Dynamic Days South America 2010 – International Conference on Chaos and Nonlinear Dynamics

Journal of Physics: Conference Series Volume 285

Sao Jose dos Campos, Brazil
26-30 July 2010

Editors:

Elbert E. N. Macau
Tiago Pereira
Antonio F. B. A. Prado
Luiz F. R. Turci
Othon C. Winter

ISBN: 978-1-61782-666-5
ISSN: 1742-6588

Printed from e-media with permission by:

Curran Associates, Inc.
57 Morehouse Lane
Red Hook, NY 12571

Some format issues inherent in the e-media version may also appear in this print version.

Printed by Curran Associates, Inc. (2011)

For permission requests, please contact the Institute of Physics
at the address below.

Institute of Physics
Dirac House, Temple Back
Bristol BS1 6BE UK

Phone: 44 1 17 929 7481
Fax: 44 1 17 920 0979

techtracking@iop.org

Additional copies of this publication are available from:

Curran Associates, Inc.
57 Morehouse Lane
Red Hook, NY 12571 USA
Phone: 845-758-0400
Fax: 845-758-2634
Email: curran@proceedings.com
Web: www.proceedings.com

TABLE OF CONTENTS

PREFACE

This proceedings presents selected contributions from the participants of South America Dynamics Days 2011, which was hosted by the National Institute for Space Research (INPE), Brazil, in July 2010.

Dynamics Days was founded in 1980 and is the longest standing and most respected international series of meetings devoted to the field of dynamics and nonlinearity. Traditionally it has brought together researchers from a wide range of backgrounds – including physics, biology, engineering, chemistry and mathematics – for interdisciplinary research into nonlinear science.

Dynamics Days South America 2010 marked the beginning of the South American branch of Dynamics Days. It brought together, for the first time in South America, researchers from a wide range of backgrounds who share a common interest in the theory and applications of nonlinear dynamics. Thus, South American researchers had a forum to promote regional as well as international scientific and technological exchange and collaboration especially, but not exclusively, on problems that are particularly relevant for the development of science and technology in the South American region. Furthermore, the conference also brought together prominent scientists from around the world to review recent developments in nonlinear science.

This conference comprised plenary invited talks, minisymposia, contributed talks and poster sessions. The articles that are compiled here were chosen from among the works that were presented as contributed talks and posters. They represent a good selection which allows one to put issues that were discussed during the conference into perspective.

It is possible to evaluate the success of an initiative by using several indices. In relation to attendees, the conference had 311 participants from 22 countries, who presented 341 works. About 86% of the participants came from South American countries. These figures allow one to classify this Dynamics Days conference as that with the greatest number of attendees ever.

Finally, we would like to express our gratitude to all the participants for their presentations, discussions, and remarkable interactions with one another. The tireless work undertaken by all the members of the International Advisory Committee and the Organizing Committee must also be recognized. We also wish to express our deep appreciation for the Scientific Societies and Research Support Agencies which supported the conference and provided all the resources which were necessary to make this idea of a South American Dynamics Days come true.

Elbert E N Macau, Tiago Pereira, Antonio F B A Prado, Luiz F R Turci, and Othon C Winter
Editors

CAPES
TAM

International Advisory Committee

Organizing Committee

Antonio Carlos Roque da Silva
USP – Universidade de São Paulo
Ribeirão Preto – SP – Brazil

Antonio F Bertachini de Almeida Prado (Co-chair)
INPE – Instituto Nacional de Pesquisas Espaciais
São José dos Campos – SP – Brazil

Arturo C Marti
Facultad de Ciencia
Montevideo – Uruguai

Carlos Leopoldo Pando Lambruschini
Benemérita Universidad Autónoma de Puebla
Puebla – Mexico

Edson Denis Leonel
UNESP – "Júlio de Mesquisa Filho"
Rio Claro – SP – Brazil

Elbert E N Macau (Chair)
INPE – Instituto Nacional de Pesquisas Espaciais
São José dos Campos – SP – Brazil

Gerard Olivar Tost
Universidad National de Colombia
Manizales – Colombia

Hamilton Varela
USP – Universidade de São Paulo
São Carlos – SP – Brazil

Hilda Cerdeira (Co-chair)
IFT – Instituto de Física Teórica
São Paulo – SP – Brazil

Iberê Luiz Caldas
USP – Universidade de São Paulo
São Paulo – SP – Brazil

José Manoel Balthazar
UNESP – "Júlio de Mesquisa Filho"
Rio Claro – SP – Brazil

Silvina Ponce Dawson
Universidad de Buenos Aires
Buenos Aires – Argentina

Vivian M Gomes
INPE – Instituto Nacional de Pesquisas Espaciais
São José dos Campos – SP – Brazil

Realization

Promotion

Sponsorship

CAPES

CNPq
Conselho Nacional de Desenvolvimento Científico e Tecnológico

FAPESP

ICTP

TRANSPORTADORA AÉREA OFICIAL

All papers published in this volume of *Journal of Physics: Conference Series* have been peer reviewed through processes administered by the proceedings Editors. Reviews were conducted by expert referees to the professional and scientific standards expected of a proceedings journal published by IOP Publishing.

Dynamic Days South America 2010 IOP Publishing
Journal of Physics: Conference Series **285** (2011) 012001 doi:10.1088/1742-6596/285/1/012001

A parametrically modulated Hénon map

Gabriela A. Casas and Paulo C. Rech
Departamento de Física, Universidade do Estado de Santa Catarina, 89223-100 Joinville, Brazil

E-mail: gabriela.a.casas@gmail.com

Abstract. In this work we consider a situation in which the two parameters of a Hénon map are linearly modulated by the output of another Hénon map, whose parameters are constant in time but can be adjusted. More specifically, here we numerically investigate modifications in basins of attraction of coexisting states, and shift in the location of critical points in bifurcation diagrams of a Hénon map, by virtue of periodic parametric modulation from another Hénon map.

1. Introduction

The Hénon map [1] is a well-known two-dimensional discrete-time dynamical system given by

$$\begin{aligned} z_{t+1} &= c - z_t^2 + dw_t, \\ w_{t+1} &= z_t, \end{aligned} \quad (1)$$

where z_t, w_t represent dynamical variables, c, d are the nonlinearity and the dissipation parameters, and $t = 0, 1, 2, \ldots$ is the discrete time. It was proposed in 1976 as a model to the Poincaré section of the continuous-time Lorenz system [2] governed by three autonomous nonlinear ordinary first-order differential equations, and have been extensively investigated in the last years. Apart from its theoretical importance, some applications are possible. It can be used as a model to a four-level CO_2 laser with modulated losses in the limit of high dissipation [3], and its parameter-space can be isomorphic to a two-dimensional parameter-space of a two-level continuous-time model of a loss-modulated CO_2 laser [4].

Dynamical systems as the Hénon map are usually characterized by a set of parameters that control their behavior and, consequently, give rise to changes in their dynamics. Most of the investigations consider these parameters constant, or allow to vary on a range of unchanging values. However, in the Nature the parameters governing the dynamics of a system frequently vary in time, and few are reports concerning it. Only in some previous investigations, some forms of modulation were applied on the parameters of the Hénon map. For example, in Ref. [5] a low-dissipative Hénon map (parameter dissipation $d = 0.9$) with coexisting period-1 and period-3 orbits was considered, and a harmonic modulation was applied on the parameter c. It was shown that, depending on the modulation, the period-3 branch can be expanded, retracted, or even suppressed. In Ref. [6] the parameter c of a high-dissipative Hénon map ($d = 0.3$) is modulated by a parametric modulation based on deterministic pseudorandom dynamics. This time it was shown that the resulting dynamics can be aperiodic and not chaotic, *i.e.*, aperiodic with a negative largest Lyapunov exponent.

In this paper we modulate both parameters, c and d, in Hénon map (1), using the solution of another Hénon map. With this aim, we make

$$c = e + fx_t, \quad d = e + fy_t, \tag{2}$$

where e, f are constant parameters, and x_t, y_t are the solution of a Hénon map given by

$$\begin{aligned} x_{t+1} &= a - x_t^2 + by_t, \\ y_{t+1} &= x_t. \end{aligned} \tag{3}$$

The result of the use of the modulation (2) in Eqs. (1) is the four-dimensional system given by

$$\begin{aligned} x_{t+1} &= a - x_t^2 + by_t, \\ y_{t+1} &= x_t, \\ z_{t+1} &= (e + fx_t) - z_t^2 + (e + fy_t)w_t, \\ w_{t+1} &= z_t, \end{aligned} \tag{4}$$

which can be regarded as a coupling of two maps of the plane: the Hénon map defined by the two first equations in (4), or by Eqs. (3), and the modulated Hénon map, defined by the two last equations in (4). Note that the two first equations do not depend on the two last equations. In other words, we have in (4) an unidirectional coupling of two Hénon maps, namely a master-slave system. System (4) maybe can be used to model the physical situation where we consider the coupling of two four-level CO_2 lasers with modulated losses in the limit of high dissipation. A similar modulation procedure was used in Ref. [7], to investigate the behavior of a logistic map modulated by the output of another logistic map.

Our purpose here is to verify possible modifications in the basins of attraction of coexisting states, and in the bifurcation points of the Hénon map (3), under the influence of the imposed modulation. The rest of this work is organized as follows. In the next section we discuss modifications introduced in bifurcation diagrams of the paradigmatic Hénon map, like shift of bifurcation points and changes in asymptotic behaviors in phase-space, as a function of the parametric modulation. Results involving basins of attraction evolution of the Hénon map when the parameters are modulated, are presented and discussed in Sec. 3. Finally, a summary is given in Sec. 4.

2. Bifurcation Diagrams

Figure 1(a) shows the bifurcation diagram (variable x) of the Hénon map (3) for $a = 1.45$ and $-1.0 \leq b \leq 0.3$, while in Figs. 1(b) and 1(c) are shown the bifurcation diagrams (variable z) of the modulated Hénon map in (4) for the same a and b parameters, with $e = 0.25$, $f = 0.2$, and $e = 0.3$, $f = 0.2$, respectively. If $a = 1.45$ and $-1.0 \leq b \leq -0.4$, then the Hénon map orbits are finally attracted to a fixed point, after a rapid transient, as can be seen in Fig. 1(a). Otherwise, If $a = 1.45$, $-1.0 \leq b \leq -0.4$, $e = 0.25$ or $e = 0.30$, and $f = 0.2$, then the modulated Hénon map orbits converge to a period-2 attractor, as shown in Figs. 1(b) and 1(c). Therefore, when the modulated Hénon map is perturbed by a Hénon map in a fixed point orbit, it converges to a period-2 orbit. The $1 \rightarrow 2$ bifurcation point at $b = -0.4$ in the Hénon map, is suppressed in the modulated Hénon map and, consequently, the fixed point disappears over the range $-1.0 \leq b \leq -0.4$. In other words, in this case the modulation procedure destroys the period-1 orbit, with consequent enlargement of the period-2 branch in the bifurcation diagrams.

Now we kept $b = 0.4$ fixed, and vary $0 \leq a \leq 1.25$. The bifurcation diagrams for the Hénon map and the modulated Hénon map are shown in Fig. 2(a) and Figs. 2(b) where $e = 0.3$, $f = 0.2$ and 2(c) where $e = 0.5$, $f = 0.2$, respectively. When $b = 0.4$, the $1 \rightarrow 2$ bifurcation point in

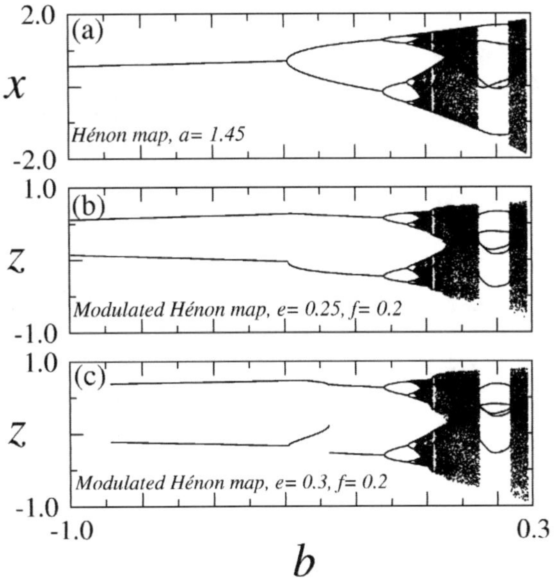

Figure 1. (a) Variable x bifurcation diagram of the Hénon map (3), for $a = 1.45$, and $-1.0 \le b \le 0.3$. (b) Variable z bifurcation diagram of the modulated Hénon map in (4), for $e = 0.25$ and $f = 0.2$. (c) Variable z bifurcation diagram of the modulated Hénon map in (4), for $e = 0.3$ and $f = 0.2$.

the Hénon map is located near to $a = 0.25$, as shown in Fig. 2(a). If $b = 0.4$, $e = 0.3$, and $f = 0.2$, then the $1 \to 2$ bifurcation occurs close to point $a = 0.125$ for the modulated Hénon map, phenomenon that can be verified in Fig. 2(b). Therefore, depending on the modulation, the period-2 branch can be enlarged with a consequent retraction of the period-1 branch. In other words, the modulation can produces a shift in the location of the bifurcation points of the Hénon map. Figure 2(c) shows again that when the modulated Hénon map is perturbed by a Hénon map in a fixed point orbit, it converges to a period-2 orbit. The $1 \to 2$ bifurcation point close to $a = 0.25$ in Fig. 2(a) disappears, together with the fixed point orbit over the range $0 \le a \le 0.25$. Hence, this time a shift on the $1 \to 2$ bifurcation point, or a suppression of the period-1 orbit may be achieved with the modulation. For all diagrams in Figs.1 and 2, the horizontal axis was divided in 1000 points, the trajectories were initialized at $(x_0,y_0)=(0.48,0.53)$ in Fig. 1(a), $(x_0,y_0)=(0.001,0.05)$ in Fig. 2(a), $(x_0,y_0,z_0,w_0)=(0.48,0.53,0.48,0.53)$ in Figs. 1(b) and 1(c), and $(x_0,y_0,z_0,w_0)=(0.001,0.05,0.001,0.05)$ in Figs. 2(b) and 2(c).

3. Basins of Attraction

Figure 3 shows the basins of attraction of the Hénon map (3) for $b = -0.3$ and $a = 1.2$. For this set of parameters we determined the periodicities in phase-space, inside the window $-3 \le x \le 3$ and $-10 \le y \le 10$, which was discretized in a grid of 750×750 points. As can be seen in Fig. 3, we have two stable dynamic equilibrium states, namely period-1 (in black color) and period-3 (in red color) states. White represents the basin of infinity. This coexistence of more than one attractor for the same set of parameters is called multistability, phenomenon that was experimentally observed in various fields, for instance, in optics [8], in biology [9], and in chemical reactions [10].

Next we investigate the suppression of multistability in the Hénon map, as an effect of the

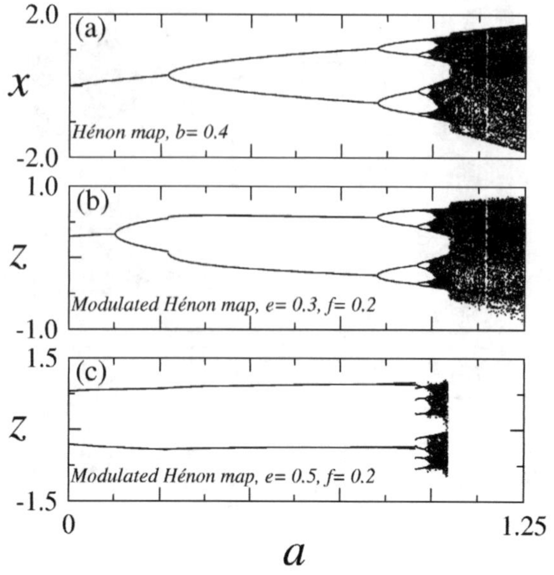

Figure 2. (a) Variable x bifurcation diagram of the Hénon map (3), for $b = 0.4$ and $0 \leq a \leq 1.25$. (b) Variable z bifurcation diagram of the modulated Hénon map in (4) for $e = 0.3$ and $f = 0.2$. (c) Variable z bifurcation diagram of the modulated Hénon map in (4) for $e = 0.5$ and $f = 0.2$.

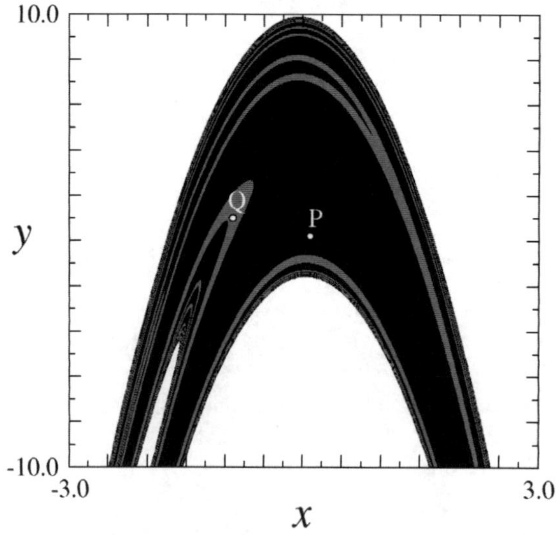

Figure 3. Basins of attraction of the Hénon map (3) for $b = -0.3$ and $a = 1.2$. Black and red colors indicate the basins of period-1 and period-3 states, respectively, and white represents the basin of the unbounded states.

modulation. As before, the periodicities in phase-space, inside the window $-3 \leq x \leq 3$ and $-10 \leq y \leq 10$, were determined in a grid of 750×750 points. Figure 4 displays two panels that refer to basins of attraction of the modulated Hénon map in (4). In the first panel, in

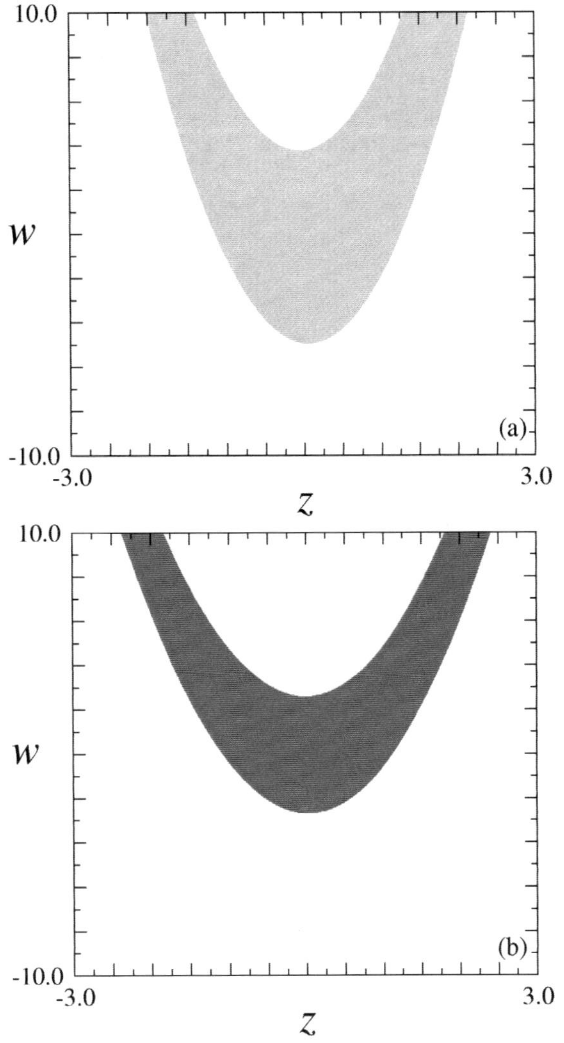

Figure 4. Basins of attraction of the modulated Hénon map in (4) for $e = 0.25$ and $f = 0.2$. (a) Initial condition (x_0,y_0)=(0.1,0.2) (point P in Fig. 3). The Hénon map in a period-1 orbit drives the modulated Hénon map to a period-2 orbit, in green. (b) Initial condition (x_0,y_0)=(-0.9,1.0) (point Q in Fig. 3). The Hénon map in a period-3 orbit drives the modulated Hénon map to a period-3 orbit, in red. As before, white represents the basin of the unbounded states.

Fig. 4(a), appears the result obtained when the modulated Hénon map with $e = 0.25$, $f = 0.2$ is perturbed by a Hénon map with $b = -0.3$ and $a = 1.2$, initialized at (x_0,y_0)=(0.1,0.2), the point P in Fig. 3, therefore a Hénon map in a period-1 orbit. We observe that this type of modulation, namely a fixed point modulation, annihilates the coexisting period-1 and period-3 attractors and their basins of attraction, and constructs a new period-2 attractor with its basin of attraction, shown in green in Fig. 4(a). In the second panel, in Fig. 4(b), is shown the result obtained when the modulated Hénon map with the same $e = 0.25$ and $f = 0.2$ parameters is perturbed by a Hénon map with the same $b = -0.3$ and $a = 1.2$ parameters, now initialized at

(x_0, y_0)=(-0.9,1.0), the point Q in Fig. 3, therefore a Hénon map in a period-3 orbit. This time we note that the modulation, namely a period-3 modulation, annihilates the coexisting period-1 and period-3 attractors with their basins of attraction, and constructs a new period-3 attractor with another basin of attraction, shown in red in Fig. 4(b).

Hence, a parameter modulation procedure of a Hénon map, by another Hénon map, is able to perform as a control, suppressing and/or generating attractors in the phase-space. In the two cases above presented two coexisting attractors were completely destroyed, while only one was generated, *i.e.*, a bistable Hénon map became monostable on account of the modulation.

There are successful experiments on attractor destruction in multistable systems. By instance, in Ref. [11] was reported an experiment involving control of multistability in an erbium-doped fiber laser with pump modulation of the diode laser, where a harmonic modulation of the diode current annihilates one or two stable limit cycles in the laser, selecting a desired state.

Similar results as the above noticed by us, can be obtained for the Hénon map using other sets of parameters. For example, if we consider $a = 1.3$ and $b = -0.24$, the result is the coexistence of two other attractors in the phase-space of the Hénon map: a period-2 and a period-6 orbits. Then, if the modulated Hénon map with $e = 0.25$ and $f = 0.2$ is perturbed by this Hénon map initialized at (x_0, y_0)=(0.1,0.2), which oscillates in a period-2 orbit, the period-6 attractor is destroyed and the modulated Hénon map becomes monostable, in a period-2 state. Otherwise, if the modulated Hénon map with same $e = 0.25$ and $f = 0.2$ parameters as above is perturbed by this same Hénon map, now initialized at (x_0, y_0)=(-0.94,1.7), which oscillates in a period-6 orbit, the period-2 attractor is destroyed and the modulated Hénon map again becomes monostable, this time in a period-6 state.

4. Summary

In this work we have numerically investigated the consequence of using a solution arising from a Hénon map, to modulate the parameters of another Hénon map. With the help of bifurcation diagrams, we have shown that the modulation can produces a shift on the location of bifurcation points, including the suppression of periodic orbits. We also have shown that the basins of attraction of coexisting states in phase-space can be modified under the influence of periodic modulations. More specifically, we have shown that bistability can be controlled, and attractors can be annihilated or/and created depending on modulation, leading the system to monostability.

Acknowledgments

The authors thank Conselho Nacional de Desenvolvimento Científico e Tecnológico-CNPq and Coordenação de Aperfeiçoamento de Pessoal de Nível Superior-CAPES, Brazilian agencies, for the financial support.

[1] Hénon M 1976 *Commun. Math. Phys.* **50** 69
[2] Lorenz E N 1963 *J. Atmos. Sci.* **20** 130
[3] Pando C L, Luna Acosta G A L, Meucci R and Ciofini M 1995 *Phys. Lett. A* **199** 191
[4] Bonatto C, Garreau J C and Gallas J A C 2005 *Phys. Rev. Lett.* **95** 143905
[5] Saucedo-Solorio J M, Pisarchik A N and Aboites V 2002 *Phys. Lett. A* **304** 21
[6] Nandi A, Bhowmick S K, Dana S K and Ramaswamy R 2009 *Chaos* **19** 033116
[7] Elhadj Z and Sprott J C 2008 *Chaos* **18** 023119
[8] Xu Q and Lipson M 2006 *Opt. Lett.* **31** 341
[9] Selivanov V A, Votyakova T V, Zeak J A, Trucco M, Roca J and Cascante M 2009 *PLoS Comput. Biol.* **5** 1
[10] Wilhelm T 2009 *BMC Syst. Biol.* **3** 90
[11] Pisarchik A N, Barmenkov Yu O and Kir'yanov A V 2003 *Phys. Rev. E* **68** 066211

Dynamic Days South America 2010 IOP Publishing
Journal of Physics: Conference Series **285** (2011) 012002 doi:10.1088/1742-6596/285/1/012002

Extracting Chaos Control Parameters from Time Series Analysis

R B B Santos[1] and J C Graves[2]

[1] Centro Universitário da FEI, Avenida Humberto de Alencar Castelo Branco 3972, 09850-901, São Bernardo do Campo, SP, Brazil

[2] Instituto Tecnológico de Aeronáutica, Praça Marechal Eduardo Gomes 50, 12228-900, São José dos Campos, SP, Brazil

E-mail: `rsantos@fei.edu.br`

Abstract. We present a simple method to analyze time series, and estimate the parameters needed to control chaos in dynamical systems. Application of the method to a system described by the logistic map is also shown. Analyzing only two 100-point time series, we achieved results within 2% of the analytical ones. With these estimates, we show that OGY control method successfully stabilized a period-1 unstable periodic orbit embedded in the chaotic attractor.

1. Introduction

Chaotic behavior is a common feature found in a wide variety of complex systems of interest to science and engineering. Owing to incomplete knowledge of the initial conditions, and to roundoff errors due to finite precision calculations, time evolution of a chaotic system is effectively unpredictable even when the underlying dynamics is strictly deterministic. Chaotic oscillations in a dynamical system may be reduced or even suppressed by chaos control techniques that disturb the system slightly. In this paper, we present a simple way to estimate the control parameters of the OGY chaos control method[1] through an analysis of the chaotic time series output by the system.

A wide class of physical and biological systems presents chaotic behavior in some range of operation. The generality of chaotic behavior is due to the fact that almost every real physical system is governed by nonlinear equations whose solutions are strongly dependent on the initial conditions at least in some part of the system parameter space[2]. In traditional applications, dynamical systems are maintained in the neighborhood of one of its stable equilibrium points. Close to one of these points, the system dynamics becomes essentially linear and no sensitive dependency on the initial conditions is observed.

There are many routes for which a physical system may become chaotic, but all of them involve a set of bifurcations that disturbs the system stability as one or more parameters are changed. A complete description of routes to chaos is outside the scope of this paper. Perhaps the most well known route to chaos is the period doubling cascade pioneered by Feigenbaum[3, 4]. The basic idea is that a system presents a cascade of subharmonic flip bifurcations until a point in which a strange attractor emerges and the dynamics becomes chaotic. Feigenbaum route to chaos is usually treated in the context of maps, particularly the logistic map.

The logistic map is a unimodal map with a single maximum point at $x = 1/2$ ruled by

$$x_{n+1} = F(x_n, r) = rx_n (1 - x_n), \quad (1)$$

and the parameter r controls the behavior of the map. It is widely known that a cascade of period doubling bifurcation beginning at $r = 3$ leads to the emergence of chaos at $r \approx 3.5699$. Apart from a trivial fixed point at $x = 0$, all other fixed points are given by $x^* = 1 - 1/r$ throughout the entire parameter range $0 < r \leq 4$.

Unidimensional maps as the logistic map are not so artificial as it may seem at first sight. Chaotic multidimensional systems in a strong dissipation regime may have just one relevant degree of freedom. In these systems, other degrees of freedom are often deemed irrelevant owing to the phase space volume contraction in those directions for which the Lyapunov exponents are negative. In the remaining relevant direction, sensitive dependence on the initial conditions holds[5].

Time series are tipically used as a means to gain insight on the general dynamics of a chaotic system, including Lyapunov exponents, and some topological characteristics[6]. However, from a control engineering perspective, this kind of information is excessive. The novelty of the proposed method is to analyze the time series to extract just the information required to control chaos, drawing attention only to the dynamics on a small neighborhood of the unstable periodic orbits.

A recent aspect related to the understanding of chaotic systems is the possibility of chaos control[7]. Chaos control techniques allows us to keep a chaotic system in an unstable periodic orbit embedded within a strange attractor with the application of small perturbations to the underlying dynamics. Due to the sensitive dependence on the actual state of the system, a small perturbation may cause dramatic changes in the system evolution[8]. Owing precisely to the chaotic behavior of the system, its control turns out to be surprisingly efficient because every small region in the phase space of a strange attractor is crossed by orbits that visit every other region in the attractor.

In this context, OGY control method is a very simple one, relying on a variation of linear control with feedback proportional to the system output[1]. The technique has been applied successfully in a variety of chaotic systems since its first experimental demonstration[9]. One of the advantages of OGY technique is that it requires little knowledge of the equations governing the system. In a codimension-1 map

$$\mathbf{x}_{n+1} = \mathbf{F}(\mathbf{x}_n, r_n), \quad (2)$$

the controls needed to stabilize a period-1 unstable periodic orbit (fixed point) are small disturbances Δr_n around the nominal value r_0 of the control parameter r:

$$\Delta r_n = r_n - r_0 = -\gamma \hat{\mathbf{n}} \cdot (\mathbf{x}_n - \mathbf{x}_0^*) \quad (3)$$

where $\hat{\mathbf{n}}$ is a unit vector normal to the Poincaré section Σ in the neighborhood of the unstable fixed point $\mathbf{x}_0^*$, and γ is a proportional gain.

Applied to the stabilization of a fixed point in the logistic map[3]

$$x_{n+1} = F(x_n, r) = rx_n (1 - x_n), \quad (4)$$

OGY control method is based on the linearized map around the unstable fixed point $x_0^* = 1 - 1/r_0$, and around the nominal control parameter r_0,

$$x_{n+1} \approx x_0^* + \alpha (x_n - x_0^*) + \beta (r_n - r_0) \quad (5)$$

that leads to a modified map

$$x_{n+1} = (r_0 - \gamma\,(x_n - x_0^*))\,x_n\,(1 - x_n) \tag{6}$$

when the system is close enough to the unstable fixed point x_0^*. Optimal control is achieved by choosing $\gamma = -\alpha/\beta$. Therefore, one must determine from a series of runs of the experiment the location of the fixed point x_0^*, and the sensitivities

$$\alpha = \left.\frac{\partial F}{\partial x}\right|_{\substack{x=x_0^*\\ r=r_0}} \tag{7}$$

and

$$\beta = \left.\frac{\partial F}{\partial r}\right|_{\substack{x=x_0^*\\ r=r_0}} \tag{8}$$

in order to apply controls on the system. It should be pointed out that this conclusion is valid even in delayed feedback control[10]. Although the feedback, in this case, does not involve explicitly the unstable fixed point x_0^*, knowledge of the location of the fixed point is still required to evaluate the sensitivities α and β.

Reconstruction of some aspects of the chaotic dynamics from time series is a subject explored through a variety of methods[6, 11, 12]. We introduce a very simple recurrence method to estimate the parameters required in order to apply the OGY method to control chaos in dynamical systems. For ease of presentation, we restrict this discussion to unidimensional maps as the logistic map, and to period one unstable periodic orbits. Therefore, it suffices to estimate the values of x_0^*, α, and β from a time series.

It must be noted that the method proposed does not attempt to reconstruct the chaotic attractor. Instead, we focus on the local dynamics of the attractor around their unstable periodic orbits. In this way, we were able to reduce substantially the volume of information needed to be extracted from the experiment and processed in order to control a chaotic system.

1.1. Our Proposal

The dynamics of a chaotic system is tipically slow around an unstable periodic orbit. Hence, we begin analyzing the behavior of a unidimensional map around one of its fixed points. A linearized version of the map around $x = x_0^*$, and $r = r_0$ may be written as

$$x_{n+1} \approx x_0^* + \alpha\,(x_n - x_0^*) + \beta\,(r_n - r_0)\,. \tag{9}$$

Fixing the value of the control parameter to the nominal value r_0, and selecting only those situations in which there is a close encounter, that is, situations in which the variation

$$\Delta_n = x_{n+1} - x_n \approx (\alpha - 1)\,(x_n - x_0^*) \tag{10}$$

is small, less than some ε, for instance. In these close encounter cases, the system is guaranteed to be in a small neighborhood of radius ε of the fixed point due to the slow dynamics of the map around x_0^*. Then, it is straightforward to show that

$$x_0^* \approx \left\langle \frac{x_n\Delta_m - x_m\Delta_n}{\Delta_m - \Delta_n} \right\rangle \tag{11}$$

where Δ_m is the corresponding variation for another close encounter in the neighborhood of the same fixed point, and the brackets indicate averaging over the several such close encounters provided by a long time series. Despite the simplicity of this recurrence method, fixed point detection may be enhanced by techniques that manipulates the probability distribution on the attractor through fixed point transforms[11].

Having an estimate for x_0^*, it is easy to show that

$$\alpha \approx 1 + \left\langle \frac{\Delta_n}{x_n - x_0^*} \right\rangle. \tag{12}$$

Finally, in order to estimate β, we vary the control parameter to a value $r_1 = r_0 + \delta r$ slightly different from r_0, and estimate the new fixed point x_1^* using equation (11). Hence, averaging over many different values of r_1 close to r_0, we discover that

$$\beta \approx (1-\alpha) \left\langle \frac{\delta x^*}{\delta r} \right\rangle = (1-\alpha) \left\langle \frac{x_1^* - x_0^*}{r_1 - r_0} \right\rangle. \tag{13}$$

The procedure outlined is very easy to perform, and we should expect to find a fair estimate of the control parameters after analyzing relatively small number of long enough time series of the output of the chaotic system. In the next section, we apply this procedure to control the very simple logistic map.

2. Results and Discussion

To demonstrate the power of this simple method, we estimate x_0^*, α, β, and γ for the logistic map with only two runs of 100 points each. In these simulations, we set $r_0 = 3.90$ and $r_1 = 3.91$, right in the middle of a heavily chaotic region, and we also fixed the radius for a close encounter to be detected in 10% of the total variation of the logistic map. Figure 1 shows a typical trajectory obtained in these very short simulations.

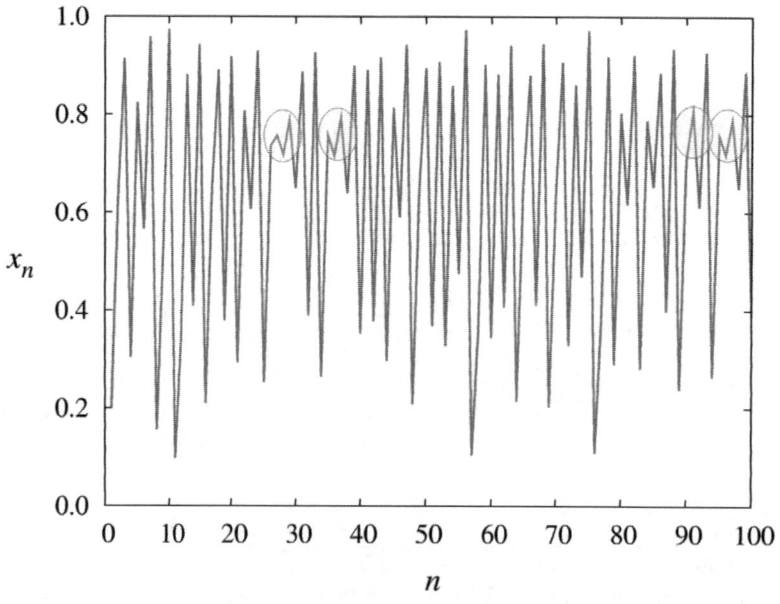

Figure 1. A 100-point chaotic trajectory in the logistic map with $r_0 = 3.90$. The circles indicate close encounters along the trajectory.

We performed the analysis proposed in section 1.1 and the results are shown in table 1. Even these short few runs were enough for a fair estimation of the parameters needed to control the chaos in the system.

Table 1. Theoretical and simulated values for the fixed point x^*, sensitivities α and β, and proportional gain γ.

Parameter	x^*	α	β	γ
Theory	0.74359	−1.90	0.1907	9.97
Simulation	0.74347	−1.86	0.1892	9.85
Error	0.016%	2.1%	0.79%	1.2%

Using the estimated parameters, we were able to stabilize a period-1 unstable periodic orbit embedded in the heavily chaotic region of the logistic map as shown in figure 2. In order to keep the orbit stabilized, additional controls amounting to only $\Delta r/r_0 = 0.074\%$ had to be continuously applied.

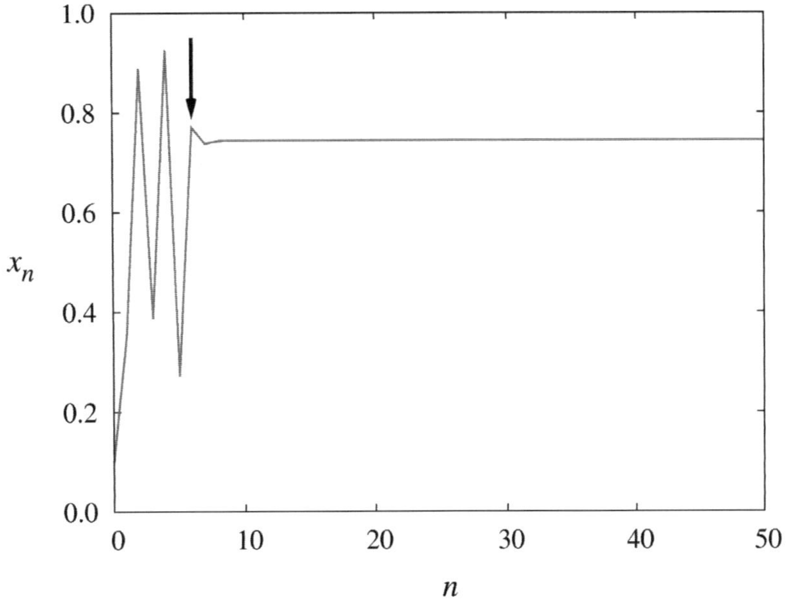

Figure 2. Stabilized trajectory in the logistic map ($r_0 = 3.90$) using OGY chaos control method with parameters estimated from time series. The arrow marks the first time that controls were applied.

3. Conclusion

We presented a simple procedure to estimate the set of parameters needed in order to control chaos. Analysis of only two very short time series of the logistic map was enough to estimate the control parameters within 2% of their analytical values. With these estimates, OGY method could successfully stabilize an unstable periodic orbit. The method applied here to stabilize a fixed point in the logistic map may be extended to stabilize a period-k unstable periodic orbit in higher dimensional systems where α is replaced by the jacobian matrix $[\alpha_{ij}] = [\partial F_i/\partial x_j]$, and β is replaced by a matrix of sensitivities $[\beta_{ij}] = [\partial F_i/\partial r_j]$.

Acknowledgments

RBBS acknowledges partial financial support from Fundação de Amparo à Pesquisa do Estado de São Paulo (FAPESP). JCG wishes to thank Fundação Educacional Inaciana Pe. Sabóia de Medeiros and Coordenação de Aperfeiçoamento de Pessoal de Nível Superior (CAPES) for financial support. We are grateful to an anonymous referee for suggestions that made the exposition more precise.

References

[1] Ott E, Grebogi C and Yorke J A 1990 *Phys. Rev. Lett.* **64** 1196
[2] May R M 1976 *Nature* **261** 459
[3] Feigenbaum M J 1978 *J. Stat. Phys.* **19** 25
[4] Feigenbaum M J 1979 *J. Stat. Phys.* **21** 669
[5] Ott E 2002 *Chaos in Dynamical Systems* (Cambridge: Cambridge University Press).
[6] Abarbanel H D I, Brown R, Sidorowich J J and Tsimring L S 1993 *Rev. Mod. Phys.* **65** 1331
[7] Macau, E E N and Grebogi C 2008 *Handbook of Chaos Control* ed E Schöll and H G Schuster (Weinheim, Germany: Wiley-VCH) pp 3-28
[8] Guckenheimer J and Holmes P 1983 *Nonlinear Oscillations, Dynamical Systems, and Bifurcations of Vector Fields* (Berlin: Springer)
[9] Ditto W L, Rauseo S N and Spano M L 1990 *Phys. Rev. Lett.* **65** 3211
[10] Pyragas K 1992 *Phys. Lett.* A **170** 421
[11] So P, Ott E, Sauer T, Gluckman B J, Grebogi C and Schiff S J 1997 *Phys. Rev.* E **55** 5398
[12] Palaniyandi P and Lakshmanan M 2005 *Phys. Lett.* A **342** 134

The effect of poisoning species on the oscillatory dynamics of an electrochemical reaction

B C Batista,[1] G C A Ferreira,[1] H Varela[1,2]
[1]Instituto de Química de São Carlos, Universidade de São Paulo, C.P. 780, CEP 13560-970, São Carlos, SP, Brazil, and [2]Ertl Center for Electrochemistry and Catalysis, GIST, Cheomdan-gwagiro 261, Buk-gu, Gwangju 500-712, South Korea.

E-mail: varela@iqsc.usp.br

Abstract. Electrochemistry is a prone field for observing complex behavior such as simple and mixed-mode oscillations, chaos and spatiotemporal pattern formation. Although modeling and numerical analysis of those systems is relatively common, analysis of the inner structure of the oscillatory region is a rather unexplored issue. We describe in this paper a numerical study of a surface catalyzed reaction that is being poisoned by a foreign adsorbing species. The effect of the poison coverage on the oscillatory period and amplitude is discussed with the help of parameter plane diagrams and mechanistic analysis. A coherent picture explaining the effect of the poison adsorption on the dynamic features is then constructed based on the obtained data.

1. Introduction

Electrochemistry has emerged as a powerful field for observing and controlling dynamic processes mainly due to the rather simple way in which the flux of one of the reactants, viz. the electric current, can be controlled [1]. Particularly in electrocatalysis, several systems present complex behavior expressed in the form of multi-stability, periodic and complex oscillations, chaos, and spatiotemporal pattern formation [2,3].

The oscillatory electro-oxidation of methanol on platinum surfaces can be cited as a model system, displaying simple harmonic potential oscillations and conventional Arrhenius-like temperature dependence [4-6], unlike other parent systems [3,7]. This system was recently modeled and numerically explored by Sauerbrei *et al.* [8] focusing on stability and bifurcation analysis. We present here the study of the dynamic behavior for a slightly modified version of this model, in which it incorporates anion adsorption among the possible chemical steps. The investigation was carried out by means of an in-depth investigation of the inner structure of the bifurcation diagrams, similarly to the two-dimensional analysis previously reported for other systems [9-11]. Although very demanding from computational point-of-view, such analysis provides a non-comparable detailed view of the dynamic structure of the model.

2. Model and preliminary analysis

The model utilized was based on that proposed by Sauerbrei *et al.* [8] described by equations 1 to 4, which represent, respectively, the time evolution of adsorbed oxides (θ_O), adsorbed HCO (θ_{HCO}), adsorbed carbon monoxide (θ_{CO}) and the electrode potential (φ). Other terms are introduced in Table 1.

$$\theta_O'(t) = k_3 F e^{k_6\varphi(t)} - k_4\theta_O(t)\theta_{CO}(t) - k_5\theta_O(t)e^{-k_6\varphi(t)} \quad (1)$$

$$\theta_{HCO}'(t) = k_1 F^2 e^{k_6\varphi(t)} - k_2\theta_{HCO}(t) \quad (2)$$

$$\theta_{CO}'(t) = k_2\theta_{HCO}(t) - k_4\theta_O(t)\theta_{CO}(t) \quad (3)$$

$$\varphi'(t) = I - k_1 F^2 e^{k_6\varphi(t)} \quad (4)$$

We introduced a modification on the parameter F, which represents the amount of free surface sites, by adding an adsorption term for a blocking species (θ_P),

$$F = 1 - \theta_O(t) - \theta_{CO}(t) - 3\theta_{HCO}(t) - \theta_P(t)$$

$$\theta_P(\varphi) = \frac{b.a.e^{k_6\varphi(t)}}{1 + a.e^{k_6\varphi(t)}} \quad (5).$$

This modification accounts for the anion adsorption as a function of the potential and can be easily deduced from thermodynamic and electrochemical assumptions [12]. The dependence is given in terms of parameters *b* and *a*, which stand for the maximum amount of poison coverage that can be achieved and the concentration of the anion in solution, respectively. Increasing either *b* or *a* leads to an increase in the amount of anion adsorbed for a given potential.

A general cycle of potential oscillations under galvanostatic control, i.e. constant applied current (*I*), may be described as follows: while the electrode potential is low the catalyst is said to be active. Gradually it becomes poisoned with adsorbed HCO and CO; as a consequence, the electrode potential rises in order to keep the overall current *I* constant. The increase in the electrode potential however leads to an increase of the oxide production which in turn enhances the oxidation of adsorbed carbon monoxide, cleaning thus the electrode surface. Accompanying the recovering of the high activity, the electrode potential starts decreasing again to keep the current, the amount of oxides decrease, and the whole cycle starts anew.

All these steps are related to the specific reactions taking place and that were incorporated in equations 1-5 by the kinetic law of mass action. Parameters used are listed in Table 1.

Table 1. Characteristic parameters for the simulations

Rate	*Meaning*	*Value*
k_1	Methanol adsorption	4
k_2	CO production	10
k_3	Oxide production	1
k_4	CO and oxides reaction	2
k_5	Oxide reduction	6
k_6	Faradaic term	15
I	Applied current	variable

Figure 1 brings at its left side a collection of potential time series obtained for poison concentrations ranging from 0 to 0.5 and maximum coverage set at 0.2. The curves were normalized so as to display the same amplitude. On the right side the behavior of carbon monoxide coverage is presented for those extreme concentration conditions. It is seen that the eye striking feature of the anion presence is to decrease the oscillation period. It is interesting to note however that only the rising part of the curves seems to be affected, leading to the observed loss of harmonicity. For galvanostatic experiments, the rising part of the curve is mechanistically associated with the blockage of active catalyst sites, particularly by carbon monoxide (CO). The right-side picture shows that this may also be the case for the system under investigation. It can be seen that the amount of carbon monoxide produced at the surface is greatly reduced when the anion effect is incorporated. This particular issue will be discussed in connection with experiments in a forthcoming paper [13]. In the next we explore different facets of the bifurcation diagrams.

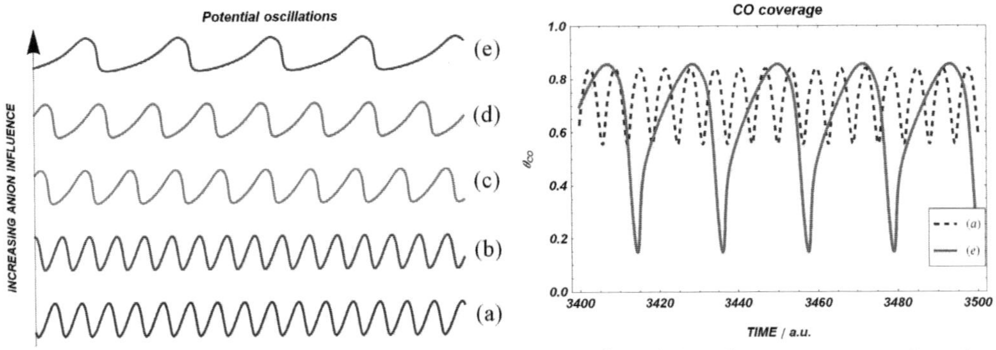

Figure 1. Left: potential time series obtained with equations 1-5 and parameters as given in Table 1 with $b = 0.2$ and (a) $a = 0, I = 0.14$; (b) $a = 0.01, I = 0.11$; (c) $a = 0.05, I = 0.09$; (d) $a = 0.1, I = 0.07$; (e) $a = 0.5, I = 0.05$. Right: CO coverage time series obtained for (a) $a = 0, I = 0.14$; and (e) $a = 0.5, I = 0.05$.

3. Parameter plane diagrams

The study of individual temporal series led to the recognition of an effect of the poisoning species on the oscillation period. Since oscillations are found to happen for differing values of parameters when different amounts of poison are present, which could cause misleading conclusions, we endeavored on a large scale analysis by constructing parameter plane diagrams and thus gaining a more representative picture of this system. Figure 2 display two-dimensional regular diagrams discriminating oscillatory (light coded) and non-oscillatory (dark-blue coded) domains for five distinct concentrations of blocking species. Diagrams were constructed by collecting a potential time series for a given set of I and b, and color was attributed in such a way that light ones accounted for the presence of oscillations. Only simple periodic oscillations were found for the whole spectra of parameters studied.

Two main aspects accompanying the increase of the poison concentration are clearly discernible in Figure 2. First, the values of current for which oscillations are found decrease with increasing poison influence, which is to say higher values of parameters a and b. Second, the region of parameters where the phenomenon is observed also shrinks for the same kind of perturbation. As the poison adsorbs on the surface inhibiting all surface catalyzed reactions their velocity should be reduced. As a consequence higher values of applied current cannot be sustained by the poisoned system explaining thus the shift of the oscillatory region towards the lower values of applied current I needed to support oscillations.

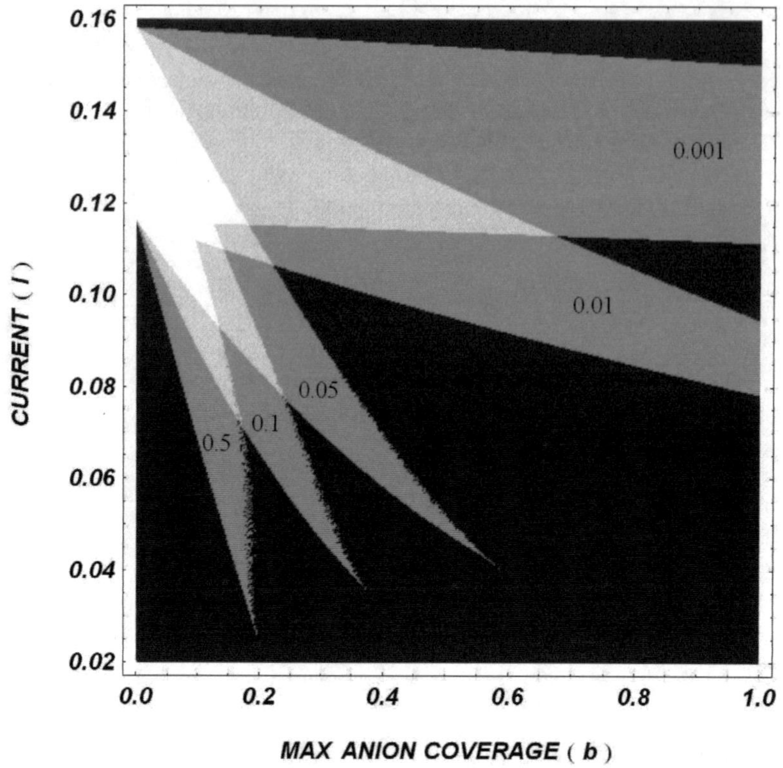

Figure 2. Bifurcation diagrams in the *b* versus *I* plane for five distinct values of anion concentration, *a*.

Deeper understanding of the underlying processes can be achieved if information of the inner structure of the parameter diagrams is available. Figure 3 brings such a study: each diagram contains 400x400 points accounting for the oscillation period obtained from numerically calculated stationary time series of the electrode potential. Each point was colored according to the oscillation period, non-oscillatory solutions were denoted in white. It is important to note that the white regions correspond to presence of stable steady states.

The general features discussed for Figure 2 are also found in Figure 3. More important is the period behavior that now arises. For any individual graph it can be seen that the increase of the maximum amount of poison that can be adsorbed leads to an increase of the oscillation period. Comparison between pictures for rising values of anion concentration reveals that large areas of the diagrams are covered with blue and pink which also indicate that period is large.

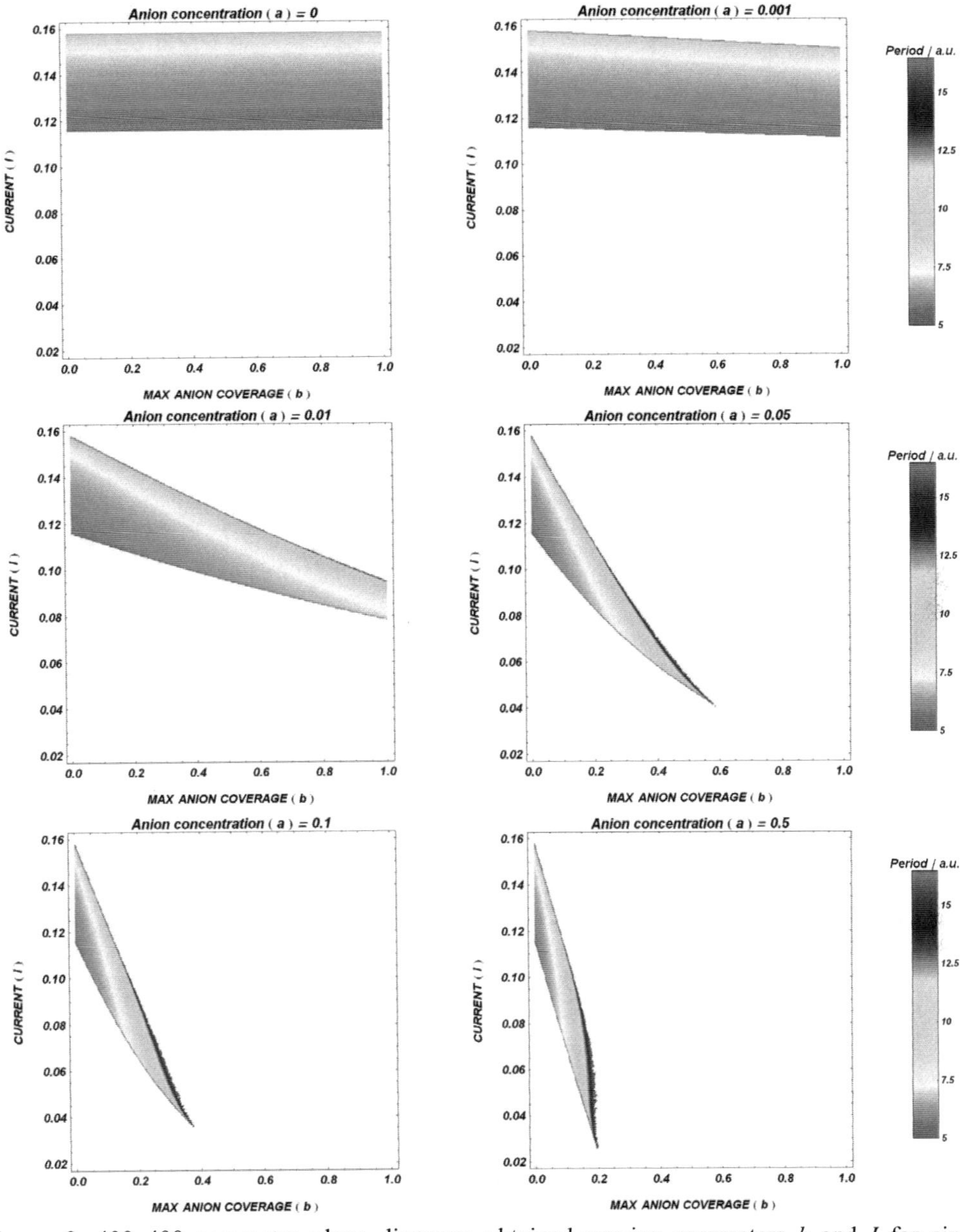

Figure 3. 400x400 parameter plane diagrams obtained varying parameters b and I for six distinct values of anion concentration as specified in the graph. The colorscale represents the temporal period associated with the oscillations found for each set of parameters.

A comprehensive summary of the data contained in the structured bifurcation diagrams is given in Figure 4. It displays the mean value of the period for five anion concentrations as function of the maximum amount of blocking species and its computation was done by summing up all values of period for a given value of b and dividing the result by the total of constituents in the sum.

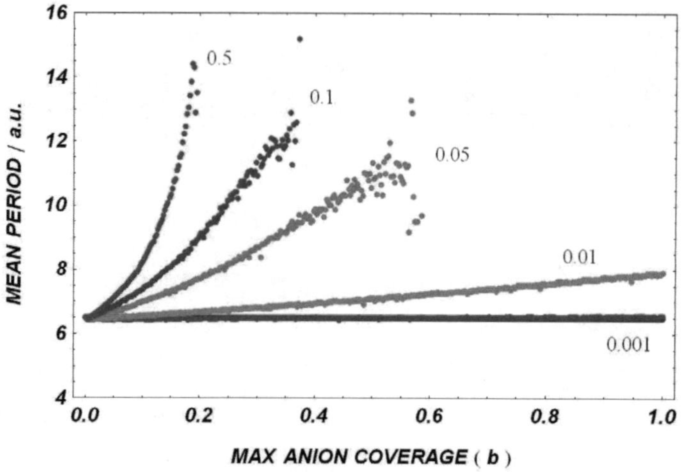

Figure 4. Mean temporal period of oscillations as function of the maximum anion coverage. The graph was constructed with data displayed in Figure 3.

4. Numerical analysis

Altogether, the obtained data reveals a coherent scenario for the anion effect: there is a general decrease in the period of oscillations when the poisoning species influence is increased. In order to take advantage of that large amount of information we shall now take a grander view at the system under consideration looking for an explanation of the period effect. In a complex oscillatory network, reactions can be classified according to their contribution to the oscillation period, some has a positive contribution, and some a negative contribution to the oscillation period [14]. In a recent article Kiss *et al.* [15] derived an analytical expression for the oscillatory frequency of a generic electrochemical model. As the system under consideration is quite more complicated and specific we now shall take a numerical approach to this analysis by studying how the individual reaction rates influence the oscillation period. Figure 5 displays such an examination. An individual constant was varied while the others were kept at their original values (see Table 1).

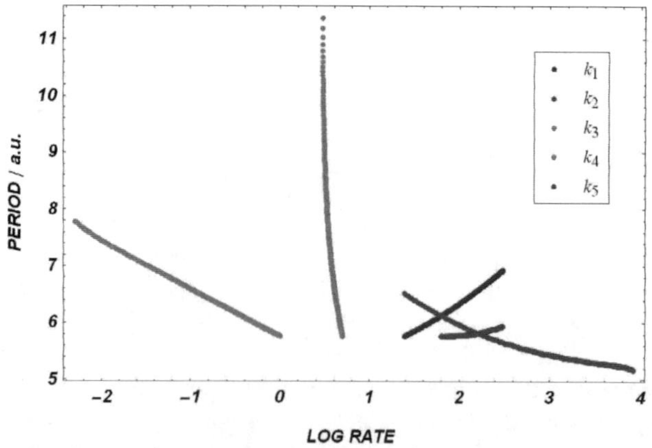

Figure 5. Dependence of the logarithm of the individual rate constants on the oscillation period. k_1 is rate of methanol adsorption; k_2 is the carbon monoxide production; k_3 is the oxide production; k_4 is the reaction between CO and oxides; and k_5 is the rate of oxide reduction.

Figure 5 shows that the formation of adsorbed carbon monoxide, the surface oxidation and the reaction between those species led to a decrease in the period of oscillations while methanol adsorption and oxide reduction caused a period increase. Particularly the rate of carbon monoxide oxidation displayed the highest effect on the oscillation period. Table 2 shows a summary of the information contained in Figure 5.

Table 2. Effect of the individual rate constants on the oscillation period

Constant	*Significance*	*Rate of change of period*
↑k_1	Methanol adsorption	+ 2.4
↑k_2	CO production	- 1.3
↑k_3	Oxide production	- 2
↑k_4	CO and oxides reaction	- 40
↑k_5	Oxide reduction	+ 0.7

It is noteworthy that the overall trend in Figure 5 and Table 2 is the decrease of the period as the rates of reaction are increased given that the period reducing steps possess a higher weight. One way of altering the individual rates all at once is, for example, to increase the temperature of the system. Electrochemical experiments aiming at understanding this effect have revealed that the oscillations do indeed follow an Arrhenius type behavior for which an increase in temperature leads to an increase in the oscillation frequency [5]. The results presented in Table 2 thus provide a mechanistic basis to understand these experimental results.

We are now at a position of explaining the anion effect in the oscillatory electro-oxidation of methanol on platinum surfaces. The general effect of the anion is to block active sites thus reducing the velocity of reactions associated with free catalyst sites. This should greatly reduce the amount of carbon monoxide produced since its production involves the highest number of metal sites. It was shown in Figure 2 that the carbon monoxide poisoning is indeed reduced when the anion influence is incorporated in the set of equations. Since the velocity of CO oxidation is directly dependent on its coverage it will also be reduced by adsorbed anions. Figure 6 brings the velocity of the reaction discussed and shows that there is indeed a decrease in its overall and mean values when the anion is added.

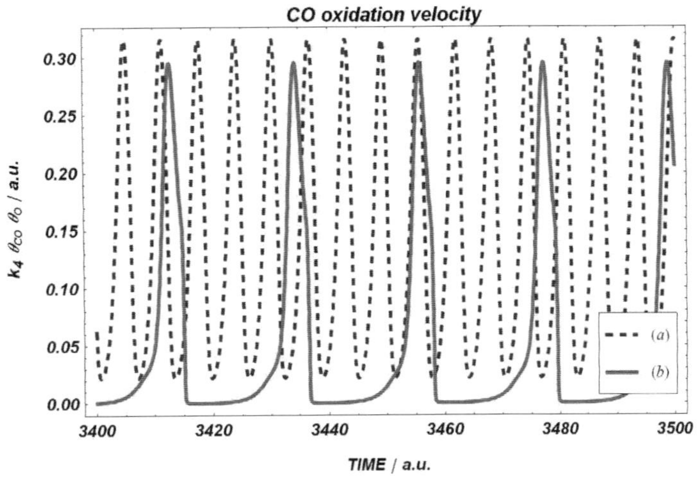

Figure 6. Left: temporal evolution of the velocity of CO oxidation for conditions given in Table 1 with $b = 0.2$ and (a) $a = 0$, $I = 0.14$; (e) $a = 0.5$, $I = 0.05$.

As given in Table 2, the CO oxidation reaction possesses the highest weight on the period of oscillations. It has been demonstrated that the poison greatly influences the reaction of CO oxidation, reducing its velocity. Therefore, reducing the CO reaction is equivalent to moving towards the left side of Figure 5 which leads to a decrease in the period of oscillations.

5. Conclusion

We investigated in this paper the impact of poisoning anions on the oscillatory electro-oxidation of methanol on platinum. The study was performed in terms of a model consisting of four ordinary differential equations. It has been demonstrated that the inclusion of a competing adsorbing species leads to a decrease in the period of oscillations and reduced rates of individual reactions. The anion effect was reasoned by its influence on the carbon monoxide oxidation reaction which was found to have the most profound weight on the oscillation frequency.

Acknowledgments

The authors acknowledge Fundação de Amparo à Pesquisa do Estado de São Paulo (FAPESP, HV # 09/07629-6) and Conselho Nacional de Desenvolvimento Científico (CNPq, BCB # 141723/2009-2, HV # 302698/2007-8) for financial support.

References

[1] Krischer K 2002 *Modern Aspects of Electrochemistry* vol 32, ed B E Conway and J O'M Bockris (New York: Kluwer Academic Press) p 1
[2] Krischer K and Varela H 2003 *Handbook of Fuel Cells: Fundamentals Technology and Aplications* vol 2, ed W Vielstich et. al. (Chichester: Wiley) p 679
[3] Sitta E, Nascimento M and Varela H 2010 *Phys. Chem. Chem. Phys.* DOI: 10.1039/c002574g
[4] Krausa M and Vielstich W 1995 *J. Electroanal. Chem.* **399** 7
[5] Carbonio E A, Nagao R, Gonzalez E R and Varela H 2009 *Phys. Chem. Chem. Phys.* **11** 665
[6] Boscheto E, Batista B C, Lima R B and Varela H 2010 *J Electroanal. Chem.* **642** 17
[7] Nagao R, Epstein I R, Gonzalez E R and Varela H 2008 *J. Phys. Chem. A* **112** 4617
[8] Sauerbrei S, Nascimento M, Eiswirth M and Varela H 2010 *J. Chem. Phys.* **132** 154901-1
[9] Bonatto C and Gallas J A C 2008 *Phil. Trans. R. Soc. A* **366** 505
[10] Freire J G, Field R J and Gallas J A 2009 *J. Chem. Phys.* **131** 044105
[11] Nascimento M A, Gallas J A C and Varela H 2011 *Phys. Chem. Chem. Phys.* **13** 441
[12] Gileadi E 1993 *Electrode kinetics for chemists, chemical engineers and material scientists* (Moskow: VCH Publishing)
[13] Ferreira G C A, Batista B C and Varela, H 2010 *in preparation*
[14] Ruoff P 1992 *J. Interdiscipl. Cycle Res.* **23** 92.
[15] Kiss I Z, Pelster L N, Wickramasinghe M and Yablonsky G S 2009 *Phys. Chem. Chem. Phys.* **11** 5720

Dynamic Days South America 2010 IOP Publishing
Journal of Physics: Conference Series **285** (2011) 012004 doi:10.1088/1742-6596/285/1/012004

Radial dependence of self-organized criticality behavior in TCABR tokamak

G. Z. dos Santos Lima[1], K. C. Iarosz[2], A. M. Batista[3], Z. O. Guimarães-Filho[4], I. L. Caldas[5], Y. K. Kuznetsov[5], I. C. Nascimento[5], R. L. Viana[6], S. R. Lopes[6]

[1]Escola de Ciência e Tecnologia, Universidade Federal do Rio Grande do Norte, 59014-615, Natal, RN, Brazil
[2]Pós-Graduação em Ciências, Universidade Estadual de Ponta Grossa, 84030-900, Ponta Grossa, Paraná, Brazil.
[3]Departamento de Matemática e Estatística, Universidade Estadual de Ponta Grossa, 84030-900, Ponta Grossa, PR, Brazil.
[4]Equipe DSC/PIIM, Université de Provence, France
[5]Instituto de Física, Universidade de São Paulo, 05508-090, SP, Brazil
[6]Departamento de Física, Universidade Federal do Paraná, 81531-990, Curitiba, PR, Brazil.

E-mail: gzampier@ect.ufrn.br;batista@interponta.com.br

Abstract. In this work we present evidence of the self-organized criticality behavior of the plasma edge electrostatic turbulence in the tokamak TCABR. Analyzing fluctuation data measured by Langmuir probes, we verify the radial dependence of self-organized criticality behavior at the plasma edge and scrape-off layer. We identify evidence of this radial criticality in statistical properties of the laminar period distribution function, power spectral density, autocorrelation, and Hurst parameter for the analyzed fluctuations.

1. Introduction

Electrostatic turbulence observed at the plasma edge tokamaks is a severe limitation to plasma confinement [1, 2]. There is considerable theoretical and experimental progress on the understanding of this turbulence, however a complete description of the observations does not have yet been achieved [3]. Thus, basic knowledge of the statistical fluctuation or driven transport fluctuations are still necessary to improve the understanding of plasma edge turbulence [4].

Quantitative investigations of the electrostatic fluctuations using spectral analysis show that drift waves are destabilized in the confining magnetic field generating a turbulent spectrum [5]. Moreover, dynamical diagnostics used to describe fluid turbulence have been applied to analyze the plasma edge turbulence, as the return-time statistics [4, 6], recurrence analyses [7] and multifractality [8]. Furthermore, the structure of intermittent plasma signals can be studied by analyzing fat tails and long-range correlations. Some of the methods to characterize the long-range correlations from experimental time series evolve autocorrelation functions, power spectral densities, and probability distribution functions (PDFs).

Recent results show analysis of electrostatic fluctuations from confined magnetically plasmas that have been described by their space and time self-organized similarity behavior (SOC) [9, 10].

Self-organized criticality (SOC) is a general property of dynamical systems with an attracting critical state, displaying scale invariance in both spatial and temporal degrees of freedom [11]. The time evolution of systems displaying SOC is characterized by the presence of self-organized structures, or avalanches, which obey a power-law statistical distribution. The paradigmatic example of a system displaying SOC is a sandpile having a constant injection of sand [8-10]. The excess mass in the surface of the sandpile is released through avalanches which were found to obey such power-law distribution. SOC has been viewed as a mechanism by which there is spontaneous emergence of complexity from simple local interactions. This encourages the search for SOC in complex systems as magnetically confined plasmas [12, 13]. In the last years, several SOC characteristics were identified in the plasma edge turbulence, as the existence of critical average gradients and profile resilience to the power-law power spectra of plasma parameters fluctuations (density, temperature, magnetic field) [14]. Many numerical models with SOC properties reproduce experimental turbulence phenomenology [3-7] and features such as transport barriers [3, 5]. SOC may be used to explain the existence of noise $1/f$ (f is the frequency) in several experimental observations [14], and the self-similar character of electrostatic fluctuations at the plasma edge [13]. Moreover, SOC gives us connection between the scale invariance in space and time.

Here, we present evidence of the self-organized criticality behavior in the electrostatic turbulence data obtained with Langmuir probes in TCABR tokamak [15]. Our analysis also reveals the radial dependence of the algorithms that characterize the observed criticality.

The paper is organized as follows: Section 2 we present the experimental setting. Section 3 treats of the analysis of the experimental results relate to self-organized criticality behavior. The last section is devote to our conclusions.

2. Tokamak data

The analyzed experiments were performed in a hydrogen circular plasma in the Brazilian tokamak TCABR [15] (major radius R=61 cm and minor radius a=18 cm). The maximum value of the plasma current is 110 kA, with duration 100 ms, the hydrogen filling pressure is 3×10^{-4} Pa, and the toroidal magnetic field is $B_T = 1.1$ T. The floating potential was measured by two Langmuir probes, poloidally separated by 0.4 cm. The probes are mounted on a movable shaft that can be displaced radially from $r = 15.0$ to 23.0 cm, with respect to the center of the plasma column. As a matter of fact, we focus on the range from 16.5 to 21.0 cm (r/a from 0.92 to 1.16) to cover the plasma edge and also the so-called scrape-off layer (SOL), i. e., the low density plasma layer comprised between the plasma column and the vessel wall.

The probe displacement occurs only for separate discharges, such that it does not disturb the plasma due to the movement of the probe. The measurements were performed at a sampling frequency of 1 MHz, and the measuring circuit has a 300 kHz bandwidth to avoid aliasing [5]. Fig. 1 shows the typical time evolution of a considered plasma discharge in TCABR. The plasma current Fig. 1(a) grows slowly in the first 85 ms. The electron density evolution indicated by Fig. 1(b) exhibits the region analysed that is a plateau with value $n_e = 1.1 \times 10^{19}$ m^{-3}. We are interested in the floating electrostatic potential signals in the radial region between inside (Fig. 1c) and outside the plasma column, which shows irregular fluctuations. We analyze data from time intervals, as the one in the window indicated by vertical lines in Fig. 1, chosen before MHD activity start increasing and changing the plasma turbulence [16]. Moreover, to study the self-organized criticality we consider the fluctuations around the mean electrostatic potential as the time series of Fig 1(d).

3. Radial dependence of self-organized criticality behavior in experimental data

In order to investigate the radial dependence of self-organized criticality on the TCABR plasma turbulence we analyze some typical SOC evidences we found in the probability density function

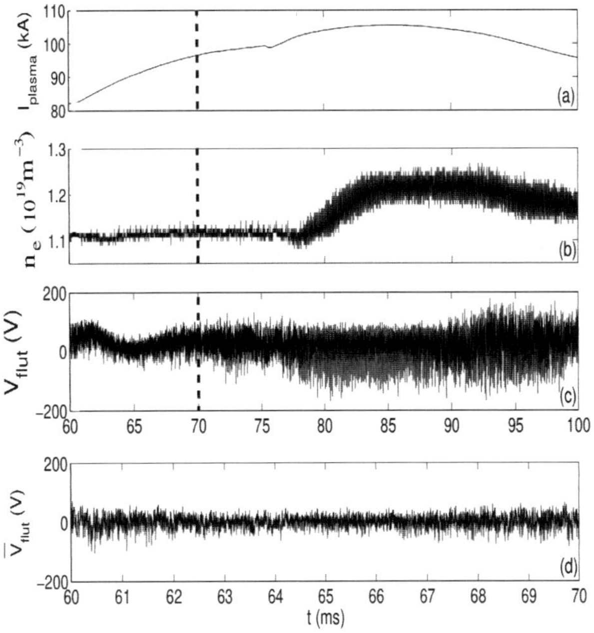

Figure 1. Time evolution of plasma discharge in TCABR tokamak. (a) Plasma current, (b) central chord plasma mean density, (c) floating electrostatic potential for a typical discharge inside the limiter (r=17 cm) and (d) its fluctuation during the analyzed time interval, from t=60 ms to 70 ms (indicated by dashed line).

(PDF) of laminar periods, auto-power frequency spectrum, autocorrelation function, and Hurst parameter on TCABR float potential fluctuations [9].

Figure 2 shows the PDFs of the laminar period duration for different radial positions. The laminar period is the time when the amplitude of the fluctuations remains below a treshold (Fig. 2 presents the PDFs for tresholds defined as $\delta = 2\sigma$, where σ is the standard deviation). Regarding Figs. 2(b) and 2(c), we can observe a linear fit that suggests a power-law decay, while in Fig. 2(a) and (d) a power-law decay is not possible to be fitted. In doing so, the PDF shows a power-law dependence around the plasma column radius.

In SOC systems, the Fourier spectra are expected to be similar to those applied to $1/f$ noise signal for a given range of frequencies [17]. This behavior can be inferred from the existence of avalanches or transport. The shape of the power frequency spectrum is characterized by a $1/f^{\alpha}$ power-law spectrum, where α is from 0 to 2 or higher separated by three regions quite different according to the theory. It has been discussed theoretically that transport events and electrostatic fluctuations at the plasma edge in magnetically confined plasmas have some characteristics of self-organized critical systems [9]. For the TCABR tokamak data measured at $r/a = 1$ (Fig. 3b) and $r/a = 1.05$ (Fig. 3c), the frequency spectrum shows two distinct regions in the frequency range with decay indexes of -1 and -4, respectively. The range involving the $1/f^4$ dependence occurs for the higher frequency part of the spectrum ($\geq$ 200 kHz). It means that small scale events are related to low power events in the obtained spectrum due to plasma background or instrumental noise. Furthermore, in Fig. 3 (b) and (c) we can observe a range with a $1/f$ dependence, this behavior has been related to the transport events linked with

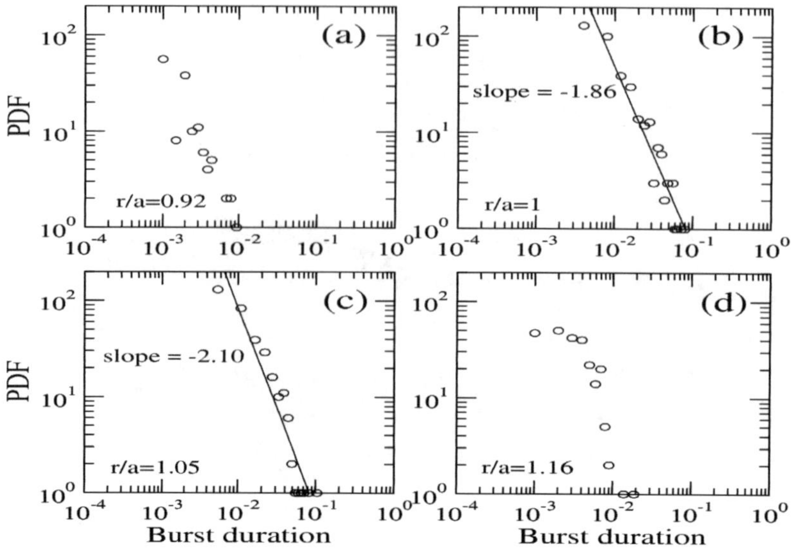

Figure 2. Probability density function of laminar periods (defined in the text), at (a) $r/a = 0.92$, (b) $r/a = 1$, (c) $r/a = 1.05$ and (d) $r/a = 1.16$.

self-organized similarity behavior. Moreover, in this figure the separation between frequency ranges are smooth. As a matter of fact, these results are in agreement with other experimental observations [14, 13]. On the other hand, Fig. 3(a) and (d) do not present regions with power-law $1/f$.

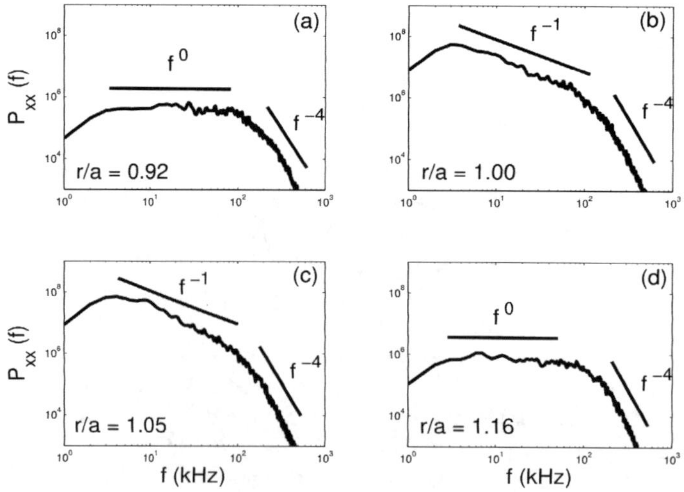

Figure 3. Power spectrum of the floating potential fluctuation at radial position (a) $r/a = 0.92$, (b) $r/a = 1$, (c) $r/a = 1.05$ and (d) $r/a = 1.16$.

The autocorrelation function (ACF) was obtained from the same shots considered in Fig. 3

is showed in Fig. 4. The width of the peak (the e-folding time of the ACF) is taken to be the decorrelation time of the local turbulence (see black dashed line). In Fig. 4(b) the e-folding time is nearby 0.05 ms [18] (at $r/a = 1$). Fig. 4(a-d) shows that the ACF in the the e-folding time depends on the radial coordinate [9]. Moreover, the autocorrelation function close to the plasma border shows an extended tail at large delay time. The existence of long-time correlation, at plasma edge, enforces the evidence of SOC behavior on TCABR plasma edge.

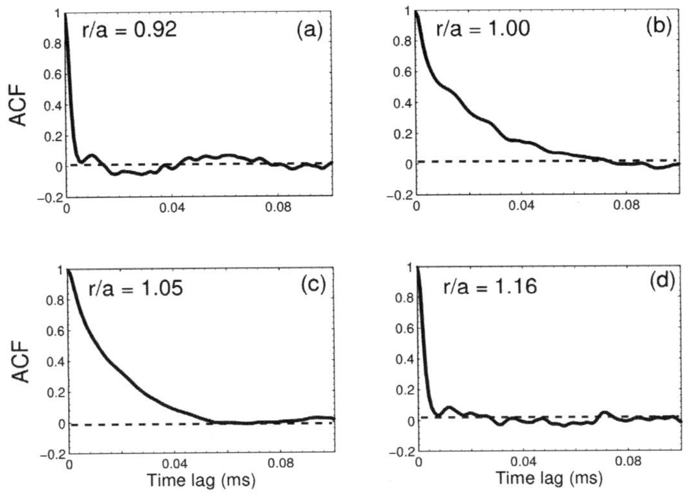

Figure 4. Autocorrelation function (ACF) of the floating potential fluctuations measured at radial position (a) $r/a = 0.92$, (b) $r/a = 1$, (c) $r/a = 1.05$ and (d) $r/a = 1.16$.

Other evidence of self-organized criticality on TCABR fluctuations is obtained through the Hurst parameter (with R/S method) [13, 19]. The parameter H is the self-similarity parameter. The value of the Hurst parameter varies between 0 and 1. For $H = 0.5$, the process is not strongly correlated. Moreover, for $0 \leq H \leq 0.5$ the time series are characterized like anti-persistent or anti-correlated, and for $0.5 < H \leq 1$ the signal is persistent or auto-correlated. Fig. 5 shows the radial dependence of the Hurst parameter of the TCABR plasma turbulence (red triangles), where we can see that the Hurst parameter decreases from the radial position $r/a = 1$ towards inside the plasma column (plasma edge). The same situation is observed for Hurst parameters that decreases in the scrape-off layer plasma according to the radial position moving away from $r/a = 1$. The variation of the TCABR Hurst values varies from 0.55 (on the scrape-off layer) to a maximum equal to 0.92 on the limiter ($r/a = 1$), indicating that the plasma turbulence, as well as plasma transport, present high self-organized behavior.

4. Conclusions

We analyzed the plasma edge turbulence in TCABR tokamak to verify the radial dependence of self-organized criticality behavior of the experimental fluctuating floating electrostatic potential, measured by Langmuir probes. The dependence of the laminar period probability density function at $r/a \approx 1$ is best recognized in a log-log plot, where the linear fit suggests a power-law decay. The frequency spectrum analysis presentes scaling behavior. Moreover, we can observe that the autocorrelation function shows an extended tail at large delay times and we also obtain a higher Hurst parameter, indicating the existence of self-organized criticality in TCABR fluctuations.

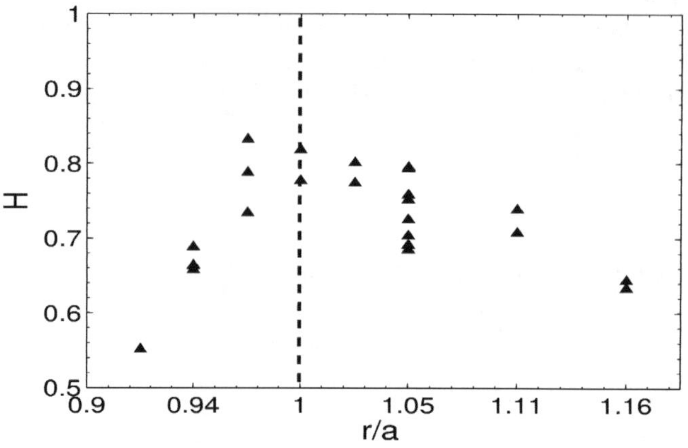

Figure 5. Radial dependence of the Hurst parameter (red triangles). The vertical black line marks the position of the plasma column ($r/a = 1$).

We have observed that our results are in agreement with SOC behavior around the radial position $r/a = 1$, as well as plasma transport mechanisms based on avalanches.

In tokamaks, the SOC radial dependence at the plasma edge reported in this article may be associated with coherent or recurrence structures propagating at this region. A theorethical description of plasma edge turbulence should predict these statistical properties and explain their influence on the anomalous particle transport in tokamaks.

Acknowledgments

This work was made possible with the help of CNPq, CAPES and Fundação Araucária (Brazilian Government Agencies).

References

[1] W. Horton 1999 Reviews of Modern Physics **71** 3 735.
[2] R. D. Hazeltine and S. C. Prager 2002 Physics Today **55** 7 30.
[3] C. Hidalgo 2004 Astrophysics and Space Science **292** 681.
[4] M. S. Baptista, I. L. Caldas, M. V. A. P. Heller, A. A. Ferreira, R. Bengtson, and J. Stöckel 2003 Physics of Plasmas **10** 5 1283.
[5] A. A. Ferreira, M. V. A. P. Heller, and I. L. Caldas 2000 Physics of Plasmas **7** 9 3567.
[6] E. G. Altmann, E. C. Silva, I. L. Caldas 2004 Chaos 14 975.
[7] Z. O. Guimarães-Filho, I. L. Caldas, R. L. Viana, I. C. Nascimento, Yu. K. Kuznetsov, and J. Kurths 2010 Physics of Plasmas **17** 0123031.
[8] C. R. Neto, Z. Guimarães-Filho, I. L. Caldas, I. C. Nascimento, Yu K. Kuznetsov 2008 Physics of Plasmas **15** 082311.
[9] Y. H. Xu, S. Jachmich, R. R. Weynants, A. Huber, B. Unterberg and U. Samm 2004 Physics of Plasmas **11** 5413.
[10] E. Spada, V. Carbone, R. Cavazzana, L. Fattorini, G. Regnoli, N. Vianello, V. Antoni, E. Martines, G. Serianni, M. Spolaore and L. Tramontin 2001 Physical Review Letters **86** 14 3032.
[11] P. Bak, C. Tang and K. Wiesenfeld 1987 Physical Review Letters **59** 4 381.
[12] P. A. Politzer 2000 Physical Review Letters **84** 1192.
[13] B. A. Carreras, B. Ph. van Milligen, M. A. Pedrosa, R. Balbín, C. Hidalgo, D. E. Newman, E. Sánchez, M. Frances, I. García-Cortés, J. Bleuel, M. Endler, C. Riccardi, S. Davies, G. F. Matthews, E. Martines, V. Antoni, A. Latten and T. Klinger 1998 Physics of Plasmas **5** 3632.
[14] F. Sattin and M. Baiesi 2006 Physical Review Letters **96** 105005.
[15] R. M. O. Galvão, Y. K. Kuznetsov, I. C. Nascimento, E. Sanada, D. O. Campos, A. G. Elfimov, J. I. Elizondo,

A. N. Fagundes, A. A. Ferreira, A. M. M. Fonseca, E. A. Lerche, R. Lopez, L. F. Ruchko, W. P. de Sá, E. A. Saettone, J. H. F. Severo, R. P. da Silva, V. S. Tsypin, R. Valencia and A. Vannucci 2001 Plasma Physics and Controlled Fusion **43** 9 1181.

[16] G. Z. dos Santos Lima, Z. O. Guimarães-Filho, A. M. Batista, I. L. Caldas, S. R. Lopes, R. L. Viana, I. C. Nascimento and Y. K. Kuznetsov 2009 Physics of Plasmas **16** 042508.

[17] M. A. Pedrosa, C. Hidalgo, B. A. Carreras, R. Balbin, I. Garcia-Cortés, D. Newman, B. van Milligen,E. Sánchez, J. Bleuel, M. Endler, S. Davies and G. F. Matthews 1999 Physical Review Letters **82** 18 3621.

[18] Ch. P. Ritz, H. Lin, T. L. Rhodes and A. J. Wootton 1990 Physical Review Letters **65** 20 2543.

[19] Hurst, H.E. 1951 Transactions of the American Society of Civil Engineers **116** 770.

Dynamic Days South America 2010 IOP Publishing
Journal of Physics: Conference Series **285** (2011) 012005 doi:10.1088/1742-6596/285/1/012005

On the effect of a parallel resistor in the Chua's circuit

Flavio Prebianca[1], **Holokx A Albuquerque**[1] **and Rero M Rubinger**[2]

[1] Departamento de Física, Universidade do Estado de Santa Catarina, 89223-100 Joinville, Brazil
[2] Departamento de Física e Química, Universidade Federal de Itajubá, 37500-903 Itajubá, Brazil

E-mail: `dfi2haa@joinville.udesc.br`

Abstract. We report a numerical bifurcation study on the Chua's circuit with parallel resistor. Through the largest Lyapunov exponent, we constructed a two-dimensional parameter space of the model. We also implemented the experimental circuit to show the similarities between the model and the experimental data. With that modification we discuss the effect of a parallel resistor in the dynamics of a Chua's circuit.

1. Introduction

Chua's circuit is a paradigm of chaotic behavior [1] and is extensively studied, with many experimental and/or theoretical works reporting important properties of this system [2, 3, 4, 5, 6, 7, 8].

In this work, we study the dynamical behavior of a Chua's circuit with parallel resistor. Our studies are based on a recent work of Braga *et al.* [9], where the authors analytically studied the governing nonlinear equations, where the nonlinearity of the diode is a cubic function, and the intrinsic inductor resistance was omitted. In our studies, we numerically study the complete set of equations, based on the Chua's circuit model [5, 6], with a piecewise-linear function for the nonlinearity, and adding a resistor in parallel with the inductor. Our aim is to obtain the global bifurcation behavior constructing two-dimensional parameter spaces of the model with the largest Lyapunov exponent method [5, 6, 10, 11, 12]. We also realized the experimental circuit, and we obtained the experimental phase portraits (attractors) for three parameters of the system.

In Fig. 1, we present the Chua's circuit with parallel resistor R_P. The dynamical variables are i_L, the current across the inductor, v_1 and v_2, the voltages across the capacitors C_1 and C_2, respectively. The set of equations that describes the circuit in Fig. 1 is given by

$$\begin{aligned}
\dot{v}_1 &= \frac{dv_1}{dt} = (v_2 - v_1)/(RC_1) - i_d(v_1)/C_1,\\
\dot{v}_2 &= \frac{dv_2}{dt} = (v_1 - \alpha v_2)/(RC_2) + i_L/C_2, \qquad (1)\\
\dot{i}_L &= \frac{di_L}{dt} = -v_2/L - i_L(r_L/L),
\end{aligned}$$

where

$$i_d(v_1) = m_0 v_1 + \tfrac{1}{2}(m_1 - m_0)\left(|v_1 + b| - |v_1 - b|\right),$$

and $\alpha = (R + R_P)/R_P$. R_P is the parallel resistor which modifies the standard Chua's circuit ($\alpha = 1$) [2, 5, 6].

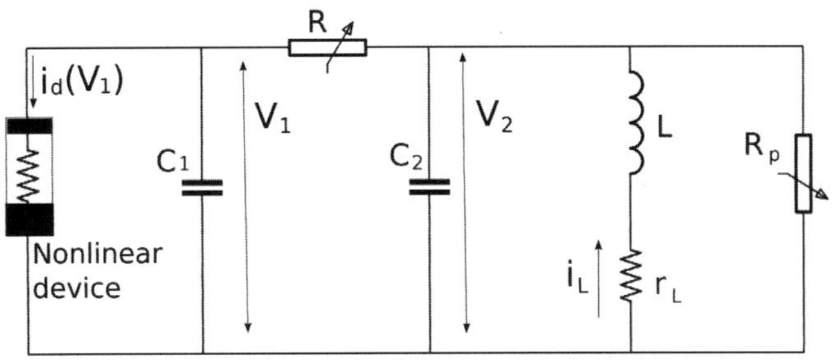

Figure 1. Scheme of the Chua's circuit with parallel resistor R_P. The dynamical variables $v_1(t)$, $v_2(t)$, and $i_L(t)$ correspond to the voltage across the capacitor C_1, the voltage across the capacitor C_2, and the current through the inductor L, respectively. The nonlinear element is the Chua's diode, with $i_d(v_1)$ nonlinear characteristic.

This paper is organized as follows: Sec. 2 is devoted to the results obtained from the numerical studies of Eqs. (1), Sec. 3 is devoted to the results obtained from the experimental studies, and the summary is given in Sec. 4.

2. Numerical Results

The numerical study carried out in this work consists of to calculate the largest Lyapunov exponent, numerically solving the Eqs. (1) using the fourth-order Runge-Kutta method with fixed time step for each pair of parameters (R, r_L). We tested the time step from 10^{-4} up to 10^{-1} and the system behavior remained the same, but to decrease the asymptotic convergence time to the attractor, we used the time step equal to 10^{-1}. The ranges of parameter values were discretized in a mesh of 500×500 points equally spaced. For each pair of parameters (R, r_L) we used the same initial condition $(x, y, z) = (0.1, 0.1, 0.1)$, and to calculate the largest Lyapunov exponent we firstly tested the convergence time of the exponent for many pairs (R, r_L), looking at the exponent behavior with the iteration time. The tests showed that 5×10^5 iterations are the optimal convergence time to estimate the largest Lyapunov exponent in most cases. We identify for each largest Lyapunov exponent a color, varying continuously from black (zero exponent), up to yellow (positive exponent). Blue color identifies the divergence cases of Eqs. (1).

Figure 2 shows the two-dimensional parameter space for the parameters (R, r_L) in Eqs. (1), with $\alpha \neq 1$, i.e., $R_P = 1600\Omega$. Black regions represent stable behaviors (fixed-points or periodic), and the yellowish regions represent unstable behaviors (chaotic). The blue color represents the divergence of Eqs. (1), and due to be large the convergence time of the largest Lyapunov exponent in some regions of the parameter space (Fig. 2), the black region marked with FP is also a fixed-point region. The white color represents fixed points for regions with small convergence time of the Lyapunov exponent. Inside the chaotic regions, we can observe the existence of immersed periodic structures, represented by the black regions inside of the yellowish region. Those periodic structures have a different accumulation boundary, where the periodic structures accumulate in a fixed-point region (marked with FP in Fig. 2) as the periodicity increases by period-adding cascade towards that region. Recent works [11, 12] reported accumulation boundaries characterized by periodic regions with periods equal to the period-adding factor, i.e., if the period-adding cascade increases the periodicity of the structures by a factor 2, the accumulation boundary is a period-2 region. However, in our case, the accumulation boundary is a fixed-point region.

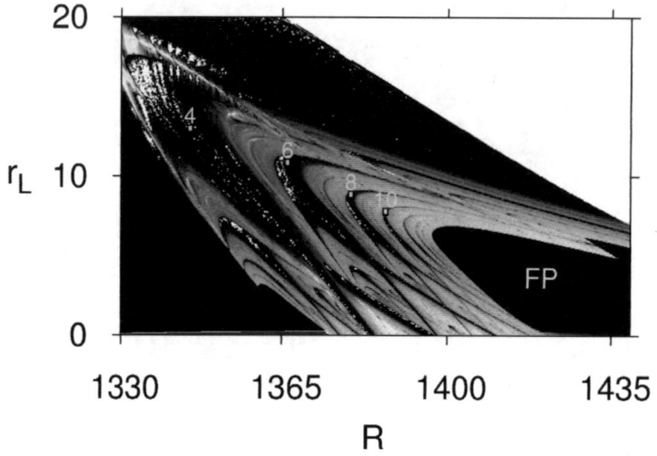

Figure 2. Parameter space diagram associating colors to different λ ranges (see the text). The numbers are the lowest periodicity of the structures. Here, R and r_L are in resistance units, Ω.

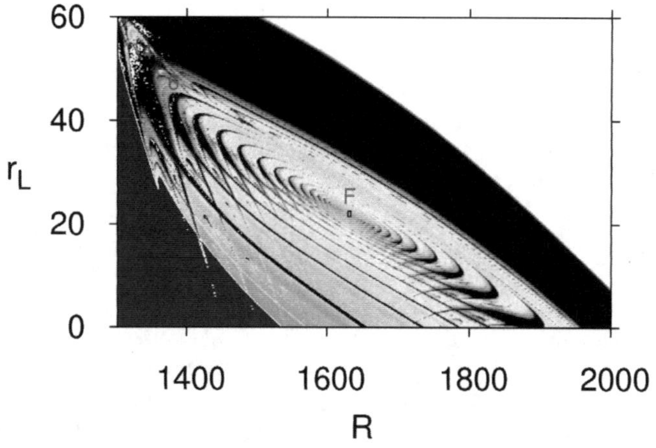

Figure 3. Parameter space of system (1), with $\alpha = 1$, associating intensity colors to different λ ranges (see the text). The numbers 4 and 6 are lowest periodicity of the structures. R and r_L are in resistance units, Ω. The periodic structures organize themselves in a single spiral structure that coils up around the focal point F, as reported in Ref. [6].

To discuss another effect of the parallel resistor R_P in the Chua's circuit, we show in Fig. 3 the parameter space of Eqs. (1) for $\alpha = 1$. In that configuration, the circuit, Fig. 1, is the standard Chua's circuit [6]. We observe periodic structures organizing themselves in a single spiral structure that coils up around a chaotic focal point in the parameter space. That behavior seems to be a common feature presented in various systems [6, 10] with homoclinic origin, and recently an experimental evidence of spirals in a chaotic electronic circuit was reported [13]. Comparing Figs. 2 and 3, we can observe the effect of the R_P resistor in the circuit's dynamics. The spiral structure was destroyed, and the remaining periodic structures reorganized in an accumulation boundary formation, as previously above described.

Figure 4 shows the attractors for parameter values represented by two green marks labeled by the numbers 4 and 6 in Figs. 2 and 3. For each attractor we numerically solved the Eqs. (1) using the fourth-order Runge-Kutta method with fixed time step equal to 10^{-1}, initial condition $(x, y, z) = (0.1, 0.1, 0.1)$, and with 5×10^5 iterations discarding the first 4.5×10^5 iterations. We observe similarities regarding periodicity between the pair of attractors belonging to distinct parameter spaces. Although the spiral has disappeared the first structures with lower periods survived.

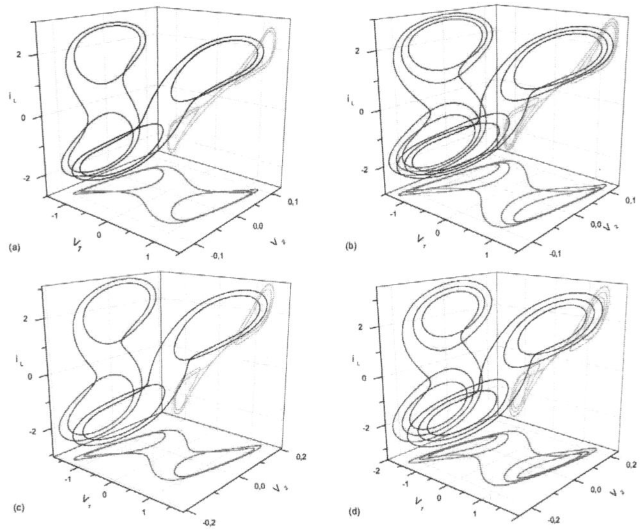

Figure 4. Three-dimensional attractors generated by Eqs. (1) for different values of the parameters R and r_L. (a) and (b) periodic attractors for the parameter values represented by two green marks labeled by the periods 4 and 6 in Fig. 2. (c) and (d) periodic attractors for the parameter values represented by two green marks labeled by the periods 4 and 6 in Fig. 3. The axes v_1 and v_2 are in voltage units (Volts). The axis i_L is in current units (Amperè).

3. Experimental results

We experimentally implemented the circuit shown in Fig. 1 using the standard Chua's circuit as base platform, and adding a potentiometer R_P in series with the inductor L. The component values for the standard Chua's circuit are the same reported in Ref. [2]. For the measurements we use a digital oscilloscope Tektronix TDS 2024B, four channels, 200 Mhz, and 2 GS/s. For the standard Chua's circuit, we use the *inductorless* implementation [14], where the inductor L is simulated by a additional circuit. In that configuration, the state variable i_L can be replaced by another voltage variable, so we measure three voltages as state variables of the circuit. To evaluate the state variable i_L, we use the inverse transformation obtained by the Kirchhoff's laws in the simulated inductor. For details about the *inductorless* implementation and the transformations, please see Refs. [2, 14].

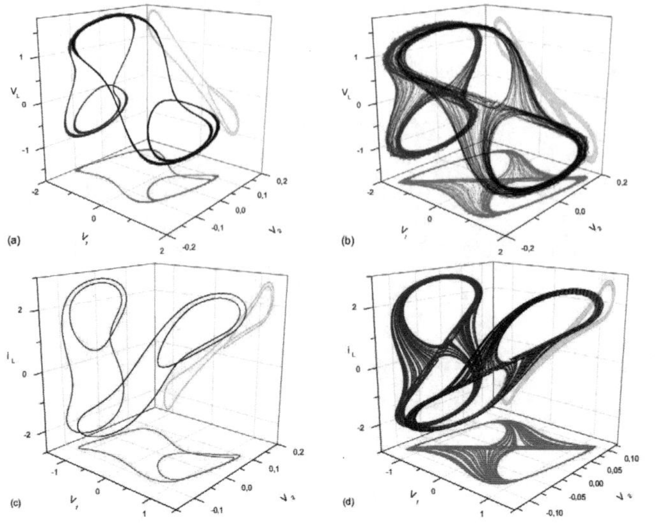

Figure 5. Three-dimensional attractors generated by experiment, (a) and (b), and generated by Eqs. (1), (c) and (d), for different values of the parameters R, r_L, and R_P. (a) $R = 1362\Omega$, $r_L = 1.2\Omega$, and $R_P = 2217\Omega$; (b) $R = 1375\Omega$, $r_L = 1.0\Omega$, and $R_P = 2261\Omega$. (c) $R = 1352.22\Omega$, $r_L = 7.244362\Omega$, and $R_P = 1600\Omega$; (d) $R = 1355.88\Omega$, $r_L = 5.335915\Omega$, and $R_P = 1600\Omega$. The axes v_1, v_2 and v_L are in voltage units (Volts). The axis i_L is in current units (Amperè).

Figure 5 shows two experimental attractors and two corresponding numerical attractors. For the experimental attractors, Figs. 5(a) and 5(b), the parameters are (a) $R = 1362\Omega$, $r_L = 1.2\Omega$, and $R_P = 2217\Omega$; (b) $R = 1375\Omega$, $r_L = 1.0\Omega$, and $R_P = 2261\Omega$. For the numerical attractors, Figs. 5(c) and 5(d), the parameters are (c) $R = 1352.22\Omega$, $r_L = 7.244362\Omega$, and $R_P = 1600\Omega$; (d) $R = 1355.88\Omega$, $r_L = 5.335915\Omega$, and $R_P = 1600\Omega$. The topological similarities for both experimental and numerical attractors, Figs. 5(a) with 5(c), and Figs. 5(b) with 5(d), are evident despite the difference between experimental and numerical parameters.

In Fig. 6 we show evidences of Shilnikov-type homoclinic chaos, with two different global turns, as reported in Ref. [4] for other modified Chua's circuit. Figs. 6(a) and 6(c) are the 3D attractors, and Figs. 6(b) and 6(d) are the experimental time series for the v_1-variable, respectively. These attractors clearly show that a double or a triple spikes repeats with intermediate intervals of large amplitude oscillations.

4. Summary

A two-dimensional parameter space, using the largest Lyapunov exponent codified in a continuous range of colors, for the Chua's circuit with parallel resistor is reported. With that modification, we observe the disappearance of the spiral structure and the emergence of a fixed points region. A new accumulation boundary sequence is reported. In the experimental implementation of the circuit, we observe a good agreement between theoretical and experimental attractors. Two experimental chaotic attractors are observed, showing experimental evidences of Shilnikov-type homoclinic chaos in that circuit.

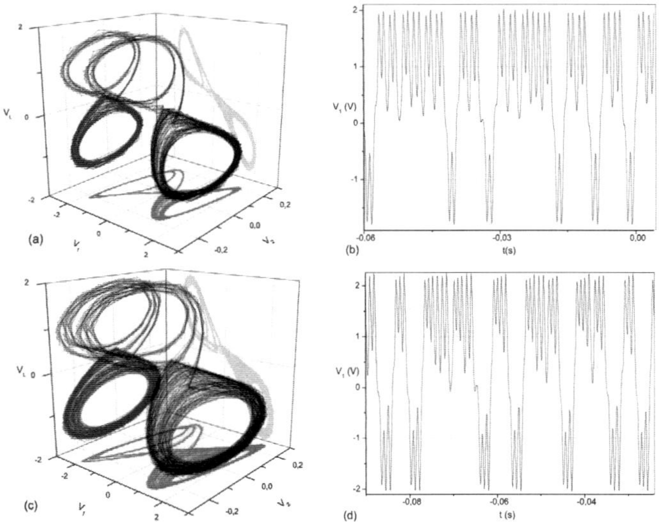

Figure 6. Three-dimensional attractors and time series generated by experiment for different parameter values R, r_L, and R_P. (a) and (b) for $R = 1384\Omega$, $r_L = 8.0\Omega$, and $R_P = 2218\Omega$. (c) and (d) for $R = 1434\Omega$, $r_L = 1.5\Omega$, and $R_P = 2238\Omega$. The axes are in voltage units (Volts).

Acknowledgements
The authors would like to acknowledge the Brazilian agencies Fundação de Apoio à Pesquisa Científica e Tecnológica do Estado de Santa Catarina-FAPESC, and Conselho Nacional de Desenvolvimento Científico e Tecnológico-CNPq, for the financial support.

References
[1] Chua L O 1994 *J. Circ. Sys. Comp.* **4** 117
[2] Albuquerque H A, Rubinger R M, Rech P C 2007 *Physica D* **233** 66
[3] Maranhão D M, Baptista M S, Sartorelli, Caldas I L 2008 *Phys. Rev. E* **77** 037202
[4] Chakraborty S, Dana S K 2010 *Chaos* **20** 023107
[5] Albuquerque H A, Rubinger R M, Rech P C 2008 *Phys. Lett. A* **327** 4793
[6] Albuquerque H A, Rech P C 2010 *Int. J. Circ. Theor. Appl.* http://dx.doi.org/10.1002/cta.713, in press
[7] Shilnikov L P 1993 *IEEE Trans. Circ. Sys. I* **40** 784
[8] Medrano-T R O, Baptista M S, Caldas I L 2005 *Chaos* **15** 033112
[9] Braga D C, Mello L F, Messias M 2009 *Math. Probl. Engin.* **2009** 149563
[10] Gallas J A C 2010 *Int. J. Bif. Chaos* **20** 197
[11] Bonatto C, Gallas J A C 2007 *Phys. Rev. E* **75** 055204(R)
[12] Bonatto C, Gallas J A C 2008 *Philos. Trans. R. Soc. London, Ser. A* **366** 505
[13] Stoop R, Benner P, Uwate Y 2010 *Phys. Rev. Lett.* **105** 074102
[14] Tôrres L A, Aguirre L A 2000 *Electronics Lett.* **36** 1915

Dynamic Days South America 2010 IOP Publishing
Journal of Physics: Conference Series **285** (2011) 012006 doi:10.1088/1742-6596/285/1/012006

Hyperbolic kaleidoscopes and Chaos in foams and Hele-Shaw cell

A P B Tufaile[1], A Tufaile[1], G Liger-Belair[2]

[1]Escola de Artes, Ciências e Humanidades da Universidade de São Paulo, R. Arlindo Bettio, 1000, 03828-000, São Paulo, Brazil

[2]Laboratoire d'OEnologie et Chimie Appliquée, UPRES EA 2069, URVVC, Faculté de Sciences de Reims, Moulin de la Housse, B. P. 1039, 51687 Reims, Cedex 2, France

E-mail: atufaile@usp.br

Abstract. Liquid foams have fascinating optical properties, which are caused by the large number of light refractions and reflections by liquid films and Plateau borders. Due to refraction and reflection at the interfaces, the direction of the rays leaving a Plateau border can vary greatly for the same incident angle and a small positional offset. A close look in some configurations of the Plateau borders or liquid bridges reveals the existence of some triangular patterns surrounded by a complex structure, and these patterns bear a resemblance to those observed in some systems involving chaotic scattering and multiple light reflections between spheres. Provided the optical properties of the sphere surfaces are chosen appropriately, fractals are natural consequences of multiple scattering of light rays in these cavities. The cavity acts as a hyperbolic kaleidoscope multiplying the scattering of light rays generating patterns related to Poincaré disks and Sierpinski gaskets in comparison to linear kaleidoscopes. We present some experimental results and simulations of these patterns explained by the light of the chaotic scattering.

1. Introduction

Foams have been intensively investigated for many years as natural phenomena that are prominent in everyday life. The propagation of light in foams has received attention, and a number of patterns are generated spontaneously due to the reflection and refraction of light. One of these patterns is observed by the formation of some triangular images inside liquid bridges in a single layer of bubbles between two plates, known as the Hele-Shaw cell. By inspection, these patterns bear a resemblance to those observed in some systems involving chaotic scattering and multiple light reflections between spheres [1]. These patterns can be obtained using mirrored spheres, and basically the image obtained is due to the fact that the different spheres are mirrored in each of the other spheres giving rise to multiple mirror images [2]. Similar systems presenting fractal-like patterns have been studied by some authors [3, 4, 5].

The use of fractal characteristics may have a potential to control and localize light in a different way from the conventional optical devices, and lead to new engineering applications of light spreading devices such as optical links [6]. For the case of liquid bridges and Plateau borders, the curvature of liquid/air interface acts as scatterer, and the direction of the rays leaving the liquid bridges can vary greatly for the same incident angle with only a small positional offset, due to refraction and reflection at the interface, and consequently the occurrence of chaotic scattering is also possible. In this paper we experimentally observed these fractal-like patterns of chaotic scattering of light in hyperbolic kaleidoscopes and foams.

The goal of this paper is to describe the existence of these triangular patterns in foams and their relation with the amazing images obtained from the chaotic scattering of light in spheres and spherical shells. We are going to look how these patterns observed in spheres are related with fractal dimension. After that, we associate some aspects of chaotic scattering of cavities with negative curvature with patterns observed in hyperbolic kaleidoscopes. Some dynamical features of the light scattering of these hyperbolic optical elements can be explored using non-Euclidean geometry and recurrence equations. In addition to this, we look at some aspects of Kolmogorov-Sinai entropy in soft billiards in order to throw some light in similar systems with light scattering in the presence of reflection and refraction.

2. Experimental Apparatus and methods

We have used different setups, involving Hele-Shaw cell, reflective spheres and hyperbolic prims. Some images were obtained in a transparent Hele-Shaw cell, the setup I, consisting of two plain parallel Plexiglas plates separated by a small gap (20 x 20 x 0.2 cm^3). The cell contains only air and an amount of commercial dishwashing liquid (V = 30 cm^3). This liquid is manufactured by Bombril, and is used without dilution. The essential surfactant is Linear Alkylbenzene Sulfonate (LAS). The surface tension is γ = 25 dyne/cm, and the density of this detergent is ρ = 0.95 g/cm^3, with refractive indices of liquid n_l = 1.333, and n_g = 1.0 for the air, see reference [7] for more details. For the case of reflective spheres and spherical shells, the configurations were constructed with two sets of reflective surfaces, the setup II is a set of three reflective spheres.

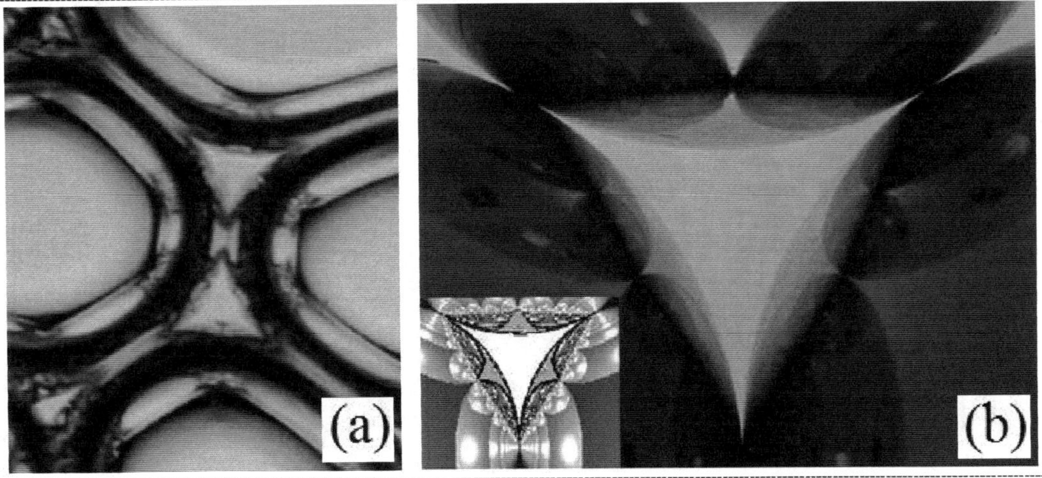

Figure 1. (a) The liquid bridges observed in a Hele-Shaw cell and some intriguing triangular patterns. (b) Image obtained from three Christmas ball ornaments similar to those obtained from liquid bridges in Hele-Shaw cells. Inset, simulation obtained with ray tracing technique for a cavity inside three spheres showing the triangular pattern.

The setup III is a set of three reflective spherical shells in order to construct hyperbolic kaleidoscopes. A small sphere was placed between the spherical shells as the real object of a hyperbolic kaleidoscope. We also have constructed hyperbolic prisms with refractive index n_p = 1.32, in order to obtain images of multiple reflections of light in "solid" foams using a beam from a diode laser (650 nm, 5 mW), that we call setup IV. The light source used when photographing setups I, II and III was a 30W fluorescent lamp, with a diffusive screen and backlighting technique. We have used some routines written in C and commercial software of ray-tracing image (Mathematica) in order to simulate the light scattering in spheres and spherical shells.

3. Triangular patterns

In figure 1, there are some examples of these patterns obtained with our experimental apparatus. In figure 1(a), the liquid bridges observed in a Hele-Shaw cell have scattered the light forming some intriguing triangular patterns. These images are compared to the image obtained from three Christmas ball ornaments in figure 1(b). The inset of figure 1(b) is the simulation obtained with ray tracing technique for a cavity inside three spheres showing the same triangular pattern. The liquid bridges shown in figure 1(a) act as curved refractive-reflective surfaces which change with time. For example, in figures 2(a)-(b), the foam gets thinner over time, leading to a variation of these patterns with time.

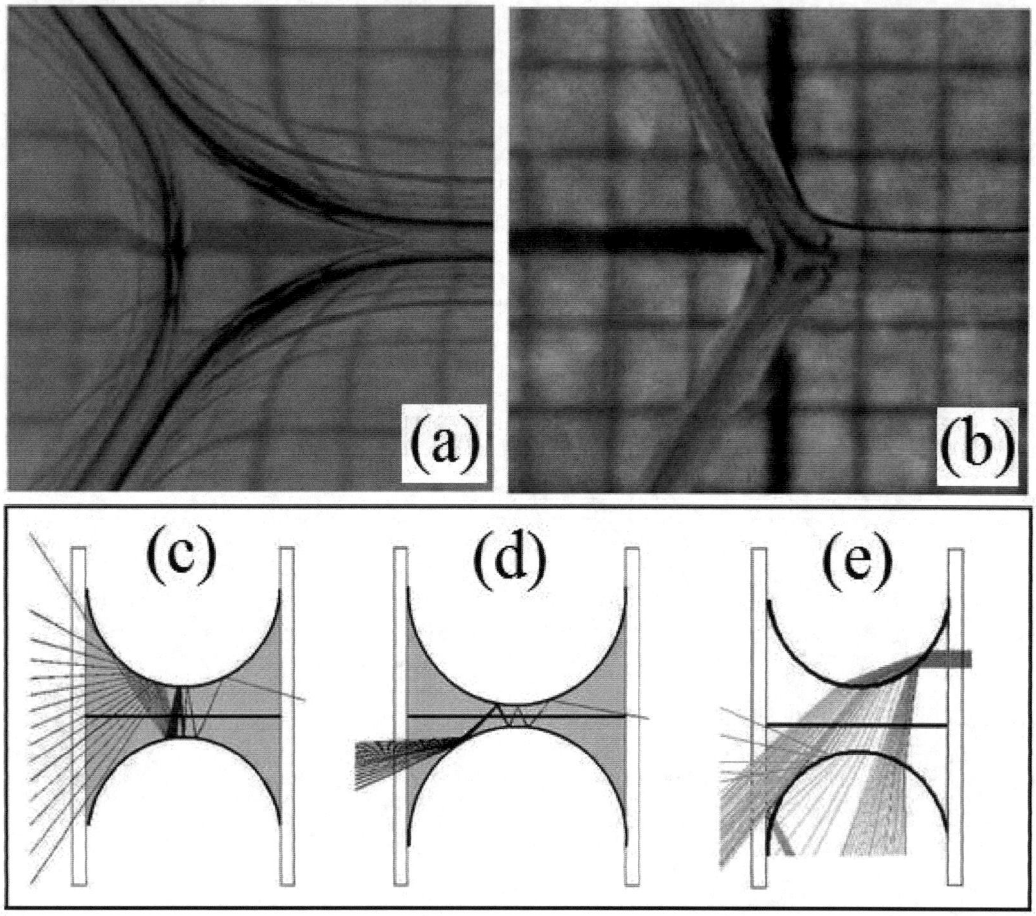

Figure 2. (a)-(b) The effects of the light refraction in the liquid bridges can be observed by the distortion of the frame behind the Hele-Shaw cell. From (a) to (b) we can observe the difference provided by the liquid drainage. (c)-(d) Sketch of the rays paths in pure reflective cavities, and (e) the rays paths in liquid bridges.

We can observe the net result of the foam drainage by the optical deformation of a square grid behind the cell. Assuming that the liquid wets perfectly the plates, the liquid bridge boundary is made up of arcs of circles. For a pure reflective case of figures 2(c)-(d), the geometry of the cross-section of spherical shells affects its scattering properties. Taking into account the refraction and reflection, the light rays with only a small difference in the incident position impacting the liquid/air interface may cause different light paths, as it is shown in figure 2(e), and consequently the occurrence of chaotic scattering is also possible. Moreover, we can see that light reflection is much stronger in the pure reflective case than in the liquid bridges and the Plateau borders, which implies that it is mainly the multiple reflections, which attribute to the generation of the triangular patterns. Quite generally, these patterns are related to some geometrical features of the set up and the fractal dimension of the chaotic repeller formed by the cavity between the spheres. We will discuss some of these features in the next section.

4. Fractal dimension and kaleidoscopes

The fractal mirror images can be explored using optical billiards, which consist of cavities between polished, reflective surfaces of spheres. Exploiting multiple reflections of light between the spheres, a variety of fractal optical effects can be observed, and some interesting properties of these optical systems were summarized in reference [8].

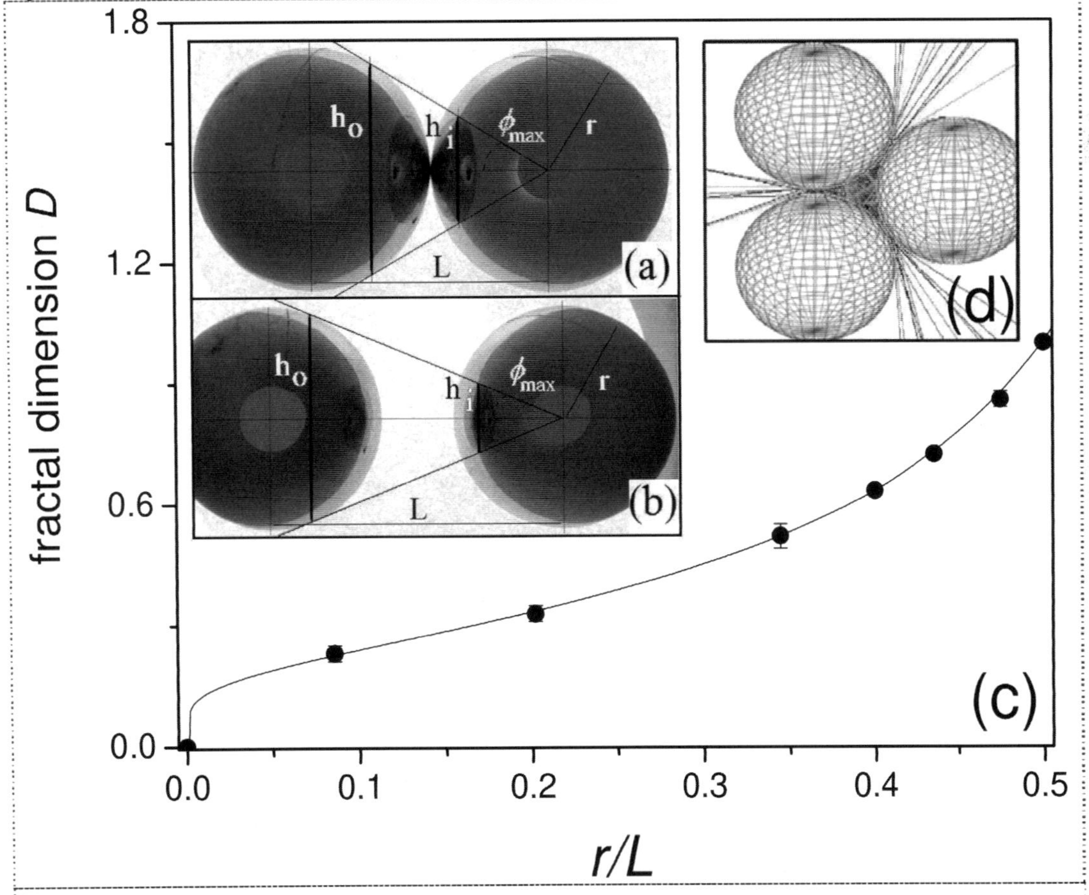

Figure 3. (a) The lateral reduction is related to the size of the image h_i, and the apparent size of the sphere h_o given by the semiangle ϕ_{max}. Example of $r/L = 0.5$. (b) Example of the lateral reduction for the case of $r/L = 0.4$. (c) The fractal dimension obtained with equation (7) compared to numerical values obtained from references [3, 9, 10]. (d) Multiple scattering of a pencil of trajectories in three sphere system for the case of $r/L = 0.49$, with the fractal dimension close to 1.

A three dimensional optical billiard can be considered as a discrete time dynamical system mapping a vector describing an incident ray at the *nth* reflection from the scatterer to the $(n + 1)th$ reflection, and these reflections create the mirror images. These mirror images show a fractal structure, whose fractal dimension increases with the number of spheres in a cluster as well as with the ratio of the radius of each sphere, r, to the distance between two spheres, L. The fractal dimension D has been discussed and calculated numerically for the case of relatively small r/L ratios by Eckhardt [9], while Korsch and Wagner [3] studied the fractal dimension D of mirror images for larger values of r/L ratios. We have to define the lateral reduction factor p, which is characterized by the ratio involving the image of the object and the object, creating the effect of lateral reduction. Assuming large angle of reflection, and taking into account the distance L between two identical spheres with radius r, the lateral reduction is defined by the height of the image h_i divided by the height of the part of the object h_o, figures 3(a)-(b), given by $p = h_i/h_o$, and it is approximately given by $p = [2(L/r-1)]^{-1}$ [3]. The rays reaching the surface of the object from the center of the mirrored sphere must lie within a cone of semiangle ϕ_{max}, given by $sin(\phi_{max}) = r/L$, which gives h_o smaller than the sphere diameter $d = 2r$. For example, for the case of spheres in contact ($r = L/2$) represented in figure 3(a), the lateral reduction is $p = 0.5$. In figure 3(b), increasing the distance between the spheres so that $r/L = 0.4$, the lateral reduction is $p = 1/3$.

For the case of three spheres the number of mirror images, $m = 2$, there is an approximate formula for the fractal dimension. The fractal sets can be considered as a limiting set of n reflections, for n going to infinite, where each reflection is the reduction of the set to $1/m$ of the previous set (See reference [10]). The scaling factor s^{-1} enables us to generalize the dimensionality of this approach. When A_n is the total measure of reflection n, it can be covered by $N(\varepsilon_n) \sim \varepsilon^{-D}$ intervals of size $\varepsilon = A_n m^{-n}$ and we obtain:

$$A_n = \varepsilon_n N(\varepsilon_n) \sim \varepsilon^{1-D} = (A_n m^{-n})^{1-D}, \tag{1}$$

and the A_n will decrease exponentially for large values of n, so that

$$A_n \sim m^{-n(1/D-1)} = e^{-\gamma n}, \tag{2}$$

with the rate γ given by

$$\gamma = (1/D-1)\ln m. \tag{3}$$

It follows that

$$D = \frac{\ln m}{\ln m + \gamma} = \frac{\ln m}{\ln(m\delta)}, \tag{4}$$

where

$$\delta = e^{\gamma} \approx \delta_n = \frac{A_{n-1}}{A_n} \tag{5}$$

is approximately the ratio between the measure of two successive reflections. For a self-similar set with scaling factor s we have $A_n = mA_{n-1}/s$, and we recover the definition of self-similar set as

$$D = -\frac{\ln m}{\ln p}, \tag{6}$$

which gives in the case of spheres:

$$D = \frac{\ln m}{\ln[2(L/r-1)]}. \tag{7}$$

In figure 3(c) is shown the fractal dimension for the case of three spheres with different *r/L* ratios. This fractal dimension is a measure of the chaoticity of the dynamics. In order to explain this system, we can represent the three spheres as a sequence of partitions formed by planar three hard disk system. The dimension of each partition reduces to zero in the limit *r/L* approaching to 0, and to unity for $r = L/2$, for the case of spheres in contact. For example, for the case of $r/L = 0.4$ the fractal dimension $D = 0.631$. In this way, for three spheres the fractal dimension ranges from 0 to 1 depending on the *r/L* ratio.

The existence of chaotic scattering was obtained numerically by Poon *et al.* [5] in a similar system, in which a three-disk system presents chaotic saddle with Wada property, and depending on the angle of incidence, it is conjectured the existence of trajectories trapping the light rays. However, in real systems it is not possible to observe this kind of trajectories because the reflectance of the sphere surface is not perfect. In addition to this, the observed images are created by the light rays which enter from one side of the cavity and exit at the opposite side. In systems like these, there are no trapped orbits, but there are orbits in which the light ray will leave the scattering region only after many collisions. José *et al.* [11] have reported the existence of stable and unstable manifolds in the elastic particle scattering from two hard disks. It is important to note that, two orbits starting out close together on the same side of the stable manifold may have their last hits on different disks and hence leave the scattering region in a quite different direction. This sensitiveness to initial condition can be observed in figure 2(c). The closer to the stable manifold, the grater is the difference of the directions of the light rays. Consequently, the light rays in the three sphere system are scattered by a cavity with properties very close to those proposed by the theory of chaotic scattering.

5. Hyperbolic geometry and Sierpinski gasket

At this point it is interesting to understand the geometric structure of the images obtained from the experiment. These multiple reflections are related to some properties of kaleidoscopes. For the case of regular three-mirror kaleidoscope, it creates an image that repeats one triangular pattern continuously, as it is shown in figure 4(a), hence producing a faceted pattern that fills the entire field of view, with a uniform tilling of the plane with triangles. For the case of the cavity between three spheres, the image produced shows a three-dimensional appearance in comparison with a regular kaleidoscope. We can consider that the cavity acts as a hyperbolic kaleidoscope, multiplying and distorting the images of some object placed at the center of the cavity.

The coding of generalized straight lines on curved surfaces or geodesics on a surface of negative curvature can be found in reference [12]. In our case, this coding represents the tessellation of triangles forming the border of the sequence of reflections from the initial conditions on a spherical surface. This approach involves the Fuchsian groups, which do for non-Euclidean geometry what crystallographic groups do for Euclidean geometry. In order to give a hint of this kind of image distortion, we can explore some aspects of non-Euclidean geometry. In general, the mathematics of kaleidoscopes in *N* dimensions is the study of finite groups of orthogonal *N* x *N* real matrices that are generated by reflection matrices. In that way, patterns observed in kaleidoscopes are related to Möbius transformations such as reflection symmetries, even in a non-Euclidean space. The three mirrored spheres could be an analogy to a stereographic projection of a regular kaleidoscope. Such stereographic projection is the Poincaré hyperbolic disk, as the image obtained experimentally by us in figure 4(b), with hyperbolic tensor metric [13]:

$$ds^2 = \frac{dx^2 + dy^2}{(1 - x^2 - y^2)}. \qquad (8)$$

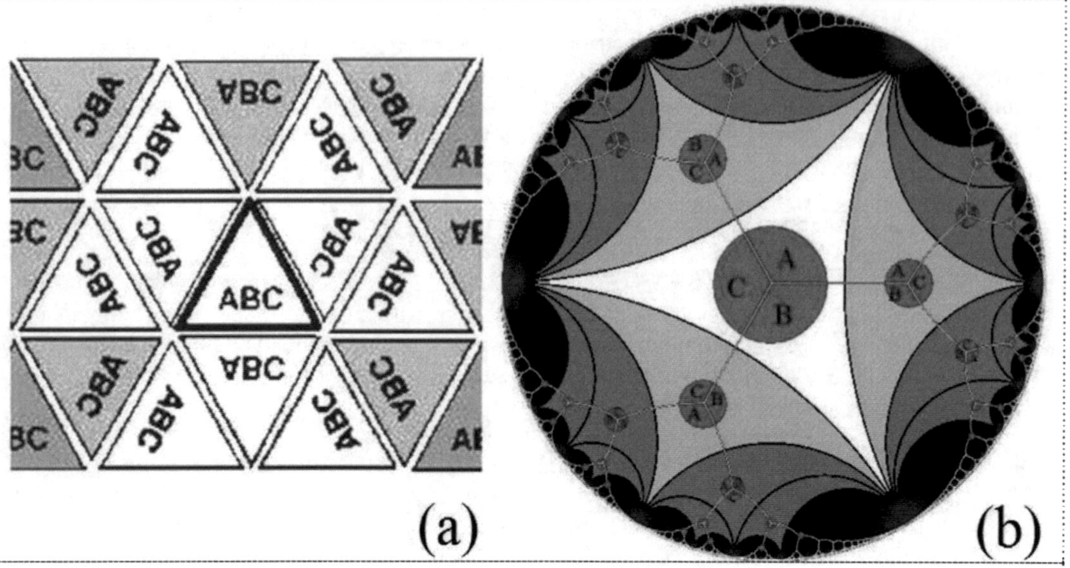

Figure 4. (a) Diagram of the reflection of a regular three-mirror kaleidoscope. (b) Poincaré disc model and lines connecting the reflected object forming the Serpinski-like gasket.

Using this metric, the lines are circular arcs that cut the boundary of the disk at right angles, and the angles in this model are the same as the Euclidean angle, the lines are badly distorted. We can observe these effects of the kaleidoscope placing an object at the center of the cavity (figure 5). Some of these transformations, called isometries, are indicated by the reflections and rotations of the letters ABC in the Poincaré disk model, as it is seen in figure 4(b). The parallel lines extending to the infinity in the kaleidoscope of plane mirrors are transformed in accumulation points at the border of the Poincaré disk. The regions ABC related to different initial conditions of the multiple reflections form a fractal-like pattern. Notice that as the lines separating each region approaches the border of the disk, the image repeats in finer and finer detail, approaching infinity.

The outstanding aspect of these images (figures 4(b) and 5(a)-(b)) is the representation of the Sierpinski gasket in the reflections [4]. However, it is not an exact Sierpinski gasket because it is distorted by the curvature of the sphere surfaces, consequently the sections are only triangular in shape instead of being true triangles, and the vertices are lost in the multiple reflections that arise where two spheres touch.

To illustrate a similar Sierpinski gasket effect in a three-sphere system, we have obtained the image of figure 5(a). Using ray trace technique, we have simulated three spherical shells touching each other. Inside the cavity between these spherical shells, there is a small sphere acting as object of this kaleidoscope. The bright and the contrast were adjusted to enhance the sequence of reflections of this small sphere. Figure 5(b) reproduces the image of a hyperbolic kaleidoscope with a small sphere acting as the object placed in the cavity. The configuration of this hyperbolic kaleidoscope correlates with ray tracing arrangements of spheres of different sizes.

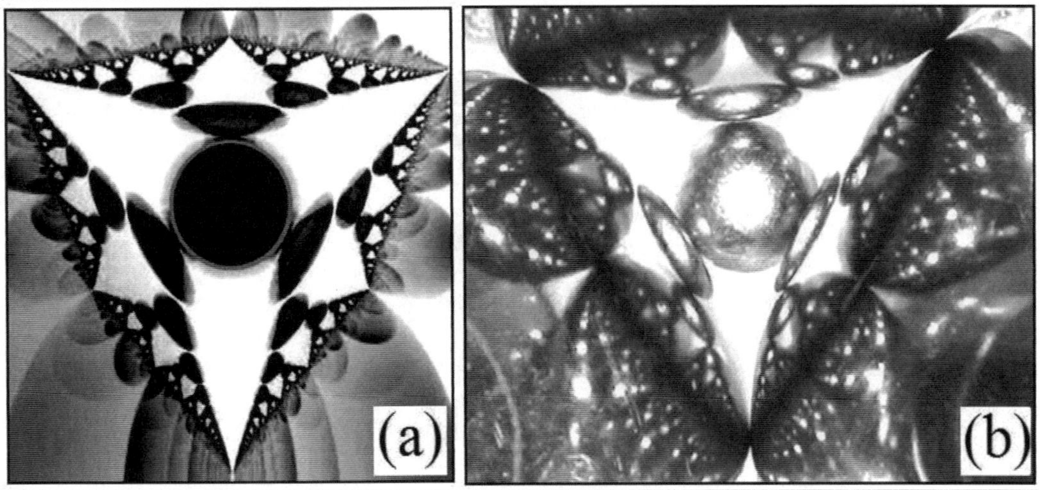

Figure 5. (a) Simulation of a three spherical shells acting as a hyperbolic kaleidoscope using a ray-tracing software and (b) image of a hyperbolic kaleidoscope using three spherical shells and a small sphere placed at the center (setup III).

6. Reflection and Refraction

The analogy between chaotic scattering in dynamical systems and light scattering in liquid bridges is based on the fact that in both cases there is a bounded area in which light rays or particles bounce back and forth for a certain time. For example, in figure 6(a), we present a configuration of liquid bridges in a Hele-Shaw cell containing three triangular zones with multiple reflections and refractions. We have compared one of these regions (figure 6(b)) with the pure reflective case of a similar configuration using reflective shells (figure 6(c)). In both images we can observe the existence of a distorted triangle surrounded by a toothed edge. We can observe some differences between these two images. In figure 6(b) there are black regions in the edge of these triangles, because most part of the light was refracted and reflected internally, while in figure 6(c) there is no black region, and the edges are well defined.

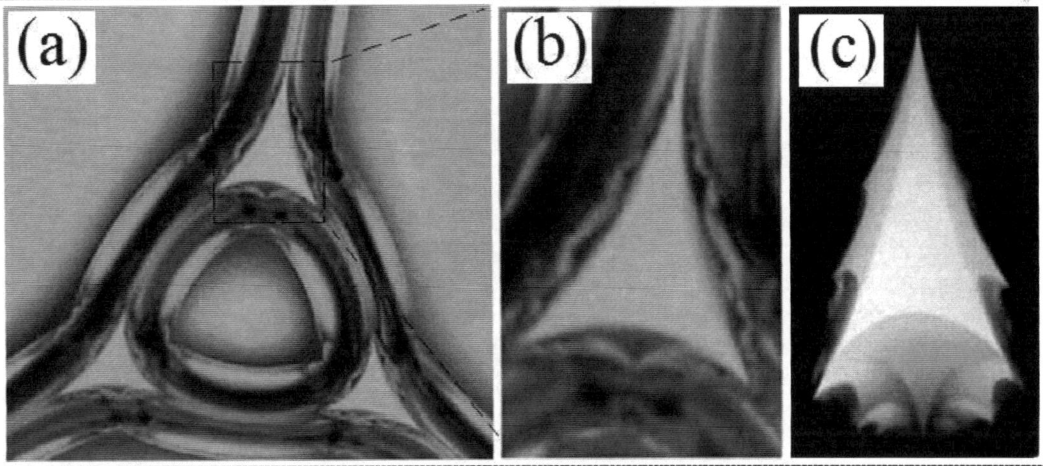

Figure 6. (a) Foam in a Hele-Shaw cell with three regions with patterns of light scattering (setup I). (b) Blow up of one of these triangular patterns. (c) Pure reflective case of a similar configuration (setup II).

The high curvature shapes of the liquid bridges can cause incident light to be scattered in many different directions through reflection, in addition to refraction [14]. Given an arbitrary frame of reference, any incoming ray can be identified by its direction θ_i and by the impact parameter b_i. The impact parameter is proportional to the angular momentum $I_i = kb_i$, where k is the wave number. The geometrical relationship between the incident and outgoing ray is given by Snell's law

$$n_i \sin\theta_i = n_t \sin\theta_t, \tag{9}$$

where n_i, n_t are the refractive indices of two media, θ_i is the angle of incidence and θ_t is the angle of refraction.

For the case of a gas/liquid interface the law of reflection states that $\theta_i = \theta_r$ (reflected), and when $n_i > n_t$ there exists the total internal reflection if the angle of incidence is above the critical angle, θ_c, and the reflectivity at normal incidence, R_{rf}, are given by

$$\begin{aligned} \theta_c &= \arcsin\left(\frac{n_g}{n_l}\right), \\ R_{rf} &= \left(\frac{n_l - n_g}{n_g + n_l}\right)^2. \end{aligned} \tag{10}$$

The fraction of the light that is reflected is given by the Fresnel reflection coefficient, a value which depends on the indices of refraction of liquid and gas (n_l, n_g) and also on the incident angle of the light. A fraction of the light entering the liquid will also be reflected inside the liquid. According to Suffern [4], the Sierpinski gaskets obtained by reflection are result of reflections at nearly normal incidence. For the case of gas/liquid interfaces, the reflectivity decreases the observation of Sierpinski gaskets, because the reflectivity of this interface at normal incidence is about 2 %, in contrast to the case of polished gold, about 50% in white light [15]. Due to the curvature and physical properties, the liquid bridges can be seen as hyperbolic prisms.

The mapping relating the incoming coordinates between successive collisions described in reference [16] is given by:

$$\theta_o = \theta_i, \tag{11}$$

$$a_o = \begin{cases} a_i, (if \sin^2(\theta_i) > 1 - U), \\ a_i + 2\arccos\left[\dfrac{(\sin\theta_i)}{\sqrt{1-U}}\right], (otherwise), \end{cases} \tag{12}$$

where the o subscripted stands for output and we called this mapping of a map.

Equations (12) represent the cases of reflection and refraction of the particle, in which U is the finite potential height of the interior of all disks with radius r measured in units of kinetic energy. We also have the light rays interacting with the scatterers in the plane centered at sites of a triangular lattice, with distance $L = 2\sqrt{3}$, in which we have the T map with the relations between (θ_o, a_o), (θ_i, a_i) and (b_o, β_o) [16]:

$$T = \begin{cases} R(\sin\beta_i - \sin\theta_o) = M\sin(a_i + \theta_o) + N\cos(a_i + \theta_o), \\ R(\cos\beta_i - \cos\theta_o) = -M\cos(a_o + \theta_o) + N\sin(a_i + \theta_o), \\ \pi + b_i = a_o + \theta_o + \beta_o. \end{cases} \tag{13}$$

The location of the scatterers is determined by a pair (*N*, *M*). The output of this model is used in the computations of the entropy of section 8.

7. Hyperbolic Prism acting as a hyperbolic kaleidoscope

We have used three spheres, touching each other, as a mold to make hyperbolic prisms. The material has a refractive index equal to 1.32 and it scatters some of the light passing inside, so we can have the image of the light beam inside the prism. The three spheres mold corresponds to the bubbles of the foam. As in the liquid bridges, we have a material more refractive than the air in the bubbles. Similarly to the case of hyperbolic kaleidoscopes, the total internal reflection of light can be observed in hyperbolic prisms. For an incoming laser beam we can observe the quasi-self-similarity of the reflections as it is shown in figure 7(a).

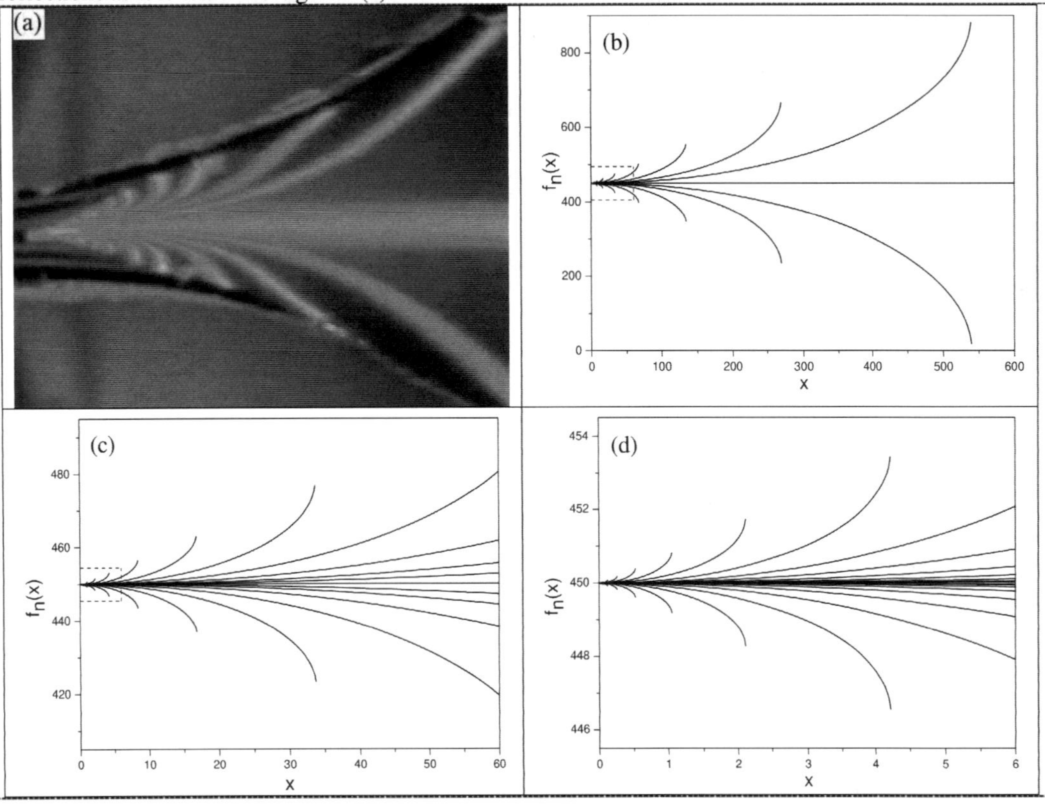

Figure 7. (a) Image of a laser beam injected in a hyperbolic prism with multiple internal reflections (b) Plot of recurrence equation obtained from hyperbolic prism. (c) Zooming the region inside the dashed lines of figure 7(b) we can observe a similar pattern. (d) Another magnification of the middle left region inside the dashed lines of figure 7(c) contains a small copy of the previous pattern.

The projection of a straight line in a spherical surface is a circle in the plane determined by the line and the center of the spherical surface. In any other plane, the image is an ellipse. Considering the Möbius transformations due to multiple reflections discussed previously, we have obtained a recurrence equation given by fitting an elliptical curve to the image of Fig. 7(a):

$$f_n(x) = \sqrt{-P_1 x^2 + \frac{P_2}{2^{2n}}} + \sqrt{P_2}\left(1 - \frac{1}{2^n}\right), \tag{14}$$

with $P_1 = 0.695(2)$ and $P_2 = 203{,}100(200)$, using pixel units. Using the symmetry of the pattern of Fig. 7(a), the recurrence equation $f_n(x)$ represents the upper side of figure 7(a), while $-f_n(x)$ represents the down side. We can define a sequence in which each ellipse is a reflection of the preceding one, as it is shown in figure 7(b). This pattern appears approximately identical at different scales. For example, figures 7(c)-(d) are magnifications of the middle left of the previous pattern. This result demonstrates the existence of patterns due Möbius transformations in a hyperbolic prism acting as a reflecting prism, similar to those observed in foams.

8. The Kolmogorov-Sinai entropy

The question at this point is: is there chaos in a system with refraction and reflection? Due to refraction and reflection at the interfaces, the direction of the rays leaving the liquid bridges can vary greatly for the same incident and only a small positional offset, and consequently there is the possibility of chaotic scattering, but how to characterize this chaotic effect? In the case of foam, it is not possible to define the exit between the interfaces in a similar way as in the case of spheres, because of the transparency of the foam. So we have to explore other ways to characterize this dynamical system.

In systems with escape, the rate of escape of particles from an open system is expressed in terms of the sum of the positive Lyapunov exponents and the Kolmogorov-Sinai entropy on the repeller, it can be found in reference [17]. Using similar maps (equations (12) and (13)) as mentioned above, some models of physical systems present chaotic scattering involving both reflection and refraction [16]. Such physical systems represent the propagation of rays passing from one medium to another, such as composite billiards [18], or soft billiard systems [16]. Baldwin has provided an empirical formula for the Kolmogorov-Sinai entropy h [16]:

$$h = \ln\left[1 + \frac{U}{r_l^2} + 1.29\left(\frac{U}{r_l^2}\right)^{0.4}\right], \tag{15}$$

where, again U is a finite potential height defined by the refractive index n, and r_l is the radius of each scatterer. Equation (15) is valid for the general case of a geometrical optical system, where a light ray travels through a lattice of circular scatterers with refractive index $n = (U - 1)^{1/2}$, in a medium where $n = 1$. For the pure reflective case, $U = 1$. The signature of chaos is obtained by the computation of the Kolmogorov-Sinai entropy h. The Kolmogorov-Sinai entropy h is the measure of the information loss in N-dimensional phase space [19]. The phase space is divided in cells of size $\{\varepsilon\}$, and takes the measurements at time increments $\{\tau\}$. The existence of finite positive values of h indicates the existence of chaotic behavior [19]. Using equation (15) for the pure reflective case the value of h is around 1.98. For the liquid case, we have used $U_l = n^2 - 1$, due to the fact that the refractive index of the scatterer is smaller than the index of the medium, so that $U_l = 0.777$. In both cases, we have used $r_l = 0.5$, meaning that each scatterer (sphere or bubble) touch its nearest neighbors. For the case of the presence of refraction and reflection, we have obtained h around 1.81. This value of entropy demonstrates that a system with refraction and reflection presents chaotic behavior. Comparing both entropy values, we can notice the both case have the same order of magnitude, but the system with refraction and reflection has the entropy value (1.81) which is a little bit smaller than the entropy of the pure reflective case (1.98).

It is interesting to note that using an approach based on determining optical absorption lengths for foams of different values of liquid fraction and the transport theory, some authors [20, 21] determined the fraction f on the path that a ray of light spends in the liquid phase of foam. They have obtained that though $1/f$ decayed linearly with most values of liquid fraction, for intermediate values of liquid fraction $1/f$ is lower than predicted by random sampling. In comparison to the values of the Kolmogorov-Sinai entropy, these results indicate there is the possibility of h even bigger than those obtained by equation (15) for the case of a light ray traversing a 2D periodic foam in a random walk with a $1/f$ decay.

9. Conclusion

The most interesting feature of all the ideas mentioned in this paper is the concept of chaotic scattering of light in foams based on hyperbolic geometry. In doing so, we have presented the analogy between chaotic scattering and the effects of light rays in liquid bridges observed in Hele-Shaw cells. Some of the main aspects of image formation in three reflective spherical shells are compared to the case of hyperbolic kaleidoscopes using the Poincaré disk model. This three-mirrored sphere system can be regarded as a chaotic system where a wave or a particle is scattered by a cavity composed by surfaces with negative curvature. In that way, the incoming wave undergoes successive Möebius transformations, such as translations, rotations, inversions, and reductions. The effects of the refraction and reflection of the light rays were studied using some properties of soft billiards. In addition to this, even a hyperbolic prism behaves like a hyperbolic kaleidoscope, because the pure case of reflection can be observed in the hyperbolic prism, and consequently foams retain properties of ordinary billiards. The existence of finite positive values of Kolmogorov-Sinai entropy is an indicative that light can be channeled through the network of Plateau borders.

Acknowledgments

This work was supported by Conselho Nacional de Desenvolvimento Científico e Tecnológico (CNPq), and Instituto Nacional de Ciência e Tecnologia de Fluidos Complexos (INCT-FCx) and FAPESP.

References

[1] Sweet D, Ott E, Yorke J A 1999 Nature (London) **399** 315
[2] Sweet D, Zeff B W, Ott E, Lathrop D P 2001 Physica D **154** 207
[3] Korsch H J and Wagner A 1991 Computers in Phys. **5** 497
[4] Suffern K G 2000 The Visual Computer **16(7)** 379
[5] Poon L, Campos J, Ott E, Grebogi C 1996 Int. J. Bifurcation and Chaos **6** 251
[6] Harayama T, Davis P 1998 Optics Letters **23** 1426
[7] Tufaile A and Tufaile A P B 2008 Phys. Lett. A **372** 6381
[8] Aguirre J, Viana R L, Sanjuán M A F 2009 Rev. of Modern Phys. **81** 333
[9] Eckhardt B 1987 J. Phys. A: Math. Gen. **20** 5971
[10] Korsch H J, Jodl H, Hartmann T 2008 Chaos: a program collection for the PC (Berlin: Springer Verlag)
[11] José J V, Rojas C, Saletan E 1992 Am. J. of Phys. **60** 587
[12] Stadlbauer M 2004 Fundamenta Math. **182** 221
[13] Weisstein E W Poincaré Hyperbolic Disk From MathWorld http://mathworld.wolfram.com/PoincaréHyperbolicDisk.html
[14] van der Net A, Blondel L, Saugey A, Drenckhan W 2007 Colloids and Surfaces A **309** 159
[15] Hecht E 1997 Optics (New York: Wesley)
[16] Baldwin P R 1988 Physica D **29** 321
[17] Gaspard P and Nicolis G 1990 Phys. Rev. Lett. **65** 1693
[18] Baryakhtar V G, Yanovsky V V, Naydenov S V, Kurilo A V 2006 J. Experim. and Theor. Phys. **103** 292
[19] Çambel A B 1993 Applied Chaos Theory – A Paradigm for Complexity (San Diego: Academic Press) 154
[20] Gittings A S, Bandyopadhyay R, Durian D J 2004 Europhys. Lett. **65** 414
[21] Sun Q C, Hutzler S 2007 Coll. Surf. A **309** 182

Evaluating strong measurement noise in data series with simulated annealing method

J. Carvalho[1],F. Raischel[2],M. Haase[3],P. Lind[1,2]

[1]Departamento de Física, Faculdade de Ciências da Universidade de Lisboa, 1649-003 Lisboa, Portugal
[2]Center for Theoretical and Computational Physics, University of Lisbon, 1649-003 Lisbon, Portugal
[3]Institute for High Performance Computing, University of Stuttgart, D-70569 Stuttgart, Germany

E-mail: raischel@cii.fc.ul.pt

Abstract. Many stochastic time series can be described by a Langevin equation composed of a deterministic and a stochastic dynamical part. Such a stochastic process can be reconstructed by means of a recently introduced nonparametric method, thus increasing the predictability, i.e. knowledge of the macroscopic drift and the microscopic diffusion functions. If the measurement of a stochastic process is affected by additional strong measurement noise, the reconstruction process cannot be applied. Here, we present a method for the reconstruction of stochastic processes in the presence of strong measurement noise, based on a suitably parametrized ansatz. At the core of the process is the minimization of the functional distance between terms containing the conditional moments taken from measurement data, and the corresponding ansatz functions. It is shown that a minimization of the distance by means of a simulated annealing procedure yields better results than a previously used Levenberg-Marquardt algorithm, which permits a rapid and reliable reconstruction of the stochastic process.

1. Introduction

Physical systems often can be resolved on two different time scales, observing slowly varying macroscopic motion while much faster microscopic interactions are perceived as an effective noisy driving. An adequate description of these systems can be obtained by a Langevin equation, which yields a deterministic drift term and a stochastic diffusion term. With respect to this description, reconstructing the drift and diffusion terms of an unknown process in time or scale from a data set enables one to enhance the predictability of that process. Recently, a nonparametric method of reconstruction based solely in the measured time series has been suggested [1]. The method has been applied successfully to data sets from fields such diverse as turbulent fluid dynamics [2], human movement [3], financial data [4], climate indices [5, 6], and to electroencephalographic recordings from epilepsy patients [7, 8]. For the case of finite time steps [9, 10], which can occur when the data is not available at sufficiently high sampling rates, recent improvements have been suggested [11, 12, 13].

However, any experimental setup will record additional measurement noise when recording a data series, which cannot be easily distinguished from the intrinsic dynamical noise. The presence of measurement noise can destroy the Markovian properties of the data and can lead to incorrect estimates for the drift and diffusion terms [18, 20].

It has been suggested to separate the measurement noise from the dynamics of the measured variable using different predictor models or schemes for noise reduction [14, 15]. Recently, another approach has been suggested, which allows reconstructing the properties of stochastic time series affected by exponentially correlated Gaussian noise, using algebraic properties of the normal distribution [16]. Here, we revisit an approach presented in Refs. [17, 18], which minimizes the functional distance between

Published under licence by IOP Publishing Ltd

characteristic functions extracted from the measurement data and corresponding terms calculated from a suitably parametrized ansatz for the drift and diffusion functions. We show that using a Simulated Annealing (SA) [19] method for the optimization yields results that are superior to the ones obtained previously by means of the Levenberg-Marquardt (LM) method.

We start in Sec. 2 by describing the general framework for extracting the evolution equation of a signal spoiled by strong measurement noise, together with the amplitude of said measurement noise. In Sec. 3 both LM and SA methods are addressed and compared. Section 4 concludes the paper.

2. Langevin approach for data sets with measurement noise

We consider a one-dimensional Langevin process $x(t)$ defined as

$$\frac{dx}{dt} = D_1(x) + \sqrt{D_2(x)}\Gamma_t, \tag{1}$$

where Γ_t represents a Gaussian δ-correlated white noise $\langle \Gamma_t \rangle = 0$ and $\langle \Gamma_t \Gamma_{t'} \rangle = 2\delta(t - t')$. The Kramers-Moyal (KM) functions $D_1(x)$ and $D_2(x)$ are defined as

$$D_n(x) = \frac{1}{n!} \lim_{\tau \to 0} \frac{1}{\tau} M_n(x, \tau) \tag{2}$$

for $n = 1, 2$, where D_1 describes the deterministic drift and D_2 corresponds to a diffusion process. An estimate of the n-th order conditional moments of the data $M_n(x, \tau)$ can be extracted directly from the measured time series as

$$\hat{M}_n(x_i, \tau) = \langle (x(t + \tau) - x(t))^n \rangle|_{x(t)=x_i}, \tag{3}$$

the hat indicating that they are calculated from the measured data $x(t)$ directly, whereas the brackets $\langle\rangle$ denote averaging over suitable parts of the time series – or ensemble averages, if available – as described in Refs. [1, 4, 6, 17]. Thus, a nonparametric reconstruction of the stochastic process can be achieved.

In the following, we will consider a time series generated by integrating Eq. (1) with drift and diffusion coefficients assumed to be linear and quadratic forms, respectively,

$$D_1(x) = d_{10} + d_{11}x \tag{4a}$$

$$D_2(x) = d_{20} + d_{21}x + d_{22}x^2. \tag{4b}$$

Though we concentrate on the particular expressions for D_1 and D_2 given above, it should be stressed that they comprehend a large collection of different processes, such as Ornstein-Uhlenbeck processes [17]. The parameter d_{10} can always be eliminated through a simple transformation $x \to x' = x + d_{10}/d_{11}$. For the numerical examples in the remainder of the text, we will use the parameters $d_{10} = 1, d_{11} = -1, d_{20} = 1, d_{21} = -1, d_{22} = 1$.

We now consider the case that the signal of $x(t)$ is affected by a Gaussian δ-correlated measurement white noise, which leads to the series of observations

$$y(t) = x(t) + \sigma\zeta(t) \tag{5}$$

where σ denotes the amplitude of the measurement noise. As Figs. 1a and 1b show, although the trend of $x(t)$ is still visible in the presence of measurement noise, it is intuitively difficult to distinguish between the intrinsic dynamic noise and the measurement noise. Likewise, Fig. 1c illustrates that for increasing σ one obtains broader probability density functions $P(y)$: the standard deviation of these distribution increases with the measurement noise, whereas the mean value remains constant.

In the presence of measurement noise, the conditional moments yield non-zero finite values at $\tau = 0$ and thus the limit $\lim_{\tau \to 0} \frac{1}{\tau}\hat{M}_n(x, \sigma \neq 0, \tau)$ does not exist [18]. Likewise, the requirements for Markovianity of the measured time series $y(t)$ may not be satisfied [20].

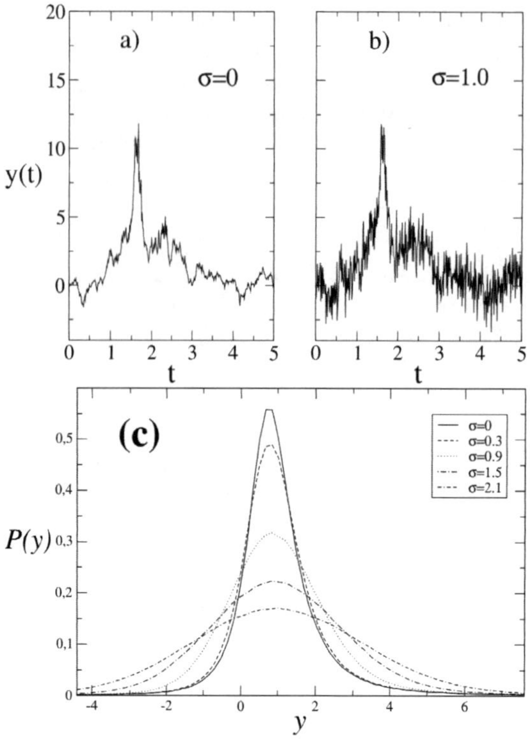

Figure 1. Langevin time series with different measurement noise strengths. Here we show a time series **(a)** without measurement noise, and **(b)** with strong measurement noise. In **(c)** the probability density function $P(y)$ of the series with measurement noise (see Eq. (5)) is shown. In all cases, the assumed time series $x(t)$ without measurement noise uses the coefficients $D_1(x) = 1 - x$ and $D_2(x) = 1 - x + x^2$, when integrating Eq. (1).

However, the two conditional moments [1, 4, 6, 17, 18] $\hat{M}_n(y_i, \tau)$ $(n = 1, 2)$ can be derived taking $y(t)$ instead of $x(t)$ in Eq. (3). Figure 2 shows plots of both conditional moments as a function of τ for different strengths of the measurement noise. The linear dependence of M_1 and M_2 on τ makes it still possible to obtain an estimate for the 'spoiled' KM-coefficients:

$$\hat{D_n}(y) = \frac{\hat{M}_n(y, \tau_2) - \hat{M}_n(y, \tau_1)}{n!(\tau_2 - \tau_1)} . \tag{6}$$

Within this description, an estimate for the amplitude of the measurement noise yields $\sigma \approx \sqrt{\frac{\hat{M}_2(\mu, 0)}{2}}$ (see Ref. [17, 21]), where μ is the average value of $y(t)$ data points in the time series. As shown in Fig. 2c, such an estimate is not valid for sufficiently strong measurement noise, i. e. $\sigma \gtrsim 0.5$.

The coefficients D_1 and D_2 cannot be correctly estimated, though. Figure 3 shows how the estimated parameters $\hat{d}_{ij}$ deviate from the 'true' values d_{ij} of the process given by Eq. (4) as measurement noise increases. Notice that for $\sigma = 0$ (left vertical axis in each plot of Fig. 3) the estimated parameter values are approximately correct.

To correctly derive the drift and diffusion coefficients $D_1(x)$ and $D_2(x)$ when σ is strong, we consider the measured conditional moments $\hat{M}_1(y_i, \tau)$ and $\hat{M}_2(y_i, \tau)$. Since these conditional moments depend

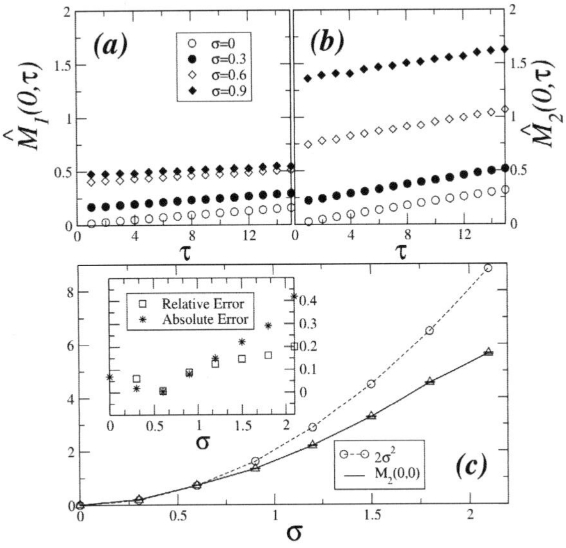

Figure 2. Conditional moments **(a)** $\hat{M}_1(y_i, \tau)$ and **(b)** $\hat{M}_2(y_i, \tau)$ as a function of τ, for bin $y_i = 0$ and different measurement noise strengths σ. The same $x(t)$ as in Fig. 1 was used. **(c)** First estimate of the measurement noise (see text).

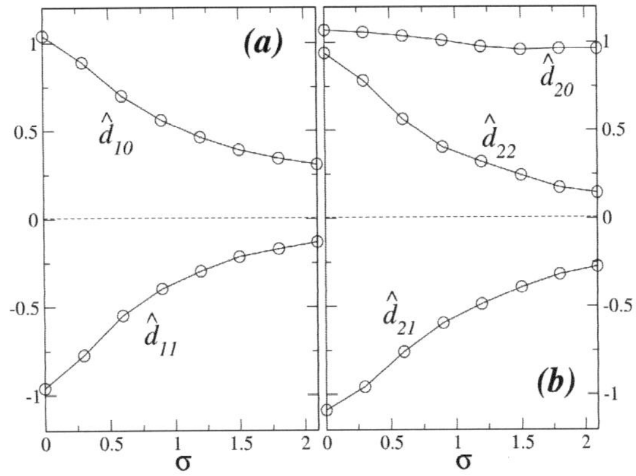

Figure 3. Noise dependence of functions $\hat{D}_1(y)$ and $\hat{D}_2(y)$ (see text and Eq. (6)). The underlying Langevin time series $x(t)$ without noise is the same as in Fig. 1.

in a non-trivial way on both time τ and amplitude y_i, we approximate them up to first order on τ:

$$\hat{M}_1(y_i, \tau) = \langle y(t+\tau) - y(t) \rangle|_{y(t)=y_i} = \tau \hat{m}_1(y_i) + \hat{\gamma}_1(y_i) + \mathcal{O}(\tau^2), \tag{7a}$$

$$\hat{M}_2(y_i, \tau) = \langle (y(t+\tau) - y(t))^2 \rangle|_{y(t)=y_i} = \tau \hat{m}_2(y_i) + \hat{\gamma}_2(y_i) + \sigma^2 + \mathcal{O}(\tau^2), \tag{7b}$$

where $y(t)$ is taken in the range $y_i \pm \Delta y/2$ for each bin i, and Δy depends on the binning considered.

While the functions $\hat{m}_i$ and $\hat{\gamma}_i$ ($i = 1, 2$) are obtained explicitly for each bin value y_i, the decisive aspect of this approach is the fact that the ansatz functions m_i and γ_i can be calculated from the drift and diffusion coefficients D_1, D_2 and the measurement noise distribution f_σ as follows [18]:

$$\gamma_1(y) = \int_{-\infty}^{+\infty} (x-y)\bar{f}_\sigma(x|y)dx \tag{8a}$$

$$\gamma_2(y) = \int_{-\infty}^{+\infty} (x-y)^2\bar{f}_\sigma(x|y)dx \tag{8b}$$

$$m_1(y) = \int_{-\infty}^{+\infty} D_1(x)\bar{f}_\sigma(x|y)dx \tag{8c}$$

$$m_2(y) = 2\int_{-\infty}^{+\infty} [(x-y)D_1(x)+D_2(x)]\bar{f}_\sigma(x|y)dx, \tag{8d}$$

where $\bar{f}_\sigma(x|y)$ is the probability for the system to adopt the value x if a measured value y is observed.

The problem we then solve is to parametrize D_1,D_2 and the noise strength σ and to determine the set of parameters that minimize the functional distance F between the measured functions $\hat{m}_i$, $\hat{\gamma}_i$ and their counterparts m_i, γ_i defined as

$$F = \frac{1}{M}\sum_{i=1}^{M}\Big[\frac{(\hat{\gamma}_1-\gamma_1(y_i))^2}{\sigma^2_{\hat{\gamma}_1}(y_i)} + \frac{(\hat{\gamma}_2-\gamma_2(y_i)-\sigma^2)^2}{\sigma^2_{\hat{\gamma}_2}(y_i)} + \frac{(\hat{m}_1-m_1(y_i))^2}{\sigma^2_{\hat{m}_1}(y_i)} + \frac{(\hat{m}_2-m_2(y_i))^2}{\sigma^2_{\hat{m}_2}(y_i)}\Big], \tag{9}$$

where the summation extends over all M bins, and $\sigma_{\hat{\gamma}_1}(y_i)$ is the error associated to function $\hat{\gamma}_1$ at the value y_i (and similarly for $\sigma_{\hat{\gamma}_2}$, $\sigma_{\hat{m}_1}$ and $\sigma_{\hat{m}_2}$, all of them taken directly from the data only).

After computing the functions $\hat{\gamma}_1$, $\hat{\gamma}_2$, we start from the initially estimated set of values for the parameters and iteratively improve the solution by seeking lower values of F, until convergence is attained. In the following section we use two different methods for minimizing the functional F, namely the LM method [18] and SA method. We will show that for our purposes SA is significantly better than LM.

3. Comparing two optimization methods: Levenberg-Marquardt and Simulated Annealing

In this section, we consider the parameters $\sigma, d_{11}, d_{20}, d_{21}$ and d_{22} which we denoted as p_k with $k = 1, \dots, 5$, respectively. For the Levenberg-Marquardt procedure one computes the derivatives of F with respect to the parameters $\mathbf{p} = \{p_k\}$, and uses them to define the coefficients $\beta_k = \frac{1}{2}\frac{\partial F}{\partial p_k}$ and $\alpha_{kl} = \sum_{j=1}^{4}\sum_{i=1}^{M}\frac{1}{\sigma_j^2(y_i)}\frac{\partial f_j(y_i,\mathbf{p})}{\partial p_k}\frac{\partial f_j(y_i,\mathbf{p})}{\partial p_l}$, where $f_j(y_i,\mathbf{p})$ is the j-th summand in the bracketed inner sum defining F, Eq. (9). The descent δp_l is computed from $\sum_{l=1}^{M}\alpha'_{kl}\delta p_l = \beta_k$, which uses a damped version of the Gaussian matrix:

$$\alpha'_{kl} = \begin{cases} (1-\lambda)\alpha_{kl}, k=l \\ \alpha_{kl}, k \neq l \end{cases} \tag{10}$$

The parameter λ is updated during the optimization. The procedure stops after sufficient convergence has been achieved. For details, see Ref. [22]. We generated several data sets from Eq. (1) with different 'input' measurement noise amplitudes $0 < \sigma_I \leq 1.2$, in order to compare the reconstructed parameters to the actual values.

The optimization results are shown in Fig. 4, triangles indicating the first estimate for the parameter values (see Sec. 2), solid lines indicating the true values used to generate the data, and squares indicating the value after LM-optimization.

While the parameters are well estimated, particularly the magnitude of measurement noise and the drift parameters d_{10} and d_{11}, it was found that LM sometimes does not converge, or converges to a local

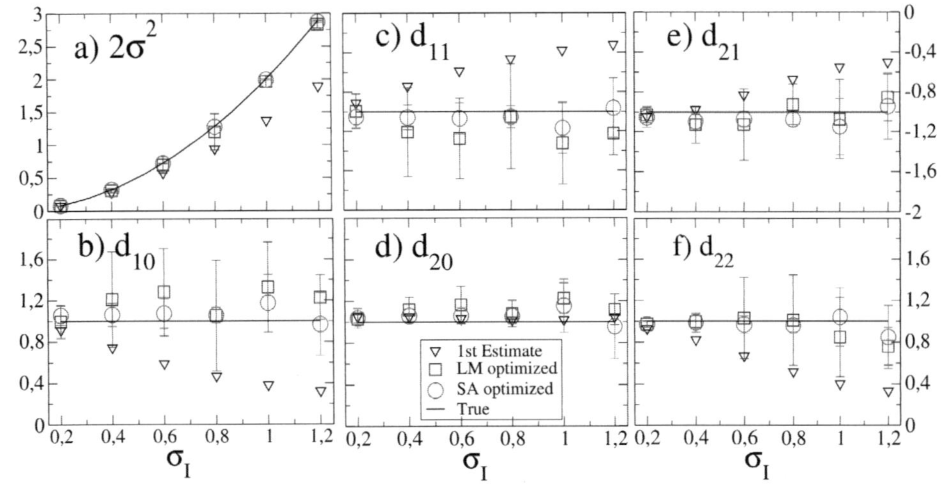

Figure 4. Comparison of the optimized parameters values (bullet) with the first estimate and the true values for different input measurement noise strengths σ_I: **(a)** $2\sigma^2$, **(b)** d_{10}, **(c)** d_{11}, **(d)** d_{20}, **(e)** d_{21}, **(f)** d_{22}. The measurement noise is correctly extracted, as well as the parameters representing the drift coefficient $D_1(x)$ and the diffusion coefficient $D_2(x)$. SA results are significantly better than LM results (see text). Error bars represent the maximum and minimum values encountered in 10 realizations.

minimum, the latter fact becoming visible when comparing the results after starting from different initial values. Both these shortcomings required that the LM optimization results were drawn as the best results of a series of attempts with different starting values. The intent to overcome these shortcomings led us to attempt the SA method (squares in Fig. 4).

Simulated annealing is a probabilistic global optimization method. Starting from initial guesses for the parameter vector it proceeds with a step in a random direction. The step is accepted immediately if the energy function at the new position F_{new} (see Eq. (9)) is lower than the previous energy F_{old}, or it is accepted with a probability given by a Boltzmann factor $\exp(-\Delta F/kT)$, where $\Delta F = F_{\text{new}} - F_{\text{old}}$. Otherwise the step is refused and a new parameter step is chosen. This procedure corresponds to the motion of thermal atoms in an attractive potential. Gradual cooling is then achieved by reducing the temperature parameter, thus annealing the test particle in the minimum of the energy landscape.

Figure 4 shows that both optimization methods determine the value of input measurement noise σ_I correctly in all cases, which we find remarkable given the comparable magnitude of measurement noise and the time series $x(t)$ for $\sigma \approx 1$. Furthermore, it can be seen that both LM and SA estimate the parameters d_{1k} for the drift and d_{2l} for the diffusion coefficient correctly; as expected [18], the drift coefficients are determined with higher precision. In general, the SA routine provides more accurate results than LM.

In the LM case, for stronger measurement noise amplitudes, namely for $\sigma > 1.2$, the algorithm is sometimes stuck in a local minimum of the function F, leading to unreliable coefficients d_{ik}. The same set of 10 time series with 10^6 data points was used for each method, but in the LM case, not all cases showed local minima sufficiently close to the true parameter values. Therefore, in our comparison, we overestimate the best performance of the LM method, by considering only the best five cases out of the set of 10 time series (squares in Fig. 4).

Figure 5a shows a plot of the energy, i.e. the distance

$$dF = F(\mathbf{p}) - F_0(\mathbf{p}_0) \tag{11}$$

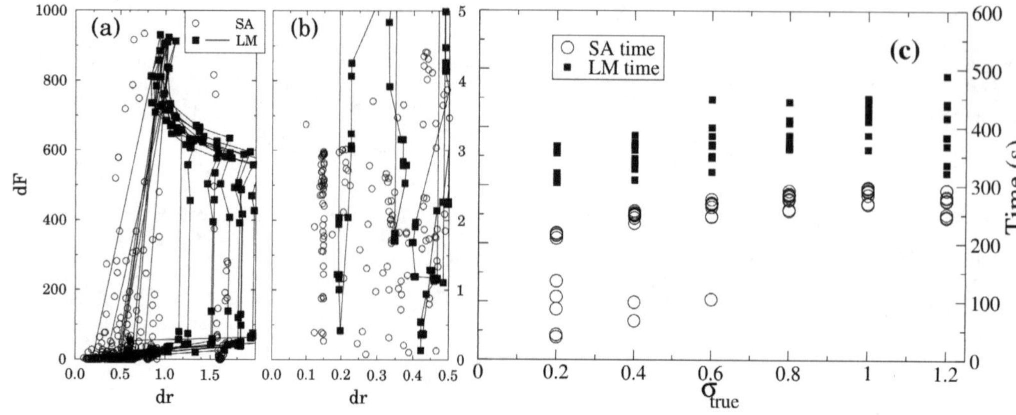

Figure 5. **(a)** Plot of the relative energy in comparison to the true value, Eq.(11), as a function of the Euclidean distance $dr = ||\mathbf{p} - \mathbf{p}_0||_2$ of the parameters from the true parameter values $\mathbf{p}_0$ found with the SA method (circles) and with the LM method (squares) for $\sigma = 1.2$. **(b)** Close-up of (a) near the minimum of F. For better visibility, not all SA-points are shown in (a) and (b). **(c)** Execution time (in seconds) as a function of σ for LM (squares) and SA (circles).

where $F_0(\mathbf{p}_0)$ is the value found when taking the true parameter values, as a function of the Euclidean distance $dr = ||\mathbf{p} - \mathbf{p}_0||_2$ of the parameters from the true parameters values. Apparently, the energy is strongly asymmetrical in at least one of the parameters, which is in agreement with Ref. [18]. Also, the existence of local minima, which apparently attract the LM algorithm, is confirmed. Figure 5b shows a close-up near the lowest minimum to emphasize the higher accuracy of the final result for SA (circles) compared with LM (squares). In this example, it can be seen that the LM algorithm does not converge to the global minimum but deviates shortly before.

Another argument in favor of the SA algorithm is that the execution time approximately 100 s lower than for the LM method, as shown in Fig. 5c. Additionally, whereas the LM algorithm needs to be applied to 10 realizations of the time series in order to produce a reasonably accurate result, the SA algorithm was found to converge to the global minimum in almost all instances, implying that fewer realizations are needed for averaging.

4. Discussion and Conclusions

We describe the improvement of a nonparametric procedure to extract measurement noise in empirical stochastic series with strong measurement noise by applying a simulated annealing approach for the optimization step. Simulated annealing accurately extracts the strength of measurement noise and the values of the parameters representing drift and diffusion. These parameters fully describe the evolution equation for the measured quantity. SA produces more reliable and accurate results than a previously used Levenberg-Marquardt algorithm, with the additional benefit of being significantly faster.

Nonparametric reconstruction of the Langevin Eq. (1) from measured stationary data sets merely requires that the process exhibits Markovian properties and fulfills the Pawula theorem [6], whereas the measurement noise needs to be uncorrelated. The Pawula constraint can be relaxed extending the analysis to Lévy noise [23, 24]. It should be pointed out that the method presented here relies solely on the Markovian properties of the underlying, undisturbed process $x(t)$, and does not require the measured process $y(t)$ to be Markovian.

An extension of this work to multidimensional Langevin time series will be essential for assessing the complexity behind EEG time series or earthquake data. It is likely that we may combine this denoising approach with the method of eigendirections presented in [25].

Acknowledgments
The authors thank D. Kleinhans, M. Wächter, J. Peinke and B. Lehle for useful discussions. They also wish to thank *Fundação para a Ciência e a Tecnologia* (FCT) for the *Ciência 2007* support (PGL), for the fellowship SFRH/BPD/65427/2009 (FR), and for the BII student support (JC). To FCT and *Deutscher Akademischer Austauschdienst* (DAAD) the authors express their gratitude for support through the bilateral cooperation DREBM/DAAD/03/2009 between Portugal and Germany. Finally, they are indebted to the organizers and participants of Dynamics Days 2010 South America for useful discussions in an inspiring atmosphere.

References

[1] R. Friedrich and J. Peinke, Phys. Rev. Lett. **78**, 863 (1997).
[2] A.P. Nawroth, J. Peinke, D. Kleinhans, R. Friedrich, Phys. Rev. E **76**, 056102 (2007).
[3] A. van Mourik A, A. Daffertshofer, P. Beek, Biological Cybernetics **94**, 233(2006).
[4] R. Friedrich, J. Peinke and Ch. Renner, Phys. Rev. Lett. **84**, 5224 (2000).
[5] C. Collette and M. Ausloos, Int. J. Mod. Phys. C **15**, 1353 (2004).
[6] P.G. Lind, A. Mora, J.A.C. Gallas and M. Haase, Phys. Rev. E **72**, 056706 (2005).
[7] J. Prusseit and K. Lehnertz, Phys. Rev. E **77**, 041914 (2008).
[8] D. Lamouroux, and K. Lehnertz, Phys. Lett. A **373** 3507 (2009).
[9] M. Ragwitz and H. Kantz, Phys. Rev. Lett. **87**, 254501 (2001).
[10] R. Friedrich, Ch. Renner, M. Siefert, and J. Peinke, Phys. Rev. Lett. **89**, 149401 (2002).
[11] D. Kleinhans, R. Friedrich, A. Nawroth, J. Peinke, Phys. Lett. A **346**, 42-46 (2005).
[12] J. Gottschall and J. Peinke, New J. Physics **10**, 083034 (2008).
[13] C. Anteneodo and R. Riera, Phys. Rev. E **80**, 1 (2009).
[14] H. Kantz and T. Schreiber, *Nonlinear Time Series Analysis*, (Cambridge University Press, Cambridge, England, 1997).
[15] H.D.I. Abarbanel, R. Brown, J.J. Sidorowich, L.S. Tsimring, Rev. Mod. Phys. **65** 1330 (1993).
[16] B. Lehle, "Analysis of stochastic time series in the presence of strong measurement noise", to appear in Phys. Rev. E (2011).
[17] F. Boettcher, J. Peinke, D. Kleinhans, R. Friedrich, P.G. Lind, M. Haase, Phys. Rev. Lett. **97** 090603 (2006).
[18] P. G. Lind, M. Haase, J. Peinke, D. Kleinhans, and R. Friedrich, Phys. Rev. E **81**,1 (2010).
[19] S. Kirkpatrick, C.D. Gelatt, and M.P. Vecchi, Science **220**, 671 (1983).
[20] D. Kleinhans, R. Friedrich, M. Wächter and J. Peinke, Phys. Rev. E **76**, 041109 (2007).
[21] M. Siefert, A. Kittel, R. Friedrich and J. Peinke, Europhys. Lett. **61** 466 (2003).
[22] W.H. Press, B.P. Flannery, S.A. Teukolsky and W.T. Vetterling, *Numerical Recipes* (Cambridge University Press, Cambridge, 1992).
[23] R. Friedrich, J. Peinke and M.R.R. Tabar, *Complexity in the view of stochastic processes* in *Springer Encyclopedia of Complexity and Systems Science* (Springer, Berlin, 2008).
[24] S. Siegert and R. Friedrich, Phys. Rev. E **64**, 041107 (2001).
[25] V. Vasconcelos, F. Raischel, M. Haase, D. Kleinhans, J. Peinke and P. G. Lind, in preparation (2011).

Stabilizing equilibrium by linear feedback control for controlling chaos in Chen system

V A Costa[1] **and G A González**[2, 3]

[1] Departamento de Ciencias Básicas, Facultad de Ingeniería (UNLP), La Plata, Argentina.

[2] Departamento de Matemática, Facultad de Ingeniería (UBA), Buenos Aires, Argentina.

[3] Consejo Nacional de Investigaciones Científicas y Técnicas (CONICET), Argentina.

vacosta@ing.unlp.edu.ar, ggonzal@fi.ub.ar

Abstract. Stabilization of a chaotic system in one of its unstable equilibrium points by applying small perturbations is studied. A two-stage control strategy based on linear feedback control is applied. Improvement of system performance is addressed by exploiting the ergodicity of the original dynamics and using Lyapunov stability results for control design. Extension to the not complete observability case is also analyzed.

1. Introduction

It is well known that one of the Control of Chaos objectives is to suppress the chaotic dynamical behaviour naturally arising in a given system. Since the beginning of the nineties, much work has been developed on this field [1, 2, 3]. An appropriate question was addressed. Given a chaotic system: how a desired time-periodic motion with improved performance can be achieved by applying only small perturbations on some accessible system parameter or system variable? The problem was firstly approached in [4, 5] as follows: a) determination of some of the unstable low-period (or even steady-state) orbit that are embedded in the chaotic attractor, b) choice of one which yields improved system performance and, c) application of small control action to stabilize this already existing orbit. This method, -the OGY method- described in detail in the discrete-time case, applies to the given continuous-time chaotic system by means of Poincaré section.

Not much later, several authors concentrated on controlling chaos in continuous-time systems without using Poincaré discretization and resorted to the well developed machinery of Modern Control Theory. It is known that fundamental features of controlling chaos like taking full account of the special aspects of chaotic motion and applying only small perturbations have been often neglected by them. For example, with the objective of suppressing chaotic behaviour, in references [6, 7, 8, 9] linear feedback control is explored while in reference [10], a controller based on a PI regulator control is addressed. However, these approaches only take care of local stabilization.

This work concentrates on Chen system but the approaches presented here are straightly applicable to other chaotic systems like Lorenz, Chua, Rossler, etc. Our purpose is to stabilize the system in one of its (unstable) equilibrium points by using linear feedback control. Improvement of system performance is addressed by exploiting the ergodicity of the original dynamics and using Lyapunov stability results for control design [11].

2. Purpose and method

Let us assume that we have a chaotic dynamical system described by:

$$\dot{X} = F(X)$$

and that *E* is one of its unstable equilibrium points embedded in its chaotic attractor.

Our aim is controlling chaos by applying feedback control. For this purpose, we will construct a linear feedback control, depending on a gain parameter *k*, to stabilize the system in the equilibrium point *E*. The following facts will be of relevance:

- due to ergodicity of the free system, (almost) every trajectory initialized in the strange attractor reaches a chosen *E*-neighbourhood, $B(E,\delta)$;
- fixed *k* such that locally asymptotic convergence is guaranteed, the corresponding attraction region of the controlled system may be estimated as a set Ω_k.

We look for a *k*-depending control law such that $B(E,\delta) \subset \Omega_k$ and which remains bounded by a desired fixed bound. The control strategy consists in making the free system run till it reaches the neighbourhood $B(E,\delta)$. At the time that the trajectory reaches the neighbourhood $B(E,\delta)$, the feedback control is activated and it is kept so for all future times. Therefore, once the trajectory enters Ω_k, it will stay there for ever while the feedback control will be kept under the desired bound.

Chen dynamical system [12, 13] is described by:

$$\begin{aligned} \dot{x}_1 &= a(x_2 - x_1) \\ \dot{x}_2 &= (c-a)x_1 + cx_2 - x_1x_3 \\ \dot{x}_3 &= -bx_3 + x_1x_2 \end{aligned} \tag{1}$$

being x_1, x_2 and x_3, the state variables and *a, b* and *c*, positive real constants. For $a=35$, $b=3$ and $c=28$, it has a chaotic attractor and its unstable equilibrium points are $E_1 = (0,0,0)$, $E_2 = (\sqrt{63},\sqrt{63},21)$ and $E_3 = (-\sqrt{63},-\sqrt{63},21)$ (see Figure 1).

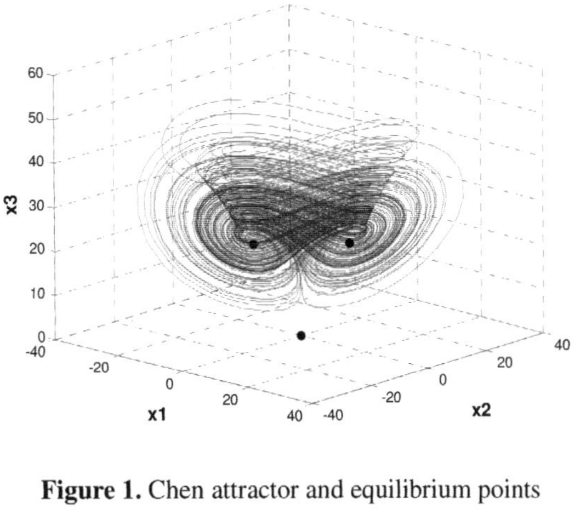

Figure 1. Chen attractor and equilibrium points

As in [6], we assume that all the state variables are observable and that each system equation may be affected by an additive control, i.e.

$$\begin{aligned} \dot{x}_1 &= a(x_2 - x_1) + u_1 \\ \dot{x}_2 &= (c-a)x_1 + cx_2 - x_1x_3 + u_2 \\ \dot{x}_3 &= -bx_3 + x_1x_2 + u_3 \end{aligned} \tag{2}$$

Let us choose $E=E_2$ and introduce the following linear controls:

$$u_1 = -k(x_1 - \sqrt{63}),\ u_2 = -k(x_2 - \sqrt{63}),\ \text{and}\ u_3 = -k(x_3 - 21).$$

From the stability analysis on the linearization of the controlled system (2), local convergence is assured if $k \geq 4.22$. By means of Lyapunov function construction [11], we estimate the region of attraction of system (2) as being an ellipsoid centered in E, Ω_k. In Table 1, the lengths of the major and the minor ellipsoid axis, $s_M(k)$ and $s_m(k)$ respectively, for some values of k are displayed.

Table 1. Estimated region of attraction. $s_M(k)$: length of the ellipsoid major axis, $s_m(k)$: length of the ellipsoid minor axis

k	$s_M(k)$	$s_m(k)$
7	0.3196	0.0459
10	0.7344	0.1485
20	2.8445	0.9300
30	6.195	2.65
50	15.387	8.73
100	28.399	20

To state the strategy we must choose k that verifies:

$$(x_1(t), x_2(t), x_3(t)) \in \Omega_k \Rightarrow \|(u_1(t), u_2(t), u_3(t))\|_2 \leq U \tag{3}$$

being U, the desired fixed control bound.

As

$$\|(u_1(t), u_2(t), u_3(t))\|_2 \leq k\|(x_1(t), x_2(t), x_3(t)) - E\|_2 \leq k \cdot s_M(k), \tag{4}$$

(3) is verified by k such that

$$k \cdot s_M(k) \leq U. \tag{5}$$

Due to ergodicity, every sphere $B(E,\delta)$ with $\delta > \Delta$ will be visited at any time, by (almost) every trajectory of system (1), initialized in the strange attractor. From numerical experience, we estimate $\Delta=0.82$. On the other hand, convergence is guaranteed if $B(E, \delta) \subset \Omega_k$ which is valid if $\delta \leq s_m(k)$. So, we need to choose δ such that

$$\Delta \leq \delta \leq s_m(k). \tag{6}$$

Hence, let us fix k and δ according to (5) and (6). The algorithm consists of two stages. In the first stage, the system runs freely (control not activated). Let t_f the instant of time at which its trajectory reaches $B(E, \delta)$. The second stage begins at time $t=t_f$ at which the feedback control is activated. Note that, differently from the OGY method [4], the control is kept activated for all $t>t_f$.

3. Results and extension

We emphasize that under the stated requirements, trajectories convergence and *U*-bounded controls are formally proved, for (almost) initial conditions in the strange attractor.

Note that the smaller is δ, the smaller *k* value may be set what yields to control effort reduction. However, a great reduction on δ will probably result on a drastic increase of the waiting time (first stage time). Besides, in general, a too small *k* delays too much convergence in the second stage. Therefore, control parameters values must be chosen respecting a compromise between control effort, and total convergence time (that is, the sum of the two stages times).

Conservative feature of our estimations must be pointed out. This is put in evidence by simulations. Let us see two examples. In both of them, the system is initialized in *(-20,-20, 30)*. Choosing $\delta=2$ and $k=30$, convergence is guaranteed and $\|(u_1(t), u_2(t), u_3(t))\|_2 \leq 185.7$ is predicted. In Figure 2, the corresponding states and controls signals are displayed. Note that the control bound is meaningfully less than the predicted one. By instead, setting $k=7$, convergence is verified although it is not theoretically proven (see Figure 3).

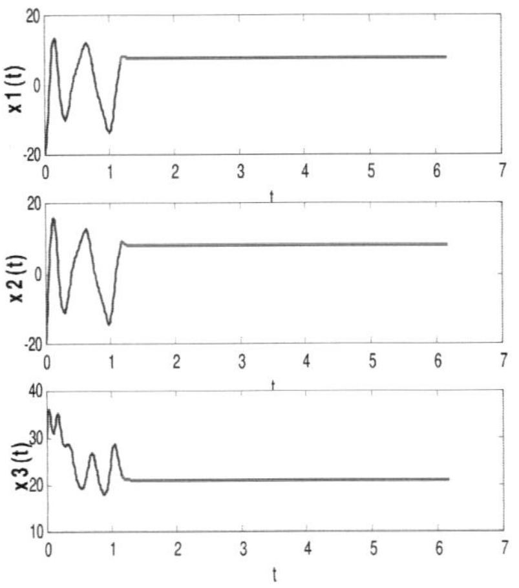

Figure 2.1. States.

— control not activated

— control is activated *(k= 30)*

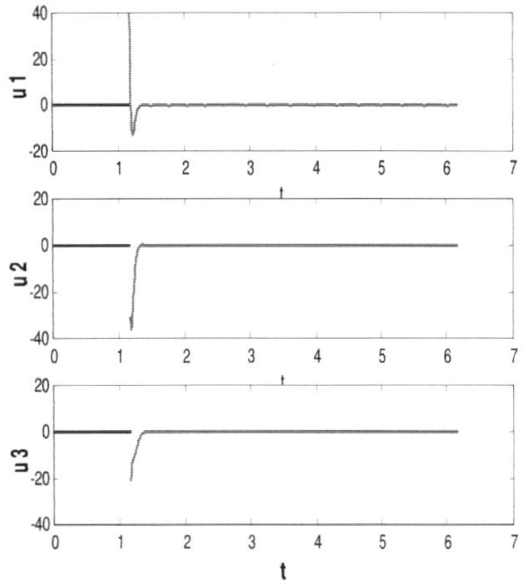

Figure 2.2. Controls.

— control not activated

— control is activated *(k=30)*

We wonder if this methodology applies when not all states are observable. Suppose that we have system $\dot{X} = F(X)$ but only the output *y* is at our disposal, being $y=CX$. The objective is to make the output converge to $y_E=CE$. We implement a two stages-algorithm as in the complete observability case, save that: *i)* the criteria for control activation is $|y - y_E|<\delta$, *ii)* the condition for control activation must be verified at every time. This is because in this case, we do not have estimated region of attraction.

In spite of this limitation, we obtain experimental evidence of the algorithm success. Let us show it by applying the algorithm to Chen system. The output of system (1) is described by $C=(0,1,0)$. Setting in (2) the linear feedback control: $u_1=0$, $u_2=- k(y-y_E)$ and $u_3=0$, the system output locally converges to y_E *if* $k \geq 6.059$. The parameters values are chosen as $\delta = 2$ and $k=10$ so we predict $\|(u_1(t), u_2(t), u_3(t))\|_2 \leq 20$. Output convergence and control bound are verified through simulation. For example, with initial condition given by *(5,-15, 40),* the resulting output and control are shown in Figure 4.

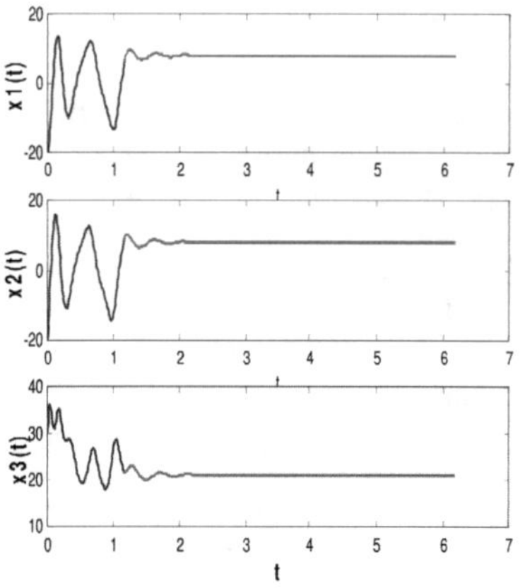

Figure 3.1. States.

— control not activated

— control is activated *(k= 7)*

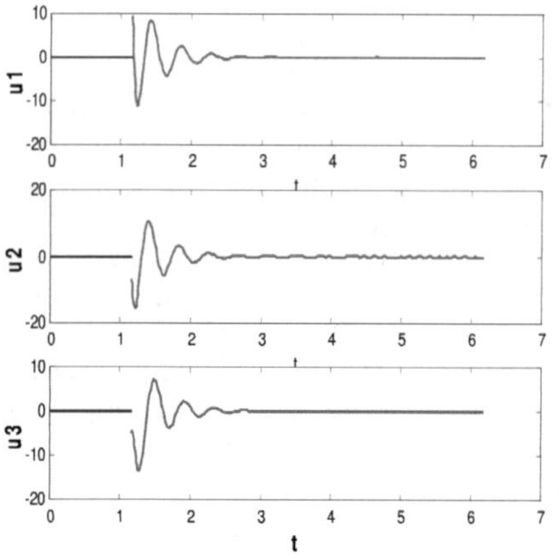

Figure 3.2. Controls.

— control not activated

— control is activated *(k= 7)*

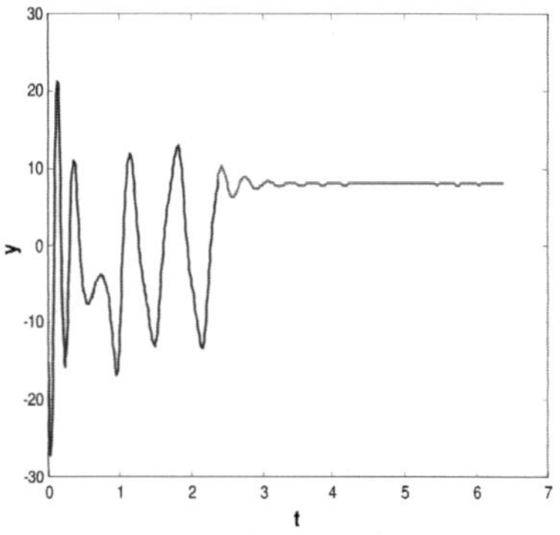

Figure 4.1. Output.

— control not activated

— control is activated *(k=10)*

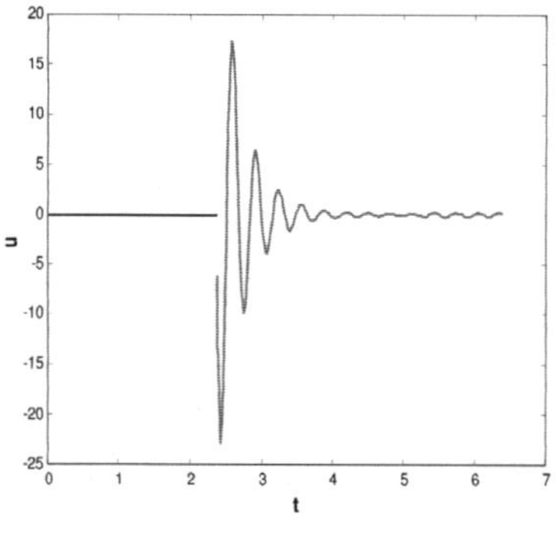

Figure 4.2. Controls.

— control not activated

— control is activated *(k=10)*

4. Discussion and conclusions

Chen system has been stabilized while considering fundamental features on controlling chaos. Hence, some progress with respect to previous works [6, 7, 8, 9] has been made.

As in OGY method, the on-line implementation only requires data on system linearization. But, for control design, extra system information is needed (for estimation of the region of attraction) to choose control parameters which guarantee convergence. Then, fixing these values, not only convergence but also no "kicking"' of the trajectory out of the neighbourhood of the equilibrium point is assured. On the other hand, simulated results show us that our theoretical estimation may be too conservative. This drawback will be object of future investigation as well as the extension of these ideas to other plausible situations like stabilization of periodic orbits, controllability restrictions, system affected by noise, etc.

We have also considered the case of uncompleted observability as in [10]. Bounded controls (under the desired bound) have been achieved by our approach. Of course, issues about compromise between convergence times and control bound restrictions also arise here. The theoretical proof of convergence or any other property of the control strategy promise interesting research.

Acknowledgments

The authors thank the reviewers for their suggestions to improve the readability of the paper.
This work has been partly supported by UBACyT, Progr. 2008-2010 and Progr. 2010-2012.

References

[1] Boccaletti S, Grebogi C, Lai Y C, Mancini H and Maza D 2000 The control of chaos: theory and applications *Physics Reports* **329** 103-197.

[2] Chen G, Dong X 1998 From Chaos to Order: Perspectives, Methodologies and Applications, Singapore: World Scientific.

[3] Fradkov A L and Evans R J 2005 Control of chaos: Methods and applications in engineering. *Annual Reviews in Control* **29** 33-56.

[4] Ott E, Grebogi C and Yorke J 1990 Controlling Chaos *Phys. Rev. Lett.* **64** 1196.

[5] Romeiras F, Grebogi C, Ott E and Dayawansa W P 1992 Controlling Chaotic Dynamical Systms *Physica D: Nonlinear Phenomena* **58** 165-192

[6] Yassen M T 2003 Chaos control of Chen chaotic dynamical system, *Chaos, Solitions & Fractals* **15** 271-283.

[7] Wang X, Tian L and Yu L 2006 Linear Feedback Controlling and Synchronization of the Chen's Chaotic System, *International Journal of Nonlinear Science* **2** no.1 43-49.

[8] Cui-Xiang L and Jia-Shi T 2008 Equilibrium points and bifurcation control of a chaotic system, *Chinese Physics B* **17** No.1 135-38.

[9] Plienpanich T, Niamsup P and Lenbury T 2005 Controllability and stability of the perturbed Chen chaotic dynamical system *Applied Mathematics and Computation* **171** 927-947.

[10] Jiang G P and Wang S P 2003 Stabilizing Unstable Equlibrium of Chaotic Systems using PI regulator *Proceedings of IEEE International Symposium on Circuits and Systems* **III** 64-67.

[11] Khalil H 2001 Nonlinear Systems *Prentice Hall* 750.

[12] Chen G and Ueta T 1999 Yet another chaotic attractor *International Journal of Bifurcation and Chaos* **9** 7 1465-1466.

[13] Zhou T, Tang Y and Chen G 2004 Chen´s attractor exists *International Journal of Bifurcation and Chaos* **14** 9 3167-3177 .

Connectivity and Nestedness in Bipartite Networks from Community Ecology

Gilberto Corso[1], A. I. Levartoski de Araujo[2], Adriana M. de Almeida[3]

[1] Departamento de Biofísica e Farmacologia, Centro de Biociências, Universidade Federal do Rio Grande do Norte, UFRN - Campus Universitário, Lagoa Nova, CEP 59078 972, Natal, RN, Brazil
[2] Instituto Federal de Educação, Ciência e Tecnologia do Ceará Av. Treze de Maio, 2081 - Benfica CEP 60040-531 - Fortaleza, CE, Brazil
[3] Departamento de Botânica, Ecologia e Zoologia, Centro de Biociências, Universidade Federal do Rio Grande do Norte, UFRN - Campus Universitário, Lagoa Nova, CEP 59078 972, Natal, RN, Brazil

E-mail: corso@cb.ufrn.br

Abstract. Bipartite networks and the nestedness concept appear in two different contexts in theoretical ecology: community ecology and islands biogeography. From a mathematical perspective nestedness is a pattern in a bipartite network. There are several nestedness indices in the market, we used the index ν. The index ν is found using the relation $\nu = 1 - \tau$ where τ is the temperature of the adjacency matrix of the bipartite network. By its turn τ is defined with help of the Manhattan distance of the occupied elements of the adjacency matrix of the bipartite network. We prove that the nestedness index ν is a function of the connectivities of the bipartite network. In addition we find a concise way to find ν which avoid cumbersome algorithm manupulation of the adjacency matrix.

1. Introduction

The last decade have experienced an explosion of papers on network theory [1]. Despite the fact that some of most popular networks are bipartite the works on bipartite networks are a minority. A very popular bipartite network is the actors collaboration network [2] where actors and movies are two sets and elements of these two sets have connections every time an actor participate in a movie. However, this bipartite network is presented in a simplified form where movies do not take part. The usual actor network has actors as nodes and a link between two actors is established every time they collaborate in the same movie. Of course there is also the movie network, but it is not so popular. An other well known network that comes from a projection of a bipartite graph is the cientific colaboration network [3].

An important application of bipartite networks are the interaction networks that comes from community ecology. Plants and their pollinators or animals and their parasites are examples of interaction networks in community ecology. We call specialist a species that interact with one or few species and generalist a species that interact with many. Nestedness is investigated in these networks characterizing the generalist-specialist balance in communities [4]. In nested interaction matrices, generalist species interact with many species, while a specialist interacts

Published under licence by IOP Publishing Ltd

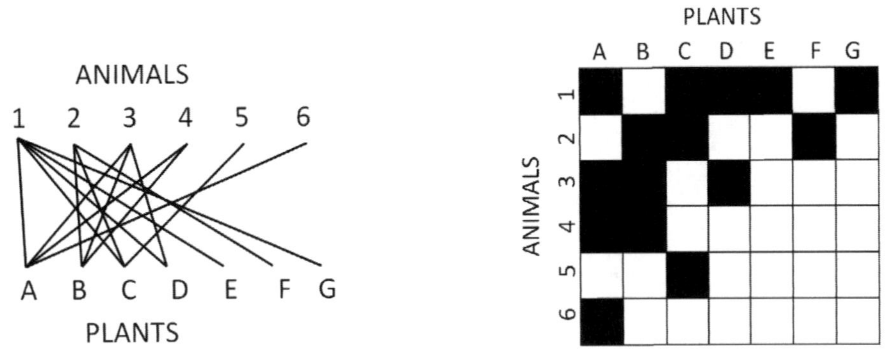

Figure 1. A sketch of a bipartite network of plants and animals, the flower-polinator network is the standard example. In the left we show the usual bipartite representation while in the right the lattice-like representation.

with few ones. In nested matrices, species are organized in such a way that specialist species only interact with more generalist ones. Ecologists suggest that nestedness patterns can distinguish between mutualist (e.g.. pollinator and plants) and antagonist (animals and parasites) networks [4]. In addition, the nestedness degree is important in the study of species co-evolution [5].

Bipartite networks have also been used in island biogeography [6], a very traditional branch of community ecology. Given a set of islands (from an archipelago) and a set of species, every time a species is found on an island a link between elements of these two sets is established, originating a bipartite network. In general, larger islands will support more species. But one can pose the question: species present in small islands will always be present in larger ones? A rough answer to this question is yes. This problem was studied for many archipelagos and several taxa (birds, lizards, beetles, etc) [7]. This pattern of species-poor sites composing a subset of species-rich sites is called a nested structure. Since forest remnants can be considered islands in a sea of anthropogenic disturbed landscape, conservation studies have revived some aspects of island biogeography in last three decades [8].

In figure 1 we show two representations of an interaction network between a set of plants i and animals j. We sketch a plant-polinator network just to fix the ideas, in fact the theoretical framework developed is valid for any bipartite network. In the lattice format the full spaces indicates the presence of interactions while the blank spaces the absences of interactions. The same information is also show in a graph representation. In this case the mathematical object that describes this phenomenon is a bipartite network which establishes links between two sets: animals and plants. The main quantity that characterizes the network is the connectivity of animals and plants, $k_A(i)$ and $k_P(j)$, which count the number of interactions of animals and plants.

The main objective of this manuscript is to show that a recent index of nestedness ν introduced in the literature [9] depends only in the connectivity of the bipartite network. The work is organized as follows: in Section 2 we outline the meaning of nestedness in community

Figure 2. Two adjacency matrices with the same occupation, the matrix in the left is more nested than the matrix in the right.

ecology and introduce other indices used by the specialists in the area; in Section 3 we present the nestedness index ν, its main mathematical properties and derive a relation between ν and the interactions of animals and plants; finally in Section 4 we conclude the work.

2. The nestedness concept

Nestedness is not a straightforward concept, and this fact leads to misunderstandings in the literature as well as a proliferation of estimators [10] (this situation is similar to the definition of diversity, another key concept in community ecology that also shows several conflicting definitions and estimators [11]). To visualize nestedness from the adjacency matrix it is common in this area to rank rows and columns of the matrix. In this way the matrix usually show in the lattice format of fill and empty squares, present a decreasing connectivity where the large number of connections is packed in the upper left corner.

The oldest and most popular nestedness estimator is the "temperature" of Atmar and Patterson [12]. This index is constructed using a median line in the ranked adjacency matrix. This line equally separates holes and full cells in the matrix (zeros and ones). The "temperature" index τ is a measure of the dispersion of holes in respect to the median. There are other indices [13, 14] that basically count the number of vacancies in the matrix. The difficult point of these indices is that they are defined by cumbersome algorithms that impede further analytical developments.

In figure 2 we show two adjacency matrices to help to clarify the nestedness idea. The two matrices have the same occupation $\rho = 7/16$, the occupation is defined by the number of ones divided by the total number elements of the matrix. In both cases the sum of the total marginals (connectivities k) is ranked as we have previously explained. The matrix in the left is more nested than the matrix in the right because it has more holes. In this matrix the occupied elements are more distant from the upper left corner, and we will use this property to define the nestedness index in the next section.

3. The nestedness coefficient and the connectivity

We investigate nestedness using a nestedness index $\nu = 1 - \tau$ developed in [9]. This estimator is based on distances on the adjacency matrix of the network. The intuitive idea behind this nestedness index comes from the dispersion of ones and zeros through the adjacency matrix. As exposed in the previous section we start packing the adjacency matrix of the observed interactions. A highly nested matrix is the one that, after packing, present a minimal mixing of ones and zeros. We estimate nestedness using the average distance d of the matrix:

$$D = \frac{1}{N} \sum_{k=1}^{N} d_{i,j}(k) \tag{1}$$

for N the number of ones in the matrix and $d_{i,j} = i + j$ the Manhattan distance. To normalize the temperature index τ, two artificial matrices are used: the perfectly nested matrix and the equiprobable random matrix. These two matrices have the same sizes L_1, L_2 and occupancy ρ of the adjacency matrix, and they work as benchmarks to properly define the nestedness index. In other words, we parametrize τ with help of the distance D of D_{nest} and D_{rand}. The first is the average distance related to a completely nested matrix and the second to the random matrix. τ is defined as follows:

$$\tau = \frac{D - D_{nest}}{D_{rand} - D_{nest}} \tag{2}$$

A zero τ corresponds to a state of minimal disorder (zero temperature), where all elements are perfectly nested. Conversely, $\tau = 1$ corresponds to a state of equiprobable randomly dispersed matrix elements (high temperature). The main advantage of τ is that it is posed in an analytical way. Because of that we have already discussed the relation between nestedness and abundance [15] and now we intend to work the connection between nestedness and connectivity.

A puzzling question when studying the nestedness pattern is the following: is it possible that nestedness can be defined only by the connectivity of matrix elements? An answer to this question will strongly help to demystify the nestedness concept. Using our index expressed in (1) and (2) we answer positively to this question. We start with the definition of D as expressed above:

$$D = \frac{1}{N} \sum_{i,j} d_{i,j} \tag{3}$$

where we emphasize that the sum is taken over the N occupied elements. As the Manhattan distance of each element is $(i + j)$ and the matrix elements in this equation may be written in the following form:

$$D = \frac{1}{N} \sum_{i,j} a_{i,j}\,(i + j) = \frac{1}{N}(\sum_{i,j} a_{i,j}\, i + \sum_{i,j} a_{i,j}\, j) \tag{4}$$

where the elements $a_{i,j}$ assume the values 1 or 0 for occupied and absent elements. Using the definition of $k_A(i) = \sum_j a_{i,j}$ and $k_P(j) = \sum_i a_{i,j}$ we conclude:

$$D = \frac{1}{N}(\sum_{i} i\, k_A(i) + \sum_{j} j\, k_P(j)). \tag{5}$$

This formula sets that the distance D depends only on the connectivity of animals and plants. As the nestedness $\nu = 1 - \tau$ that depends on D we have proved that nestedness rely on the connectivities of animals and plants. In addition Equation (5) enable ones to find ν without complex algorithmic manipulation of the bipartite network. In this way Equation (5) shows the nestedness in a compact form, this situation is quite distinct of other algorithms that calculate nestedness that are based on elaborated algorithmic manipulation of the bipartite netowork [12, 13, 14].

4. Final Remarks

We use the metric concept of nestedness previously developed, to prove that the nestedness index is completely determined by the connectivity of the bipartite network. Indeed, we prove a simple statement that can simplify the theoretical framework of community ecology. Rigorously speaking, the nestedness index is not a necessary concept, the connectivity is enough to characterize the nestedness of a bipartite network. Another important aspect of this work concerns the algorithmic work to find the nestedness index. The Patterson and Atmar index or the Brualdi index cited in Section 2 have a computational complexity much larger that our index. Indeed ν can be easily calculated by Equation (5).

It is worth to note that Equation (5) gives that the sum of the distances D is a function of plants and animal connectivities, that means: $D = D(k_A(i), k_P(j))$. The usual way of considering connectivity in network theory [1, 16] is employing the ditribution of connectivity $P(k)$ that is the integral of the connectivity $k(l)$. As D is not a constant, but a function of the connectivities, we have not proved that the nestedness is a function of $P(k)$. Therefore the present work analitically advances the original paper [15] that defines a nestedness metric, but does not provides any hint of the relation of this metric with the conectivities of the bipartite network.

We believe that a well defined nestedness estimator can be useful to investigate order and structure in bipartite networks such as the metabolic network of a cell or social webs like actors bipartite network [16]. To conclude, nestedness is a subject of high current interest in biogeography, community ecology and conservation biology. The *Web of Science* shows more than 300 papers on the subject in the last 14 years. In this way, this work will be of interest to researchers in a great multitude of areas, as well as to physicists working with patterns in bipartite networks.

Acknowledgements

The authors gratefully acknowledge the financial support of CNPq (Conselho Nacional de Desenvolvimento Científico e Tecnológico, Brazil). We also thanks to Thomas Lewinsohn and Paulo Inacio Prado for the encouraging discussions in theoretical ecology.

References

[1] Costa L F Rodrigues F A Travieso G Villas Boas P R 2007 Characterization of complex networks: A survey of measurements *Adv. Phys.* **56** 167–242
[2] Jiang Y Xu T He D 2004 A self-adaptative model of actor-collaboration network *Int. J. Mod. Phys. B* **18**, 2604–2607
[3] Newman M E J 2001 Scientific collaboration networks *Phys. Rev. E* **64** 016131.
[4] Lewinsohn T M Prado P I Jordano P Bascompte J and Olesen J M 2006 Structure in plant-animal interaction assemblages *Oikos* **113** 174–184
[5] Bascompte J Jordano P and Olesen J M 2006 Asymmetric coevolutionary networks facilitate biodiversity maintenance *Science* **312** 431–433
[6] Mac Arthur R H and Wilson E O 1967 *The theory of Island Biogeography* Monographs in Population Biology. Princeton Univ. Press.
[7] Darlington P J 1957 *Zoogeography: the geographical distribution of animals* Wiley, New York.
[8] Fleishman E Betrus C J Blair R B MacNally R. and Murphy D 2002 Nestedness analysis and conservation planning: the importance of place, environment, and life history across taxonomic groups *Oecologia* **133** 78–78
[9] Araujo A I L Corso Gilberto Almeida A M and Lewinsohn T M 2010 An analytical approach to the measurement of nestedness in bipartite networks *Physica A* **389** 1405–1411
[10] Almeida-Neto M Guimarães Jr P R Lewinsohn T M 2007 On nestedness analyses: rethinking matrix temperature and anti-nestedness. *Oikos* **116** 716–722.
[11] Magurran A E 2003 *Measuring Biological Diversity* Blackwell Pub.
[12] Patterson B D and Atmar W 1986 Nested subsets and the structure of insular mammalian faunas and archipelagos *Biol. J. Linn. Soc.* **28** 65–82

[13] Brualdi R A and Sanderson J G 1999 Nested species subsets, gaps and discrepancy *Oecologia* **119** 256–264
[14] Almeida-Neto M, Guimarães P, Guimarães Jr P R, Loyola R D and Ulrich W 2008 A consistent metric for nestedness analysis in ecological systems: reconciling concept and measurement *Oikos* **117** 1227–1239
[15] Araujo A I L., Almeida A M, Cardoso M Z and Corso G 2010. Abundance and nestedness in interaction networks *Ecological Complexity Preprint* doi:10.1016/j.ecocom.2010.02.004
[16] Albert R and Barabási A L 2002 Statistical mechanics of complex networks *Rev Mod Phys* **74** 47–97

Evidence of transport barrier in TCABR tokamak with high MHD activity

G. Z. dos Santos Lima[1], F. A. Marcus[2], I. L. Caldas[3], Yu. K. Kuznetsov[3], I. C.Nascimento[3]

[1]Department of Science and Technology: Federal University of Rio Grande do Norte, 59014-615, Natal, RN, Brazil

[2] Department of Naval Architecture and Ocean Engineering, University of São Paulo, 05508-970 - São Paulo, SP, Brazil

[3]Institute of Physics, University of São Paulo C.P. 66318, 05315-970 São Paulo, S.P., Brazil

E-mail: gzampier@ect.ufrn.br; ibere@usp.br

Abstract.
In this work we analyze the particle transport at the plasma edge during TCABR tokamak discharges with high MHD activity. The interpretation of this transport as chaotic, in a quasi integrable Hamiltonian system formed by the plasma flow and the drift waves, predicts its dependence on a confinement parameter, proportional to the difference between the plasma flow and the drift wave phase velocities. For the analyzed discharges, we observe a particle transport decrease where the confinement parameter has a maximum. In the considered quasi integrable description, this can be interpreted as an evidence of a localized transport barrier.

1. Introduction

Plasmas have been magnetically confined in tokamaks to investigate the possibility of generating energy from controlled thermonuclear fusion [1, 2]. One of the main problems to confine plasma in tokamaks is the anomalous loss of particles at the plasma edge [3]. It is believed that this transport of particles is related to the plasma turbulence due to drift waves [4, 5, 6].

Understanding the transport of particles in the tokamaks has a fundamental importance to improve plasma confinement [7, 8]. Recent theoretical studies [9] and experimental results in tokamaks [10] showed that transport barriers may arise at plasma edge, under some condiction, increasing the plasma confinement [6].

Unlike most tokamaks, in TCABR the frequency spectra of electrical and magnetic fluctuations have a peculiar partial superposition [11, 12]. Moreover, in some electrical discharges the MHD activity can enhance and modulate the electrostatic turbulence at the plasma edge [7, 8], coupling and synchronizing with the electrostatics fluctuations [12]. Furthermore, for plasma discharges with high MHD activity, the transport is high where a resonance condition is satisfied [7, 8], i.e, where the drift velocity associated to the radial electric field and toroidal magnetic field components is equal to the drift wave phase velocity.

Some transport properties in TCABR tokamak discharges, with high MHD activity, have been interpreted as a consequence of the chaotic particle motion due to drift waves propagating in the plasma edge with $\vec{E} \times \vec{B}$ poloidal flow. This chaotic motion has been described by a two wave quasi integral Hamiltonian model, which predicts a transport dependence on a confinement

Published under licence by IOP Publishing Ltd

Journal of Physics: Conference Series **285** (2011) 012010 doi:10.1088/1742-6596/285/1/012010

parameter, proportional to the difference between the plasma flow and the drift wave phase velocities [9, 13]. In this work, for the analyzed discharges, we present an evidence of a particle transport decrease where the confinement parameter has a maximum. In the considered quasi integrable description, this can be interpreted as an indication of a transport barrier at the plasma edge.

The remainder of the paper is organized as follows: Section 2 is a brief description of the TCABR experimental setup. Section 3 contains the experimental results about the relation between the drift and wave phase velocity difference and the barrier transport. This is followed by Section 4, consisting of our analysis conclusion.

2. Experimental setup

The experiments analyzed were performed in a hydrogen circular plasma in the TCABR tokamak [13], which has the major radius equal to $R = 61$ cm and minor radius equal to $a = 18$ cm (shown in Fig.1). The maximum value of the plasma current is around 110 kA, with duration 100 ms, as well as the hydrogen filling pressure presents 3×10^{-4} Pa, and toroidal magnetic field is $B_T = 1.1$ T. The fluctuations was measured by two Langmuir probes [Fig.1], poloidally separated by 0.4 cm. The probes are mounted on a movable shaft that can be displaced radially from $r = 15.0$ to 22.0 cm, with respect to the center of the plasma column. We take the range from 16.5 to 21.0 cm in order to cover the entire region of the plasma edge. The probe displacement occurs only for separate discharges in order to not disturb the plasma confinement and stability. The measurements were performed at a sampling frequency of 1 MHz.

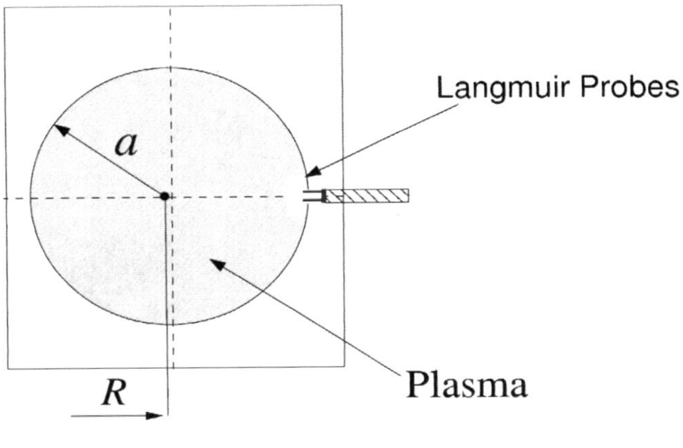

Figure 1. Cross section of the TCABR tokamak showing the Langmuir probes and plasma column.

During TCABR tokamak discharges with high MHD activity, the electrostatic plasma edge turbulence and particle transport are strongly influenced by the MHD activity. This high MHD activity can be obtained in two different regimes of TCABR discharges. In one regime, the enhancement of MHD activity was observed only after the activation of a biasing electrode while, in the other regime, this enhancement occurs spontaneously without activation of the electrode. In this article, we considered the second TCABR regime.

The evolution of a plasma discharge (number 18352) in TCABR is depicted in the Fig.2. The plasma current [Fig.2(a)] grows slowly for the first 85 ms, decaying slowly after that. The electron density evolution, indicated in the Fig.2(b), exhibits an evolution with a plateau level reach around $n_e = 1.2 \times 10^{19}$ m^{-3} in the region analyzed between vertical lines. The amplitude of the magnetic oscillations, shown in Fig.2(c), starts increasing at $t = 78$ ms and stays high until $t = 100$ ms. The time evolution of the ion saturate current is shown in Fig.2(d). The vertical black lines indicate the time interval with high MHD activity.

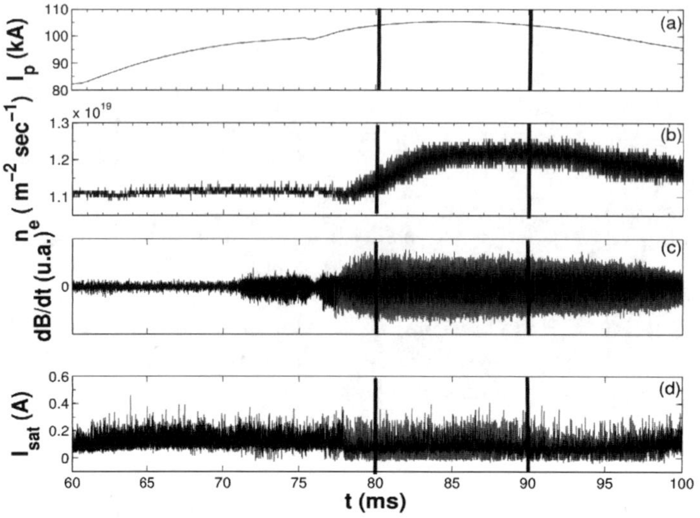

Figure 2. Time evolution of plasma discharge in TCABR tokamak. (a) Plasma current, (b) central chord plasma mean density, (c) magnetic oscilation, (d) ion current saturation. The vertical black lines indicate the analyzed time interval with high MHD activity.

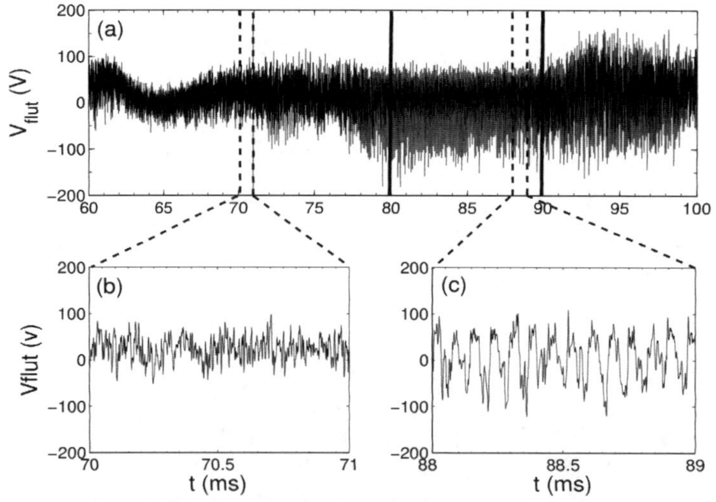

Figure 3. Time evolution of floating electrostatic potential fluctuation (a). Amplified time intervals with low MHD activity (b) and with high MHD activity (c) (dashed vertical lines). The vertical black lines indicate the analyzed time interval with high MHD activity.

The fluctuation we particularly studied was the floating electrostatic potential and a representative example is depicted in the Fig.3(a). The MHD activity change significantly the floating electrostatic potential as we present in the Figures 3(b) and 3(c); two diferent time intervals with low and high MHD activity respectively. All analyzed data were chosen during time intervals represented by the window indicated by full vertical lines (see Fig.3). We choose this time interval window before MHD activity starts to increase and changing the plasma turbulence [7, 10, 12].

3. Evidence of transport barrier in TCABR tokamak: Results and Discussion

Experimental results presented in this paper, on the drift and phase velocities difference and transport barrier at TCABR plasma edge and scrape-off layer [10], give us important evidence for the Hamiltonian model presented in Ref.[6, 9]. In this model there is a function, the trapping profile (U), proportional to the difference between the dominant wave phase velocity and the drift flow velocity, that determines the chaotic particle transport in the radial direction [14]. Thus, where the trapping profile ($U(x)$) is zero a resonance condition exists and a high radial particle transport is predictable. In contrast, where $dU/dr = 0$ the particle flux should be low.

To search for these theoretical predictions, we analyze data from several discharges (18346 t0 18379) and drew in the Fig.4 the radial profile of the averages phase velocity (v_{ph}), the drift velocity ($v_{\vec{E}\times\vec{B}}$) and the difference between these velocities, proportional to the experimental trapping profile, i.e, $U(r) \propto \frac{1}{B_\phi}\frac{dV_{flut}}{dr} - v_{ph}$.

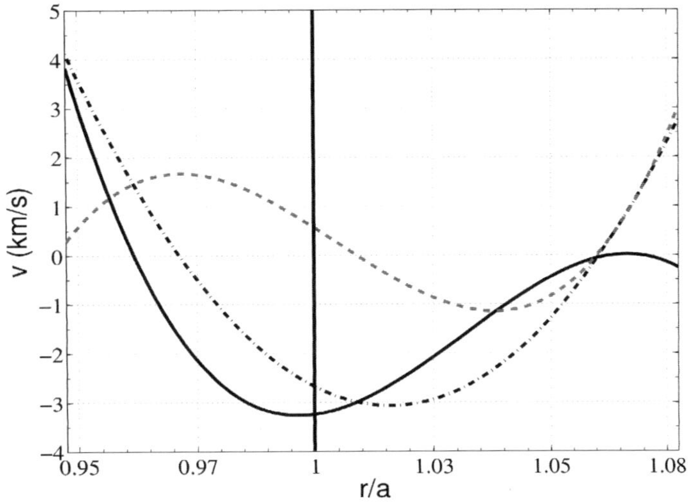

Figure 4. Radial profile of the averages phase (black line) and drift (dot-dashed line) velocities, and the difference between these velocities (dashed line)

Figure 5 shows the radial profiles of the experimental particle transport (triangle up) and the difference between phase and drift flow velocities $U(r)$ (dashed line). We observe a particle transport decay in the region where the difference between velocities has a local maximum ($dU(r)/dr = 0$) at ($r/a = 0.97$), corresponding to a shearless point. This decay indicates a possible transport barrier at this position, as predicted in [6, 9] and observed in [10]. Thus, the localized transport decay may be an indication of a transport barrier localized on the shearless region (indicated by a vertical arrow in Fig.5).

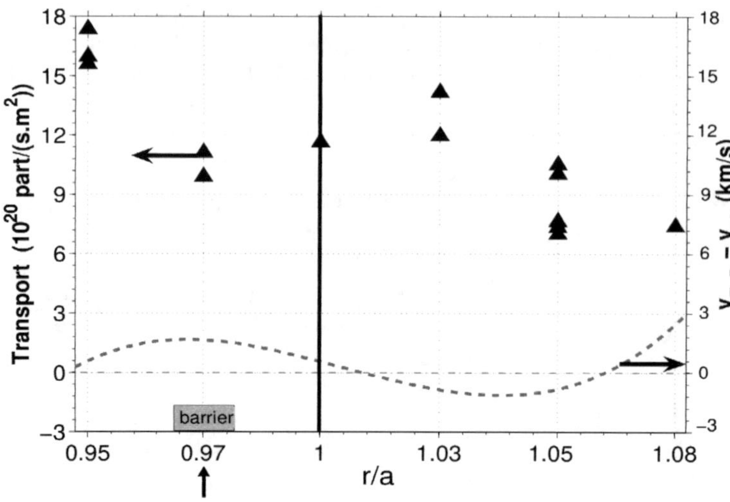

Figure 5. Radial profile of particle transport and the difference between the phase and flow velocities. Transport barrier localized on the shearless region on r/a =0.97(indicated by the vertical arrow).

Our transport barrier prediction is based on radial profiles obtained by analyzing probe data from different radial positions in a sequence of similar discharges. This prediction should be further confirmed in new experiments with simultaneous probe measurements in different radial positions. Moreover, the uncertainty in Figs. 4 and 5 profiles justify new measurements to better precise the barrier localization.

4. Conclusion

Analysing the radial profile of turbulence and transport of particles of the TCABR plasma discharges with high MHD activity, we found evidence that the particle transport decreased where the difference between phase and drift velocities had a local maximum. Although the experiment was not performed to verify any theoretical model, we used a chaotic two wave Hamiltonian model to interpret this result as an indication of the existence of a transport barrier created in the radial position where the drift flow shear vanishes.

Acknowledgments

This work was made possible with the help of FAPESP, CNPq, CAPES and UFRN.

References

[1] Jeffrey F 1990 *Plasmas Physics and Fusion Energy* (Cambridge)
[2] Gross R A 1984 *Fusion Energy* (John Wiley and Sons)
[3] Hazeltine R D and Prager S C 2002 *Physics Today* **55** 30
[4] Wesson J 2004 *Tokamaks* (Oxford University Press, 3rd Edition)
[5] Hidalgo C 2004 *Astrophysics and Space Science* **292** 681
[6] Horton W 1999 *Review of Morden Physics* **71** 735
[7] dos Santos Lima G Z and et al 2009 *Physics of Plasmas* **16** 042508
[8] Guimaraes-Filho Z O and et al 2010 *J. Phys.: Conf. Series* **246** 012014
[9] Marcus F A and et al 2008 *Physics of Plasmas* **15** 112304

[10] dos Santos Lima G Z 2009 *Análise da Turbulência do Plasma no Tokamak TCABR* (Ph.D. thesis, Intitute of Physics, University of São Paulo)
[11] Heller M V A P and et al 2007 *Journal of Plasma Physics* **73** 295
[12] Guimaraes-Filho Z O and et al 2008 *Physics of Plasmas* **15** 062501
[13] Nascimento I C and et al 2005 *Nuclear Fusion* **45** 796
[14] Horton W 1985 *Plasma Physics and Controlled Fusion* **27** 937

Dynamic Days South America 2010 IOP Publishing
Journal of Physics: Conference Series **285** (2011) 012011 doi:10.1088/1742-6596/285/1/012011

Blocking Radial Diffusion in a Double-Waved Hamiltonian Model

Caroline G. L. Martins[1, 4], F. A. Marcus[2], I. L. Caldas[3] and R. Egydio de Carvalho[1]

[1] UNESP-Univ Estadual Paulista; Instituto de Geociências e Ciências Exatas; Departamento de Estatística, Matemática Aplicada e Computação, Av. 24A, 1515, 13506-900 Rio Claro , SP, Brazil.
[2] Universidade de São Paulo; Departamento de Engenharia Naval e Oceânica 05508-970 São Paulo, SP, Brazil.
[3] Universidade de São Paulo; Instituto de Física 05315-970 São Paulo, SP, Brazil.
[4] Instituto Tecnológico da Aeronáutica; Departamento de Física 12228-900 São José dos Campos, SP, Brazil.

E-mails: carolinegameiro@gmail.com, albertus@if.usp.br, ibere@if.usp.br, regydio@rc.unesp.br

Abstract A non-twist Hamiltonian system perturbed by two waves with particular wave numbers can present Robust Tori, barriers created by the vanishing of the perturbing Hamiltonian at some defined positions. When Robust Tori exist, any trajectory in phase space passing close to them is blocked by emergent invariant curves that prevent the chaotic transport. We analyze the breaking up of the RT as well the transport dependence on the wave numbers and on the wave amplitudes. Moreover, we report the chaotic web formation in the phase space and how this pattern influences the transport.

1. Introduction

A Hamiltonian model with two waves was introduced by W. Horton [1, 4] to describe drift waves, originated by particles drift proportional to $\vec{E} \wedge \vec{B}$ in nonuniform plasmas, propagating in a magnetic toroidal and an electric radial fields. The model has been applied in many works as to investigate the influence of the equilibrium electric and magnetic fields on the radial transport as to analyze experimental results [2, 3].

Robust Tori [5-7] are dynamical barriers present in Hamiltonian models that survive intact under the action of generic perturbations [8]. Our proposal is to show that the two-wave Hamiltonian model contains RT, for certain parameters set, and even if other waves are included in the system they do not break up themselves. This is an important fact since the creation of barriers in Hamiltonian systems has been a remarkable subject in physics, mainly in Hamiltonians used for plasma studies [9, 10].

We are going to consider the behavior of an integrable Hamiltonian H_0 under the effect of a small amplitude perturbation described as,

$$H(x, y, t) = H_0(x) + \varepsilon H_1(x, y, t) \tag{1}$$

where ε is the control perturbation parameter. We also consider the transport along the "axis x", called radial transport [6, 7], due to the waves, in poloidal sections of the tokamak. The Hamiltonian describing that transport is [1],

$$H(x,y,t) = H_0(x) + \sum_n A_n \sin(k_{xn} x)\cos(k_{yn} y - \omega_n t) \tag{2}$$

where $H_0(x)$ is an unperturbed integrable Hamiltonian perturbed by a collection of waves. The coordinates x and y are the canonical momentum and its conjugate coordinate respectively. In this work we consider only two waves.

The canonical transformation mediated by the generating function, $F(x,y',t) = x\,(y' - u_1.t)$ from where $y = \frac{\partial F(x,y',t)}{\partial x} = y' - u_1 t$, $x' = \frac{\partial F(x,y',t)}{\partial y'} = x$, $\frac{\partial F(x,y',t)}{\partial t} = -x u_1$ and $H(x,y,t) = H(x',y',t) + \frac{\partial F(x,y',t)}{\partial t}$, leads the Hamiltonian from Eq. (2) in the following equation,

$$H(x,y,t) = H_0(x) - u_1 x + A_1 \sin(k_{x1}x)\cos(k_{y1}y) + A_2 \sin(k_{x2}x)\cos(k_{y2}(y-ut)) \tag{3}$$

where $u = (\omega_2 / k_{y2}) - (\omega_1 / k_{y1})$ is the difference of the phase velocities between the two waves, while $u_1 = (\omega_1 / k_{y1})$ is the phase velocity of the first wave, (A_1, A_2) and (k_{x1}, k_{y1}) are the amplitudes and the wave numbers of the first and second waves respectively.

An important characteristic of the unperturbed Hamiltonian system is its rotation number,

$$\Omega(x) = \frac{dH_0(x)}{dx} \tag{4}$$

because depending if $\Omega(x)$ is monotonic or not the system is twist or non-twist [11, 12, 13] respectively. Many models are described by non-twist Hamiltonians, for instance orbits in particle accelerators [14], plasma wave heating [15] and fluid dynamics [16]. The unperturbed Hamiltonian $H_0(x)$ that we will consider here presents a non-twist profile and is giving by,

$$H_0(x) = Ax^3 + Bx^2 + Cx - u_1 x \tag{5}$$

where A, B and C are constants numbers.

2. Hamiltonian approach for one wave

At first, we consider the Hamiltonian of Eq. (3) with only one wave, choosing $A_2 = 0$, so that,

$$H(x,y,t) = Ax^3 + Bx^2 + Cx - u_1 x + A_1 \sin(k_{x1}x)\cos(k_{y1}y) \tag{6}$$

We note that the perturbed Hamiltonian vanishes when $\cos(k_{y1}y) = 0$ or $\sin(k_{x1}x) = 0$. However, looking at the equations of motion,

$$\dot{x} = A_1 k_{y1} \sin(k_{x1} x) \sin(k_{y1} y)$$
$$\dot{y} = \left[3Ax^2 + 2Bx + C - u_1\right] + A_1 k_{x1} \cos(k_{x1} x) \cos(k_{y1} y) \qquad (7)$$

we can see that for $\cos(k_{y1} y) = 0$ the motion still exists in x and y coordinates. So, the RT will appear only in the cases where the perturbation and also the motion in at least one coordinate vanish [6]. Our interest is about the radial transport, then we will analyze the case when $\sin(k_{x1} x) = 0$. For this condition, the perturbation and the motion in the coordinate x vanish (see Eqs. (6) and (7)). However, the motion still exists along the coordinate y. There will be straight lines with *x=constant*, corresponding to the RT, at $x = \frac{n_1 \pi}{k_{x1}}$; for integers n_1.

Figure 1 shows the Poincaré section for the one-wave Hamiltonian, with islands that look like cells flatted by straight lines. These lines, in black, are the RT previously mentioned. The Hamiltonian of Eq. (7) presents only one wave and the system is integrable, i.e. there is not any chaotic orbit.

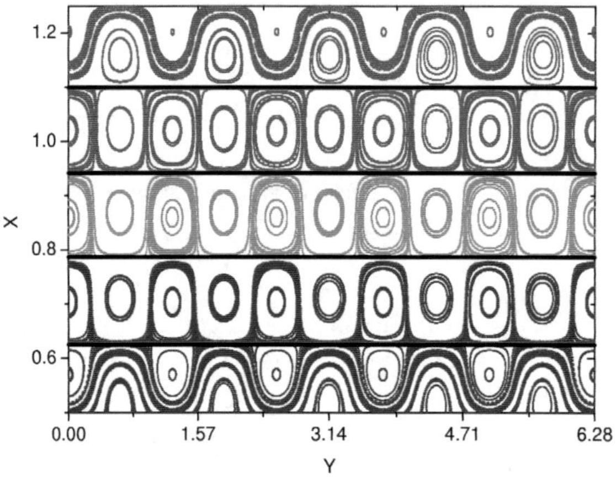

Figure 1 – Poincaré section for the Hamiltonian of Eq. (7) with one wave, $A_2=0$, and A = -2; B = 5.25 and $(C-u_1)$ = -4.5

3. Two-waves Hamiltonian approach

The addition of another wave in the Hamiltonian of Eq. (7) will break the integrability of the system and we get this through the equation,

$$H(x, y, t) = Ax^3 + Bx^2 + Cx - u_1 x + [A_1 \sin(k_{x1} x) \cos(k_{y1} y) + A_2 \sin(k_{x2} x) \cos(k_{y2}(y - ut))] \qquad (8)$$

The equations of motion are given by,

$$\dot{x} = A_1 k_{y1} \sin(k_{x1} x) \sin(k_{y1} y) + A_2 k_{y2} \sin(k_{x2} x) \sin(k_{y2}(y - ut))$$
$$\dot{y} = \left[3Ax^2 + 2Bx + C - u_1\right] + A_1 k_{x1} \cos(k_{x1} x) \cos(k_{y1} y) + A_2 k_{x2} \cos(k_{x2} x) \cos(k_{y2}(y - ut)) \qquad (9)$$

We can see that the perturbation in square brackets in Eq. (8) vanishes for $\cos(k_{y1}y) = \cos(k_{y2}(y-ut)) = 0$, but looking at the equations of motion above, Eq. (9), we note that the motion does not vanish with only this condition analogously as mentioned in section 2.

On the other hand, for $\sin(k_{x1}x) = \sin(k_{x2}x) = 0$, there will be lines with *x=constant* at $x = \frac{n_1\pi}{k_{x1}} = \frac{n_2\pi}{k_{x2}}$ for integers n_1 and n_2. Consequently, the relation between the wave numbers should satisfy $k_{x2} = m.k_{x1}$. If *m* is an integer, all RT will remain intact, otherwise only some percentage will survive. If the wave numbers obey the condition above, all RT will remain intact under the addition of the second wave blocking the radial transport.

Fig. 2 shows two different situations for the two-wave Hamiltonian model of Eq. (8). Fig. 2(a) shows the Poincaré section for the case already described in the literature [1, 2, 3, 4]. Adding the second wave, the integrability of the system is broken and there are chaotic orbits spread along the phase space, i.e. the particles move along the radial coordinate *x* forming a chaotic web. The colours blue, green and red represent trajectories of initial conditions given on different regions in the phase space and they show the mixing radial transport done by the particles. There are no barriers in the phase space along the *axis y*. Fig. 2(b) shows the Poincaré section for the particular case $k_{x2} = m.k_{x1}$, and the RT (lines in black) remain intact under the addition of the second wave. We see that there is no mixture of the s blue, green and red, showing that the RT are blocking the radial diffusion.

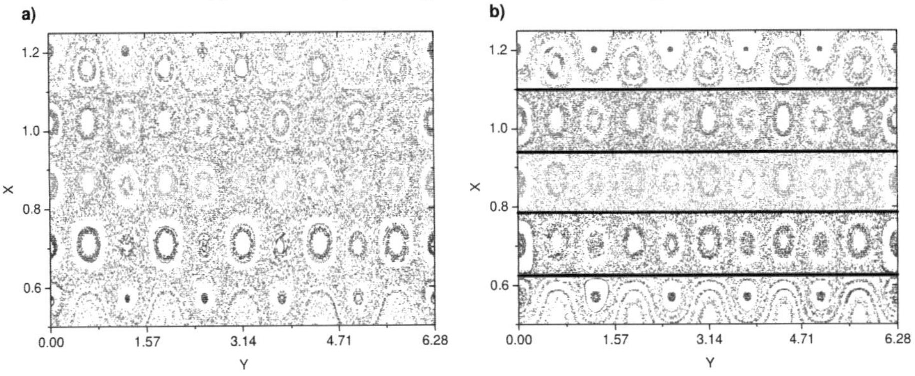

Figure 2 - Poincaré sections for the Hamiltonian with two waves of Eq. (8); **a)** for $k_{x2} \neq m.k_{x1}$ without RT. **b)** for $k_{x2} = m.k_{x1}$ with RT in black.

4. Diffusion coefficient and robust tori

The onset of chaos in the phase space takes place when the second wave is added and consequently the transport can be evaluated by calculating the diffusion coefficient from the orbits. The calculation of the local diffusion coefficient, for finite time, is done considering the non-dimensional equation [2]:

$$D = \frac{1}{2tN}\sum_{i=1}^{N}[x_i(t) - x_i(0)]^2 \qquad (10)$$

where *N=1000* is the total number of initial conditions, which are uniformly distributed through the grid *x:[0.50; 1.25]* and *y:[0; 2π]*, and *t=150* is, in fact, the number of iterations for each initial condition.

In Fig. 3(a) we show the dependence of the radial diffusion coefficient on the wave number k_{x2}. We used in the numerical simulations the following parameters $k_{x1}=20;\ k_{y1}=3$. As we can see, the lowest values of the diffusion occur for $k_{x2}=m.k_{x1}=20$ for $m=1$ and for $k_{x2}=m.k_{x1}=40$ for $m=2$. The radial diffusion coefficient tends to zero for all m integer, but it is not zero for small non-integer m.

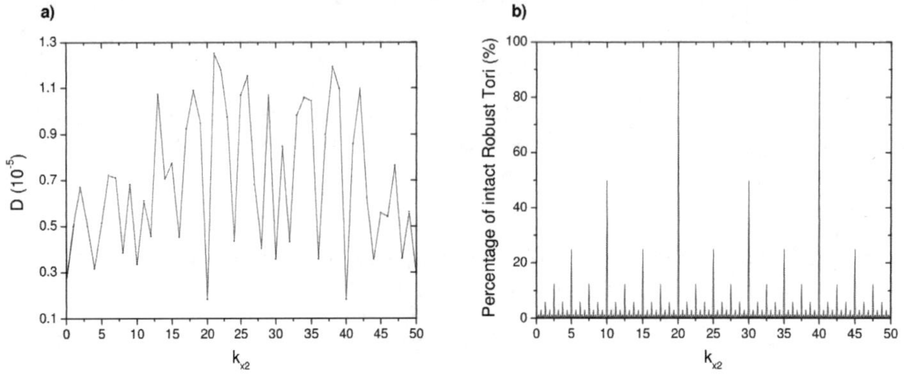

Figure 3 – Comparison between **a)** Diffusion coefficient and **b)** Percentage of intact robust tori.

The behavior of the radial diffusion coefficient is explained from Fig. 3(b), which shows the percentage of intact remaining RT after the addition of the second wave, considering an initial amount of 40 RT in the phase space. For the wave numbers $k_{x2} = 20$ and $k_{x2} = 40$ all RT still exist because the wave numbers satisfy the particular solution we present here, $k_{x2} = m.k_{x1}$ with m an integer. However, we can observe that a portion of RT also exists when m is a non-integer.

The existence of RT affects directly the particles diffusion in the radial direction. For the particular solution introduced here, the radial diffusion coefficient is zero because all RT are preserved. However, for some intermediate values of k_{x2} some RT still remain intact, decreasing the radial transport.

5. Chaotic web and the breaking up of the robust tori

In order to analyze the breaking up of the RT and the formation of a chaotic web in the phase space we performed a colour map to indicate how many iterations are necessary for the initial conditions to reach the reference line $x=0.9408$ for different values of (A_2/A_1). A grid of initial conditions is used with y in the range $[0, 2\pi]$ and x in the range $[0.6283, 0.9408]$.

Figure 4 shows the colour maps for different values of (A_2/A_1). We emphasize that the particular solution for the RT preservation was not used in this case. We used $k_{x2} \neq m.k_{x1}$ to show how the RT are broken by the time we increase the rate for the amplitudes of the two waves.

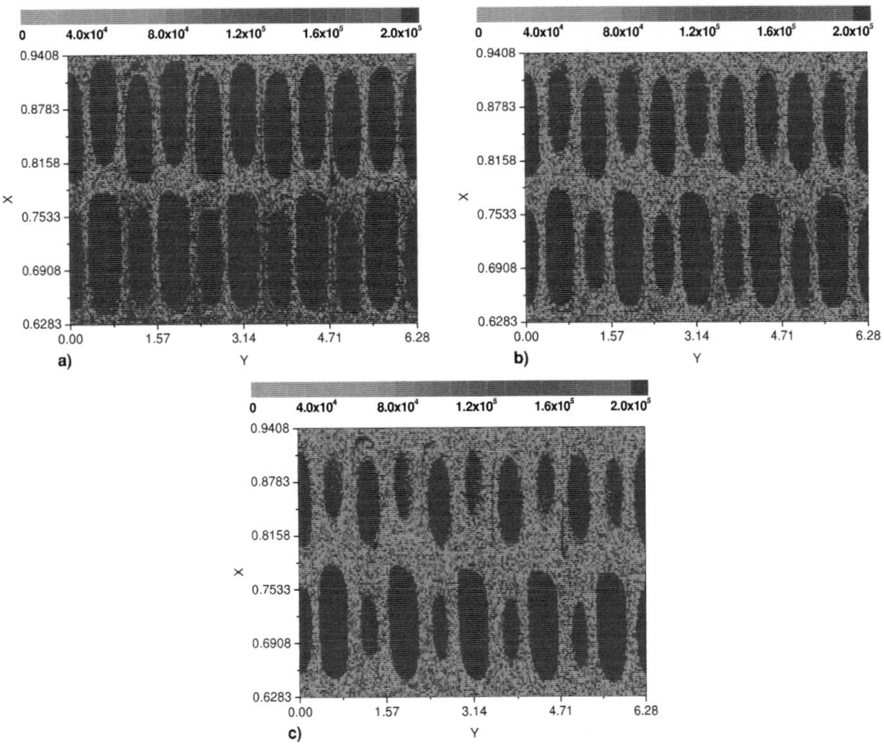

Figure 4 – Colour map showing how many iterations are necessary for the initial conditions to reach the reference line $x=0.9408$ for **a)** $(A_2/A_1)=0.05$ **b)** $(A_2/A_1)=0.1$ **c)** $(A_2/A_1)=0.15$

A trajectory with a chosen initial condition is considered trapped if the number of iteration to reach the reference line is higher than $6\text{x}10^5$ and the colour that corresponds to this situation is blue. As can be seen in Fig. 4(a), with $(A_2/A_1)=0.05$, all RT are destroyed and a chaotic web appears in the phase space. The chaotic web is created around the hyperbolic fixed points, doing a channel for the radial transport. As we increase the amplitudes rate, the chaotic web also increases, as can be seen in Fig. 4(b) so that a large number of trajectories achieve the reference line with only a few iterations. The enlargement of the chaotic web is clearer when we increase the rate of the amplitudes to $(A_2/A_1)=0.15$, as is showed in Fig. 4(c), and the area covered by the colour blue, which corresponds to the trapped trajectories, decreases.

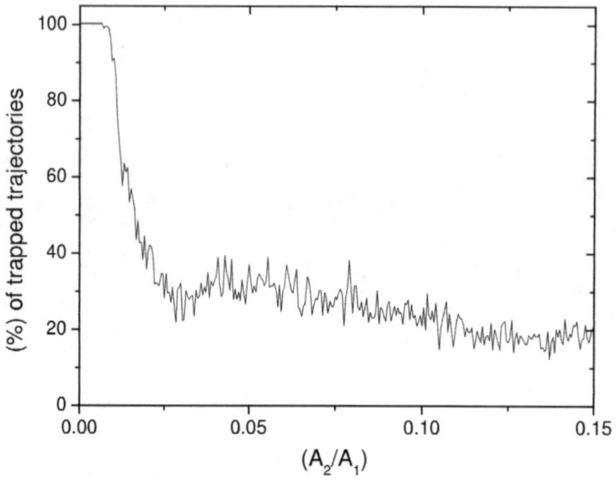

Figure 5 – Percentage of the trapped trajectories in the RT located at $x=0.62831853$ as a function of the rate of the amplitudes (A_2/A_1) of the first and the second wave.

In order to determine how many trajectories (generated by chosen initial conditions) have been trapped in the line with $x=0.62831853$ formed by one robust torus, we count how many dots in blue exist in this line. We depict in Fig. 5 the behavior of the breaking up of the robust torus. How was expected for the integrable case (with only one wave), $(A_2/A_1)=0$, all trajectories are trapped, showing the presence of the robust torus. After the breaking up of the robust torus, the number of trapped trajectories in $x=0.62831853$ decreases, with an oscillatory behavior.

6. Conclusions

The onset of chaos in the phase space takes place when the second wave is added and consequently the destruction of the manifolds of the hyperbolic fixed points creates a chaotic web in phase space, forming a channel for the radial transport. The increase of the wave amplitudes is the responsible for the enlargement of the chaotic web, improving the radial transport.

Previous studies [9, 17] have shown the importance of decreasing the radial transport induced by drift waves to improve the plasma confinement in tokamaks. It is also reported that similar Hamiltonians to the one presented in this paper have been used to study transport but only few works were dedicated to control chaos in these systems [2, 10].

Robust Tori are dynamical barriers that block the chaotic radial transport and a particular solution that preserves all RT in phase space is presented. We point out that the particular two wave solution presented in this paper can be extended for many waves and the non perturbed Hamiltonian H_0 does not influence the formation of RT along the *x-direction*. The condition to introduce RT can be rewritten as $k_{x(n+1)} = m_n . k_{xn}$ with n an integer ϵ [1, (N-1)], where (N-1) is the number of waves considered. The coefficients m_n have to obey the condition, $\prod_{n=1}^{N-1} m_n = \text{integer number}$. The multiplication of the coefficients m_n has to be an integer because all radial wave numbers k_{xn} have to be a multiple of the first radial wave number k_{x1}, to keep intact all RT in phase space.

Acknowledgments

The authors thank the Brazilian scientific agencies FAPESP, CNPq and CAPES for financial support.

References

[1] Horton W 1985 *Plasma Phys. Controlled Fusion* **27** 937
[2] Marcus F A, Caldas I L, Guimarães-Filho Z O, Morrison P J, Horton W, Kuznetsov Yu K and Nascimento I C 2008 *Phys. Plasmas* **15** 112304
[3] Kroetz T, Marcus F A, Roberto M, Caldas I L, Viana R L and da Silva E C 2009 *Computer Physics Communications* **180** 642
[4] Horton W 1981 *Plasma Phys.* **23** 1107
[5] Egydio de Carvalho R, Martins Caroline G L and Favaro G M 2009 *Braz. J. Phys.* **39** 606
[6] Martins Caroline G L, Egydio de Carvalho R, Caldas I L and Roberto M 2010 *J. Phys. A* **43** 175501
[7] Martins Caroline G L, Marcus F A, Caldas I L and Egydio de Carvalho R 2010 *Physica A* **389** 5511.
[8] Egydio de Carvalho R and Favaro G M 2005 *Physica A* **350** 173
[9] Escande D F 1985 *Phys. Rep.* **121** 165
[10] Ciraolo C, Chandre C, Lima R, Vittot M, Pettini M, Figarella C, and Ghendrih P 2004 *J. Phys. A* **37** 3589
[11] del-Castillo-Negrete D, Greene J M and Morrison P J 1996 *Physica D* **91** 1
[12] Howard J E and Humpherys J 1995 *Physica D* **80** 256
[13] Egydio de Carvalho R and Ozório de Almeida A M, 1992 *Phys. Lett. A* **162** 457
[14] Symon K R and Sessler A M 1956 *Proc. Int. Conf. on High-Energy Accelerators and Instrumentation* CERN - Geneva, p. 44
[15] Howard J E, Lichtenberg A J, Lieberman M A and Cohen R H 1986 *Physica D* **20** 259
[16] del-Castillo-Negrete D and Morrison P J 1993 *Phys. Fluids A* **5** 948
[17] Lichtenberg A J and Lieberman M A 1993 *Regular and Stochastic Motion*, Springer-Verlag, NY

Magnetic Field Line Escape: Comparison with Mean Free Path

Caroline G. L. Martins[1, 2], M. Roberto[2], I. L. Caldas[3] and R. Egydio de Carvalho[1]

[1] UNESP-Univ Estadual Paulista; Instituto de Geociências e Ciências Exatas; Departamento de Estatística, Matemática Aplicada e Computação, Av. 24A, 1515, 13506-900 Rio Claro , SP, Brazil.
[2] Instituto Tecnológico da Aeronáutica; Departamento de Física 12228-900 São José dos Campos, SP, Brazil.
[3] Universidade de São Paulo; Instituto de Física 05315-970 São Paulo, SP, Brazil.

E-mails: carolinegameiro@gmail.com, marisar@ita.br, ibere@if.usp.br and regydio@rc.unesp.br

Abstract In the present work, we determine the fraction of magnetic field lines that reach the tokamak wall leaving the plasma surrounded by a chaotic layer created by resonant perturbations at the plasma edge. The chaotic layer arises in a scenario where an integrable magnetic field with reversed magnetic shear is perturbed by an ergodic magnetic limiter. For each considered line, we calculate its connection length, i.e. the number of toroidal turns that the field lines complete before reaching the wall. We represent the results in the poloidal section in which the initial coordinates are chosen. We also estimate the radial profile of the fraction of field lines, for different temperatures, whose connection lengths are smaller than the electron collisional mean free path.

1. Introduction

Plasma-wall interaction is one of the critical issues for the development of an energy source based on nuclear fusion [1-3]. Efforts have been made to study particle transport in tokamaks outer of the plasma column to verify how this transport affects the tokamak wall. If the heat and particle loadings are too spatially localized, the sputtering processes which occur in the wall may release contaminants into the plasma column deteriorating the overall confinement quality [4].

Many techniques have been developed to modify the spatial distribution of the particles and of the heat transport in the outer region of the tokamak in order to improve the plasma confinement. One of these techniques uses an ergodic magnetic limiter that creates resonant magnetic perturbations at the plasma edge and generates a chaotic layer near the tokamak wall [5-7].

Another way to improve the confinement is to modify the plasma current to create an equilibrium with reversed magnetic shear from a non-monotonic plasma current density. In this case an internal barrier arises and the particle transport is reduced [8, 9].

A recent work has shown that the connection lengths are comparable to the electron collisional mean free path for tokamaks with divertors [10]. In our work, we are going to study the patterns produced by the escaping chaotic field line near the tokamak wall [11, 12] for an equilibrium with reversed magnetic shear perturbed by an ergodic magnetic limiter. The escaping of the field line is analyzed for a range of control parameters and the connections lengths are calculated and compared to the electron collisional mean free path in order to estimate the relevance of the field line transport. The comparison is made for different radial positions and different temperature values.

2. Hamiltonian Approach

In this section we introduce a Hamiltonian to describe the plasma equilibrium with a reversed shear magnetic field in the tokamak perturbed by an ergodic magnetic limiter (EML). The equilibrium Hamiltonian, H_0, has a toroidal symmetry whose canonical time is the associated toroidal angle ϕ (See Figure 1). This integrable non-perturbed Hamiltonian corresponds to an analytical solution of the non-linear Grad-Shafranov equation [13].

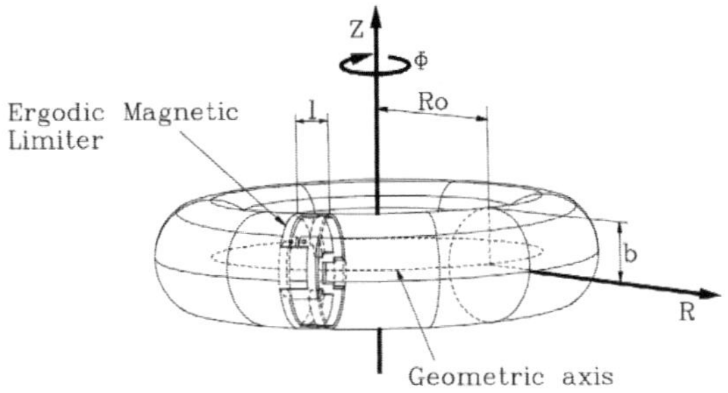

Figure.1 - Basic geometry of the tokamak with the ergodic magnetic limiter scheme.

The field lines geometry is described in non-orthogonal polar-toroidal coordinates (r_t, θ_t, ϕ_t) [13], which are related to the polar local coordinates (r,θ) by this relation:

$$r_t = r\left[1-\frac{r}{R_0'}\cos\theta+\left(\frac{r}{2R_0'}\right)\right]^{1/2} \text{ and } \sin\theta_t = \sin\theta\left[1-\frac{r}{R_0'}\cos\theta+\left(\frac{r}{2R_0'}\right)^2\right]^{-1/2} \quad (1)$$

where R_0' is the Shafranov shift. This coordinate system has been introduced to evidence the toroidal effects in the geometry of equilibrium field. Magnetic surfaces are characterized by nested surfaces of r_t = *constant*, for which the non-monotonic safety factor [14] reads as,

$$q_c(r_t)=q_c(a)\frac{r_t^2}{r^a}\left[1-\left(1+\beta'\left(\frac{rt}{a}\right)^2\right)\left(1-\left(\frac{rt}{a}\right)^2\right)^{\gamma+1}\right]^{-1} \quad (2)$$

with $q_c(a)=(I_p a^2)/(I_e R_0^2)$, where I_p is the total plasma current, I_e is the total current of the toroidal magnetic system, a is the plasma radius, R_0 is the position of the magnetic axis and the parameters $\gamma=0.78$ and $\beta=3.0$, with $\beta'=\beta(\gamma+1)/(\beta+\gamma+2)$, describes the plasma current profile.

We have chosen $q \cong 5$ at the plasma edge with $R_0=0.61\ m$ and $a=0.18\ m$, which are the TCABR parameters [15]. R_0 is the major radius and a is the plasma radius.

The perturbing Hamiltonian H_1 is created by an ergodic magnetic limiter that consists of N_r rings of length l located symmetrically along the toroidal direction of the tokamak (see Figure 1). These current rings may be considered as slices of a pair of external helical windings located at $r_t = b$, which conduct a current I_h in opposite senses for adjacent conductors. The windings are described by the following winding law: $u_t = m_0[\theta_t + \alpha\sin(\theta_t)] - n_{0t}\phi_t$, where (m_0, n_0) are the poloidal and toroidal mode numbers, respectively [14]. A tunable parameter α is introduced in such way that the variable u_t is constant along a field line.

After an appropriated canonical transformation, the unperturbed Hamiltonian H_0 is given in terms of the action J, corresponding to the toroidal normalized flux, and ϑ the poloidal angle canonically associated to J. The perturbing Hamiltonian H_1 is a function of J, ϑ and t, and it is represented by a Fourier expansion of delta-kicks due to the ergodic magnetic limiter [14]. The perturbation creates chaotic magnetic field lines at the plasma edge, and the Hamiltonian function that describes the perturbed magnetic field lines is:

$$H(J,\vartheta,t) = H_0(J) + \varepsilon \sum_{m=0}^{2m_0} H_m^*(J)\cos(m\vartheta - nt) \sum_{k=-\infty}^{+\infty} \delta\left(t - \frac{2\pi}{N_r}k\right) \tag{3}$$

where H_m^* are the Fourier coefficients and the perturbation strength is given by,

$$\varepsilon = -2\left(\frac{l}{2\pi R_0}\right)\left(\frac{I_h}{I_e}\right) \tag{4}$$

The value of epsilon is very small because in tokamaks we have $l << 2\pi R_0$ and $I_h << I_e$. The details of this derivation can be found in [14].

3. Connection Lengths

For a magnetic field line with an initial position given by (J, ϑ) in the considered Poincaré section ($\Phi = 0$), its connection length $N_{cl}\,(J, \vartheta)$ is the number of toroidal turns completed before the lines reaches the tokamak wall. In our case we can coincide the ergodic limiter with the tokamak wall at the same radius, for instance $r_t = b$ (corresponding to $J=0.055$). The field line is considered lost when it reaches this radial position. The non-uniformity at the plasma edge, due to chaos, can be described by the field line connection length distribution.

Figure 2(a) shows the distribution of the connection length values $N_{cl}(J, \vartheta)$ numerically calculated for a grid of 200 x 1000 initial conditions of (J, ϑ). The color scale in the figure indicates the N_{cl} values from 1 to 500 toroidal turns. In this figure, the lines trapped inside the islands are represented by pink points. We consider the perturbation parameter $I_h/I_p=0.11$, and the perturbation mode number as $(m_0, n_0) = (4,1)$. In Figure 2(a) initial conditions with different colors indicate that the field lines with large and small connection lengths are densely mixed, as found in reference [12]. Figure 2(b) shows the fraction of lines that escape toward the wall, as a function of the initial position J, relative to the total number of field lines lost to the wall.

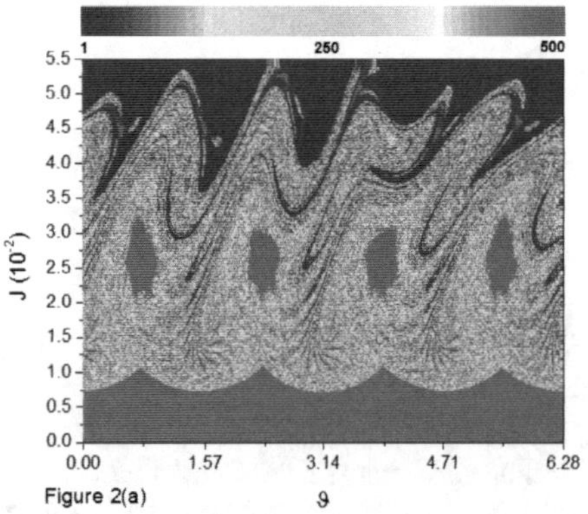

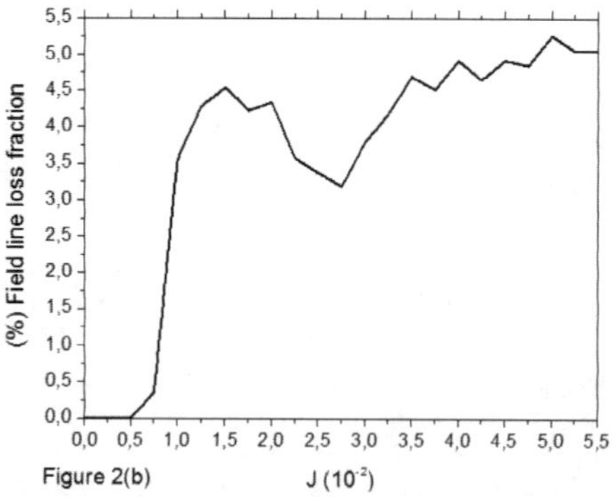

Figure 2 - **(a)** Connection lengths for $q(a)=5$. Scale from 1 to 500 toroidal turns. **(b)** Percentage of lines that escape toward the wall.

As we can see in Figs. 2(a) and 2(b), the escaping of the field lines is higher at the tokamak edge than in the plasma core. We also note the existence of four islands of stability (four elliptic fixed points) at the position $J \approx 0.025$. This fact leads to the decreasing of the field lines loss to the wall, because of the stickiness around the magnetic islands, which traps the magnetic field lines and increases their connection lengths. Moreover, these islands are surviving dimerized islands typical of plasma equilibrium with a non-monotonic spatial safety factor profile as the one considered in this work.

4. Comparisons Between Mean Free Path (MFP) and Connection Lengths

The electron mean free path λ can be estimated by the formula [16]

$$\lambda = 2.5\text{x}10^{17} \frac{T_e^2}{n_e \ln(\Lambda)} \qquad (5)$$

where the temperature is in eV and the density is in m^{-3}. To obtain λ in Eq. (5), we use the plasma density $n_e = 5.0x10^{18}$ (m^{-3}) , $ln(\Lambda)=15.2-0.5ln(n_e/10^{20})+ln(T_e/1000)$ and different values of electron temperature T_e. Table 1 presents the values obtained for λ, considering temperatures from 10 to 100 eV. The ratio $\lambda/2\pi R_0$ is equivalent to the mean free path in units of toroidal length and $2\pi R_0$ is the toroidal circumference length.

Te (eV)	λ (m)	($\lambda/2\pi R_0$)
10	0,4137	0,1078
20	1,5642	0,4081
30	3,4113	0,8900
40	5,9351	1,5485
50	9,1226	2,3801
60	12,9641	3,3824
70	17,4518	4,5533
80	22,5795	5,8912
90	28,3416	7,3946
100	34,7336	9,0623

Table 1 – Mean free paths (λ) and the ratio $(\lambda/2\pi R_0)$, from $T_e=10$ *to* $T_e=100$ eV.

Figure 3 shows the calculated connection lengths for the field lines with initial coordinates (J, ϑ). The actions are in the interval J = *[0,030; 0,055]* and the scale represent values from 0 to λ (fig. 3a) and from 0 to 2λ (fig. 3c). In Figs. 3b and 3d we also present the action profiles of the fraction of lines lost with connection lengths smaller than λ (in fig. 3b) and 2λ (in fig. 3d).

Close to the plasma edge, the field lines have smaller connection lengths than in the plasma core, as one can verify in Figs. 2 and 3. The reason for this behavior is the existence of magnetic islands showed in Figure 2(a) (pink circles surrounded by colorful structures). These islands create a chaotic region with stickiness in their neighborhood trapping the magnetic field lines and, consequently, increasing the number of toroidal turns they complete before reaching the wall, i.e., increasing the connection lengths in the region [12]. As we are interested in the peripheral region of the plasma, we choose the range J = *[0,030; 0,055]*, which corresponds to the plasma edge.

Near the wall there are purple regions in Figure 2(a) corresponding to field lines with low connection lengths, which are comparable to the mean free paths analyzed above. In order to make visible the structures in the distributions, we adjusted the connection length scales, N_{cl}, varying from 0 to 1 and N_{cl} from 0 to 2, as we can see in Figs. 3(a) and 3(c), respectively. Figure 3(b) shows that the number of the field lines that escapes with less than one mean free paths increases for larger values of J. In figure 3(d) we present the same analysis but for two mean free paths. The mean free path increases with the temperature, therefore the fraction of lost field lines in such displacement also increases. Looking at Figs. 3(b) and 3(d) we see that the percentage of lost field lines, with connection length shorter than one and two mean free paths respectively, corresponds to a maximum around 3,5% of the whole field lines that escape to the wall.

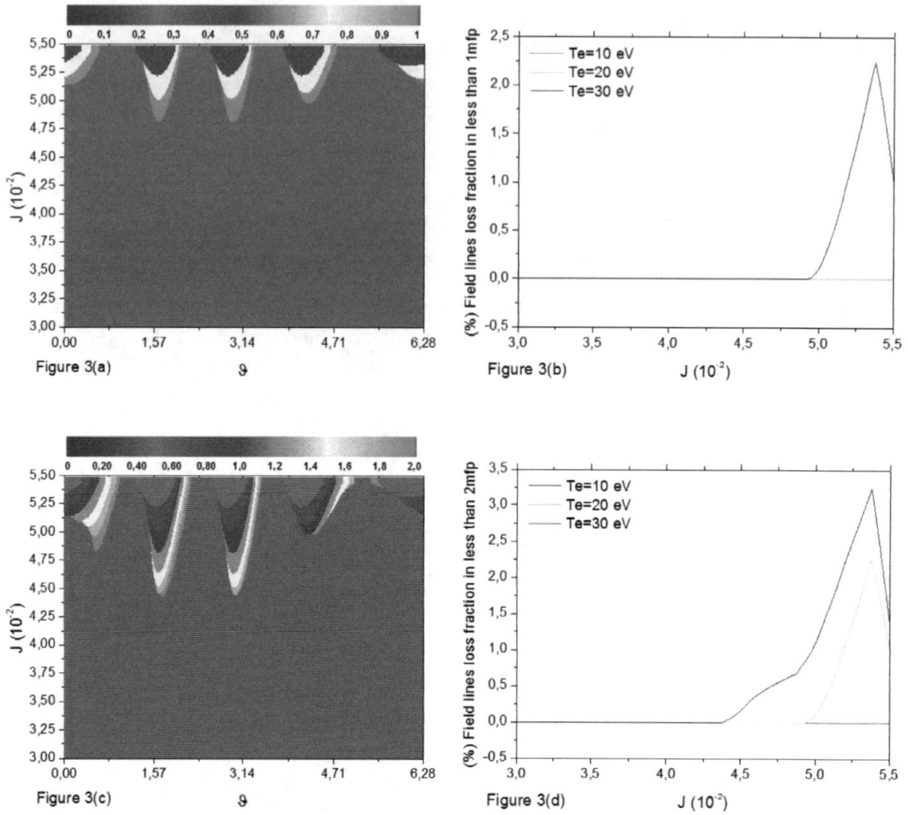

Figure 3 - **(a)** Connection lengths for $q(a)=5$ and a scale with colors indicating N_{cl} from 0 to 1. **(b)** Percentage of lost field lines with $N_{cl} < \lambda$. **(c)** Connection length with $q(a)=5$ and a scale with colors indicating N_{cl} from 0 to 2 **(d)** Percentage of lost field lines with $N_{cl} < 2\,\lambda$.

Figures 4(a) and 4(c) show the field lines that are lost during a displacement of less than one and two collisional mean free paths respectively, as a function of the variable action J, for electron temperatures from 40 to 100 eV.

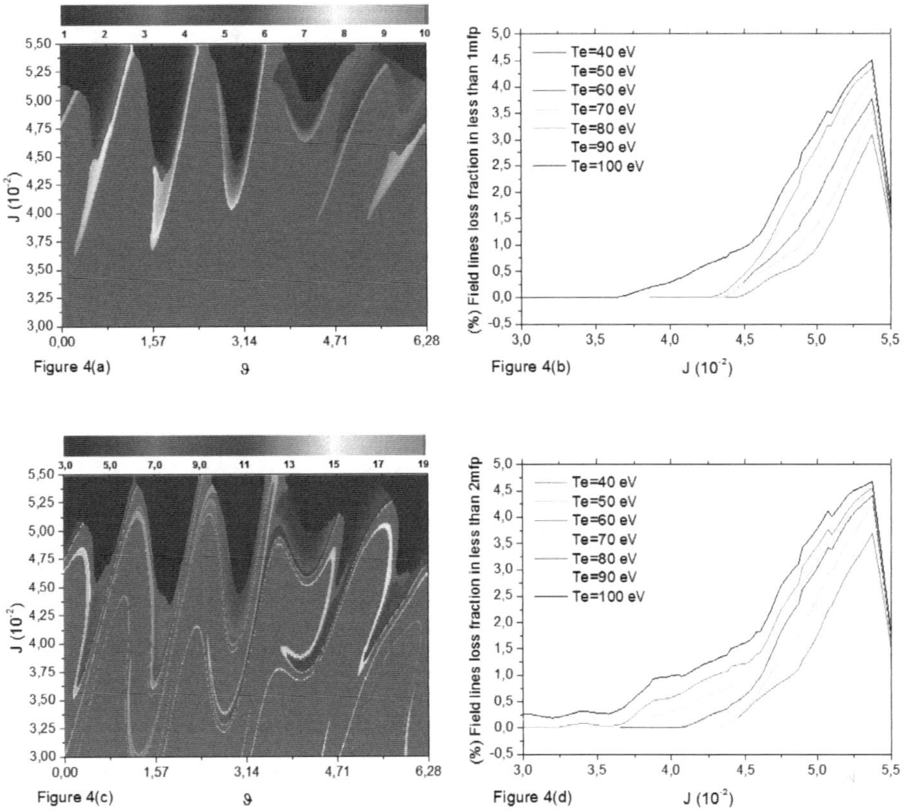

Figure 4 - **(a)** Connection lengths for $q(a)=5$. The scale indicates colors representing N_{cl} from *1* to 10. **(b)** Percentage of lost field lines with $N_{cl} < 1\ \lambda$. **(c)** Connection length with $q(a)=5$ for N_{cl} scale from 3 to 19. **(d)** Percentage of field lines loss fraction in less than 2 collisional mean free paths.

In Figs. 4(a) and 4(c), we adjust the connection length scales to $N_{cl} = 1$ until $N_{cl} = 10$ and $N_{cl} = 3$ until $N_{cl} = 19$, respectively. Figures 4(b) and 4(d) shows that the fraction of lost field lines increases with the temperature and the maximum percentage of lost field lines fraction increases for large values of T_e. Accordingly, the maximum values are 4,5% in Fig. 3(d) and 3,5% in Fig. 4(d).

5. Conclusions

We calculate the spatial connection length distribution for magnetic field lines in tokamaks with a reversed magnetic shear perturbed by an ergodic magnetic limiter. For this equilibrium, the stickiness around the dimerized magnetic islands traps magnetic field lines and increase their connection lengths.

We show that most of the magnetic field lines located inside the chaotic region describe long distances, which are much higher than the electron collisional mean free path, before reaching the tokamak wall. However, the fraction of the field lines lost to the wall with connection lengths shorter that the electron collisional mean free path increases with the electron temperature and also increases for large values of the action variable J. So, we expect that a significative fraction of field lines escape to the wall completing a toroidal excursion with a length shorter than the mean free path.

The comparison between the connection lengths obtained in this work and the electron mean free path indicates that the time for the escaping of the field line should be relevant to estimate the particle confinement time.

Acknowledgments
The authors thank the Brazilian scientific agencies FAPESP, CNPq and CAPES for financial support.

References

[1] Hazeltine R D and Prager S C 2002 *Physics Today* **55** 7 30
[2] Hidalgo C 2004 *Astrophysics and Space Science* **292** 681
[3] Horton W 1999 *Reviews of Modern Physics* **71** 3 735
[4] Parker R et al 1997 *J. Nucl. Mater* **241** 243
[5] Ghendrih Ph, Grosman A and Capes H 1996 *Plasma Phys. Control. Fusion* **38** 1653
[6] Silva E C, Caldas I L, Viana R L 2001 *IEEE Trans. On Plasma Science* **4** 617-631
[7] Silva E C, Caldas I L, Viana R L, Sanjuán M A F 2002 *Phys. Plasmas* **9** 4917
[8] Levinton F M et al 1995 *Phys. Rev. Lett.* **75** 4417
[9] Connor J W, Fukuda T, Garbet X, Gormezano C 2004 *Nuclear Fusion* **44** R1-R49
[10] Fenstermacher M E et al 2008 *Phys. Plasmas* **15** 056122
[11] Silva E C, Roberto M, Portela J S E, Caldas I L, Viana R L 2006 *Nuclear Fusion* **46** S192-S198
[12] Kroetz T, Roberto M, Silva E C, Caldas I L, Viana R L 2008 *Phys. Plasmas* **15** 092310
[13] Kucinski M Y and Caldas I L 1987 *Z Naturforsch, A: Phys. Sci.* **42** 1124
[14] Roberto M, Silva E C, Caldas I L and Viana R L 2004 *Phys. Plasma* **11**(1) 214
[15] Nascimento I C et al 2007 *Nucl. Fusion* **47** 1570
[16] Kotov V, Reiter D and Kukushkin A S 2007 *Report Julich* **4257** in *www.eirene.de*

Dynamic Days South America 2010 IOP Publishing
Journal of Physics: Conference Series **285** (2011) 012013 doi:10.1088/1742-6596/285/1/012013

Electrical implementation of a complete synchronization dynamic system

C Gonçalves [1] and L G Neto [1]

[1] Universidade de São Paulo
Avenida Trabalhador São-Carlense, 400 CEP 13566-590 São Carlos-SP

E-mail: cristhiane.goncalves@usp.br

Abstract. This work presents an electrical implementation of complete synchronization systems, proposing a master/slave synchronization of two identical particle-in-a-box electronic circuits, exhibiting a rich chaotic behaviour. This behaviour was measured, and also emulated, and the results were compared. Just a few works in literature describe experimental measurements of chaotic systems. The master/slave electronic circuits employed have a very simple electronic implementation and results show a complete synchronization of the system.

1. Introduction

Synchronization of chaotic systems is a phenomenon that occurs when chaotic oscillators are coupled, or when a chaotic oscillator drives another one. In several situations, these systems can exhibit exponential divergence of the trajectories of two identical chaotic systems started with nearly the same initial conditions. However, synchronization of coupled or driven chaotic oscillators is a phenomenon well established experimentally, and well understood theoretically. It has been found that chaos synchronization is a rich phenomenon wich presents a variety of forms.

Synchronization between dynamical systems is being extensively explored in electronic circuits [1, 2]. There are many types of circuits with potential applications to generate chaotic signals, like Chua's circuit [3] and Lorenz-based chaotic circuit [4]. Pecora and Carroll proposed the drive-response scheme and determined conditions of synchronization by means of conditional Lyapunov exponents[5].

Dynamical systems can exhibit several types of synchronization, among them: complete synchronization; generalized synchronization, phase synchronization and lag synchronization.

Complete synchronization occurs when there is coincidence of states of two interacting subsystems, as described by equation (1):

$$x_d(t) = x_r(t) \tag{1}$$

This regime appears only if the both interacting systems are identical [6].

Generalized synchronization has also been described in the literature [7], and introduced for drive-response systems, as the presence of a relation between the states of response and drive, thus:

$$x_r(t) = \Im[x_d(t)] \tag{2}$$

Considering the behavior of two unidirectionally coupled chaotic systems:

$$\dot{x}_d = H(x_d, g_d) \tag{3}$$

$$\dot{x}_r = G(x_r, g_r) + \varepsilon P(x_d, x_r) \tag{4}$$

Where $x_{d,r}$ are the state vectors of the drive and response systems, respectively, and H and G define the vector fields of these systems, g_d and g_r are the controlling parameters, P represents the coupling term, and ε is the scalar coupling parameter.

The generalized synchronization (GS) regime can be observed for two identical systems, with equal or mismatched parameters, and unidirectional coupling. In this particular case, the systems are identical, so the dimensions of drive and response oscillators are equal. The system is described as follows:

$$\dot{x}_d = H(x_d, g_d) \tag{5}$$

$$\dot{x}_r = G(x_r, g_r) + \varepsilon A(x_d - x_r) \tag{6}$$

Where $A=\{\delta_{ij}\}$ is the coupling matrix, and $\delta_{ii} \neq 0$ or 1, and $\delta_{ij}=0$, if $i \neq j$. The dynamics of the response system may be considered as the no autonomous dynamics of a modified system [7]:

$$\dot{x}_m = H'(x_m, g_r, \varepsilon) \tag{7}$$

Under the external force $\varepsilon A x_d$:

$$\dot{x}_m = H'(x_m, g_r, \varepsilon) + \varepsilon A x_d \tag{8}$$

Where $H'(x,g)=H(x,g)- \varepsilon A x_d$. The term $-\varepsilon A x_d$ brings the dissipation into the modified system.

The regime of complete synchronization is a particular case of generalized synchronization [7].

In this paper, it was investigated synchronization between two identical particle-in-a-box circuits, exhibiting a rich chaotic behavior. Although the great interest in chaotic systems, just a few works in the literature show also experimental realization of these systems [1]. The great advantage of this work is to obtain synchronism with robust coupling by using simple electronic circuits [8, 9], in a closed loop scheme. In Section 2 we present the purpose of this work. In Section 3, the particle-in-a box mechanical system is described, as well as the master-slave configuration proposed by using two particle-in-a-box identical circuits, and experimental methods employed. In Section 4, and experimental results are shown. Section 5 discusses results, and conclusions are given on Section 6.

2. Methods

The electronic circuit proposed emulates a particle-in-a-box mechanical system, described as follows, on figure 1.

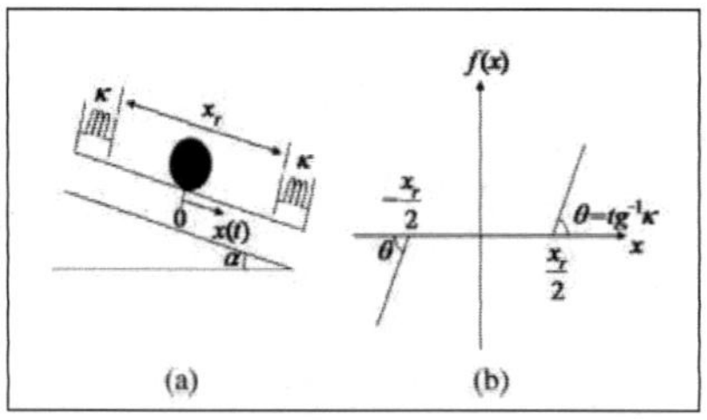

Figure 1-(a) Particle inside an oscilate box. (b)Force applied by the springs. In this system, a particle with mass m can collide with the box walls in x-direction. The walls are supported by linear springs with coefficient k.

In this system, a particle with mass m can collide the box walls in the x-direction. The term x_r is the distance between the walls. The walls are supported by linear springs with coefficient k, and the collisions between the particle and the walls are totally inelastic. The walls and springs have no mass and dimension. The contribution of the forces acting in the particle of Fig. 1(a) results in the equation:

$$m\frac{d^2x}{dt^2}+\mu\frac{dx}{dt}+f(x)=mg\sin\alpha \tag{9}$$

Where μ represents the viscous coefficient and $f(x)$ is the force applied by the springs. The angle θ represents the term $tg^{-1}k$ and the term $mg\sin\alpha$ describes the external force acting on the particle due the gravitational acceleration g.

The particle-in-a-box circuit is described on figure 2. This circuit simulates the mechanical behavior of the system of the figure 1(a).

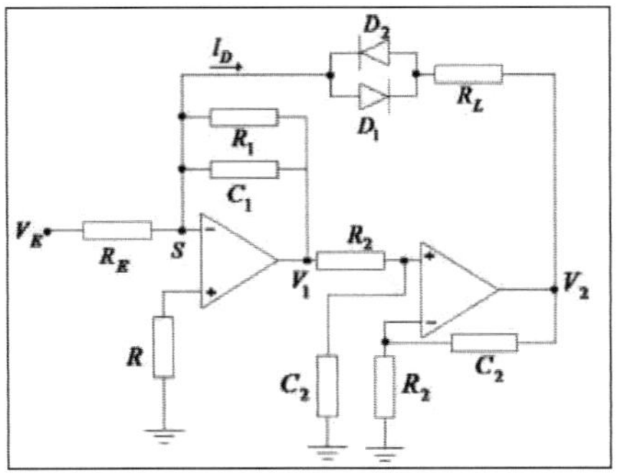

Figure 2 – Particle-in-a-box circuit. The component values are: resistors($k\Omega; R_1=47; R=R_E=R_2=10; R_L=0.51$),capacitors ($nF; C_1=C_2=12$), diodes $D_1=D_2: 2N4148$, operational amplifiers 741 or equivalents.

The particle-in-a-box electronic circuit has a sinusoidal input $V_E=V_{max}\sin(\omega t)$, and the contribution of the currents in the point S, shown in figure 2, results in the differential equation:

$$R_2C_1C_2\frac{d^2V_2}{dt^2}+\frac{R_2C_2}{C_1}\frac{dV_2}{dt}+I_D(V_2)=-\frac{V_E}{R_E} \tag{10}$$

In equation (10), the sign V_2 is the position x, and the signal is the velocity $\frac{dx}{dt}$.

The key devices of this circuit are the anti-parallel diodes D_1 e D_2 in association with the resistor R_L. They emulate the nonlinearity of this circuit, as shown on figure 3.

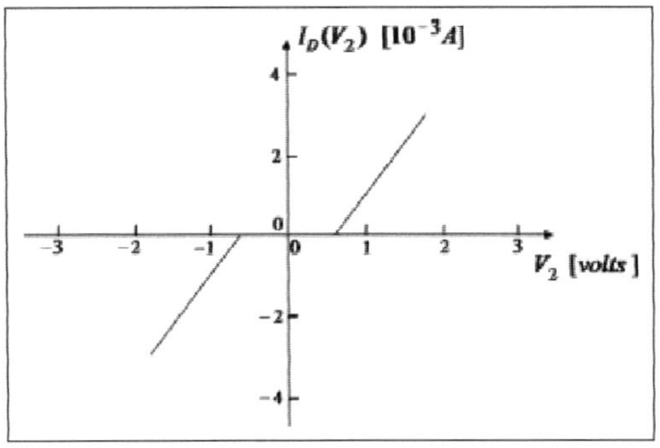

Figure 3- Current versus voltage for anti-parallel diodes D_1 and D_2. They emulate the collisions of the particle with the walls.

A master-slave configuration was proposed by using two identical systems as shown on figure 4, in a closed-loop scheme, by using R_{RE}=470Ω.

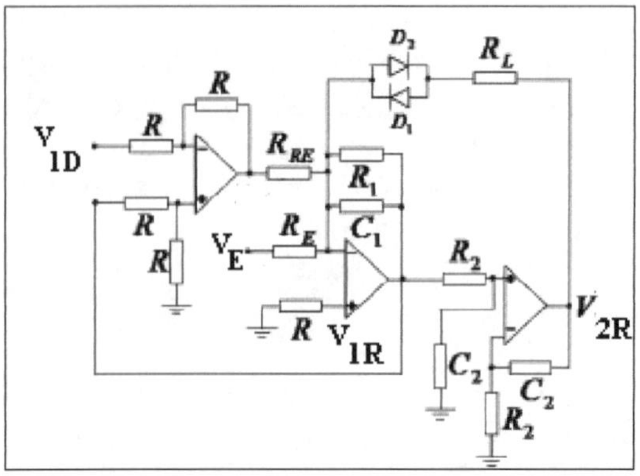

Figure 4-The circuits were applied to closed-loop scheme. The chaotic output signals V_2 obtained from the drive and response systems were taken from a digital oscilloscope and then analyzed.

A master-slave configuration was implemented by using two identical systems as shown on figure 5, in which a closed-loop scheme was applied, as described on figure 4. Measurements were taken by using a digital oscilloscope Minipa MO 2061, two function generators Minipa MGF 4201A 2MHz, a digital electronic font Minipa MPL 3303, and data were transferred to a personal computer, as shown on figure 6.

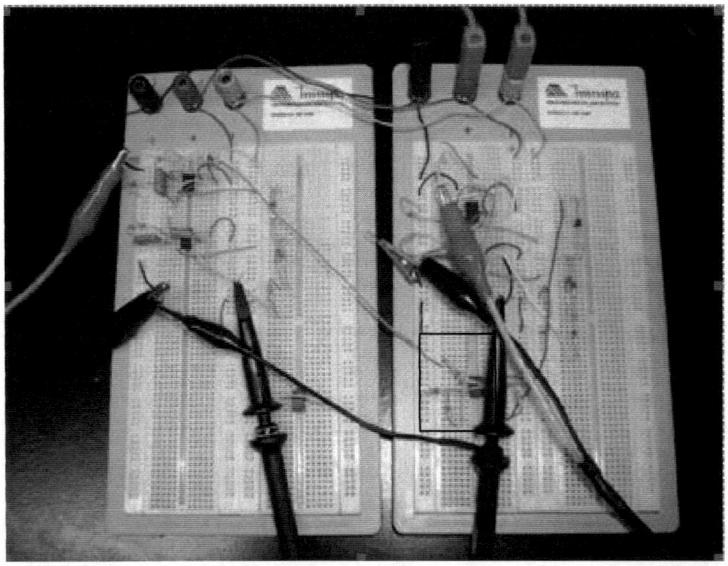

Figure 5 – Experimental realization of the proposed dynamical system. Master (left) and slave (right) particle-in-a-box electronic circuits. Detail shows the closed-loop scheme, R_{RE}=470Ω.

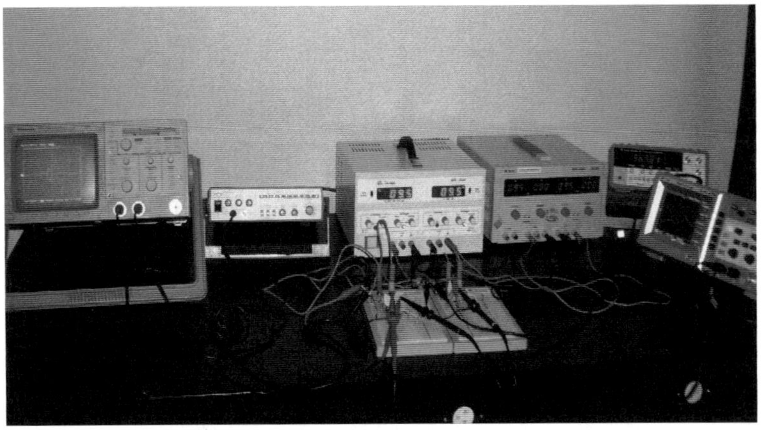

Figure 6 – Measurement of dynamical system.

3. Results

3.1. Numerical simulations results

Simulated results of synchronization between chaotic drive and response subsystems using the same parameters of experimental conditions are shown in this section.

Figure 7 exhibits the phase diagram of the particle-in-a-box circuit (V_{1d} and V_{2d}), in a frequency of f=1096Hz.

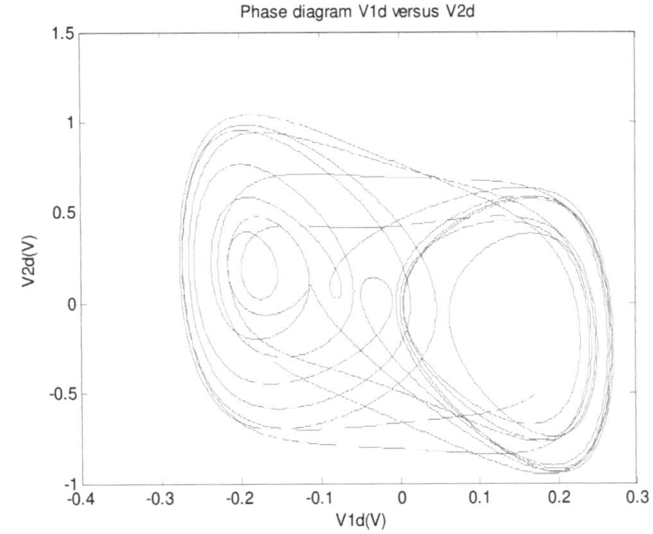

Figure 7– Numerical simulation of phase diagram of the particle-in-a-box circuit (V_{1d} and V_{2d}), by using a frequency f=1096Hz.As shown, chaotic behaviour s is exhibited.

Figure 8 exhibits the results obtained from the numerical simulation of the complete synchronization regime between output signals of drive and response subsystems.

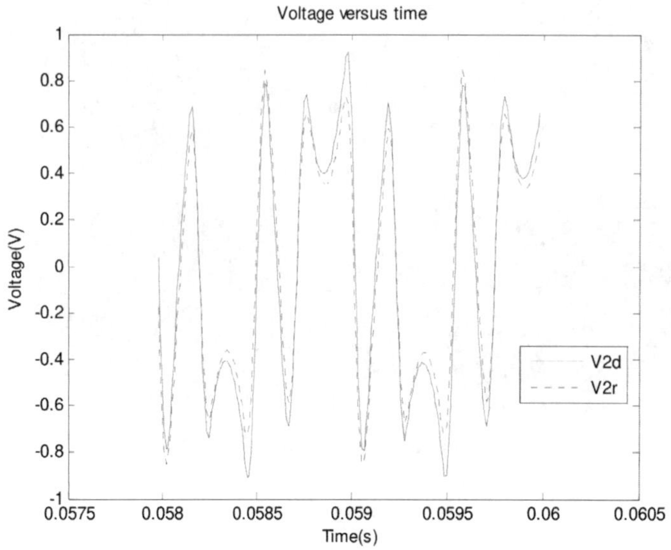

Figure 8– Numerical simulation of synchronism between output signals of drive (V_{2d}) and response (V_{2r}) subsystems. As shown, complete synchronization is exhibited.

3.2. Experimental results

Experimental results of measurements taken from a particle-in-a-box circuit are shown on this section. The periodic chaotic behavior can be obtained by the variation of the parameter of bifurcation of the circuit shown on figure 2, which controls the circuit. The variation of the parameter frequency of the input sinusoidal signal $V_E = V_{max}\sin(\omega t)$ can be used as the bifurcation parameter. Phase diagrams are shown on figure 9.

At first, the master-slave configuration of the proposed system was obtained by using two identical particle-in-a-box circuits. The circuits were applied to closed-loop scheme, by using R_{RE} =470Ω. The output signals V_2 obtained from master and slave circuits were taken from a digital oscilloscope and then analyzed. It was observed complete synchronism between drive and response signals, as shown on figure 10.

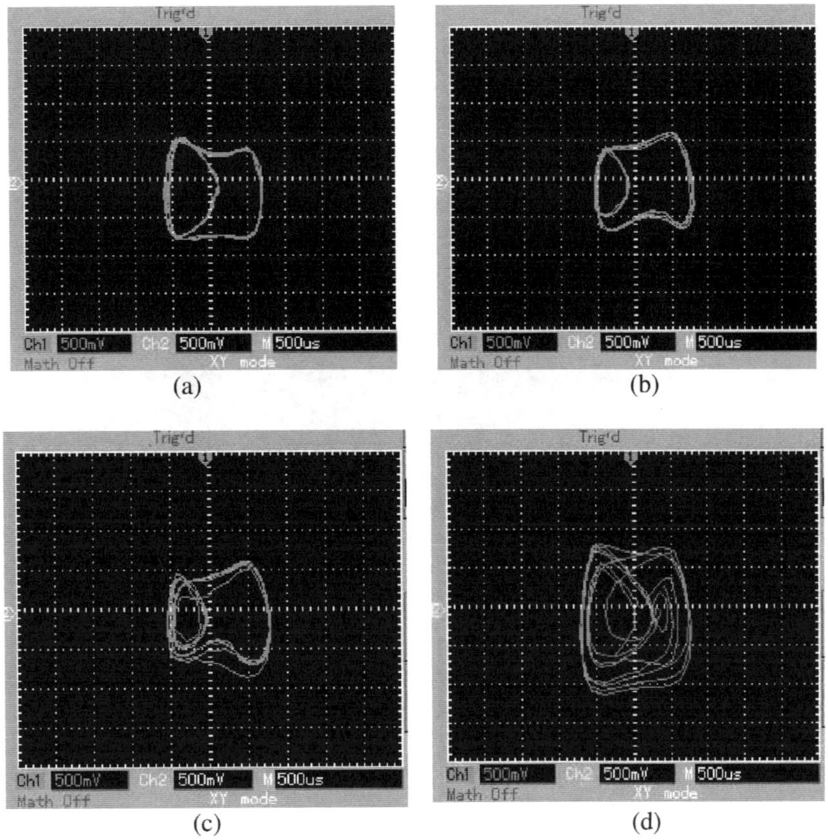

Figure 9– Phase diagrams of the particle-in-a-box circuit for the fixed amplitude 2V, and variable frequency as bifurcation parameter: (a) Period 1-*f*=958*Hz*;(b) Period 2- *f*=963*Hz;*(c) *f*=965*Hz*;(d) *Chaos f*=1096*Hz*.

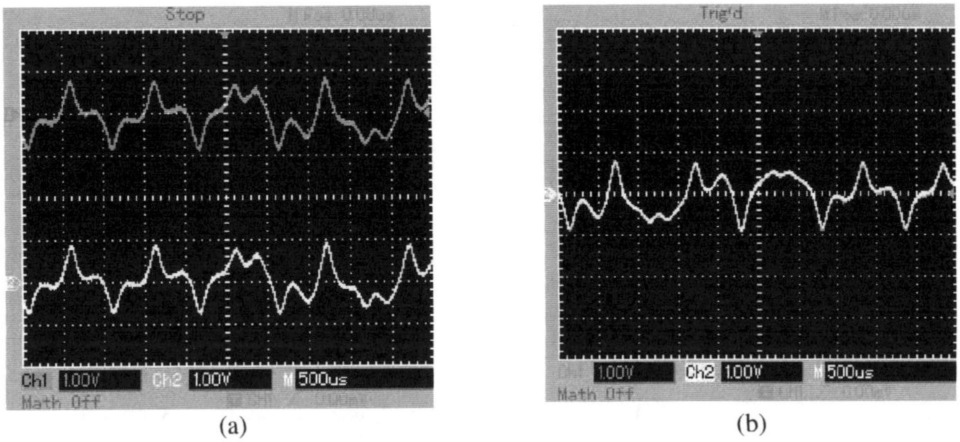

Figure 10– (a) Output signals of drive and response subsystems. (b)Complete synchronism exhibited between output signals of drive and response subsystems.

Complete synchronism can be also observed analyzing the phase diagram obtained from output signals of drive and response subsystems in a closed-loop scheme, by using R_{RE} =470 Ω. Results are shown on figure 11. Phase diagram exhibits the relation $x_d = x_r$, which means that complete synchronism was achieved.

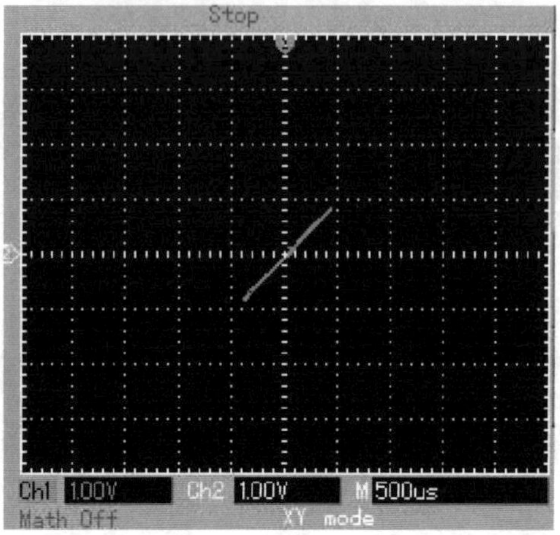

Figure 11– Phase diagram synchronism obtained from output signals of drive and response subsystems, R_{RE}=470Ω. Complete synchronism was observed, according to relation $x_d = x_r$.

Finally, a configuration with a master and two slave particle-in-a-box circuits was proposed to verify robustness of the coupling. The circuits were applied to closed-loop scheme, by using R_{RE}=470Ω. The output signals V_2 obtained from master and slaves circuits were taken from a digital oscilloscope and then analyzed. It was observed complete synchronism between drive and response signals, as shown on figure 12.

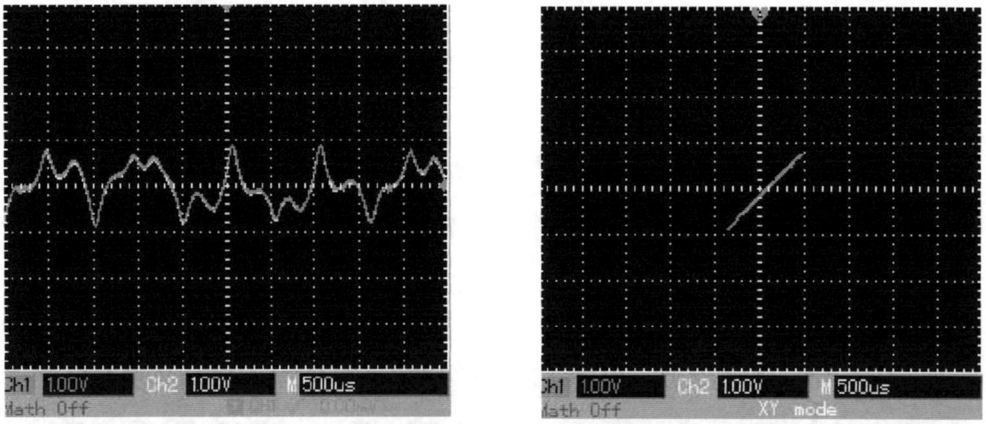

(a) (b)

Figure 12–(a) Output signals of slaves subsystems (b) Phase diagram synchronism obtained from output signals of slaves subsystems. As shown, complete synchronism between slaves subsystems was achieved. Phase diagram shows mathematical relation $x_d = x_r$.

4. Discussion

Figure 7 shows the phase diagram for an only particle-in-a-box circuit emulated according to equation (10), for the fixed amplitude 2V, and frequency $f = 1096Hz$. Figure 8 exhibits the results obtained from the numerical simulation of the complete synchronization regime between output signals of drive and response subsystems.

Phase diagrams of the particle-in-a-box circuit for the fixed amplitude 2V, by using variable frequency as bifurcation parameter, are shown on figure 9: (a) Period 1- $f = 958Hz$; (b) Period 2- $f = 963Hz$; (c) Period 4- $f = 965Hz$; and (d) Chaos- $f = 1096Hz$.

Two identical particle-in-a-box circuits were connected by using a closed-loop scheme, where R_{RE}=470Ω. They exhibit a complete synchronization regime. Figure 10 shows output signals of drive and response subsystems. Phase diagram obtained from output signals of drive and response subsystems is shown on figure 11. Both results are in agreement to mathematical relation $x_d = x_r$.

At last, a configuration with a master and two slave particle-in-a-box circuits, to verify robustness of the coupling. The circuits were applied to closed-loop scheme, by using R_{RE} =470Ω. The output signals V_2 obtained from master and slaves circuits were taken from a digital oscilloscope and then analyzed. As shown on figure 12, complete synchronism between both slaves subsystems was achieved. Phase diagram shows mathematical relation $x_d = x_r$. Experimental results showed good agreement with the simulation results.

5. Conclusion

Just a few works in literature describe experimental measurements of chaotic systems, and most employ complex electronic circuits. The master/slave electronic circuits employed in this work have a very simple electronic implementation.

Numerical results show agreement with experimental results, and show the robustness of the coupling in this system. Analyzing preliminary results, it may be concluded that the proposed system emulates the chaotic behavior of a particle-in-a-box circuit, and results show complete synchronization of the system. Due to robustness of the coupling between these systems, we intend to apply this system in chaotic secure communication. Results suggest the system could be employed in chaotic mask successfully.

6. References

[1] Grosu I, Padmanaban E, Roy P K and Dana S K 2008 *Phys.Rev. Lett.*, **100**, 23, 234102-1-4.
[2] Reagy J F, Carroll T L, Pecora L M 1994 *Phys Rev. Lett. A* , **73**, 2, 3528-31
[3] Hu Z, Chen X 2006, *Dyn. Cont.Disc.Impuls. Syst.-Series A-Math. Anal.* **13**, 489-499
[4] Cuomo K M , Oppenheim A V , Strogats S H 1993, *IEEE Trans.Circ. and Syst. II, Analog. Dig. Sig. Proc.*, **40**, 10, 626-633
[5] Pecora L M, Carroll T L 1989 *Phys.Rev.Let.* **64**, 10, 821-824.
[6] Pikovsky A S 1983 *Zeitsch. für Phys. B Cond. Mat.*, **55**, 2, 149-154.
[7] Hramov A E , Koronovskii A A 2005. *Phys.Rev.E*, **71**, 067201-04

Intriguing structures in iterative maps motivated by N-body problem

Ashok Kumar Mittal[1] and Ashok Kumar Gupta[2]

[1] Department of Physics, University of Allahabad, Allahabad, India
[2] Department of Biotechnology and Bioinformatics, Jaypee University of Information Technology, Waknaghat-173215, Solan HP, India

E-mail: mittals79@hotmail.com

Abstract. Conventional approaches to modeling any system try to incorporate increasingly realistic features into the model, thereby making it more and more complex. An opposite approach seeks to build simpler and simpler conceptual models capable of capturing some observed features of a system. This trend began with Lorenz, who simplified models of the atmosphere to obtain the Lorenz model consisting of a system of only three equations. Despite the simplicity of these equations, this system displayed surprisingly rich properties, and has been used as a conceptual model in diverse disciplines. Poincare maps help study ordinary differential equations from a qualitative perspective. Several investigators like Henon and Feigenbaum followed this simplification approach. Instead of investigating Poincare maps of realistic systems, they, along with several others, investigated simple maps for their own sake. Despite lack of realism, this approach proved to be very fruitful. A map, as simple as the Logistic map, became an important conceptual modeling paradigm. It provided a tool for understanding bifurcation routes to chaos, which were verified experimentally through various experiments in diverse fields. Coupled map lattices (CML) help explore partial differential equations (PDE). Further simplification led to the introduction of Cellular Automata (CA). These fields continue to be explored with vigor and have given rise to a rich body of knowledge, conceptually useful over a wide spectrum of disciplines. In this paper, we follow the simplification approach for modeling the N-body problem. N-body simulations, say in Gravitation, give rise to filamentary structures. Such structures are observed in the actual observed Galactic distribution. The mechanism for creation of such structures is not well understood. We present a simple iterative dynamical model, motivated by the N-body problem, which, though unrealistic, produces such filamentary structures. This model also exhibits a variety of intriguing structures. Attempts to understand these structures may lead to useful insights similar to those provided by investigations in maps, CML, CA etc.

1. Introduction

Conventional approaches to modeling any system try to incorporate increasingly realistic features into the model, thereby making it more and more complex. This approach, though useful in some ways, suffers from some drawbacks also. The cost of modeling in terms of time and money becomes a limiting factor. More important, the capacity to draw useful insights about the system becomes limited. Massive data becomes available, but getting useful manageable mental models, comprehensible by the human brain, becomes increasingly difficult.

During the last half-century an opposite trend in modeling has been advocated. This approach seeks to build simpler and simpler conceptual models capable of capturing some observed features of a system. This trend began with Lorenz, who simplified his models of the atmosphere to simpler and simpler sets of equations eventually ending with the now celebrated Lorenz model consisting of a system of only three equations [1]. Despite their simplicity, these equations displayed immensely rich properties, which despite half a century of extensive exploration, have not yet been fully revealed. Lorenz equations have been used as a conceptual model in diverse disciplines [2-20].

Poincare introduced the idea of studying ordinary differential equations from a qualitative perspective. For this purpose he introduced Poincare maps. A number of investigators like Henon [21] and Feigenbaum [22] pursued the simplification doctrine; instead of investigating Poincare maps of realistic systems, they, along with several others, investigated simple maps for their own sake. This approach was very fruitful. Logistic map became an important conceptual modeling paradigm. Once again, despite its simplicity, Logistic map offered a bewildering richness. It provided a tool for understanding bifurcation route to chaos. This route was verified experimentally through various experiments in diverse fields [23]

Just as maps were introduced to explore qualitative properties of ordinary differential equations (ODE), coupled map lattices (CML) [24] were introduced for exploring the properties of partial differential equations (PDE). Further simplification led to the introduction of Cellular Automata [25]. All these fields continue to be explored with vigor and have given rise to a rich body of knowledge, conceptually useful over a wide spectrum of disciplines.

To the best of our knowledge this simplification approach has not been applied to N-body simulation [26]. N-body simulations, say in Gravitation, give rise to filamentary structures. This is also observed in the actual observed Galactic distribution [27]. In this paper, we present a simple iterative dynamical model which produces such filamentary structures. This model exhibits a variety of intriguing structures. We believe that an attempt to understand these structures will lead to insights similar to those provided by investigations in maps, CML, CA etc.

2. A simplified model of N-body simulation

The problem of N-body simulation in 3-dimensions consists of solving a set of 6N first order ODEs. Numerical simulation requires replacement of continuous time by discrete time. Our first simplification replaces 6N ODE's by a set of N maps. The ith map gives the position of the ith particle at time t+1 as a function of the positions of all the particles at time t. Let $\vec{R}_i(t)$ denote the position of the ith particle at time t. Let $\Delta\vec{R}_i = \vec{R}_i(t+1) - \vec{R}_i(t)$ denote the change in the position of the ith particle from time t to time t+1. In the next simplification, we assume that this change is due to the sum of influences from all the other particles, so that $\Delta\vec{R}_i = \sum_{j\neq i} \Delta\vec{R}_{ij}$, where $\Delta\vec{R}_{ij} = \vec{f}(\vec{R}_j - \vec{R}_i)$. A map that we have investigated is: $\vec{f}(\vec{r}) = \frac{\alpha\vec{r}}{r^2}$. We choose the value α = 0.001. In this model the changes in the positions of the particle are made synchronously. In the next stage of simplification, we make the changes in position asynchronously. This reduces computation time. As further simplification, we pick a particle randomly and change its position according to the above rule.

Next, we make further departure from realistic modeling. Instead of picking a particle and changing its position by summing the effects of all other particles, we choose a particle and change the positions of all other particles according to a specified map. If particle i is chosen, the positions of all the other particles change according to the rule $\Delta\vec{R}_j = -\vec{f}(\vec{R}_j - \vec{R}_i)\forall j \neq i$. The ith particle can be chosen randomly.

To avoid the singularity at r = 0, we modify $\vec{f}(\vec{r})$ so that $\vec{f}(\vec{r}) = \frac{\alpha\vec{r}}{r^2+\beta^2}$. We choose N = 100000 uniformly distributed random points inside a circle of radius 1. Some of the intriguing structures observed after m = 1000 iterations for α = 0.001 and different values of β are shown in figures 1 - 8.

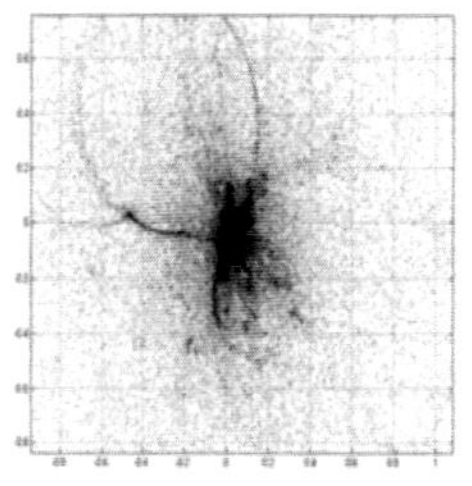

Figure 1. β = 0.0001.

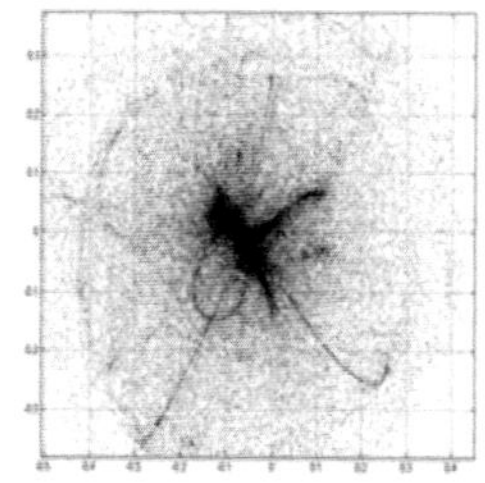

Figure 2. β = 0.0010.

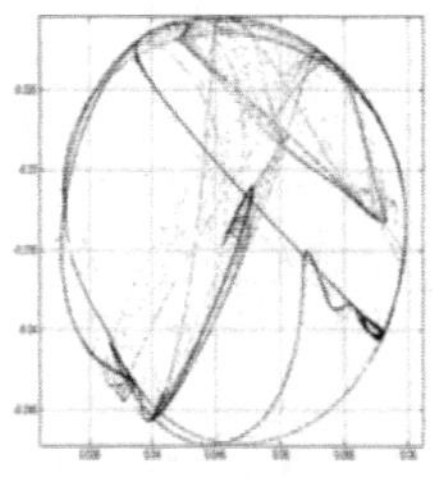

Figure 3. β = 0.0100.

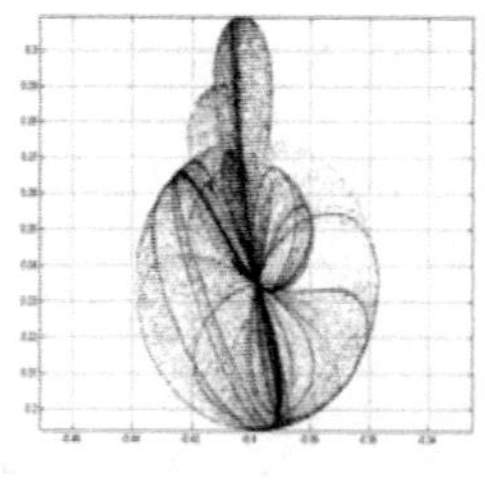

Figure 4. β = 0.0180.

These figures show that for small values of β, as β is increased, the observed structure becomes increasingly filamentary and sparse. For β = 0.018 the structure has a complex quasi 1-dimensional form. Dramatic changes take place in the observed structure for higher values of β. For β = 0.019 we find that all the particles converge to the vertices of an equilateral triangle. For β = 0.020 we find a 1-dimensional curve. For β = 0.022 a nearly circular structure is observed. For β = 0.023 and higher values, we always find that all the particles converge to a single point.

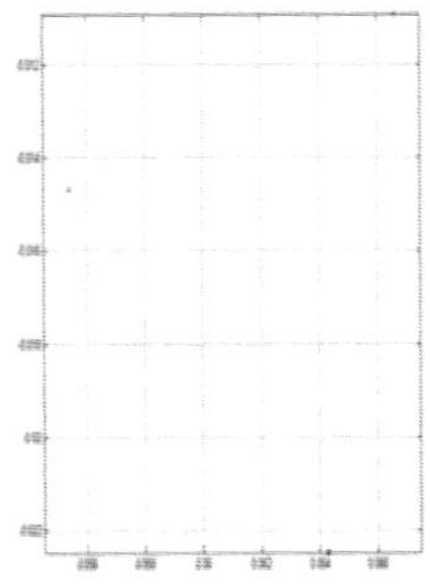

Figure 5. β = 0.0190.

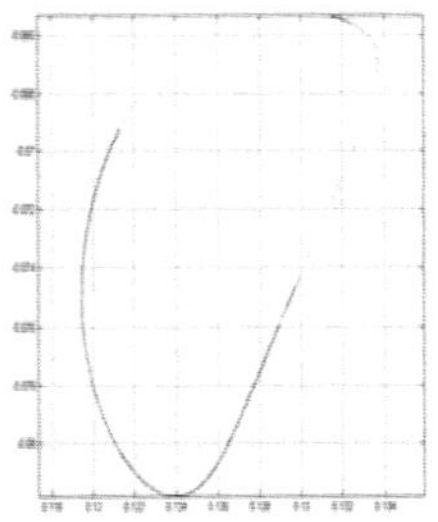

Figure 6. β = 0.0200.

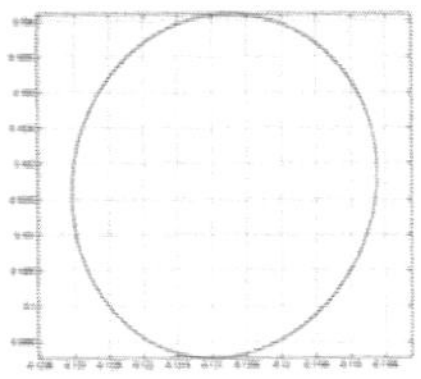

Figure 7. β = 0.0220.

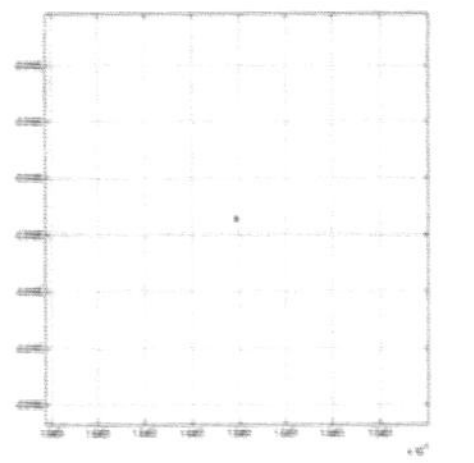

Figure 8. β = 0.0230.

Keeping β fixed at 0.019, we increased the number of iterations. We found that the points remain on the vertices of an equilateral triangle; the length of the side does not change with increase in the number of iterations. For β between 0.019 and 0.021, we find that on increasing the number of iterations sufficiently, the structure eventually converges to the vertices of an equilateral triangle, whose size does not change on increasing the number of iterations. Another intriguing observation is that this convergence is much faster if the number of points N is a multiple of 3 compared to when it is not. In one dimension, however, the number of iterations before convergence tends to decrease as the number of points N is increased.

For β = 0.018 and below, the structures did not converge even as the number of iterations was increased to 250000. For β = 0.022 the structure remained nearly circular even for m as large as 17000.

3 Analysis of the model

For N = 2, the dynamics of the separation between the points can be analyzed exactly in terms of the map f. As the only property of interest is the distance between the two points, negative values of R have the same meaning as positive values. The separation R' at time (t+1) is determined by the separation R at time t by

$$R' = \left| R - \frac{\alpha R}{R^2 + \beta^2} \right| = F(R) \tag{1}$$

This map is shown for β = 0.0200 in Figure 9.

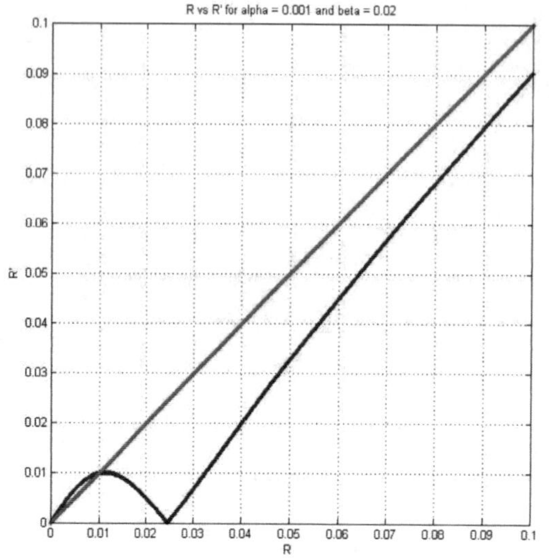

Figure 9. Plot of R' vs R is shown in blue. R' = R line is shown in red color.

The map has two fixed points at R = 0 and $R = \tilde{R} = (\frac{\alpha}{2} - \beta^2)^{1/2}$ and zeros at R = 0 and $R = R_0 = (\alpha - \beta^2)^{1/2}$. For $R < R_0$, the derivative of the map is given by

$$F'(R) = \frac{\alpha}{R^2+\beta^2} - \frac{2\alpha R^2}{(R^2+\beta^2)^2} - 1 \tag{2}$$

At R = 0, the slope of F is given by

$$F'(0) = \frac{\alpha}{\beta^2} - 1 \tag{3}$$

The magnitude of this slope becomes less than 1 for $\beta > (\frac{\alpha}{2})^{1/2} \cong 0.0224$. Hence for β greater than this critical value, R = 0 is stable and the separation between the two particles converges to zero. This explains the observation that for $\beta = 0.023$ and above, for any value of N, all the particles converge to a single point.

At $R = \tilde{R}$, the slope is given by

$$F'(\tilde{R}) = \frac{8\beta^2}{\alpha} - 3 \tag{4}$$

Therefore, $R = \tilde{R}$ is stable for $-1 < \frac{8\beta^2}{\alpha} - 3 < 1$, i.e., for $(\frac{\alpha}{4})^{1/2} \cong 0.0158 < \beta < (\frac{\alpha}{2})^{1/2} \cong 0.0224$. Clearly for this range of β, the separation between the two particles will converge to $\tilde{R}$.

For larger N, we can be sure that if all the particles lie on the vertices of a triangle of side $\tilde{R}$, then this configuration will not change shape or size on further iteration. However, can the stability of F

at $\tilde{R}$ guarantee that the N–particle system will converge to the vertices of a triangle of side $\tilde{R}$? The answer is negative, as we find empirically that for N > 3, the system converges only for $\beta > \beta_c$, where $\beta_c \approx (\frac{\alpha}{3})^{1/2}$. This observation has been verified for different values of α. In order to simplify the analysis, we consider the analogous one-dimensional system with N = 3. The state, with two particles having the same coordinate and the third particle at a distance $\tilde{R}$ from them, is an invariant state. We examine the local stability of this state. Without loss of generality we assume that in this state the coordinates of the three particles are $x_1 = 0$ and $x_2 = x_3 = \tilde{R}$. In the perturbed state, we take the coordinates of the three particles to be $x_1 = \varepsilon_1$, $x_2 = \tilde{R}+\varepsilon_2$ and $x_3 = \tilde{R}+\varepsilon_3$, where ε_i are infinitesimals.

If P_1 gets chosen as the particle which changes the positions of the other particles, then the new values of the perturbations are given by

$$\begin{bmatrix} \varepsilon'_1 \\ \varepsilon'_2 \\ \varepsilon'_3 \end{bmatrix} = \begin{bmatrix} 1 & 0 & 0 \\ 1-F'(\tilde{R}) & F'(\tilde{R}) & 0 \\ 1-F'(\tilde{R}) & 0 & F'(\tilde{R}) \end{bmatrix} \begin{bmatrix} \varepsilon_1 \\ \varepsilon_2 \\ \varepsilon_3 \end{bmatrix} \tag{5}$$

This may be written as

$$\varepsilon' = J^{(1)}\,\varepsilon \tag{6}$$

The eigenvalues of $J^{(1)}$ are:

$$\lambda_1^{(1)} = 1, \quad \lambda_2^{(1)} = F'(\tilde{R}), \quad \lambda_3^{(1)} = F'(\tilde{R}) \tag{7}$$

with corresponding eigenvectors

$$e^{(1)}{}_1 = \begin{bmatrix} 1 \\ 1 \\ 1 \end{bmatrix}, e^{(1)}{}_2 = \begin{bmatrix} 1 \\ v_2 \\ w_2 \end{bmatrix}, e^{(1)}{}_3 = \begin{bmatrix} 0 \\ v_3 \\ w_3 \end{bmatrix} \tag{8}$$

If P_2 gets chosen as the particle which changes the positions of the other particles, then the new values of the perturbations are given by

$$\begin{bmatrix} \varepsilon'_1 \\ \varepsilon'_2 \\ \varepsilon'_3 \end{bmatrix} = \begin{bmatrix} F'(\tilde{R}) & 1-F'(\tilde{R}) & 0 \\ 0 & 1 & 0 \\ 0 & 1-F'(0) & F'(0) \end{bmatrix} \begin{bmatrix} \varepsilon_1 \\ \varepsilon_2 \\ \varepsilon_3 \end{bmatrix} \tag{9}$$

which may be written as

$$\varepsilon' = J^{(2)}\,\varepsilon \tag{10}$$

The eigenvalues of $J^{(2)}$ are:

$$\lambda_1^{(2)} = 1, \quad \lambda_2^{(2)} = F'(\tilde{R}), \quad \lambda_3^{(2)} = F'(0) \tag{11}$$

with corresponding eigenvectors

$$e^{(2)}{}_1 = \begin{bmatrix} 1 \\ 1 \\ 1 \end{bmatrix}, e^{(2)}{}_2 = \begin{bmatrix} 1 \\ 0 \\ 0 \end{bmatrix}, e^{(2)}{}_3 = \begin{bmatrix} 0 \\ 0 \\ 1 \end{bmatrix} \tag{12}$$

If P_3 gets chosen as the particle which changes the positions of the other particles, then the new values of the perturbations are given by

$$\begin{bmatrix} \varepsilon'_1 \\ \varepsilon'_2 \\ \varepsilon'_3 \end{bmatrix} \cong \begin{bmatrix} F'(\tilde{R}) & 0 & 1-F'(\tilde{R}) \\ 0 & F'(0) & 1-F'(0) \\ 0 & 0 & 1 \end{bmatrix} \begin{bmatrix} \varepsilon_1 \\ \varepsilon_2 \\ \varepsilon_3 \end{bmatrix} \tag{13}$$

which may be written as

$$\varepsilon' = J^{(3)} \varepsilon \tag{14}$$

The eigenvalues of $J^{(3)}$ are:

$$\lambda_1^{(3)} = 1, \quad \lambda_2^{(3)} = F'(\tilde{R}), \quad \lambda_3^{(3)} = F'(0) \tag{15}$$

with corresponding eigenvectors

$$e^{(3)}{}_1 = \begin{bmatrix} 1 \\ 1 \\ 1 \end{bmatrix}, e^{(3)}{}_2 = \begin{bmatrix} 1 \\ 0 \\ 0 \end{bmatrix}, e^{(3)}{}_3 = \begin{bmatrix} 0 \\ 0 \\ 1 \end{bmatrix} \tag{16}$$

It is easy to see that the separation between the points P_2 and P_3 grows by a factor $|F'(\tilde{R})|$ each time the point P_1 gets chosen and by a factor $|F'(0)|$ each time the point chosen is P_2 or P_3. Since the points P_i are chosen with equal probability, it follows that the separation between P_2 and P_3 will have an overall tendency to decrease if

$$|F'(\tilde{R})||F'(0)|^2 < 1 \tag{17}$$

This inequality yields

$$\left\{ \frac{8\beta^2}{\alpha} - 3 \right\} \left\{ \frac{\alpha}{\beta^2} - 1 \right\}^2 < 1 \tag{18}$$

and holds for $\beta > \beta_c = (\frac{\alpha}{s})^{1/2}$, where s is the real root of the equation

$$3s^3 - 14s^2 + 18s - 8 = 0 \tag{19}$$

This equation has a root $s \approx 2.93$. Although the critical value of s has been obtained for N = 3 in the one dimensional case, it is remarkably close to the empirically observed value 3 for the two dimensional case with N greater than or equal to 3.

4. Conclusion

We obtained a family of iterative dynamical systems motivated by the N-body problem. This family exhibits several very intriguing properties which need to be explained. One needs to understand how filamentary structures arise, how nearly circular structures arise before all the points converge to the vertices of an equilateral triangle, how the conditions of stability can be obtained in 2 dimensions. In 3 dimensions all the points converge to the vertices of a tetrahedron, but before this convergence they settle on a nearly spherical surface. It is hoped that the type of dynamical system introduced here will produce a body of knowledge as rich as those produced by Logistic and other maps, Coupled Map Lattices, Cellular Automata etc.

References

[1] Lorenz EN, Deterministic non-periodic flows 1963 *J Atmos Sci* **20** 131-141.
[2] Mittal AK, Dwivedi S, Yadav RS 2007 Probability distribution for the number of cycles

between successive regime transitions for the Lorenz model, *Physica D* **233** 14-20.

[3] Mittal AK, Dwivedi S, Pandey AC 2005 Bifurcation analysis of a paradigmatic model of monsoon predication, *Nonlinear Processes in Geophysics* 12 707-715.

[4] Yadav RS, Dwivedi S, Mittal AK 2005 Prediction rules for regime changes and length in a new regime for the Lorenz model, *J Atmospheric Sci* **62** 2316-2321.

[5] Mehta M, Mittal AK, Dwivedi S 2003 The double cusp map for the forced Lorenz system, *Int J Bif Chaos* **13** 3029-3035.

[6] Dwivedi S, Mittal AK Forecasting the duration of active and break spells in intrinsic mode functions of Indian monsoon intraseasonal oscillations 2007 *Geophys Res Lett* **34** L16827.

[7] Dwivedi S, Mittal AK, Goswami BN 2006 An empirical rule for extended range prediction of duration of Indian summer monsoon breaks *Geophys Res Lett* **33** L18801.

[8] Palmer TN 1998 A nonlinear dynamical perspective on climate predictions *J Clim* **12**; 575-591.

[9] Palmer TN 1994 Chaos and predictability in forecasting the monsoons *Proc Ind Natl Sci Acad A* **60** 57-66.

[10] Palmer TN, Extended Range atmospheric prediction and the Lorenz model 1993 *Bull Am Meteorol Soc* **74** 49-66.

[11] Haken H, Analogy between higher instabilities in fluids and lasers 1975 *Phys Lett A*; **53**; 77-78.

[12] Malkus WVR, Non periodic convection at high and low Prandtl number 1972 *Mem Soc R Sci Liege* **4** 125-128.

[13] Yorke ED, Yorke JA 1979 Metastable Chaos: Transition to sustained chaotic behaviour in the Lorenz model *J Stat Phys* **21**; 263-277.

[14] Knobloch E 1981 Chaos in a segmented disc dynamo *Phys Lett A* **82** 439-440.

[15] Lorenz EN 1980 Noisy periodicity and reverse bifurcations, in *RHG Hellerman (Ed.), Nonlinear Dynamics*, Annal NY Acad Sci; 282-291.

[16] Pedlosky J, Frenzen, C 1980 Chaotic and periodic behavior of finite amplitude baroclinic waves, *J Atmospheric Sci* **37** 1177-1196.

[17] Pedlosky J 1972 Limit cycles and unstable baroclinic waves *J Atmosperic Sci* **29** 53-63.

[18] Brindley J, Moroz IM 1980, Lorenz attractor behavior in a continuously stratified baroclinic fluid, *Phys Lett A* **77**, 441-444.

[19] Gibbon JD 1982 Dispersive instabilities in nonlinear systems. The real and complex Lorenz equations, in *H Haken (Ed.), Chaos and Order in Nature, Synergetics*, Springer Verlag; 92-101.

[20] Gibbon JD, McGuinness 1980 A derivation of the Lorenz equations for some unstable dispersive physical systems *Phys Lett A* **77**; 295-299.

[21] Henon M 1976, A Two-Dimensional Mapping with a Strange Attractor, *Comm Mathl Physics* **50** 69-77.

[22] Feigenbaum, MJ 1979 The Universal Metric Properties of Non-Linear Transformations, *J Statistical Physics* **21** 669-706.

[23] Bai-Lin 1984 (ed): Chaos, (World Scientific, Singapore).

[24] Kaneko K 1984 Period-Doubling of Kink-Antikink Patterns, Quasiperiodicity in Antiferro-Like Structures and Spatial Intermittency in Coupled Logistic Lattice *Prog Theor Phys* **72** 480-486.

[25] Wolfram S 2002 A New Kind of Science (Wolfram Media).

[26] Aarseth SJ 2003 Gravitational N-body Simulations: Tools and Algorithms, (Cambridge University Press).

[27] Bharadwaj S, Sahni V, Sathyaprakash BS, Shandarin SF, Yess C 2000 Evidence for Filamentarity in the Las Campanas Redshift Survey, *The Astrophysical Journal*; **528** 21-29.

Dynamic Days South America 2010 IOP Publishing
Journal of Physics: Conference Series **285** (2011) 012015 doi:10.1088/1742-6596/285/1/012015

On a cellular automaton with time delay for modelling cancer tumors

K. C. Iarosz[1], C. C. Martins[1], A. M. Batista[1], R. L. Viana[2], S. R. Lopes[2], I. L. Caldas [3] and T. J. P. Penna[4]

[1]Departamento de Matemática e Estatística, Universidade Estadual de Ponta Grossa, 84030-900, Ponta Grossa, PR, Brazil
[2]Departamento de Física, Universidade Federal do Paraná, 81531-990, Curitiba, PR, Brazil
[3]Instituto de Física, Universidade de São Paulo, Caixa Postal 66316, 05315-970, São Paulo, SP, Brazil
[4]Instituto de Física, Universidade Federal Fluminense, 24210-340, Niterói, RJ, Brazil

E-mail: antoniomarcosbatista@gmail.com

Abstract. In this work we considered cellular automaton model with time delay. Time delay included in this model reflects the delay between the time in which the site is affected and the time in which its variable is updated. We analyzed the effect of the rules on the dynamics through the cluster counting. According to this cluster counting, the dynamics behavior is investigated. We verified periodic oscillations same as delay differential equation. We also studied the relation between the time delay in the cell cycle and the time to start the metastasis, using suitable numerical diagnostics.

1. Introduction

The cancerous tumors dynamics, including their growth, propagation, and treatment, is one of the major problems in mathematical biology [1]. The interest for the problem has led to the formulation of numerous growth models [2], which have been proposed in order to analyze one or several basic features, such as the metastasis [3], the lack of nutrients [4], the competition for resources and the cytotoxic activity made by the immune response [5].

In human cancers, the mean cell division cycle time is a relevant measure due to the difference that we find in patients and the region affected by cancer. The range average cycle time varies from 2 days to several weeks with a mean of approximately 6 days [6]. Cell division cycle can be divided into 4 distinct phases, such as G_1-phase, DNA synthesis or S-phase, G_2-phase and mitosis or M-phase. The duration of S-phase, G_2-phase and M-phase is relatively short, most cancerous cells in a patient's tumor are in G_1-phase. Bertalanffy and collaborators [7] investigated the cell cycle of Ehrlich tumor cell populations. Ehrlich tumors are neoplastic cell populations frequently employed in cancer research, due to the fast proliferation. They transplanted primary Ehrlich ascites tumors to a great variety of mice, and observed that the S-phase duration seems to center around 9 hours, G_2-phase may be around an hour and the mitosis ranged from 30 to 75 minutes. Some studies showed that most Ehrlich tumor cell lines pass through at least a brief G_1 phase which lasts about 3 to 6 hours. Besides the experimental evidence in DNA synthesis, there is an equation used to described the fraction of the Ehrlich cell cycle [8].

Experiences have shown the effect of various chemotherapeutic agents on metastasis, and the relationship between cell cycle and tumor growth. One element that has received attention in recent years is the selenoprotein. Mice injected with this element presented a pronounced reduction in tumor progression and metastasis in colon and prostate [9]. Moreover, experimental observations have shown that selenium can inhibit cell cycle progression in human prostate cancer [10].

Continuous and discrete models are types of approaches used to describe the cancerous growth. The mathematical modelling of cancer growth and therapy can be done using many tools, as ordinary and partial differential equations [11]. However, discrete models such as cellular automata have been used as well chiefly, because they can be formulated in a rather simple way by specifying the types of cells involved and their evolution through rules connecting the cells and their neighbors [12].

Cellular automata are prototypes of spatially extended dynamical systems, that present discrete space and time, as well thus the state variables take on a finite set of discrete values [13]. They have been used on self-organization studies in dynamical systems, as cancerous cells proliferation [14], traffic flow [15], neural network [16], ecosystems [17], and spatial pattern formation [18].

From cellular automata models, one is able to investigate how global (macroscopic) behavior of cell assemblies as tumors are affected by changes in local (microscopic) properties of the cell interaction with its surrounding medium [19]. One macroscopic property of tumors is the metastasis, or the ability to migrate to other parts of the body through the lymphatic system or the bloodstream, causing the appearance of secondary tumors in other organs. The mechanisms triggering the appearance of metastasis are not well-known presently, but one possible factor involved is some change in the interaction between the cells at the microscopic level [20, 21].

In recent papers, cellular automata models have been considered to model aspects of tumor growth and therapy [22] and the presence of immune surveillance [23], as well as to study the role of acidity in tumor growth [24]. Tumor-induced angiogenesis has also been modelled using a cellular automaton approach [25]. A two-dimensional stochastic cellular automata model was proposed to describe avascular solid tumor growth, taking into account both the competition between cancer cells and normal cells for nutrients, space and a time-dependent proliferation of cancerous cells [26].

In this paper we are to investigate the role of time delay in the cell interaction as a triggering factor for metastasis, by using a modified version of a cellular automaton model for cancer growth proposed by Qi and collaborators [23]. This model also takes into account other microscopic propertiesl, as the proliferation of cancer cells, the cytotoxic effect of the immune system, and the mechanical pressure inside the tumor; so as to reproduce the Gompertz growth of cancer tumors.

In the Qi and collaborators model the cancer cell mitosis is instantaneous, what is obviously a simplifying assumption [23]. Accordingly, we introduced a time delay in our modified model as a parameter representing the time it takes for a cancer cell to undergo mitosis [27, 28]. As we shall see, on varying this time delay it is possible to trigger the formation of metastases. This is consistent with similar findings for delay differential equation models of tumor evolution. Hence, we claim that the same outcome would be exhibited by other similar cellular automata models of cancer growth, so that our result is quite general in spite of having been observed in a particular model.

This paper is organized as follows: in Section 2 we introduce the cellular automaton with emphasis on the time delay in proliferation. Section 3 shows the results obtained with the varying of the cellular automaton model parameters that influence the dynamics behavior using the cluster counting. The last section presents the conclusions.

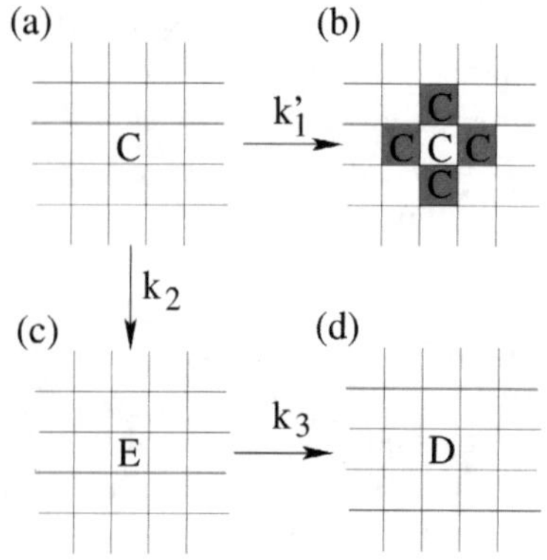

Figure 1. (a) to (b) describes the possibility of proliferation at one of the shadowed sites occupied by normal cells, (a) to (c) denotes the cytotoxic process and (c) to (d) the complex is replaced by a dead cell.

2. Cellular automaton with time delay

In this study we insert a time delay in the cellular automaton [23] in order to analyze the dynamics behavior of cancerous cells in a tissue. Then, let the cancerous (abnormal) cells, the dead cancerous cells, the effector (cytotoxic) cells (macrophages, etc) and the complexes produced by the cytotoxic process be respectively represented by C, D, E_0, and E. Fig. (1) shows the processes considering the proliferation or dissolution of a cancerous cell. As a matter of fact this process can be depicted by the following reactions:

$$C \xrightarrow{k_1'(t)} 2C, \tag{1}$$

$$C + E_0 \xrightarrow{k_2} E \xrightarrow{k_3} E_0 + D, \tag{2}$$

$$D \xrightarrow{k_4} \text{normal}. \tag{3}$$

Reaction (1) describes the proliferation of cancerous cells at a time t, with

$$k_1'(t) = k_1 \left(1 - \frac{N_c}{\phi}\right), \tag{4}$$

where $k_1(t)$ is the proliferation rate of cancerous cells, t is the time, N_c is the total number of cancerous cells and ϕ is a constant, thus N_c reaches the maximum ϕ. We can see by Fig. 1(a) to (b) that the cell C located at site will divide into two and one them will occupy the original position and the other will randomly invade one of the four neighbor sites primarily occupied by normal cell. The first reaction in (2) denotes the cytotoxic process, in that reaction (2) a single effector binds to one abnormal cell in a time. In doing so, there is not cancerous cell proliferation, as it is showed in Fig. 1(b), and there is a probability of a cytotoxic process (Fig. 1c). The second reaction of (2) depicts the dissolution of complexes. In this case the cell died in accordance with a determined probability (Fig. 1d). Equation (3) describes the dissolution of dead cells. The values adopted for the parameters k_1, k_2, k_3 and k_4 are listed in reference [23]. To overcome the lack of the experimental data for k_4, it is possible to consider the value in the wide range (0.1-0.4).

The density of cancerous cells $\rho(t)$ describes the effect of the mechanical pressure on cancer development,

$$\rho(t) = \frac{N'}{R^2}, \tag{5}$$

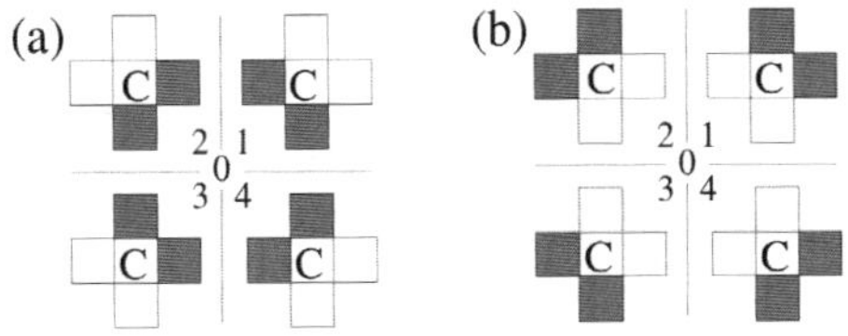

Figure 2. The number 0 indicates the center of the square lattice, where 1,2,3 and 4 denote the four quadrants. When $\rho(t) \leq \rho_c$ the second daughter cell will occupy one of the two shadowed sites with equal probability, as it is showed in (a), and (b) for $\rho(t) > \rho_c$.

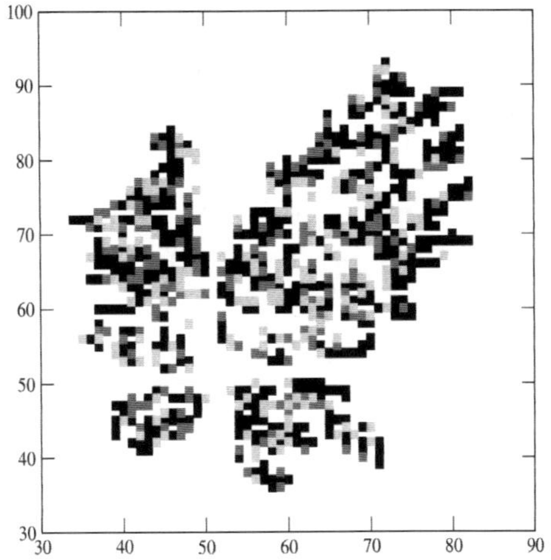

Figure 3. (Color online) The shape of a tumor $k_1 = 0.74$, $k_2 = 0.2$, $k_3 = k_4 = 0.4$, $\rho_c = 3.85$, $\phi = 10^3$ and $t = 50$. Black squares represent cancerous cells, red squares are complexes and green squares are the dead cancer cells. The initial configuration is only 5 cancerous cells in the central part of the lattice.

where $N' = N_c + N_E + N_D$, $R = (\sum R_{ij})/N'$ and R_{ij} depicts the distance from the site (i, j) occupied by a cancerous cell to the origin. Considering a critical value of $\rho(t)$, $\rho(t) = \rho_c$, we have that if $\rho(t) \leq \rho_c$, the second daughter cell resulting from the proliferation can only occupy one of the inside nearest neighboring sites occupied by normal cell (Fig. 2a), if $\rho(t) > \rho_c$, the second daughter cell may invade outside nearest neighboring sites with normal cell (Fig. 2b). This way we used equal probability of the second daughter to occupy one of the two possible neighboring sites.

The shape of the tumor is shown by Fig. 3 for $k_1 = 0.4$, $k_2 = 0.1$, $k_3 = k_4 = 0.35$, $\rho_c = 3.7$, $\phi = 10^3$ and $t = 50$. We consider at $t = 0$, 5 cancerous cells in the central part of the square lattice 101×101 and all of the remaining sites are occupied by normal cells. The black, red and green colors denote C, E and D, respectively. The white region represents the normal tissue and we can see that the tumor is irregular at each lattice site

We include a time delay τ in proliferation rate of cancerous cells $k_1'(t-\tau)$ and density of cancerous cells $\rho(t-\tau)$, in order to consider the delay between the time at which a cancerous cell commences mitosis and the time at which the daughter cell is produced.

Fig. (4a) shows the curves of N_c versus t for $\tau = 0$ (black line) and $\tau = 5$ (red line). It is seen that the curves reach their average maxima simultaneously. Thus, results given by the model fit the relevant observed cancer growth curves. The model may describe the Gompertz growth of a cancer [29], as a matter of fact its algebraic is

$$V = V_0 \exp\left\{\frac{A}{B}[1 - \exp(-Bt)]\right\}, \tag{6}$$

where V is the volume of the tumor, A and B are parameters, t is the time and V_0 denotes the initial volume. The parameters are obtained as a result of experimental data.

At $\tau = 5$ an periodic oscillation of amplitude occurs, furthermore the value of τ to occur oscillatory behavior of N_c depends on the network size. Due to this fact, we can see in the Fig. (4b) a periodic oscillation for a network with $N = 1301$ considering $\tau = 850$. In a model of delay differential equations proposed by Banerjee and Sarkar [30] it was presented the existence of periodic solutions, that is, tumor levels oscillated around a fixed point in absence of treatment. This phenomenon is known as Jeff's Phenomenon [31], and it has been observed clinically.

The value of τ may cause oscillations due to the competition between k_1 and k_2. Moreover, there is a critical value of τ for which the network presents $N_c = 0$ after a transient time, i.e., the proliferation of cancerous cells ceases.

Fig. (4b) shows that the amplitude of the oscillations increases when we increase the network size. This behavior occurs due to the fact that ϕ may be increased, and N_c reaches the maximum ϕ. The amplitude of the oscillations does not increase when the network size increases and the ϕ is maintaned constant.

In Fig. (5a) we plot each value of the time series of number of cancerous cells versus a time-delayed version, by plotting $(N_c(t), N_c(t-1))$. This is called a delay-coordinate reconstruction, or delay plot. At $\tau = 0$ a small and irregular oscillation of the values of N_c can be observed in Fig. (5a) (black line), whereas for $\tau = 6$ the red line reproduces a regular behavior. Fig. (5b) and (5c) show the power spectrum analysis for $\tau = 0$ and $\tau = 6$, respectively. The power spectrum analysis is used to detect periodic behavior, indeed for $\tau = 6$ (Fig. 5c) there is a periodic behavior with a value of frequency around 0.045. By the way, oscillatory behavior occurs in delay differential equation due to a Hopf-bifurcation. The delay in differential equation model for tumor growth is used to study the effects and interactions between tumor cells and immune cells [30][32][33].

3. Cluster

In each cell growth pattern, a cluster is defined as any set of interconnected cells which is physically isolated from any other group of cells on the pattern. We use the Hoshen-Kopelman method to study the number of clusters of the cancerous cells in the cellular automaton. Hoshen and Kopelman introduced in 1976 a breakthrough algorithm for cluster analysis in percolation phenomena [34]. Cluster size distribution of cellular aggregates was also used to characterize the growth patterns of established normal and cancer cell lines, cultured in mono-layer and collagen gel [35].

Fig. (6) shows the time evolution of the number of clusters η for two different values of the proliferation rate of cancerous cells k_1. As we observe the maximum average value of η increases with the rate k_1, although the time for achieving maximum decreases. By the way, when a time delay is included it is possible to occur a periodic oscillation in the values of η. However, the main result is the difference in the values of number of the cancerous cells varying the rate k_1. In preliminary injection experiments with a mouse bronchogenic carcinoma [36], a known metastasizable tumor, it was shown that metastasizability is exhibited by viable tumor cells in clumps and not by discret single cells.

In many cancer cases, the cancer spreads from the one site to another physiological organ via the blood stream or via the lymphatic system. This processes is known as metastasis [37]. Tumor metastasis consists of a series of discrete biological processes that move tumor cells from

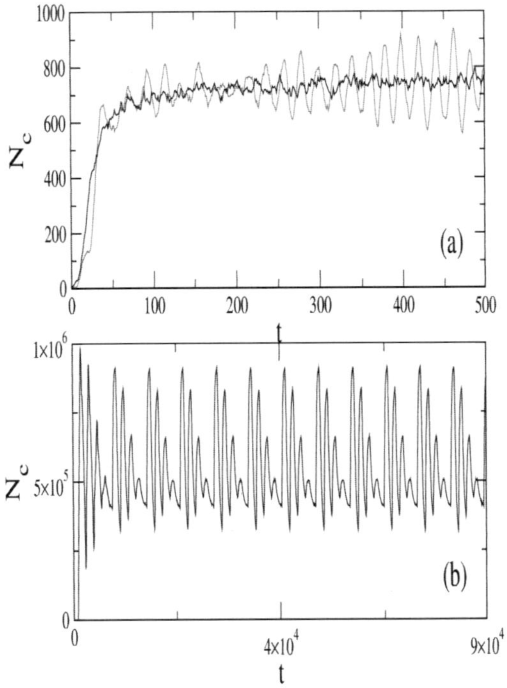

Figure 4. (Color online) The time evolution of N_c for $k_1 = 0.74$, $k_2 = 0.2$, $k_3 = k_4 = 0.4$ and $\rho_c = 3.85$. (a) 101×101, $\phi = 10^3$, $\tau = 0$ (black line) and $\tau = 5$ (red line), (b) 1301×1301, $\tau = 850$ and $\phi = 1.7.10^6$.

the primary site to a distant location [38]. When tumor cells metastasize, the new tumor is called a secondary or metastatic tumor. Metastatic tumors are very common in the late stages of cancer. It is sometimes the case where the secondary tumor is one that is fatal. Cells that metastasize are basically of the same kind as those in the original tumor. We considered that the metastasis occurs when the value η is greater than a critical value η^*.

Fig. (7) represents the probability of a cytotoxic process versus proliferation rate of cancerous cells exhibiting a gray region where the metastasis occurs. For this reason, by Fig. (7), the metastasis only can occur for values of k_1 and k_2 in the gray region when the cellular automaton presents $\eta > \eta^*$, so that we fix η^* equal 50.

Let us fix η at a value, say $\eta = \eta^* = 50$, and vary the parameter k_1 to analyze the behavior of the time to start the metastasis T_m, that is, the time to the number of clusters attains the critical value η^*. We analyze the values of the T_m for different seeds of the random number generator due to the fact the proliferation be probabilistic, then we consider several time evolutions with different initial random seeds to verify the behavior of $T_m^{(i)}$ at each seed i. This probabilistic behavior may be related to dependence of the cancerous growth on the manner and speed with which it adapts to changes in its surroundings and composition.

Fig. (8a) shows that the values of $T_m^{(i)}$ behave asynchronously, on the other hand, in Fig. (8c) we can see spikes with amplitudes higher than the case without time delay in proliferation. Moreover, increasing τ the spikes increase its amplitude and quantity. In fact, if we plot the distribution it is possible to observe that the time to start the metastasis T_m increases with the time delay, in accordance with the distribution in Fig. (8d) which has a minor area and a shift

Journal of Physics: Conference Series **285** (2011) 012015 doi:10.1088/1742-6596/285/1/012015

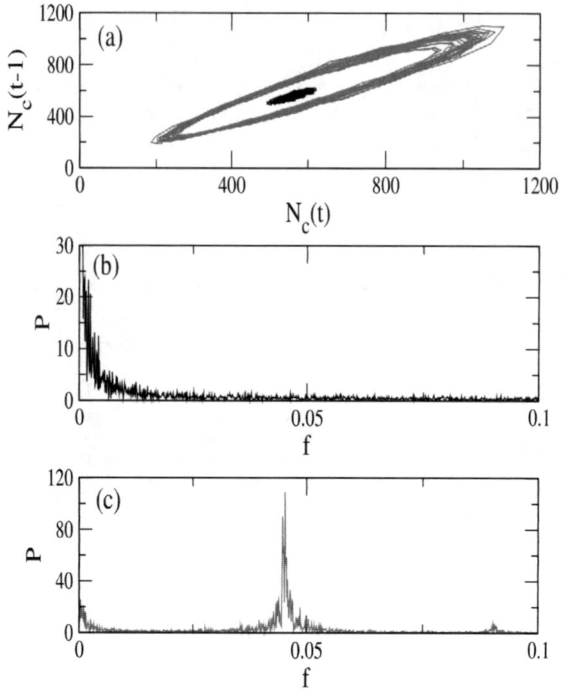

Figure 5. (Color online) We consider $k_1 = 0.74$, $k_2 = 0.2$, $k_3 = k_4 = 0.4$ and $\rho_c = 3.85$. (a) Delay plot of the number of cancerous cells showing a non regular behaviour for $\tau = 0$ (black line) and a regular for $\tau = 6$ (red line), as well as the figures (b) and (c) show the respective power spectrum.

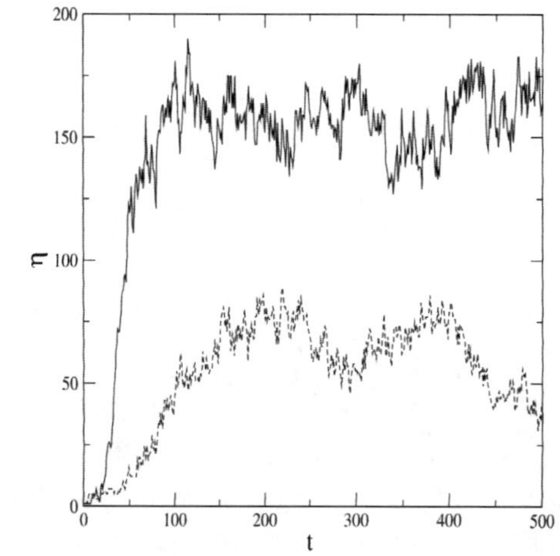

Figure 6. Time series of the number of clusters for $N = 101$, $\tau = 0$, $\phi = 1000$, $\rho_c = 3.85$, $k_2 = 0.2$, $k_3 = k_4 = 0.4$, $k_1 = 0.74$ (continuous line) and $k_1 = 0.4$ (dashed line).

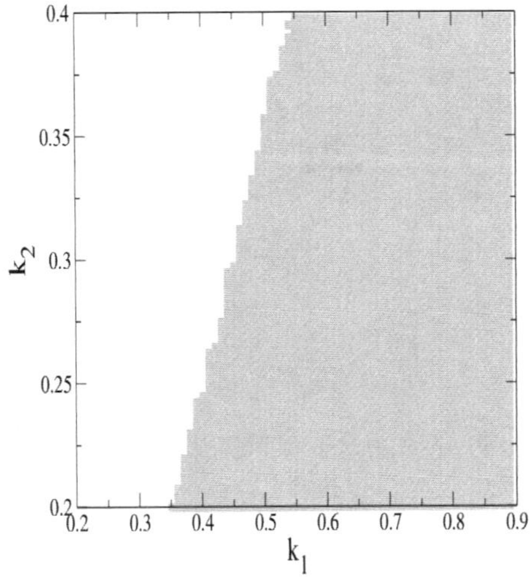

Figure 7. k_2 versus k_1 for $N = 101$, $\phi = 1000$, $\rho_c = 3.85$, $k_3 = k_4 = 0.4$ and $\eta^* = 50$. Gray region represents the values which metastasis occurs ($\eta > \eta^*$).

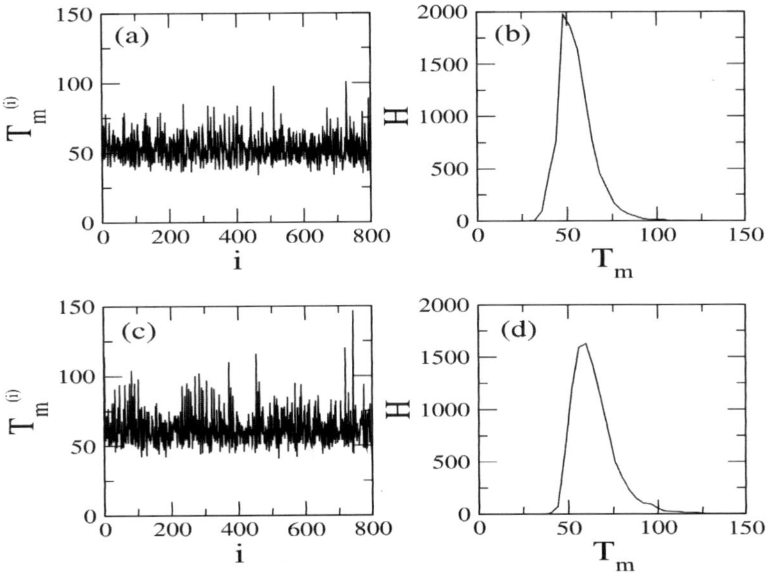

Figure 8. $T_m^{(i)}$ versus i and distribution for (a) and (b) with $\tau = 0$, (c) and (d) with $\tau = 10$. We considered $N = 101$, $\phi = 1000$, $\rho_c = 3.85$, $k_3 = k_4 = 0.4$, $k_2 = 0.2$ and $k_1 = 0.5$.

to right compared with distribution of the Fig. (8c), as a result of major values of $T_m^{(i)}$. In order to obtain a time delay in the metastasis it is necessary to consider a time delay in proliferation.

In Fig. (9) we have plotted the average time to start the metastasis versus the time delay, where our results for the average was taken by considering different initial condition profiles. We can observed the tendency of the time to start the metastasis increases when the time delay increases. This behaviour may occur as a result of chemotherapeutic agents. There are agents

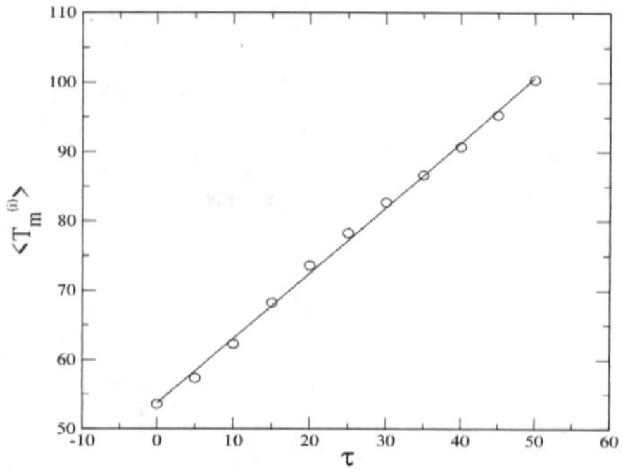

Figure 9. Average of $T_m^{(i)}$ varying the time delay, where we considered $k_1 = 0.5$, $k_2 = 0.2$, $k_3 = k_4 = 0.4$, $\rho_c = 3.85$ and $\phi = 10^3$. The solid line is a least-squares with slope≈1.

that inhibit cell cycle, as well as they are used to increase the time to start the metastasis [10].

4. Conclusions

In this paper we studied some aspects of the cancerous growth dynamics displayed by a cellular automaton, also we analyzed the metastasis starting time from a primary to a secondary cellular automaton varying parameters of the system. With a time delay in proliferation we observed periodic behavior of the number of the cancerous cells for some parameter values. This behavior may explain periodic tumors whose growth is are uncorrelated with the administration of chemotherapy, more commonly referred to as Jeff's Phenomenon.

We considered the number of clusters as a critical parameter to start the metastasis. We have found that the metastasis starting time increases when a time delay in proliferation is introduced in the model. The results suggested that the metastasis starting time increases when cancerous cell division is not instantaneous.

The model with a time delay introduced in proliferation has more complex dynamics than the model without time delay, such as oscillatory behavior and time delay in the metastasis, as well as it reproduces same periodic oscillations as delay differential equation.

Acknowledgments

This study was possible by partial financial support from the following Brazilian government agencies: CNPq, CAPES and Fundação Araucária. We acknowledge A. P. G. da Silva for useful suggestions.

References

[1] Adam J A and Bellomo N, 1997 A Survey of Models for Tumor-Immune System Dynamics (Birkhäuser, Boston).

[2] Menchón S A and Condat C A, *Cancer growth: Predictions of a realistic model*, 2088 Physical Review E **78** 022901.

[3] Pinho S T R, Freedman H I and Nani F, *A chemotherapy model for the treatment of cancer with metastasis*, 2002 Mathematical and Computer Modelling **36** 773.

[4] Scalerandi M, Romano A, Pescarmona G P, Delsanto P P and Condar C A, *Nutrient competition as a determinant for cancer growth*, 1999 Physical Review E **59** 2.

[5] Cattani C and Ciancio A, *Qualitative analysis of second-order models of tumorimmune system competition*, 2008 Mathematical and Computer Modelling **47** 1339.

[6] Wilson G D, McNally N J, Dische S, Saunders M I, Des Rochers C, Lewis A A and Bennet M H, *Measurement of cell kinetics in human tumours in vivo using bromodeoxyuridine incorporation and flow cytometry*, 1988 British Journal of Cancer **58** 423.

[7] Bertalanffy F D, Schachter R, Ali J and Ingimundson J C, *Mitotic rate and doubling time of intraperitoneal subcutaneous Ehrlich ascites tumor*, 1965 Cancer Research **25** 685.
[8] Baserga R, *The relationship of the cell cycle to tumor growth and control of cell division: A Review*, 1965 Cancer Research **25** 5 581.
[9] Hatfield D L, Yoo M -H, Carlson B A and Gladyshev V N, *Selenoproteins that function in cancer prevention and promotion*, 2009 Biochimica et Biophysica Acta **1790** 1541.
[10] Venkateswaran V, Klotz L H and Fleshner N E, *Selenium modulation of cell proliferation and cell cycle biomarkers in human prostate carcinoma cell lines*, 2002 Cancer Research **62** 2540.
[11] Wheldon T E, 1988 Mathematical Models in Cancer Research, Taylor and Francis.
[12] Moreira J and Deutsch A, *Cellular automaton models of tumor development: A critical review*, 2002 Advances in Complex Systems **5** 247.
[13] Wolfram S, 1986 Theory and Applications of Cellular Automata, World Scientific.
[14] Mallet D G and De Pillis L G, *A cellular automata model of tumorimmune system interactions*, 2006 Journal of Theoretical Biology **239** 334.
[15] Gao D, Jiang R, Hu S -X, Wang B -H and Wu Q -S, *Cellular-automaton model with velocity adaptation in the framework of Kerner's three-phase traffic theory*, 2007 Physical Review E **76** 026105.
[16] Copelli M, Roque A C, Oliveira R F and Kinouchi O, *Physics of psychophysics: Stevens and Weber-Fechner laws are transfer functions of excitable media*, 2002 Physical Review E **65** 060901(R).
[17] Manor A and Shnerb N V, *Origin of pareto-like spatial distributions in ecosystems*, 2008 Physical Review Letters **101** 268104.
[18] Suzudo T, *Spatial pattern formation in asynchronous cellular automata with mass conservation*, 2004 Physica A **343** 185.
[19] Kansal A R, Torquato S, Harsh G R, Chiocca E A, and Deisboecke T S, *Simulated brain tumor growth dynamics using a three-dimensional cellular automaton*, 2000 Journal of Theoretical Biology **203** 367.
[20] Quaranta V, Weaver A M, and Cummings P T, *Mathematical modeling of cancer: The future of prognosis and treatment*, 2005 Clinica Chimica Acta **357** 173.
[21] Gatenby R A, Gawlinski E T, *The glycolytic phenotype in carcinogenesis and tumor invasion: Insights through mathematical models*, 2003 Cancer research **63** 3847.
[22] Duchting W and Vogelsaenger T, *Recent progress in modelling and simulation of three-dimensional tumor growth and treatment*, 1985 Biosystems **18** 79.
[23] Qi A -S, Zheng X, Du C -Y and An B -S, *A cellular automaton model of cancerous growth*, 1993 Journal of Theoretical Biology **161** 1.
[24] Patel A A, Gawlinsky E T, Lemieux S K and Gatenby R A, *A cellular automaton model of Early tumor growth and invasion: the e4ects of native tissue vascularity and increased anaerobic tumor metabolism*, 2001 Journal of Theoretical Biology **213** 315.
[25] Anderson A R A and Chaplain M J, *Continuous and discrete mathematical models of tumor-induced angiogenesis*, 1998 Bulletin of Mathematical Biology **60** 857.
[26] Reis E A, Santos L B L and Pinho S T R, *A cellular automata model for avascular solid tumor growth under the effect of therapy*, 2009 Physica A **388** 1303.
[27] Byrne H M, *The effect of time delays on the dynamics of avascular tumor growth*, 1997 Mathematical Biosciences **144** 83.
[28] Cui S and Xu S, *Analysis of mathematical models for the growth of tumors with time delays in cell proliferation*, 2007 Journal of Mathematical Analysis and Applications **336** 523.
[29] Steel G G, 1977 Growth Kinetics of Tumors, Oxford: Clarendon Press.
[30] Banerjee S and Sarkar R R, *Delay-induced model for tumorimmune interaction and control of malignant tumor growth*, 2008 Biosystems **91** 268.
[31] Thomlinson R, *Measurement and management of carcinoma of the breast*, 1982 Clinical Radiology **33** 481.
[32] Villasana M and Radunskaya A, *A delay differential equation model for tumor growth*, 2003 Mathematical Biology **47** 270.
[33] Yuan Z, Hu D and Huang L, *Stability and bifurcation analysis on a discrete-time system of two neurons*, 2004 Applied Mathematics Letters **17** 1239.
[34] Hoshen J and Kopelman R, *Percoltion and cluster distribution. I. Cluster multiple labeling technique and critical concentration algorithm*, 1976 Physical Review B **14** 8 3438.
[35] Mendes R L, Santos A A, Martins M L and Vilela M J, *Cluster size distribution of cell aggregates in culture*, 2001 Physica A **298** 471.
[36] Watanabe S, *The metastasizability of tumor cells*, 1954 Cancer **7** 2 215.
[37] Evans C W, The metastatic Cell, 1991 Behaviour and Biochemistry, Chapman and Hall, London.
[38] Steeg P S, *Tumor metastasis: mechanistic insights and clinical challenges*, 2006 Nature Medicine **12** 8 895.

Dynamic Days South America 2010 IOP Publishing
Journal of Physics: Conference Series **285** (2011) 012016 doi:10.1088/1742-6596/285/1/012016

The upper and lower solution method for nonlinear fourth-order boundary value problem

Amir El-Haffaf
Université d'Oran.Faculté des Sciences
Département de Mathématiques
B.P. 1524 Es-Senia. Oran. Algerie.
e-mail: elhaffaf@yahoo.fr
elhaffaf1@univ-oran.dz

Abstract. In this paper , we study the existence of a solution for a fourth order boundary value problem

$$x^{(4)}(t) + f(t, x(t), x'(t), x''(t)) = 0, \qquad 0 < t < 1$$
$$x(0) = x'(0) = x''(1) = x^{(3)}(0) = 0$$

Where $f \in C([0,1] \times IR^2, IR)$, and $f \in C([0,1] \times IR^3, IR)$. By placing certain restrictions on the nonlinear term f , we prove the existence of at least one solution to the boundary-value problem with the use of lower and upper solution method and Schauder fixed-point theorem. The construction of lower or upper solutions is also presented.

1. Introduction Recently, fourth-order boundary value problems have been considered in many papers.for example, Ma[20] studied the existence of positive solutions for the fourth-order boundary value problem

$$x^{(4)}(t) = \lambda f(t, x(t), x''(t)) = 0, \qquad 0 < t < 1$$
$$x(0) = x'(0) = x''(1) = x^{(3)}(1) = 0$$

Under some suplinear semipsitone conditions.The boundary conditions considered here are a special case of the $(p, n-p)$ right focal boundary conditions, in which $n = p = 2$.
Extensive research has been done on focal boundary value problems. For some recent results, we refer the reader to the papers by Agarwal and O'Regan[1],Anderson[2],Henderson and Kaufman[15]. The fourth-order boundary value problems have been studied by many authors, for example yang[22], Kosmatov[18],J.R.Graef,Chen Qian and Bo yang[9].Most of them studied the existence of positive solutions for the equation

$$x^{(4)}(t) = g(t) f(x), \qquad 0 < t < 1$$
$$x(0) = x'(0) = x''(1) = x''(1) = 0 \qquad or \qquad x(0) = x(1) = x'(0) = x'(1) = 0$$

Recent contribution to the literature on the existence of multiple positive solutions include to the papers of J.Henderson,J.R.Graef and Bo.yang[10],J.Henderson and S.K.Ntouyas[17].

The method of lower and upper solution is extensively developed for lower order equations with linear and nonlinear boundary conditions. But there are only a few applications to higher-order ordinary differential equations. The lower and upper solution method is employed to acquire existence results about some boundary value problems with some monotonic or quasi-monotonic nonlinear term f which is not dependance on any lower-order derivatives.

2. Preliminaries

This paper is organised as follows. In section 2, we present two lemmas that will be used to prove the mains results. In section3, we obtain some existence results for the existence of solution for two class nonlinear fourth-order boundary value problem. The method used here it is based on Schauder fixed-point theorem and lower and upper solution method. The construction of lower or upper solution is also presented.

The purpose is to study the existence of solution for two nonlinear fourth-order boundary value problems

$$\begin{aligned} &x^{(4)}(t) + f(t, x(t), x''(t)) = 0, \qquad 0 < t < 1 \\ &x(0) = x'(0) = x''(1) = x^{(3)}(0) = 0 \end{aligned} \tag{2.1}$$

and

$$\begin{aligned} &x^{(4)}(t) + f(t, x(t), x'(t), x''(t)) = 0, \qquad 0 < t < 1 \\ &x(0) = x'(0) = x''(1) = x^{(3)}(0) = 0 \end{aligned} \tag{2.2}$$

Definition 2.1. We call $\alpha, \beta \in C^2[0,1] \cap C^4[0,1]$ lower and upper solution of problem (2.1), respectively, if

$$\begin{aligned} &\alpha^{(4)}(t) + f(t, \alpha(t), \alpha''(t)) \geq 0, \qquad 0 < t < 1 \\ &\alpha(0) = \alpha'(0) = \alpha''(1) = 0; \qquad \alpha^{(3)}(0) \geq 0 \end{aligned}$$

$$\begin{aligned} &\beta^{(4)}(t) + f(t, x(t), x''(t)) \leq 0, \qquad 0 < t < 1 \\ &\beta(0) = \beta'(0) = \beta''(1) = 0; \qquad \beta^{(3)}(0) \leq 0 \end{aligned}$$

Denote by $G(t,s)$ the Green's function of the following BVP

$$\begin{aligned} &-x^{(3)}(t) = 0, \qquad 0 < t < 1 \\ &x(0) = x'(0), \qquad x''(1) = 0 \end{aligned} \tag{2.3}$$

Then

$$G(t,s) = \begin{cases} \frac{1}{2}t^2, & 0 \leq t \leq s \leq 1, \\ st - \frac{1}{2}s^2, & 0 \leq s \leq t \leq 1, \end{cases} \tag{2.4}$$

Lemma 2.2. For $G(t,s)$ defined by (2.4), the following holds:

(1) $\dfrac{G(t,s)}{G(s,s)} \leq 1$ for $t, s \in (0,1)$,

(2) $\dfrac{G(t,s)}{G(s,s)} \geq C > 0$ for $t, s \in (0,1)$

The proof of lemma (2.2) is very straightforward and therefore omitted.

Let $\ell = \int_0^1 G(s,s)s\,ds$. It is easy to see that $\ell > 0$.

3. Main results

We consider (2.1) under the assumption that $f:[0,1]\times R^2 \to R$ is continuous. We give some lemmas which indicate some restrictions on the nonlinear term and let us constructlower or upper solutions.

Lemma 3.1 If there exist a constant $\mathrm{A} \geq 0$ such that

$$f(t,p,q) \leq \mathrm{A}, \text{ for } 0 \leq t \leq 1, \quad \ell \mathrm{A}C \leq p \leq \ell \mathrm{A}, \quad -\mathrm{A} \leq q \leq 0,$$

Then problem (2.1) has an upper solution.

Proof. Setting $u(t) = -x''(t)$. Then the Problem (2.1) is equivalent to

$$\begin{gathered} u'(t) = f(t,(Tu)(t),-u(t)); \quad 0 < t < 1, \\ u(0) = 0 \end{gathered} \tag{3.1}$$

Where $(Tu)(t) = \int_0^1 G(t,s)u(s)ds$ and $G(t,s)$ is the Green's function defined by (2.4)

it is clear that the restriction on f guarantee that $\psi(t) = \mathrm{A}t$ satisfies the first order equation

$$\begin{gathered} \psi'(t) + f(t,(T\psi)(t),-\psi(t)) \geq 0 \quad 0 < t < 1 \\ \psi(0) = 0 \end{gathered}$$

This shows that $\beta(t) = \mathrm{A}\psi(t)$ is an upper solution of problem (2.1).

Lemma 3.2. If there exist a constant $\mathrm{B} \leq 0$ such that

$$f(t,p,q) \geq \mathrm{B}, \text{ for } 0 \leq t \leq 1, \quad \ell \mathrm{B} \leq p \leq \ell \mathrm{B}C, \quad 0 \leq q \leq -\mathrm{B},$$

Then problem (2.1) has a lower solution.

Proof. Setting $\varphi(t) = \mathrm{B}t$. Then it is clear that the restriction on f guarantee that $\varphi(t) = \mathrm{B}t$ satisfies

$$\begin{gathered} \varphi'(t) + f(t,(T\varphi)(t),-\varphi(t)) \leq 0 \quad 0 < t < 1 \\ \varphi(0) \leq 0 \end{gathered}$$

This shows that $\alpha(t) = \mathrm{B}\varphi(t)$ is a lower solution of problem (2.1).

Remark 3.3 Explicitly, we can write the upper solution $\beta(t)$ and lower solution $\alpha(t)$ of

Problem (3.1), $\beta(t) = -\frac{\mathrm{A}}{24}t^4 + \frac{\mathrm{A}}{4}t^2$, $\alpha(t) = -\frac{\mathrm{B}}{24}t^4 + \frac{\mathrm{B}}{4}t^2$.

Theorem 3.4. Suppose there exist two constants $A \geq 0 \geq B$, $A \geq |B|$ such that

$$f(t,p,q) \leq A, \text{ for } 0 \leq t \leq 1, \quad \ell AC \leq p \leq \ell A, \quad -A \leq q \leq 0 \tag{3.2}$$

$$f(t,p,q) \geq B, \text{ for } 0 \leq t \leq 1, \quad \ell B \leq p \leq \ell BC, \quad 0 \leq q \leq -B \tag{3.3}$$

If $f(t, p, q)$ is increasing in p, then problem (2.1) has a solution $x(t)$ such that $\alpha(t) \leq x(t) \leq \beta(t)$, $\beta'''(t) \leq x'''(t) \leq \alpha'''(t)$, $t \in [0,1]$, where

$$\alpha(t) = B\int_0^1 G(t,s)s\,ds, \qquad \beta(t) = A\int_0^1 G(t,s)ds$$

Proof. From the conditions (3.2) and (3.3), lemma 3.1 and lemma 3.2 conbined yield that the problem (2.1) has an upper solution $\beta(t)$ and a lower solution $\alpha(t)$.
Set $\varphi' = -\alpha'''$, $\psi' = -\beta'''$, then $\varphi'(t) = B \leq A = \psi'(t)$. On the other hand, $A \geq 0 \geq B$ implies $\alpha'''(t) \geq \beta'''(t)$. The fact that $G(t,s) \geq 0$ and taking into account boundary condition yield $\alpha(t) \leq \beta(t)$, for $0 \leq t \leq 1$.
Define now the folowing operators D, F, where
$D : domD = \{u \in C^1(0,1) \cap C[0,1] \to C[0,1] : u(0) = 0\}$ is derivate operator such that $(Du)(t) = u'(t), t \in (0,1)$, and $F : C[0,1] \to C[0,1]$ is a continuous operator such that

$$Du = Fu, \quad u \in dom(D), \text{ where } F \text{ is defined as} \tag{3.4}$$

$$(Fu)(t) = f(t, T(\min\{\psi, \max\{u,\varphi\}\}), -(\psi, \max\{u,\varphi\}) \tag{3.5}$$

If u^* is a solution of (3.4), then $\varphi(t) \leq u^*(t) \leq \psi(t)$, $t \in [0,1]$. Consequentely, $u^*(t)$ is a solution of (3.1). Furthemore, $u^*(t) = (Tu^*)(t)$ is a solution of (2.1) satisfying $\alpha(t) \leq u^*(t) \leq \beta(t)$. In fact, if $\varphi(t) > u^*(t)$, there exist $t_1 \in [0,1]$ such that $\varphi(t_1) > u^*(t_1)$. As $u^*(0) = -\alpha''(0) = \varphi(0) = 0$, by the continuity of u^* and φ, there exist an interval (t_2, t_3) such that $\varphi(t_2) = u^*(t_2)$ and $\varphi(t) > u^*(t)$, for $t \in (t_2, t_3)$. Therefore, for $t \in [t_2, t_3]$,

$$(Fu^*)(t) = f(t, T(\min\{\psi, \max\{u^*, \varphi\}\}), -\varphi)$$

Let $k(t) = u^*(t) - \varphi(t)$, by the monotonocity of f and T, one has

$$\varphi'(t) = -\alpha'''(t) \leq f(t, \alpha(t), \alpha''(t)) = f(t, T\varphi(t), -\varphi(t)), \quad t \in [0,1]$$

$$u^{*\prime}(t) = f(t, T(\min\{\psi, \max\{u^*, \varphi\}\}), -\varphi(t)), \quad t \in [t_2, t_3],$$

then $k'(t) \geq 0$, $t \in [t_2, t_3]$. Thus $k(t_2) = 0$, implies $k(t) \geq 0$, $t \in [t_2, t_3]$. Namely $\varphi(t) \leq u^*(t)$, $t \in [t_2, t_3]$. This contradicts our supposition. Thus, $\varphi(t) \leq u^*(t)$, $t \in [0,1]$.
The proof of $\psi(t) \geq u^*(t)$, $t \in [0,1]$ is quite similar of $\varphi(t) \leq u^*(t)$ and therefore omitted.
Let now, $H : C[0,1] \to [0,1]$, defined by

$$(Hu)(t) = \int_0^t (Fu)(s)ds, \quad t \in [0,1]$$

It is clear that H is a continuous operator and a fixed point of H is a solution of Problem (3.4).
Set $M_A = \{\theta \in C[0,1] : \|\theta\| \le A\}$, it is not difficult to see that $\psi, \varphi \in M_A$.
We prove now, that $H : M_A \to M_A$ is a completely continuous operator. From the condition (3.2), one has

$$|(Hu)(t) - (Hu)(s)| = \left|\int_s^t (Fu)(r)dr\right| \le A|t-s|$$

That is $\{H(M_A)\}$ is equi-continuous and bounded uniformly. Form Arzela-Ascoli theorem, we assert that $H : M_A \to M_A$ completely continuous operator, Schauder fixed-point guarantee that H has a fixed point $u^* \in M_A$. Therefore $x^*(t) = \int_0^1 G(t,s)u^*(s)ds$ is a solution of Problem (2.1) such that $\alpha(t) \le x^*(t) \le \beta(t), \quad \alpha'''(t) \le x^{*'''}(t) \le \beta'''(t), \quad t \in [0,1]$.

Consider now Problem (2.2). Assume that $f : [0,1] \times R^3 \to R$ is continuous. Lemmas similar to lemmas 3.1 and 3.2 can be obtained. It is not difficult to see that the equation of problem (2.2) is equivalent to

$$u'(t) = f(t, (Tu)(t), \Gamma(u)(t), -u(t)); \qquad 0 < t < 1,$$
$$u(0) = 0$$

where $(Tu)(t) = \int_0^1 G(t,s)u(s)ds$, $(\Gamma u)(t) = \int_t^1 u(s)ds$.

Lemma 3.5 If there exist a constant $A \ge 0$ such that

$$f(t,p,q,r) \le A, \text{ for } 0 \le t \le 1, \quad \frac{p}{\ell}, 2q \in [0,A], \quad r \in [-A, 0] \quad ,$$

Then problem (2.2) has an upper solution.

Lemma 3.6 If there exist a constant $B \le 0$ such that

$$f(t,p,q,r) \ge B, \text{ for } 0 \le t \le 1, \quad p \in [\ell B, 0], \quad 2q, r \in [0, -B] \quad ,$$

Then problem (2.2) has a lower solution.

Lemmas 3.5 and 3.6 can be proved similarely as lemmas 3.1 and 3.2 and so omitted. An argument similar to the one in theorem 3.4 provides the following result about Problem (2.2).

Theorem 3.7 Suppose there exist two constants $A \ge 0 \ge B, \quad A \ge |B|$ such that

$$f(t,p,q,r) \le A, \text{ for } 0 \le t \le 1, \quad \frac{p}{\ell}, 2q \in [0,A], \quad r \in [-A, 0] \tag{3.6}$$

$$f(t,p,q,r) \ge B, \text{ for } 0 \le t \le 1, \quad p \in [\ell B, 0], \quad 2q, r \in [0, -B] \tag{3.7}$$

If $f(t,p,q,r)$ is increasing in each of p and q, then problem (2.2) has a solution $x(t)$ such that $\alpha(t) \le x(t) \le \beta(t), \quad \beta'''(t) \le x'''(t) \le \alpha'''(t), \quad t \in [0,1]$, where

$$\alpha(t) = B\int_0^1 G(t,s)sds, \quad \beta(t) = A\int_0^1 G(t,s)ds, \quad \ell = \frac{1}{8}.$$

Example 3.8 Consider the problem

$$\begin{aligned} &x''''(t) + \frac{1}{8}\left[t + e^{x(t)} + x''(t)\right] = 0 \\ &x(0) = x'(0) = x''(1) = 0, \quad x'''(0) = 0 \end{aligned} \tag{3.8}$$

Let $f(t,p,q) = \frac{1}{8}\left[t + e^{p} + q\right]$, it is easy to check that $\mathrm{A} = 4$, $\mathrm{B} = 0$ satisfy the bounday conditions of problem (2.1), respectively, therefore,

$$\alpha(t) = 0, \quad \beta(t) = 4\int_0^1 G(t,s)sds$$

are lower and upper solutions of (3.8). By theorem 3.4, Problem (3.8) has a positive solution x^* such that $0 \le x^*(t) \le 4\int_0^1 G(t,s)sds$.

Example 3.9 Consider the problem

$$\begin{aligned} &x''''(t) + \frac{1}{3}\left[t + \ln(1 + x(t) + (x'(t))^2 - x''(t)\right] \\ &x(0) = x'(0) = x''(1) = 0, \quad x'''(0) = 0 \end{aligned} \tag{3.9}$$

Let $f(t,p,q,r) = \frac{1}{3}\left[t + \ln(1+p) + q^2 - r\right]$, it is easy to check that $\mathrm{A} = 1$, $\mathrm{B} = 0$ satisfy the boundary conditions of problem (2.2) , respectively, therefore,

$$\alpha(t) = 0, \quad \beta(t) = \int_0^1 G(t,s)sds$$

are lower and upper solutions of (3.9). By theorem3.7, Problem (3.9) has a solution x^* such that

$$0 \le x^*(t) \le \int_0^1 G(t,s)sds.$$

Acknowledgments. The author would like to thank the anonymous referees for their valuable suggestions

References

[1] Agarwal R.P. and D.O'Regan 2001 Multiplicity results for singular conjugate focal and (n,p) problems, J. Diff. Eq. 170 ,142-156.

[2] Anderson D.R. 2003 Green's functions for third-order generalized right focal problem, J. Math. Anal. Appl. 288, 1-14.

[3] Bai Z., 2008 Existence of solutions for some third-order boundary-value problems, Elect J. Diff. Eq. 25, 1-6.

[4] Deimling K. 1985 Nonlinear Functional Analysis (Berlin: Springer).

[5] Dhage B.C. 2003 Local fixed point theory for the sum of two operators in Banach spaces (Fixed Point Theory) vol 4, 1 pp 49-60.

[6] Dhage B.C. Henderson J, 2004 Existence theory for nonlinear functional boundary value problems, Elec. J. Qualit. Theo. Diff. Eq.1, 1-15.

[7] Dugundji J. and Granas A. 1982, Fixed Point Theory, (Warsaw :Monograph Math)

[8] Fend W., webb J.R.L. 1997 Solvability of a m-point boundary value problems with nonlinear grouth, J.Math.Anal.Appl. 212 ,467-480.

[9] Graef J.R., Qian C. Yang B. 2004 Positive solutions of three point boundary value problem for nonlinear differential Equations, Proc. Dyn. Syst Appl. 4, 431-438.

[10] Graef J.R., Henderson J, Yang B. 2007 Positive Solution For A Nonlinear Higher Order boundary value Problem, Elect. J. Diff Eq. 45, 1-10.

[11] Granas A., Guenther R.B, Lee J.W.1991 Some general existence principles for Caratheodory Theory of nonlinear differential equations, J.Math.Pures et Appl. 70 , 267-276.

[12] Gupta C.P. 1992 Solvability of a three-point nonlinear boundary value problem for a second order ordinary differential equations, J. Math.Anal.Appl.168 , 540-551.

[13] C.P.Gupta, 1995 Solvability of an m-point nonlinear boundary value problem for a second order ordinary differential equations, J. Math.Anal.Appl.189, 575-584.

[14] J. Henderson, 1995 Boundary value problems for functional differential equations (Singapore: World Scientific)

[15] Henderson J. Kaufman E.R, 1997 Multiple positive solutions for focal boundary problems, Commun. Appl. Anal. 1, 53-60.

[16] Henderson J. and Hudson W. N.,1996 Eigenvalue problem for nonlinear functional differential equations, Comm. Appl. Nonlin. Anal. 3, 51-58.

[17] Henderson. J, Ntouyas S.K., 2007 Positive solution for systems of nth-order three-point nonlocal boundary value problems, Elec. J. Qualit. Theory. Diff. Eq. 18 ,1-12.

[18] Kosmatov N. 2004 Countably many solutions of a fourth-order boundary value problem, Elec. J. Qualit. Theory. Diff. Eq. 12, 1-15.

[19] Liu B.,2002 Positive solutions for a nonlinear three-point boundary value problem, Comp. Math. Appl. 44, 201-211.

[20] Ma R.,2003 Multiple positive solutions for semipositone fourth-order boundary value problem, Hiroshima Math. J.33, 217-227.

[21] Sun Y. P., 2004 Nontrivial solution for a three-point boundary value problem, Elect J. Diff. Eq. 11 , 1-10.

[22] Yang B., 2005 Positive solution for a fourth-order boundary value problem, Elect J. Diff. Eq. 3, 1-17.

Dynamic Days South America 2010 IOP Publishing
Journal of Physics: Conference Series **285** (2011) 012017 doi:10.1088/1742-6596/285/1/012017

Characterization of hyperchaotic states in the parameter-space of a modified Lorenz system

Marcos J. Correia and Paulo C. Rech

Departamento de Física, Universidade do Estado de Santa Catarina, 89223-100 Joinville, Brazil

E-mail: marcorreia18@yahoo.com.br

Abstract. In this paper we report a new four-dimensional autonomous system, constructed from a Lorenz system by introducing an adequate feedback controller to the third equation. We use a numerical method that considers the second largest Lyapunov exponent value as a measure of hyperchaotic motion, to construct a two-dimensional parameter-space color plot for this system. Different levels of hyperchaos are represented in this plot by a continuously changing yellow-red scale. Practical applications of this plot includes, by instance, walking in the parameter-space of hyperchaotic systems along suitable paths.

1. Introduction

Historically, hyperchaos was first presented by Rössler [1] to characterize a chaotic system with at least two positive Lyapunov exponents. Some years after, the existence of hyperchaos was experimentally confirmed [2, 3, 4]. The dynamics of a hyperchaotic system is expanded in two or more directions simultaneously, resulting in a more complex attractor than a chaotic attractor, which has only one positive Lyapunov exponent. This expansion of the dynamics, happening at the same time in at least two directions makes hyperchaotic systems to have better performance in many chaos based fields, including technological applications, when compared to chaotic systems. For instance, hyperchaotic systems, due to the much more complicated structure of the attractors, can be used to improve the security in chaotic communication systems, where a chaotic signal is used to mask the message to be transmitted. It was shown that a message masked by a chaotic system is not always secure [5].

Hyperchaotic systems are common in many fields such as nonlinear circuits [6, 7], secure communications [8, 9], lasers [10, 11], colpitts oscillators [12], control [13, 14, 15], synchronization [16, 17, 18], and in a quantum cellular neural network [19]. Hyperchaos is generally present in high-dimensional nonlinear dynamical systems [20], and its occurrence is related with the loss of shadowing of chaotic trajectories on account of a strong form of non-hyperbolicity called unstable dimension variability [21]. A standard way to investigate the dynamics of a hyperchaotic system is by modeling them with differential equations. If we prefer to use autonomous first order ordinary differential equations to model a hyperchaotic system, we need to consider at least four these equations. In a hyperchaotic four-dimensional dissipative system, there is only one possibility to the Lyapunov exponents spectrum: two are positive, one is null, and one is negative.

In this paper we consider the magnitude of the second largest Lyapunov exponent to numerically characterize the points with hyperchaotic behavior in a two-dimensional parameter-space of a dynamical system modeled by a set of four nonlinear autonomous first order ordinary differential equations. Each point of this parameter-space is painted with a color that indicates the level of hyperchaos of the point. Here we report specific results obtained for a system constructed by us, from a Lorenz system, by introducing an adequate feedback controller to the third equation. This new controlled system is given by

$$\begin{aligned} \dot{x} &= -\sigma(x-y), \\ \dot{y} &= rx - y - xz, \\ \dot{z} &= xy - bz + w, \\ \dot{w} &= -a(x^2+w), \end{aligned} \tag{1}$$

where x, y, z, w represent dynamical variables, and $\sigma > 0$, $r > 0$, $b > 0$, and $a > 0$, are parameters.

The paper is organized as follows. In Sec. 2 we discuss some basic properties of this new system, and show that the trivial equilibrium point can be unstable. In Sec. 3 numerical results involving Lyapunov exponents spectrum and phase-space portraits are presented. Finally, the paper is summarized in Sec. 4.

2. Some Properties of the System (1)

The divergence of the vector field (1) is given by

$$\frac{\partial \dot{x}}{\partial x} + \frac{\partial \dot{y}}{\partial y} + \frac{\partial \dot{z}}{\partial z} + \frac{\partial \dot{w}}{\partial w} = -(\sigma + b + 1 + a), \tag{2}$$

from where we conclude that the system (1) is always dissipative and, therefore, that the phase-space contracts volumes as the time increases. As a consequence, all the bounded system trajectories finally settle onto an attractor in a four-dimensional phase-space.

The equilibrium points of the system (1) are calculated by doing $\dot{x} = \dot{y} = \dot{z} = \dot{w} = 0$, that is, by solving the set of coupled equations

$$\begin{aligned} -\sigma(x-y) &= 0, \\ rx - y - xz &= 0, \\ xy - bz + w &= 0, \\ -a(x^2+w) &= 0, \end{aligned} \tag{3}$$

for x, y, z, and w. the origin $P_0 = (0,0,0,0)$ is an equilibrium point. The Jacobian matrix for system (1) at P_0, denoted by J_0, is given by

$$J_0 = \begin{pmatrix} -\sigma & \sigma & 0 & 0 \\ r & -1 & 0 & 0 \\ 0 & 0 & -b & 1 \\ 0 & 0 & 0 & -a \end{pmatrix},$$

and the characteristic equation, calculated using $\det(J_0 - mI) = 0$, where m represents the eigenvalues and I is the standard 4×4 identity matrix, is

$$(m+a)(m+b)[m^2 + (1+\sigma)m + (1-r)\sigma] = 0, \tag{4}$$

from where we obtain the eigenvalues

$$m_1 = -a, \quad m_2 = -b, \quad \text{and} \quad m_\pm = \frac{-\sigma - 1 \pm \sqrt{(\sigma-1)^2 + 4r\sigma}}{2}.$$

The origin P_0 is a stable equilibrium point if the real part of its corresponding eigenvalues is negative. Take into account that $a > 0$, $b > 0$, and $\sigma > 0$, we conclude that m_1, m_2, and m_+ are always less than zero. Otherwise, for $r > 1$, $m_+ = (-\sigma - 1 + \sqrt{(\sigma-1)^2 + 4r\sigma})/2$ is greater than zero. Therefore, P_0 is an unstable saddle-node point for $r > 1$, a necessary condition to the occurrence of chaos, hyperchaos, and limit cycles in system (1).

3. Some Numerical Results

Figure 1 shows two parameter-space plots displaying different dynamical behaviors for

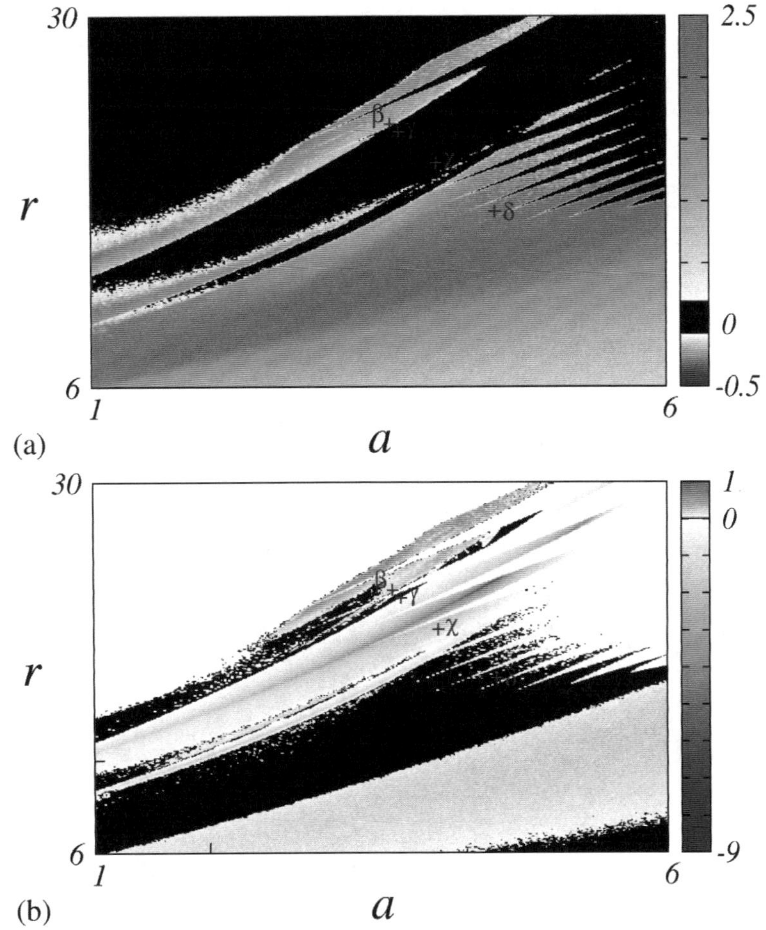

Figure 1. Regions of different dynamical behaviors in (a,r) parameter-space of system (1), for $b = 8/3$, $\sigma = 10$. (a) The largest Lyapunov exponent. (b) The second largest Lyapunov exponent. In each plot colors represent the magnitude of the Lyapunov exponent, as shown in the respective column at right. About the points β, γ, χ, and δ, see the text.

system (1). Both plots were obtained by computing Lyapunov exponents λ on a 500×500 mesh of parameters (a,r). In Fig. 1(a) is considered the largest Lyapunov exponent, while Fig. 1(b) considers the second largest Lyapunov exponent, for $b = 8/3$, $\sigma = 10$, $1 < a < 6$, and $6 < r < 30$. In these and further plots, system (1) was always integrated with a fourth-order Runge-Kutta algorithm with a fixed step size equal to 10^{-3}, and considering 5×10^5 steps to compute each exponent. Furthermore, with respect to the initial conditions, every orbit in the phase-space, for each one of the 2.5×10^5 pairs (a,r), was started from the same (x_0,y_0,z_0,w_0)=(10,10,10,10), that is, we do not follow the attractor.

Colors are associated with the magnitude of the Lyapunov exponent. Blue-white for more negative, black for zero, and red for more positive. Indeed, a positive exponent is indicated by a continuously changing yellow-red scale, while a negative exponent is indicated by a continuously changing blue-white scale. Note that the color scale of each one diagram was redefined in each individual plot: in plot shown in Fig. 1(a) $-0.5 < \lambda < 2.5$, while in Fig. 1(b) $-9 < \lambda < 1$. A negative largest Lyapunov exponent indicates a stable equilibrium point, a zero largest Lyapunov exponent indicates a stable periodic attractor (or a quasiperiodic attractor, when the second largest Lyapunov exponent is also equal to zero), and a chaotic attractor has a positive largest Lyapunov exponent.

Now we concentrate our attention in the last case, that is, when the largest Lyapunov exponent is positive. In consequence, there are only two possibilities to the second largest Lyapunov exponent (the third largest Lyapunov exponent is null, and the minor is negative): a positive value, which indicates hyperchaotic motion with two positive Lyapunov exponents, and a zero value indicating chaotic motion. Therefore, from above said, we conclude that the points pertaining to the yellow-red region in parameter-space of Fig. 1(b), are characterized by hyperchaotic motion. In other words, we numerically characterize hyperchaotic behavior in the parameter-space of the set of four nonlinear autonomous first order ordinary differential equations (1), by looking for the positive second largest Lyapunov exponent. For all set of parameters (a,r) that this exponent is greater than zero, the motion is hyperchaotic. Note that the yellow-red region in Fig. 1(b) is a subset of the region with same shading in Fig. 1(a). It is a consequence of the fact that if the second largest Lyapunov exponent is greater than zero, then the largest Lyapunov exponent also is greater than zero.

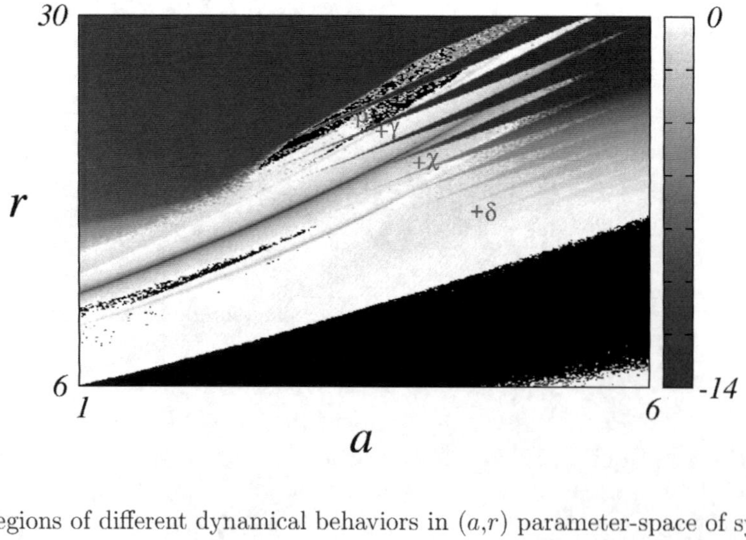

Figure 2. Regions of different dynamical behaviors in (a,r) parameter-space of system (1), for $b = 8/3$, $\sigma = 10$, when the third largest Lyapunov exponent is considered. About the points β, γ, χ, and δ, see the text.

Figure 2 shows the parameter-space plot obtained when the third largest Lyapunov exponent is considered. It is useful, together with plots in Fig. 1, to characterize all possible dynamical behaviors presented by system (1), namely equilibrium, periodicity, quasiperiodicity, chaos, and hyperchaos. For instance, a two-torus region (quasiperiodic region), for which the two largest Lyapunov exponents are null and the third is negative, is concerned with a common region painted in black in plots of Fig. 1, and painted in blue-white in plot of Fig. 2.

In Fig. 3 one sees four two-dimensional projections of typical phase-space trajectories

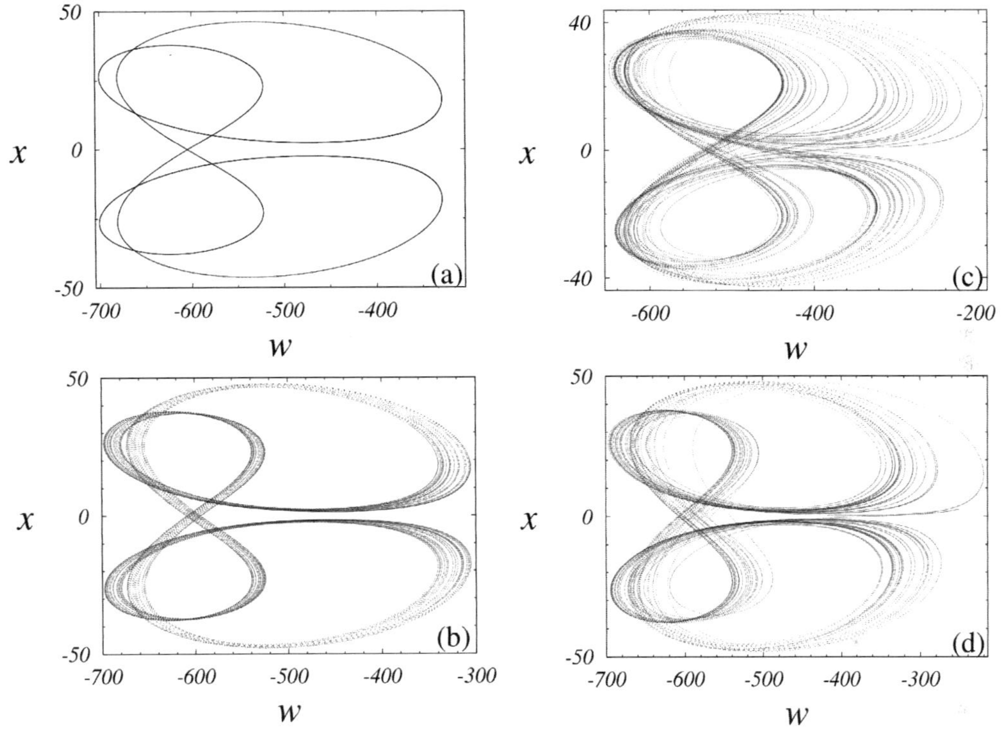

Figure 3. Phase-space trajectories of system (1) in $w-x$ plane, for different points in Figs. 1 and 2. (a) Point χ, inside a periodic region. (b) Point γ, inside a quasiperiodic region. (c) Point δ, inside a chaotic region. (d) Point β, inside a hyperchaotic region.

(attractors) in $w-x$ plane , which are result of numerical simulations for the four points, χ, γ, δ, and β, shown in the $a-r$ parameter-spaces of Figs. 1 and 2, and whose coordinates are (a,r)=(4.0,20.750), (a,r)=(3.675,22.78125), (a,r)=(4.5,17.625), and (a,r)=(3.625,23.09375), respectively. Shown are a cycle-3 periodic orbit in Fig. 3(a), a quasiperiodic orbit in Fig. 3(b), a chaotic orbit in Fig. 3(c), and a hyperchaotic orbit in Fig. 3(d). Our motivation to include these diagrams here is that they give us an opportunity to observe the structure of each type of orbit. All attractors were constructed with 20×10^3 points, and from the initial condition (x_0,y_0,z_0,w_0)=(10,10,10,10).

The complete Lyapunov exponents spectrum for points χ, γ, δ, and β is given in Table 1, where one sees that this results corroborate all said above.

Point	Attractor	λ_1	λ_2	λ_3	λ_4
χ	Periodic	0.017	-1.403	-1.562	-22.414
γ	Quasiperiodic	0.014	0.005	-0.446	-24.598
δ	Chaotic	1.848	0.012	-0.880	-27.185
β	Hyperchaotic	0.729	0.310	0.014	-25.961

Table 1. Lyapunov exponents spectrum for points χ, γ, δ, and β.

4. Summary

In summary, in this paper we have reported the usage of the second largest Lyapunov exponent value as a measure of hyperchaotic motion, to construct two-dimensional parameter-space color plots for a system modeled by a set of four nonlinear autonomous first order ordinary differential equations, namely a system constructed from the paradigmatic Lorenz system by introducing an adequate feedback controller to the third equation. The method consists in to associate colors to the magnitude of the above-mentioned Lyapunov exponent. More specifically, we have used here a continuously changing yellow-red scale to indicate a positive second largest Lyapunov exponent. It should be pointed out that is not necessary knowledge of the model equations. In the case of an experimental time series, for instance, the Lyapunov exponents spectrum can be estimated by applying techniques for nonlinear time series analysis and, in consequence, the parameter-space can be constructed in the same manner.

Acknowledgments

The authors thank Conselho Nacional de Desenvolvimento Científico e Tecnológico-CNPq and Coordenação de Aperfeiçoamento de Pessoal de Nível Superior-CAPES, Brazilian agencies, for the financial support.

[1] Rössler O E 1979 *Phys. Lett. A* **71** 155
[2] Matsumoto T, Chua L O and Kobayashi K 1986 *IEEE Trans. Circuits Syst.* **33** 1143
[3] Stoop R and Meier P F 1988 *J. Opt. Soc. Am. B* **5** 1037
[4] Stoop R, Peinke J, Parisi J, Röhricht B and Hübener R P 1989 *Physica D* **35** 425
[5] Perez G and Cerdeira H A 1995 *Phys. Rev. Lett.* **74** 1970
[6] Li Q D and Yang X S 2008 *Int. J. Circ. Theor. Appl.* **36** 19
[7] Takahashi Y, Nakano H and Saito T 2004 *IEEE Trans. Circuits Syst. II* **51** 468
[8] Li C D, Liao X F and Wong K W 2005 *Chaos Solit. Fract.* **23** 183
[9] Goedgebuer J -P, Levy P, Larger L, Chen C -C and Rhodes W T 2002 *IEEE J. Quantum Electron.* **38** 1178
[10] Vicente R, Dauden J, Colet P and Toral R 2005 *IEEE J. Quantum Electron.* **41** 541
[11] Zhang S H and Shen K 2003 *Chin. Phys.* **12** 149
[12] Cenys A, Tamasevicius A, Baziliauskas A, Krivickas R and Lindberg E 2003 *Chaos Solit. Fract.* **17** 349
[13] Yan Z 2005 *Appl. Math. Comput.* **168** 1239
[14] Zhou X, Wu Y, Li Y and Xue H 2008 *Appl. Math. Comput.* **203** 80
[15] Zhu C, 2010 *Appl. Math. Comput.* **216** 3126
[16] Zhu C 2009 *Appl. Math. Comput.* **215** 557
[17] Zheng S, Dong G and Bi Q 2010 *Appl. Math. Comput.* **215** 3192
[18] Wang C -N, Li S -R, Ma J and Jin W -Y 2010 *Appl. Math. Comput.* **216** 647
[19] Cai L, Ma X K and Wang S 2003 *Acta Phys. Sin.* **52** 3002
[20] Granemann E, Lopes S R and Viana R L 2008 *Phys. Rev. E* **78** 066206
[21] Anteneodo C, Batista A M and Viana R L 2006 *Physica D* **223** 270

A Discrete *SIRS* Model with Kicked Loss of Immunity and Infection Probability

F Paladini[1,2], I Renna[3,4] and L Renna[1,2,*]
[1] Dipartimento di Fisica dell'Università del Salento, 73100 Lecce, Italy
[2] Istituto Nazionale di Fisica Nucleare, Sezione di Lecce, 73100 Lecce, Italy
[3] ISIR, Université Pierre et Marie Curie/CNRS, F-75005 Paris, France
[4] IIT - Italian Institute of Technology, University of Genova, Italy

E-mail: luigi.renna@le.infn.it

Abstract. A discrete-time deterministic epidemic model is proposed with the aim of reproducing the behaviour observed in the incidence of real infectious diseases, such as oscillations and irregularities. For this purpose we introduce, in a naïve discrete-time *SIRS* model, seasonal variability in the loss of immunity and in the infection probability, modelled by sequences of kicks. Restrictive assumptions are made on the parameters of the models, in order to guarantee that the transitions are determined by true probabilities, so that comparisons with stochastic discrete-time previsions can be also provided. Numerical simulations show that the characteristics of real infectious diseases can be adequately modeled.

1. Introduction

The dynamics of epidemics are generally based on the *SEIR* or *SIR* models [1], in which the host population is partitioned into categories containing susceptible, infected but not yet infectious (exposed), infectious, and recovered individuals. These models have endemic asymptotically stable equilibria or the disease dies out. Because oscillations are observed in the incidence of many infectious diseases (such as measles, mumps, rubella and influenza), it is of interest to determine how oscillating solutions can arise in epidemiological models.

Usually, epidemic models are analysed in the field of continuous systems. In the deterministic framework, periodic solutions exist if the transmission rate is allowed to vary seasonally or spatial heterogeneity is included [2, 3]. It is also suggested that non linear incidence rate [4], temporary immunity (through which an individual's return to susceptibility can be delayed) [5], the introduction of a quarantine class [6], a constant removal rate of the infective individuals [7] are mechanisms that can induce oscillations in the occurrence of diseases.

Discrete models have also been used to study infectious disease dynamic. Discrete-time models are useful for several reasons. The discrete time-step approach is justified because, on one hand, it is an approximation to the continuous time case, while on the other hand, our every day life has a certain periodicity, e.g., seasonal changes. Moreover, it provides a substantial reduction of computational time, and the entire collection of tools developed for the study of maps can be advantageously used.

[*] To whom any correspondence should be addressed.

In discrete-time deterministic *SIRS* models periodic solutions are found when weak conditions on the force of infection are assumed [8], or the models parameters are not let to vary within realistic ranges of values [9], or unreasonable discrete-time steps are introduced [10]. In more realistic approaches, the interactions within the population are described by small world networks [11], and/or dynamics are described by delay difference equations [11,12]. In the small word networks approach [11], for a finite value of the disorder of the network (ranging from ordered lattices to random graphs), a transition from a stationary state (fixed point, with fluctuations) to self-sustained oscillations in the size of the infected population was found. In the time-delay approach [12], individuals spend a fixed amount of time in a generalized immune class; and oscillations occur for a specific ranges of immunity duration and infected individual exposure.
The conditions that lead to periodicity in the infectious disease was also investigated in reference [13] by a naïf discrete *SIRS* model, and it was found that the epidemic oscillates when a small fraction of individuals became not permanently immunised. In fact, the inclusion in a *SIR* model of a positive feedback, from the removed class to a susceptible one, in a very narrow range of the corresponding control parameter, has the effect of periodically enhancing the spread of disease. The smaller is the probability that a recovered becomes susceptible, generally the larger is the period of the oscillations in the infected population. This observation together with results in [11,12] suggest that oscillations can be the result of two factors: the manner in which the infection takes place, and residence time in the class of immunized.
In this paper we add to the discrete *SIRS* model of ref. [13] a seasonal variability, introduced by means of a sequence of kicks, which change periodically (*i*) the fraction of recovered individuals, which lose immunity in the sampling intervals, and/or (*ii*) the infection probability. Our goal is understanding and reproducing the oscillations of real diseases propagation. The simplicity of the model makes it possible to easily verify the correctness of above assumptions.
The outline of the paper is the following. We introduce the *SIRS* model in Sec. 2. In Sect. 3 seasonality in the loss of immunity parameter is introduced by means of a sequence of kicks. We discuss dynamical properties of the system when the infection probability is perturbed, in Sect. 4. Finally, we give conclusions in Sect. 5.

2. Epidemic *SIRS* model

We consider a population consisting of susceptible (S), infected (I) and recovered (R) individuals. As a consequence, the total population is

$$N = S + I + R. \tag{1}$$

Susceptible elements can pass to the infected state through contagion by an infected one. Infected elements can pass to the recovered state, and recovered elements can return to the susceptible state by loss of immunity. This kind of system is usually called *SIRS*, for the cycle that a single element goes over. The contagion is possible only during the S phase, and only by an I element. During the R phase, the elements are immune and not infectious.
SIRS models are excitable systems known to display relaxation oscillations in mean field or well-mixed approaches. In spatially extended versions space-time oscillations can occur, due to the interaction between neighbouring elements.
The basic discrete-time deterministic (*SIRS*) model has the form

$$\begin{aligned} S_{n+1} &= qS_n + cR_n \\ I_{n+1} &= (1-q)S_n + bI_n \\ R_{n+1} &= (1-c)R_n + (1-b)I_n \end{aligned} \tag{2}$$

where, during each sampling interval n, q denotes the probability that a susceptible avoids the infection, b is the proportion of individuals which remains infected ($0 \le b \le 1$), c ($0 \le c \le 1$) the fraction of recovered individuals, which lose immunity.

The probability q can be modelled in different ways [13]. The probability q is an arbitrary function $0 \leq q(S, I) \leq 1$ with the property $q(S, 0) = 1$. It depends on the particular form of propagation of the disease. Let us denote by p $(0 < p < 1)$ the probability of the infection effectively transmitting through a time of contact. We assume the population has a homogeneous spatial distribution and the probability per unit time of a contact between two individuals is the same for any pair. Then I_n /N (proportion of infected in a population) represents the probability that an infective individual is in contact with a susceptible, and $p\ I_n /N$ is the probability that a susceptible remains infected. As a consequence we obtain the following expression

$$q = 1 - pI / N \tag{3}$$

for the standard probability that a susceptible avoids the infection.

2.1. Periodic incidence of the disease

Let us consider the propagation of an infection on a population of constant size N in the *SIRS* model with the standard probability (3). When $c = 0$, the epidemic equations reduce to the *SIR* equations ones. So $c \neq 0$ was used throughout what follows.

By utilizing the relation (1) and inserting equation (3) in the system (2) one obtains

$$\begin{aligned} S_{n+1} &= (1 - pI_n / N)S_n + c(N - S_n - I_n) \\ I_{n+1} &= (pS_n / N + b)I_n \end{aligned} . \tag{4}$$

The system has two fixed-point:

$$S^* = N, I^* = 0, \tag{5}$$

$$S^* = N\frac{1-b}{p}, I^* = N\frac{c(p+b-1)}{p(1-b+c)}. \tag{6}$$

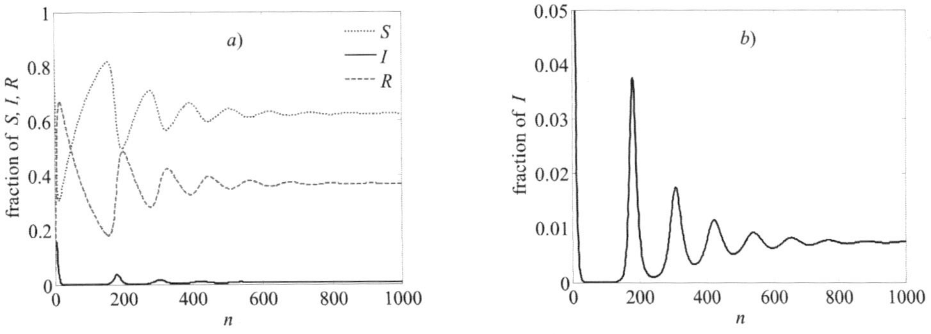

Figure 1. *a)* Fraction of susceptible, infected and recovered individuals as a function of time for $p = 0.8$, $b = 0.5$ and $c = 0.01$. An initial fraction of 0.1 infected, and the rest susceptible, was used. The system evolves through damped oscillations toward the endemic equilibrium (6). *b)* Magnification of infected population time series.

The fixed point (5) is stable if $p + b < 1$, otherwise it is unstable. Quite the opposite, the fixed point (6) is stable if $p + b > 1$, otherwise it is unstable. Thus, the fixed points change their stability on different sides of $p + b = 1$, and the system undergoes a transcritical bifurcation [13]. So, for $p + b > 1$ the dynamics correspond to that of an endemic infection, with a persistent fraction of infected individuals.

Typical time series are shown in figure 1. An initial fraction of 0.1 infected, and the rest susceptible, was used in all the results shown here.
In reference [13] it is shown that for each fixed pair of values p and b such that $p + b > 1$, there exists a value c_M of c such that oscillations are observed in the infected population, if we take $0 < c < c_M$. By reducing the value of $c < c_M$, periodic incidence of the disease is observed at higher values of time (figure 2).
These results highlight the importance of properly modelling the ''loss of immunity'' step, for observations of periodic epidemic outbreaks. A second issue is obviously the mode of infection spread between infected and healthy individuals.

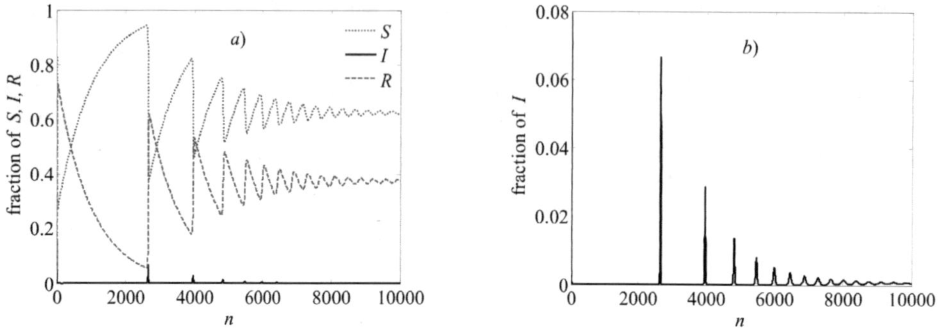

Figure 2. As in figure 1, but for $c = 0.001$.

We wish to insert some degree of variability both in the recovery parameter c and in the infection probability p. Epidemiological justifications can, for example, reside in the immunity memory response or pathogen mutation effects. As we are meanly interested in introducing some form of discontinuity in the dynamics, we merely add a perturbation in the form of a series of periodic kicks.

3. Perturbation of the recovery class

We evaluate now the hypothesis that the oscillations occur if the duration of immunity has a limited duration or if the fraction of recovered individuals suddenly change over time. In our model this can be easily simulated by perturbing the fraction of recovered individuals which lose immunity by a sequence of kicks with amplitude k and period t. We therefore assume that

$$c = c_0 + k\delta_{n,t}\,. \tag{7}$$

We start by considering

$$\delta_{n,t} = \begin{cases} 1 & \text{if } n/t = \text{integer} \\ 0 & \text{if not} \end{cases}\,. \tag{8}$$

Under these assumptions, the system's dynamics are quite different. Most notably, as the parameter k increases from 0, sustained oscillations emerge in time series. The plots of figure 3 was obtained by increasing c_0 of 0.3 one time step every 100 time steps. Calculations show also that at high values of k the amplitude of the oscillations becomes larger.
The time series is periodic, with an almost defined period (equal to t) and negligible variation in amplitude. Epidemiologically, this behaviour is qualitatively similar to the epidemic patterns typical of diseases such as influenza [14].
The mechanism, by which the oscillations emerge, is easily explained: while the fraction of the population that is infected is close to a steady state, the transition from immune to susceptible class of

a suitable number of individuals increases the possibility of infection, and consequently the size of epidemic raises. Then the system moves back toward the steady state and the process is repeated.

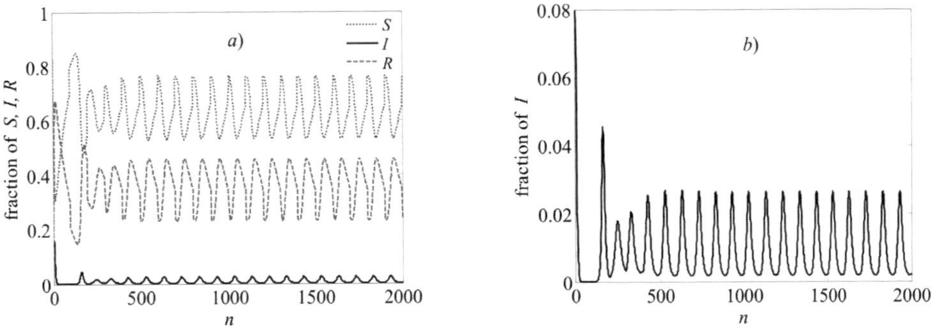

Figure 3. *a*) Fraction of susceptible, infected and recovered individuals as a function of time for $p = 0.8$, $b = 0.5$ and $c_0 = 0.01$. The fraction of recovered, which in each time-step are returned to the susceptible class, is perturbed by a sequence of kicks, as described by equation (8). Other parameters are $k = 0.3$, $t = 100$. Self-sustained oscillations develop. *b*) Magnification of infected population dynamic.

The shape of the oscillations depends on the values of c_0 and k, the period depends on t. In figure 4 *a*) simulations are shown for different values of these parameters. The number of infected individuals grows with c_0.

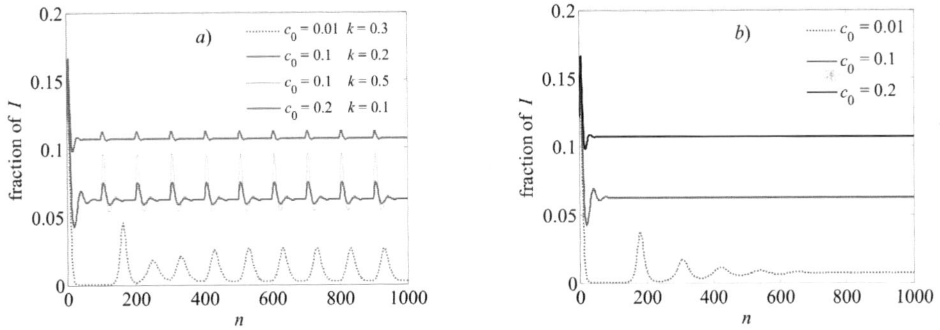

Figure 4. *a*) Fraction of infected individuals as a function of time for $p = 0.8$, $b = 0.5$, $t = 100$, and different values of c_0 and k. *b*) As in *a*), but with $k = 0$.

Increasing the value of c_0, the system returns faster to steady state corresponding to $k = 0$, so that the peaks in the size of the infected population appear separated in time.
While increasing the value of c_0 involves oscillations with peaks above the steady states, his decrease works in the opposite direction, and the oscillations are a result of a decrease in the number of infections compared to the value of the steady states, which are shown in figure 4 *b*), for a comparison. The k value determines the height of the peaks, t the oscillations period. Decreasing only t, a situation equal to that of figure 4 *a*) is restored, but with smaller oscillations periods.
The kicks can be continued for more time steps. In this case the kicking is modelled as follows:

$$\delta_{n,t,s} = \begin{cases} 1 & \text{for } n/t, n/t+1, \ldots, n/t+s-1 \; (n/t, s = \text{integer}) \\ 0 & \text{otherwise} \end{cases} \tag{9}$$

where the parameter s is the number of consecutive times that a kick works (duration of kicks). For s = 1 equation (9) becomes equivalent to equation (8) (1 kick every t).

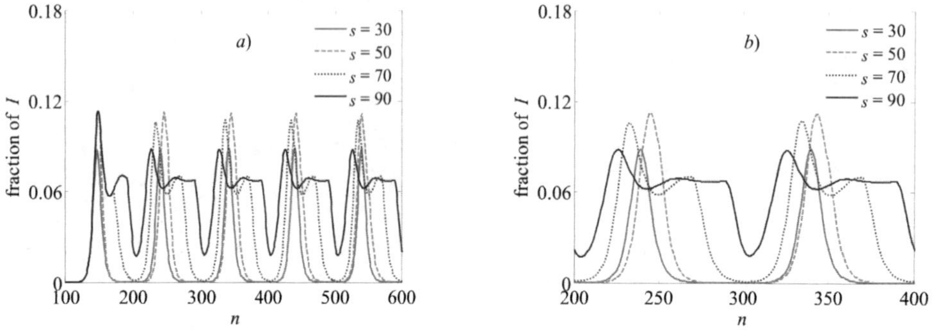

Figure 5. *a*) The dynamics of infection as a function of the number of time steps, for different values of s, when formula (9) is used for kicking. Parameters are: p = 0.8, b = 0.5, t = 100, c_0 = 0.01, k = 0.1. *b*) An enlargement.

Figures 5 illustrate the effect of kicking during s successive time steps. The dynamics of infection as a function of the number of kicks s are the following: by increasing s also the time interval of persistence of the infection increases; starting from about $s = t/2$, increasing still s a secondary minimum occurs immediately after the primary one and the oscillations can have a plateau, which stabilizes around the steady state corresponding the value $c = c_0 + k$.

4. Perturbation of the probability of infection

We now address the second hypothesis that links the occurrence of oscillations to changes in the probability of infection. For this purpose we introduce perturbations similar to the previous, but this time referring to p:

$$p = p_0 + k\delta_{n,t,s} \tag{10}$$

where $\delta_{n,t,s}$ is given by (9).

Numerical calculations with s = 1 show that the discontinuity introduced on the infection probability also produces periodic oscillations of period t (figure 6). The fraction of infected has a peak in correspondence of each kick, (i.e. when $\delta_{n,t,s} = 1$). Besides, the average number of infected elements creases with c.

At constant values of c, variations in the value of the parameter s produce two main effects: a change of peaks amplitude and a shifting or a frequency variation of the time series between the peaks. An example of this behaviour is reported in figure 7. From the epidemiological point of view, different infection patterns can be obtained qualitatively similar to the real ones.

Therefore, the duration s of the kick is an important factor, affecting the structure of the temporal evolution of the epidemic. Indeed, reversing symmetrically the duration of the kicks/no kicks time steps (with p_0 = 1.0, k = –0.2), you get again, apart from an initial phase, the pattern of figure 7.

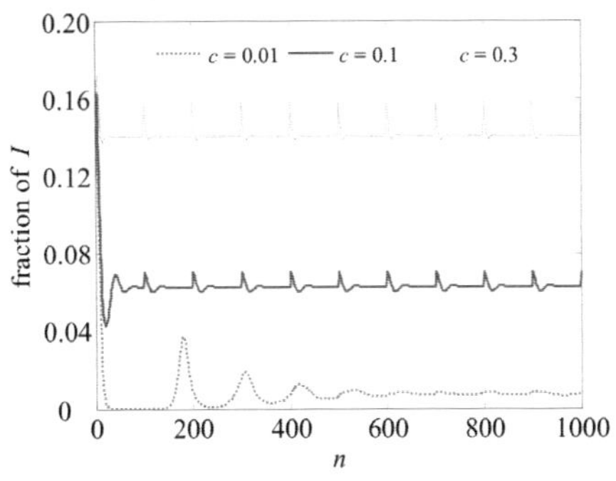

Figure 6. The dynamics of infection as a function of the number of time steps, at different values of c, when formulas (9) and (10) are used. Parameters are: $p_0 = 0.8$, $b = 0.5$, $s = 1$, $k = 0.2$, $t = 100$.

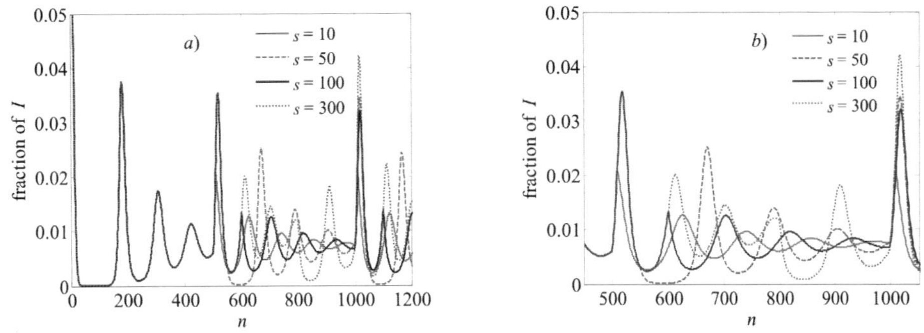

Figure 7. *a*) Time series of infected individuals for different values of s. Other parameters are as in the previous figure, except $c = 0.01$, $t = 500$. *b*) An enlargement.

The relevant fact that emerges from the preceding discussion is that the basic phenomenon that underlies on the presence of oscillations in the dynamics is the discontinuity in p. This is also confirmed by the result shown in the figure 8, where for the discontinuity the value $p = 0$ was used.

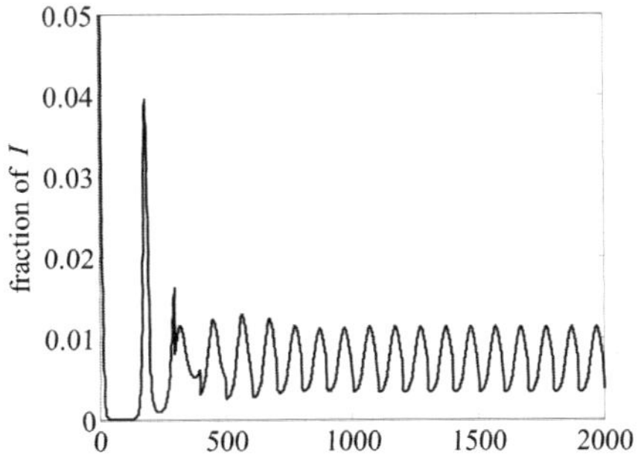

Figure 8. The infection dynamic as a function of the number of time steps. Parameters are $p_0 = 0.8$, $b = 0.5$, and 1 kick of $p = 0$ every 100 is used.

5. Conclusions

To understand the evolution of many infectious disease systems, we introduced the effects of seasonality in a discrete *SIRS* model. We have shown that the introduction of seasonal forcing in loss of immunity and in transmission rate can well explain the periodicity observed in the outbreaks of diseases.

We have quantified the effect of contact rate and immunity duration on oscillatory behaviour by modelling their variability by means of sequences of kicks, acting on the corresponding model parameters.

Epidemiologically, oscillations in the number of infected individuals in a population could be due essentially or to a mutating pathogen or to a seasonal enhancement of the infection probability. Model behaviors are qualitatively similar to the epidemic patterns typical of diseases such as influenza [14]. Moreover, they emphasize the importance of the interplay between seasonal infection and recruitment and suggest that our simulation patterns are meaningful in different epidemic scenarios.

6. References

[1] Anderson R M and May R M 1991 Infectious Diseases of humans: Dynamics and control *Oxford University Press.*

[2] Lloyd A L and May R M 1996 *J. Theor. Biol.* **179** 1.

[3] Kamo M and Sasaki A 2002 *Physica D* **165** 228.

[4] Liu W M, Hethcote H W and Levin S A 1987 *J. Math. Biol.* **25** 359.

[5] Hethcote H W, Stech H W and van den Driessche P 1981 Periodicity and stability in epidemic models: a survey, in: Busenberg S, Cooke K L (Eds.) differential equations and applications in Ecology, Epidemics and Populations Problems *Academic Press* 65-82.

[6] Hethcote H W, Zhien M A and Shengbing L 2002 *Math. Biosc.* **180** 141.

[7] Wang W and Ruan S 2004 *J. Math. Anal. Appl.* **291**.

[8] Allen L J S 1994 *Math. Biosci.* **124** 83.

[9] Ramani A, Cartea A S, Willox R and Grammaticos B 2004 *Physica A* **333** 278.

[10] Jansen H and Twizell E H 2002 *Math. And Comp. in Simulation* **58** 147.

[11] Kuperman M and Abramson G 2001 *Phys. Rev.Lett.* **86** 2909.

[12] Girvan M, Callaway D S, Newman M E J and Strogatz S H 2002 *Phys. Rev. E* **65** 031915-1.

[13] D'Innocenzo A, Paladini F and Renna L 2006 *Physica A* **364** 497.

[14] Quenel P and Dab W 1998 *Eur. J. Epidemiol.* **14** 275.

Dynamic Days South America 2010 IOP Publishing
Journal of Physics: Conference Series **285** (2011) 012019 doi:10.1088/1742-6596/285/1/012019

Cryptography with chaos using Chua's system

C H Oliveira [1] and J C Pizolato Jr [2]

[1] State University of Londrina (UEL-PR), Brazil
[2] Federal University of São Carlos (UFSCAR-SP), Brazil

E-mail: cleberho@hotmail.com

Abstract: In the last years, chaotic systems have been applied in information security. These systems have a complex and unpredictable behavior, what makes them more attractive for data cryptography applications. In this work, the chaotic behavior of signals generated by Chua's system is combined with the original information in order to obtain a safe cryptographic method. The experimental results demonstrate that the proposed scheme can be used in data cryptography applications.

[1] cleberho@hotmail.com
[2] jcpizolato@yahoo.com.br

1. Introduction

Experts in nonlinear dynamics use the word "chaos" as a technical term to refer to the irregularity and unpredictable behaviour of nonlinear deterministic system [1-2]. These systems, although having only one or two- degrees-of-freedom, can display an amazing random behaviour, from periodic up to chaotic.

For thereabout two decades, researchers have been studying the chaotic system characteristics and their use in information cryptography. In Rong He and P.G. Vaidya's work [3], the chaotic synchronization system is used in the cryptography key generation. The modified Chua's circuit was used to send audio and image information in a chaotic synchronization scheme [4]. In these two ultimate results published [3-4], there is a requirement of systems synchronization between transmitter and receiver. Nevertheless, there are proposals for cryptographic systems with no requirement of synchronization between transmitter and receiver. The algorithm proposed by Batista [5] suggests a cryptographic system based on a unidimensional chaotic system, the logistic map [1]. In it, the chaotic behavior is used for data ciphering.

This paper proposes a new algorithm of data cryptography based on the contributions from the cryptographic method presented in [3], in which the algorithm utilizes the Chua's system in order to cause a chaotic behaviour. The proposed chaotic cipher uses chaotic signals with no requirement of systems' synchronization between transmitter and receiver. Results reveal that this algorithm does not demand high capacity of computational processing.

This paper will be presented as it follows. The mathematical model of Chua's system is analyzed in section 2. The algorithm proposed is presented in section 3. Section 4 presents qualitative results through the application of the proposed cryptographic process on two pieces of information, being the first a sort of image information and the second a sort of text information. Furthermore, it displays the frequency analysis on a sort of text information. In section 5, some considerations are made on the algorithm proposed. The conclusions concerning the cryptography scheme proposed are in section 6

2. The Chua's system

The mathematical model of Chua's system [4], is described by the system of state equations that follows:

$$C_1 \frac{dvc_1}{dt} = \frac{1}{R}(vc_2 - vc_1) - g_{NR}(vc_1) \tag{1}$$

$$C_2 \frac{dvc_2}{dt} = \frac{1}{R}(vc_1) + i_L \tag{2}$$

$$L\frac{di_L}{dt} = -vc_2 \tag{3}$$

Where: v_{c1} e v_{c2} correspond to the tensions through the capacitors C_1 and C_2 respectively, i_L is the current whereby the inductor L and $g_{NR}(vc_1)$ represent a nonlinear function which defines the characteristics $v-i$ of Chua's diode, mathematically described in (4).

$$g_{NR}(v_{c1}) = G_b v_{c1} + \frac{1}{2}(G_a - G_b)(|v_{c1} + E| - |v_{c1} - E|) \tag{4}$$

The application of the transformations ($x_1 = v_{c1}/E, x_2 = v_{c2}/E, x_3 = i_L/EG, \tau = Gt/C_2$ where $E = 1\text{V}$ and $G = 1/R$), it is possible to get the system of state equations of Chua's system normalized, as follows:

$$\dot{x}_1 = \alpha(x_2 - f(x_1)) \quad (5)$$

$$\dot{x}_2 = x_1 - x_2 + x_3 \quad (6)$$

$$\dot{x}_3 = -\beta x_2 \quad (7)$$

Where: $\dot{x} = dx/d\tau, \alpha = C_2/C_1 > 0$ and $\beta = C_2/G^2 L > 0$ are the system bifurcation parameters. The nonlinear function $f(x_1)$ is mathematically described by (8):

$$f(x_1) = bx_1 + \frac{1}{2}(a-b)(|x_1 + c| - |x_1 - c|) \quad (8)$$

Where: $a = G_a/G < 0$ e $b = G_b/G < 0$.

The parameters of the system ($b = -\frac{1}{7}, a = \frac{2}{7}, \alpha = 9, \beta = 14{,}28$ and $c = 1$), were adjusted to allow the generation of the chaotic attractor in Chua's system [6-7].

3. The proposed algorithm

The original message is called plain text, the unreadable message is called ciphered and the key is called k. The process of turning a plaintext into a ciphered text is called ciphering and the inverse process is called deciphering. Since only one plaintext is ciphered at a time, the proposed algorithm is a flow ciphering system [8]. The algorithm assumes that for each unit of plaintext, there must be a key k to make it unrecognizable. In this algorithm each key k is used only once in the encoding process.

The key generator k, uses a non-periodic orbit, in which $f_p : I \to I$, through Chua's chaotic system behaviour, with the interval I and parameter P [6-7]. The interval I is divided in N sub-intervals $Ik, (k = 1,2,3,..., N)$ used to generate keys k, defined by $x_1[t], x_2[t]$ e $x_3[t]$.

The ciphering process consists of iterating f_p consecutively, starting from the initial condition (key) $x_0 \in I$, computing the number of iterations and associating each site Ik for a plaintext unit ciphering p, by using $c = (p+k)\text{mod}(256)$. The c_j deciphering is made by using the same key k, that is, associating the same sub-interval Ik with the text unit ciphered c_j, using $p = c - (k\,\text{mod}(256))$. These processes use an alphabet compounded of symbols based on the ASCII table in which the equivalence used is $0,1,...,255$.

For both the plain text p and the ciphered text $c \in \{0,1,...,255\}$. In order to cipher a plaintext unit, $c = (p+k)\,\text{mod}\,256$ is used and to decipher a ciphertext unit $p = c - (k\,\text{mod}(256))$. This way, this method is especially characterized by the simplicity of the algebraic functions involved. Security is increased through the generation of the keys k. where each key is used only once either to cipher a plaintext unit or to decipher a ciphertext unit.

3.1. Key generator

In the proposed algorithm, the key generator k for ciphering data uses the Chua's chaotic system characteristics, presented in the reference [4]. The key k, is generated from each sub-interval Ik of $x_1[t], x_2[t]$ and $x_3[t]$ in which, $k = x_1[t] + x_2[t] + x_3[t]$. Through the algebraic functions involved, a complex and unpredictable behavior is obtained. Therefore, the normalized version of Chua's system is given by the equations (9-11):

$$\dot{x}_1 = \alpha(x_2 - f(x_1)) \quad (9)$$

$$\dot{x}_2 = x_1 - x_2 + x_3 \quad (10)$$

$$\dot{x}_3 = -\beta x_2 \quad (11)$$

and by the nonlinear function (12):

$$f(x_1) = bx_1 + \frac{1}{2}(a - b)(|x_1 + c| - |x_1 - c|) \quad (12)$$

However, the parameters presented in [6-7] $b = -\frac{1}{7}, a = \frac{2}{7}, \alpha = 9, \beta = 14{,}28$ are used, and $c = 1$, in which the integration algorithm used is Euler.

Figure (1) shows a three-dimensional chaotic system $\dot{x}_1, \dot{x}_2$ and $\dot{x}_3$, which represents and attractor generated by Chua's chaotic system.

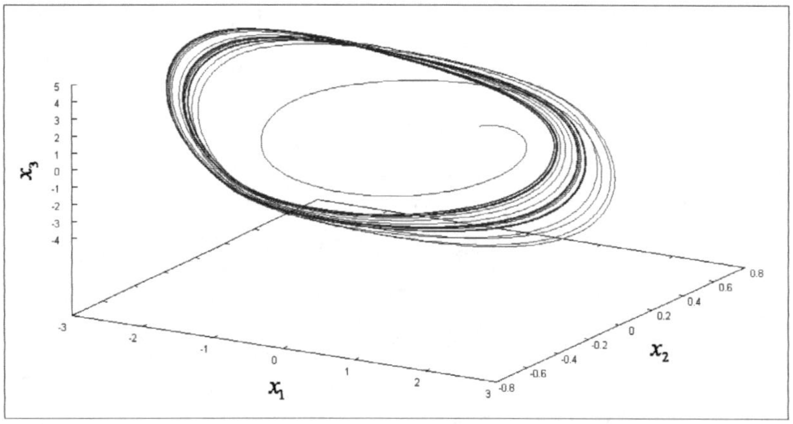

Figure 1. Attractor generated by Chua's chaotic system simulation with parameters [6-7] ($b = -\frac{1}{7}, a = \frac{2}{7}, \alpha = 9, \beta = 14{,}28$ and $c = 1$) and initial conditions ($x_1[0] = 0.2, x_2[0] = 0.6$ and $x_3[0] = 0.7001$).

Nevertheless, the attractor generated based on Chua's chaotic system, used to generate the keys k, which are defined by $x_1[t], x_2[t]$ and $x_3[t]$ are dependent on the initial conditions. Each *key* k is used only once, that is, exclusively by one text unit. Yet, for $x_1[t], x_2[t]$ e $x_3[t]$ it is determined the I interval, which becomes dependent on the amount of text units of the information. Thus, the sub-intervals Ik, of $x_1[t], x_2[t]$ and $x_3[t]$ are used to create the keys $\{k_1, k_2, k_3, ..., k_n\}$.

Figure (2) shows a key generator, which uses chaotic signals with no requirement of system synchronization between transmitter and receiver.

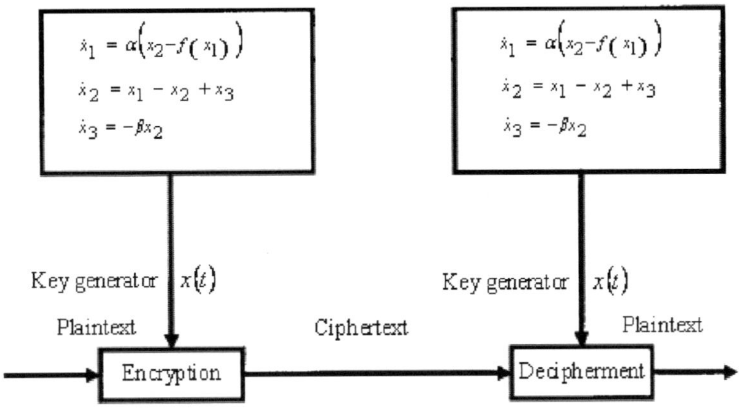

Figure 2. Ciphering / deciphering process by using Chua's chaotic system.

Based on the behavior of Chua's chaotic system, we will bring forward two examples of key generation k, used in figure (3). From the sub-interval Ik of $x_1[t], x_2[t]$ and $x_3[t]$ in this case, $t = 1.0$ we will have $x_1[t] = 0.610429$, $x_2[t] = 0.231888$ and $x_3[t] = 0.609670$, with $t = 2.0$ we will have $x_1[t] = 0.615602$, $x_2 = 0.241770$ and $x_3[t] = 0.576557$.

Therefore, we use n decimal places of the sub-interval Ik to generate a whole part, as the example, $t = 1.0$ we will have $x_1[t] = 0.610429$, $x_2[t] = 0.231888$ and $x_3[t] = 0.609670$, $t = 2.0$ we will have $x_1[t] = 0.615602$, $x_2 = 0.241770$ and $x_3[t] = 0.576557$.

The keys k are generated from $x_1[t], x_2[t]$ and $x_3[t]$ of the sub-interval Ik, in which, $k = x_1[t] + x_2[t] + x_3[t]$, for instance, $t = 1.0$ so $k = 610429 + 231888 + 609670$ however, $k = 1451987$ and $t = 2.0$, so $k = 615602 + 241770 + 576557$, however, $k = 1433929$.

3.2. Cryptographic model proposed

Based on the algorithm proposed, we present the ciphering of two plaintext units used in figure (3), where the characters "C" and "H" are represented respectively in the ASCII table by the numbers: "67" and "72". However, it is used the cryptographic method $c = (p + k)\bmod(256)$, in which, $(67 + 1451987)\bmod(256)$ and $(72 + 1433929)\bmod(256)$ so the ciphertext units referring to "C" and "H" become "22" and "145", following the ASCII table.

Figure (3) shows the ciphering process of a piece of information in which each key k is used only once to cipher a plaintext unit.

Time t	$x[t]$ $x_1[t]+x_2[t]+x_3[t]$	Keys k	Plaintext p	Ciphertext $c = (c+k)\bmod 256$
0.1	1.451987	1451987	C (67)	22
0.2	1.433929	1433929	H (72)	145
0.3	1.414838	1414838	A (65)	247
0.4	1.394711	1394711	O (79)	102
0.5	1.373548	1373548	S (83)	191

Figure 3. Ciphering process using the cryptographic model proposed.

When deciphering, the process is inverse. However, we present the deciphering of two ciphered text units "22" and "145", used in figure (4). Thus, the cryptographic method is used, $p = c - (k \bmod(256))$, in which, "$22 - (1451987 \bmod(256))$" and "$45 - (1433929 \bmod(256))$," if resulting of $p < 0$, so, $p + 256$," but the plaintext units referring to "22" and "145", become "67 = C" and "72 =H".

Figure (4) shows the deciphering process of a piece of information in which each key k is used only once to decipher a ciphered text unit.

Time t	$x[t]$ $x_1[t]+x_2[t]+x_3[t]$	Keys k	Ciphertext c	Plaintext $p = (c - (k \bmod 256))$ If $p < 0$, so $p+256$
0.1	1.451987	1451987	22	C (67)
0.2	1.433929	1433929	145	H (72)
0.3	1.414838	1414838	247	A (65)
0.4	1.394711	1394711	102	O (79)
0.5	1.373548	1373548	191	S (83)

Figure 4. Deciphering process using the cryptographic model proposed.

4. Results

This section presents qualitative results through the application of the proposed cryptographic process on two pieces of information, being the first a sort of image information and the second a sort of text information. Furthermore, it displays the frequency analysis on a sort of text information.

4.1. Appliance of the cryptographic process proposed in a sort of image information

The cryptographic method proposed was applied in a piece of image information (512 x 512 pixels) [9], as shown in figure (5).

In this section, a ciphering /deciphering process is displayed, in which the parameters defined in [6-7] and $x_1[0] = 0.1, x_2[0] = 0.2$ and $x_3[0] = 0.3$ are used to generate the keys k.

Figure 5. Cryptographic process. (a) Original image (512 x 512 pixels) [9]. (b) Ciphered image (512 x 512 pixels). (c) Deciphered image (512 x 512 pixels).

4.2. Appliance of the cryptographic process proposed in a sort of text information

Figure (6) verifies the algorithm security by showing a comparison between the sort of text information composed by information units p_j and its ciphered version c_j. The ciphering process utilizes the system of equations (9-12), with initial conditions ($x[0] = 0.2, 0.3$ and 0.1) to generate the keys k.

Concerning the figure (6), "Y – ASCII (Decimal)" represents the information units p_j and their ciphered versions c_j in decimal format (0,1,...,255), according to the ASCII table. In addition, "X – Character positions" represents the positions of the information units p_j and their ciphered version c_j respectively in positions (1,2,...,50). Therefore, for each information unit p, there is a ciphered version c as illustrated in figure (6).

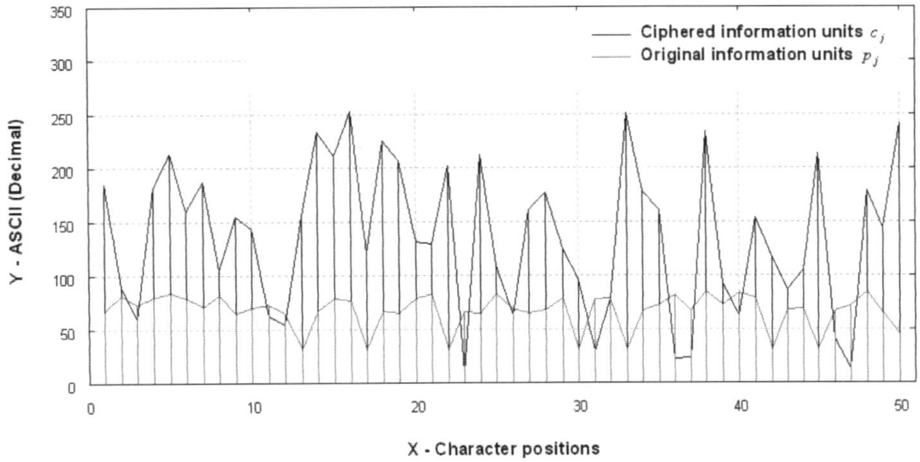

Figure 6. Graphic of a piece of information composed by information units and their ciphered version.

4.3. System security analysis by using frequency analysis

This section analyses the frequency of the ciphered text units in order to assess the algorithm security against attacks of analysis to the frequency units that emerge in ciphertexts c_j and then associate plaintext units p_j. Such test verifies a piece of information composed of 18.016 units of plaintexts p_j with 17,5 KB, in which 4.742 is the highest frequency of a plaintext unit p and 0 is the lowest frequency, in which 4.742 is the frequency difference between the plaintext units p_j.

Figure (7) shows the analysis of the information frequency composed of plaintext units p_j. Nevertheless, "Y-Frequency" displays the frequency of each plaintext unit *p*. "X-ASCII (Decimal)" represents the plaintext units p_j, in decimal format, following the ASCII table $p = \{0,...,255\}$.

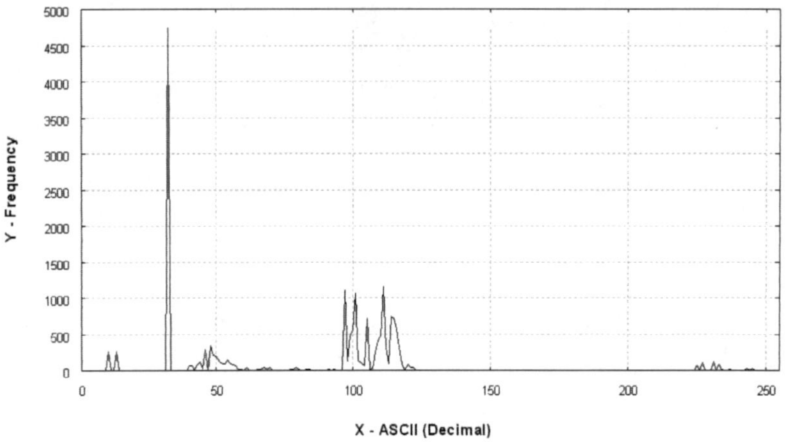

Figure 7. Illustration of the frequency analysis of a piece of information composed of plaintext units p_j, in decimal format, according to the ASCII table.

Figure (8) shows the frequency analysis of a piece of information composed of 18.016 ciphered text units c_j, with17,5 KB, referring to the information displayed in figure (7). However, "Y-Frequency" exhibits the frequency of every ciphered text unit *c*. "X-ASCII (Decimal)" represents the ciphered text units c_j in decimal format, according to the ASCII table $p = \{0,...,255\}$.

However, the ciphering was achieved through the proposed algorithm, by using the Chua's system behavior to generate keys k, with parameters defined in [6-7], and $x[0] = 0.2, 0.3$ and 0.1. Since one plaintext unit p is ciphered at a time, the proposed algorithm is assorted as a flow ciphering system.

The algorithm proposed is classified as a flow ciphering system. But, for every plaintext unit *p*, there is a ciphertext unit *c*, that is, the ciphered information will have the same size as the original one. Thus, after the ciphering process, we got 94 to the higher frequency of a ciphertext unit c and 47 the lowest frequency, being 47 the frequency difference between the ciphertext units c_j.

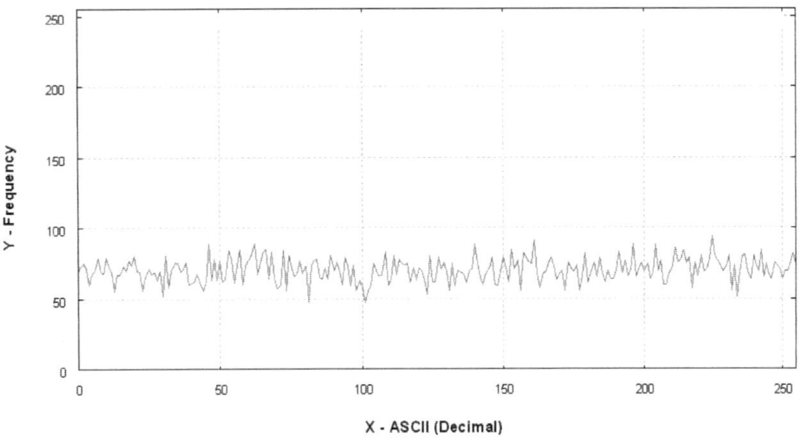

Figure 8. Illustration of the frequency analysis of a piece of information consisting of ciphertext units c_j, in decimal format, according to the ASCII table.

5. Discussion

The proposed algorithm does not require any synchronism between transmitter and receiver in the ciphering/deciphering process. Such cryptographic process does not depend on the alphabet, once it follows the ASCII chart, what implicates in higher levels of security and versatility in several applications.

6. Conclusions

This paper presents a new cryptographic algorithm by using the chaotic behavior of Chua's system to generate cryptographic keys. The proposed algorithm is grounded on the algebraic functions involved and on the complex and unpredictable behavior of the signals used in the cryptographic system. The results obtained and illustrated in section 4 reveal that this algorithm can be used for data cryptography.

Acknowledgments

The authors thank the Department of Electrical Engineering of Londrina State University and the Department of Electrical Engineering of São Carlos Federal University for their support.

References

[1] Kadanoff L P 1983 Roads to Chaos *Physics Today* p.46
[2] Gleick J 1990 Caos: A construção de uma nova ciência *Editora Campus*
[3] Rong He and P G Vaidya 1998 Implementation of chaotic cryptography with chaotic synchronization *Physical Review E.* volume 57 number 2 p. 1532-1535.
[4] G Guzman, C Cruz-Hernández, R M López-Gutierrez and E E Garcia-Guerrero 2008 Synchronization of Chua´s circuits with multi – scroll attractors: Applications to communications *Commun NonLinear Sci Numer Simulat* p. 2765-2775
[5] M S Baptista 1998 Cryptography with chaos *Physics Letters A.* no. 240 p. 50-54
[6] Madan RN 1993 Chua's circuit: A paradigm for chaos *Singapore: World Scientific*

[7] Chua LO, Komuro M and Matsumoto T 1986 The double scroll family *IEEE Trans Circ Syst I* 33(11):1072-118
[8] Singh S 2008 O livro dos códigos *Editora Record* 7ª Edição
[9] Basso E W Filtros Vetoriais para a Remoção de Ruído em Imagens Coloridas Disponível em: < www.inf.ufrgs.br/~ewbasso/works/fvrric_paper.pdf> Acesso em: 21 mar. 2011.
[10] L P L Oliveira and M Sobottka 2008 Cryptography with chaotic mixing *Chaos Solitons and Fractals*
[11] L M Pecora and Carroll T L 1990 Synchronization in chaotic systems *Phys Rev Lett.*64:821-824
[12] A Holokx A, Rech P C 2009 A parameter-space of a Chua system with a smooth nonlinearity *International Journal of Bifurcation and Chaos in Applied Sciences and Engineering* v. 19 p. 1351-1355

Dynamic Days South America 2010 IOP Publishing
Journal of Physics: Conference Series **285** (2011) 012020 doi:10.1088/1742-6596/285/1/012020

Asymptotic approach for the nonlinear equatorial long wave interactions

Enver Ramirez Gutierrez[1,4], **Pedro L Silva Dias**[1,2], **Carlos Raupp**[3,1]

[1]Department of Atmospheric Sciences, Institute of Astronomy, Geophysics and Atmospheric Sciences, University of São Paulo, Rua do Matão, 1226 - Cidade Universitária - Sao Paulo, 05508-090, Brazil.
[2]National Laboratory for Scientific Computing, Av. Getúlio Vargas, 333, Quitandinha, Petrôpolis, Rio de Janeiro, 25651-075 Brazil.
[3]Institute for Theoretical Physics, Sao Paulo State University, Rua Dr. Bento Teobaldo Ferraz, 371 Bloco II Barra Funda - Sao Paulo, 01140-070, Brazil.
[4]Earth System Science Center, Cachoeira Paulista, Rodovia Presidente Dutra, Km 40 - Sao Paulo, 12630-000, Brazil.

E-mail: `enver.ramirez@gmail.com`

Abstract. In the present work we use an asymptotic approach to obtain the long wave equations. The shallow water equation is put as a function of an external parameter that is a measure of both the spatial scales anisotropy and the fast to slow time ratio. The values given to the external parameters are consistent with those computed using typical values of the perturbations in tropical dynamics. Asymptotically, the model converge toward the long wave model. Thus, it is possible to go toward the long wave approximation through intermediate realizable states. With this approach, the resonant nonlinear wave interactions are studied. To simplify, the reduced dynamics of a single resonant triad is used for some selected equatorial trios. It was verified by both theoretical and numerical results that the nonlinear energy exchange period increases smoothly as we move toward the long wave approach. The magnitude of the energy exchanges is also modified, but in this case depends on the particular triad used and also on the initial energy partition among the triad components. Some implications of the results for the tropical dynamics are disccussed. In particular, we discuss the implications of the results for El Niño and the Madden-Julian in connection with other scales of time and spatial variability.

1. Introduction

It is common in theoretical works on large-scale tropical dynamics the use of the long wave approximation for studies related to the El Niño and/or the Madden-Julian oscillation ([1, 2]). This approximation filters out inertia-gravity waves of all wavelengths while being accurated for Kelvin and long Rossby waves. However, the long wave approximation is completely inaccurate for short Rossby waves ([10]). Furthermore, the completeness of the remaining waves has not been proven and consequently, the representation of any phenomenon on the basis of only these solutions is questionable. In the present work, an asymptotic method is used to obtain the long wave approximation as a limiting case of the shallow water equation. The difference with previous works is that the transformation of one regime to another is controlled by an external parameter ($\delta \in [0, 1]$) which is a measure of the anisotropy of the space and time scalings. The $\delta \to 1$ limit corresponds to the shallow water equation, whereas $\delta \to 0$ refers to the long wave approximation. Any value in between corresponds to intermediate realizable states. The model equations developed here are valid for both limiting cases and also for the intermediate states. The advantage of this method is that it allows a continuous approach to the long wave equations while keeping the completeness of the solutions of the shallow water equations ([11]). With this approach we have studied the resonant nonlinear wave energy exchanges in the long wave like model. We have found an effective modification by the δ parameter of the conservation laws in both the physical and spectral space. The nonlinear coupling coefficients , which is valid for resonant and off-resonant interactions also depend on δ. It was also verified by both theoretical and numerical results that the period of the nonlinear energy modulation increases as δ decreases. The magnitude of the energy exchange is also modified as a function of δ.

In section 2 the basic model equations are presented together with the anisotropic scaling procedure. Section 3 deals with the conservation of energy and potential vorticity for the anisotropic model. In section 4 the spectral representation of the model equations using the eigensolutions of the linear problem as expansion basis is presented. Section 5 deals with the nonlinear evolution equation of the spectral coefficients and also with the modification of both the nonlinear coupling and nonlinear interaction coefficient as a function of the anisotropy parameter. In section 6 numerical experiments of the reduced dynamics of a single resonant triad are presented for some selected equatorial trios. In section 7 the nonlinear resonant three wave interactions are discussed and how these are modified by the anisotropy parameter. Finally, the main findings of the work are summarized in the Conclusions section.

2. Governing equations

We start with the unforced, inviscid and nonlinear shallow water equations in the equatorial β plane in dimensional form (equation 1)

$$\partial_t u + \mathbf{v}\cdot\nabla u - \beta y v + g\partial_x H = 0 \quad (1a)$$
$$\partial_t v + \mathbf{v}\cdot\nabla v + \beta y u + g\partial_y H = 0 \quad (1b)$$
$$\partial_t H + \mathbf{v}\cdot\nabla H + H\nabla\cdot\mathbf{v} = 0 \quad (1c)$$

In equation (1), $H = \bar{H} + \eta$, $\bar{H}$ is the mean thickness of the fluid layer, η represents its perturbation and βy is the equatorial Coriolis parameter. The equations can be nondimensionalized by taking units of length and time in the range presented in equation (2).

$$L = (C/\beta)^{1/2} \approx [500, 1500]\text{km} \quad (2a)$$
$$T = \frac{1}{(C\beta)^{1/2}} \approx [8, 18]\text{hrs} \quad (2b)$$

Where $C = \sqrt{g\bar{H}} \in [10, 50]$m/s is the atmospheric baroclinic wave speed. The anisotropically scaled and nondimensionalized (δ dependant) variables are obtained as follows:

$$x = (L/\delta)x''; y = Ly''; t = (T/\delta)t'' \quad (3a)$$
$$u = Cu''; v = \delta C v''; H = (C^2/g)H''; \eta = \eta'' \quad (3b)$$

We restrict the analysis to $0 < \delta < 1$. Therefore, x'' and t'' are large scale and slow time, respectively, whereas v'' is of small magnitude. With the substitution $(x, y, t, u, v, \eta)'' \rightarrow (x, y, t, u, v, \eta)$ equations (4) are obtained.

$$\partial_t u + u\partial_x u + v\partial_y u - yv + \partial_x \eta = 0 \quad (4a)$$
$$\delta^2[\partial_t v + u\partial_x v + v\partial_y v] + yu + \partial_y \eta = 0 \quad (4b)$$
$$\partial_t \eta + u\partial_x \eta + v\partial_y \eta + (1+\eta)(\partial_x u + \partial_y v) = 0 \quad (4c)$$

3. Conserved quantities

In what follows we will show that conserved quantities for equations (4) can also be expressed as a function of δ. The energy conservation is:

$$\partial_t E + \text{div}((\eta + \frac{1}{2}\eta^2 + E)\vec{v}) = 0 \tag{5}$$

where

$$E = E(u, v, \eta; \delta) = \frac{1}{2}[(1+\eta)(u^2 + \delta^2 v^2) + \eta^2] \tag{6}$$

and the potential vorticity conservation law can also be obtained:

$$\partial_t(\frac{y + \delta^2 \partial_x v - \partial_y u}{1+\eta}) + \vec{v} \cdot \nabla(\frac{y + \delta^2 \partial_x v - \partial_y u}{1+\eta}) = 0 \tag{7}$$

In the present study we will focous on the energy conservation as it is quadratic to the lowest order. Thus, its consequenes for the nonlinear wave interactions are stronger than the imposed by the potential vorticity conservation law (see [13]).

4. Spectral representation

With $\phi = (u, v, \eta)^T$ the modal representation of the dependent variables and the spectral coefficients are:

$$\phi(x, y, t) = \sum_a \phi_a(x, y) Z_a(t); \tag{8a}$$

$$Z_a(t) = \int \phi_a^+(x, y)\phi(x, y, t) dx dy \tag{8b}$$

where $\phi_a(x, y)$ are the eigenfunctions of the linear operator $\mathcal{L}$ defined below, and $\phi_a^+(x, y)$ is the transposed-conjugated of ϕ_a

$$(\mathcal{L} - \hat{\imath}\omega_a I)\phi_a(x, y) = 0 \tag{9}$$

The asymptotic model (see equation (4)) can be written as:

$$[\partial_t \mathcal{I}(\delta) + \mathcal{L}]\phi + \mathcal{I}(\delta)\mathcal{B}(\phi, \phi) = 0 \tag{10}$$

where

$$\mathcal{I}(\delta) = \begin{pmatrix} 1 & 0 & 0 \\ 0 & \delta^2 & 0 \\ 0 & 0 & 1 \end{pmatrix}; \mathcal{L}(\phi) = \begin{pmatrix} -yv + \partial_x \eta \\ +yu + \partial_y \eta \\ \partial_x u + \partial_y v \end{pmatrix}; \mathcal{B}(\phi, \phi) = \begin{pmatrix} \vec{v} \cdot \nabla u \\ \vec{v} \cdot \nabla v \\ \nabla(\eta \vec{v}) \end{pmatrix}. \tag{11}$$

5. Evolution equation

Projection of equation (10) onto an arbitrary eigenmode ϕ_a results in:

$$[d_t \mathcal{I} + \hat{\imath}\omega_a I] Z_a(t) = 1/2 \sum_{bc} \delta_{abc} \sigma_a^{bc} \mathcal{I} Z_b^*(t) Z_c^*(t). \tag{12}$$

In equation 12, δ_{abc} indicates the interaction condition and establishes that in order for the interaction to take place, the modes must satisfy the relations $k_a + k_b + k_c = 0$ and $n_a + n_b + n_c =$ odd. Otherwise, the interaction is zero. The indentity matrix $I (= \mathcal{I}_{(\delta=1)})$;

and $\{Z_b^*, Z_c^*\}$ are the complex conjugate of $\{Z_b, Z_c\}$, respectively. The coupling coefficient σ_a^{bc} can be put as a function of δ:

$$\delta_{abc}\sigma_a^{bc}\mathcal{I}(\delta) = \int \phi_a^+(x,y)\mathcal{B}(\phi_b^*(x,y), \phi_c^*(x,y))\mathcal{I}(\delta)dxdy \tag{13}$$

Equation (12) is valid for both resonant and non-resonant nonlinear interactions. Using $\xi_a \equiv \partial_x v_a - \partial_y u_a - y\eta_a = -\hat{i}v_a/\omega_a$ and $T_a^{bc} = u_a\xi_b\xi_c$ the coupling coefficient can be written as

$$\delta_{abc}\sigma_a^{bc}\mathcal{I} = \omega_a\gamma_{abc} + (\omega_a + \omega_b + \omega_c)T_a^{bc} \tag{14}$$

As can be seen in equation (15), the interaction coefficient (γ_{abc}) is clearly modified by δ

$$\gamma_{abc} = \delta\left[\frac{(\eta_a\vec{v}_b\cdot\vec{v}_c + \eta_b\vec{v}_c\cdot\vec{v}_a + \eta_c\vec{v}_a\cdot\vec{v}_b) - T_a^{bc}}{\delta} - \delta\{T_b^{ca} + T_c^{ab}\}\right] \tag{15}$$

Taking into account the interaction ($n_a + n_b + n_c =$ odd) and resonance ($\omega_a + \omega_b + \omega_c = 0; k_a + k_b + k_c = 0$) conditions, it is possible to obtain the resonant coupling coefficient from equation (14). It can be readily seen that there is an effective reduction of the interaction frequency as δ decreases, whereas the interaction strength is also modified, increasing in absolute values for $(T_b^{ca} + T_c^{ab}) > 0$ (see equations (16)).

$$\omega_a > \delta\omega_a \text{ for } 0 < \delta < 1 \tag{16a}$$

$$\lim_{\delta\to 0}|\gamma_{abc}| > \lim_{\delta\to 1}|\gamma_{abc}| \tag{16b}$$

6. Numerics

The numerical integration scheme of equation (12) for the resonant three-wave problem uses a semi-analytic method, which assumes that the non-linear terms are constant within a $2\Delta t$ time interval. Thus, if we know the values of the expansion coefficients $Z_j, j = \{a, b, c\}$ for $[t - \Delta t, t]$, it is possible to obtain analytic solution at time $t + \Delta t$. The nonlinear terms are defined at the central time t. Zonal and meridional dependencies of the wave components, as well as their derivatives, are computed analitycally.

The problem is initialized in such a way that the most energetically active component of a particular resonant triad (the mode with the highest absolute frequency) holds most of the initial energy, i.e., the initial amplitudes ($A_j(\omega_{max}) \gg A_{i\neq j}$; defined in equation (17) below. According to [13] a more rigorous requirement to determine the most energetically active member can be obtained by using the mode with the intermediate slowness, which is the reciprocal of the phase speed). Initial phases are also consistent with the aim of maximizing energy exchange. Thus, it is requiered that $\sum \lambda_j = \pi/2$ (see [12],[15],[16]). The value of $\bar{H}(= 250\text{m})$ used here is consistent with the first baroclinic mode of the dry atmospheric dynamics.

$$Z_j(0) = A_j e^{\hat{i}\lambda_j}; j = \{a, b, c\}; \sum_j \lambda_j = \pi/2 \tag{17}$$

7. Nonlinear equatorial long-wave like interactions

In this section the modified nonlinear dynamics of the anisotropically scaled shallow water model is studied. To simplify, the analysis is on the reduced dynamics of a single resonant triad. The nonlinear energy exchange is investigated for some of the resonant triads studied in [12], but in this case, the interaction coefficient (equation 15) is used for values of $\delta = \{0.5, 1.0\}$. The first value (0.5) is a conservative estimate of the anisotropy parameter; however, the long wave like effect appears in equation (4) as the meridional momentum equation is multiplied by $\delta^2 (= 0.25)$. Meanwhile, the modification of the eigenfunctions is not such a dramatic, only of order of δ, as can be seen in [11] (using the slowness space). A proof of the asymptotic completeness of the long wave solutions can also be found in [11]. Consequently, smaller values of δ could be used as well, but in such a case, the modifications of the eigensolutions must also be considered. Therefore, intermediate values of δ are used with the aim of not to filter any of the solutions, while keeping long wave like effects. Coincidentally, for the range of values of the baroclinic wave speed 10 - 50 m/s, or the length scales 500 - 1500 km, or time scales 8 - 18 hrs; values of δ in the range of 0.3 - 0.5 are reasonable.

Each resonant triad studied is labeled by three parameters k, n, w, where k is the zonal wavenumber; n refers to the meridional quantum number and w label the wave type, say: R (Rossby); M (Mixed Rossby-Gravity); G^+ (Eastward Intertio-Gravity), G− (Westward Inertio-Gravity waves). In this way, two resonant triads are studied. The first one [(5,1,R);(1,4,G^-);(6,4,G^+)] corresponds to an interaction involving on slow mode and two fast modes, whereas the second triad [(0,1,R);(5,2,R);(5,0,M)] is a representative example of slow modes interaction.

Fig.(1) displays the nonlinear energy modulations associated with the [(5,1,R);(1,4,G);(6,4,G)] triad. Most of the initial energy is projected onto the third component, which has the highest absolute frequency and therefore, it is the most energetically active member of the triad. More precisely, the initial energy distribution is given by $A_3 = 10.0, A_{1,2} = 1.0$. On the other hand, in Fig.(2) the [(0,1,R);(5,2,R);(5,0,M)] trio is displayed, for the same two values of $\delta = \{0.5, 1.0\}$. The first component of this trio is the zonally symmetric Rossby mode, which has zero frequency and therefore, null resonant interaction coefficient. This mode is necessary to catalyze the energy exchange between the other two components. The higher the energy given to this mode, the stronger the energy exchange between the other two modes (not shown). To maximize the energy exchange, most of the initial energy is projected onto the most energetically active mode (5,0,M), with the partition $A_3 = 10.0, A_1 = 2.5, A_2 = 1.0$. It must be noted that the frequency of the third mode is only marginally larger than the frequency of the second mode (5,2,R).

As can be noted in Fig.(1) there is a slow modulation of the energy, in a period longer than the period of the phase propagation of the involved components. This slow modulation is accentuated as δ decreases as was anticipated in equations (13-15). For instance, the maximum energy exchange occurs at periods of around 5.8; 16 and 25.4 for the shallow water and at around of 6.5; 17.1 and 28 for the long-wave like approach. Although the curves for $\delta = 0$ are not displayed, the slow energy modulation occurs at periods of almost twice as the differences between the curves for $\delta = 1$ and $\delta = 0.5$. The magnitude of the energy exchange among the components is also modified. More precisely, the amplitude related to the (5,1,R) wave is increased, whereas the amplitude of the (1,4,G^-) is decreased. For this case, there is not an apparent modification of the energy associated with the most energetically active mode (6,4,G^+).

For the other triad (see Fig.(2)), despite of involving only slow modes, it is also characterized by the presence of the zonally symmetric Rossby mode that does not exchange energy with the other modes but catalyzes the energy exchange between the other two components. For the initial condition considered here, this trio is energetically less active in the sense of energy exchange than the previous one. However, there is a modification of both the period and, although less evident, the strength of the energy modulation as a function of the anisotropy parameter δ.

Thus, long wave like models tend to produce a longer nonlinear energy exchange when compared to the shallow water model. These results suggest some implications for the long term memory of the tropical region to disturbances. Physically, the long wave approach has shown to be useful to explain, in a simplified fashion, some key aspects of tropical phenomena such as El Niño and the MJO ([1, 5, 6, 2]). With the asymptotic approach used here it was possible to see that there exist differences between the shallow water and the long wave approaches under the optics of the nonlinear wave interactions (resonant or not). The close link between long wave model and simplified models of large scale tropical variability such as the El Niño or the MJO suggests that there is a potential for applications of the results of the present study. The asymptotic completeness could be useful to study the connection of the above mentioned variability with small scale disturbances such as those associated with short Rossby waves.

8. Conclusions

The shallow water model is written in a parametric form as a function of an external parameter δ. This parameter is a measure of both spatial and flow scales anisotropy and also measures the fast to slow timescales ratio. The model obtained is more general than the traditional long wave equations in the sense that it is valid for intermediate states between the shallow water and the long wave model. Using typical values of the perturbations as found in the tropical region it was possible to verify that the intermediate states are in fact realizables. For instance, values of δ between 0.3 - 0.5 are enough to produce a long wave like effect of the (order of δ^2) in the meridional momentum equation (a near geostrophic balance), and at the same time only a modification of the order of δ in the eigensolutions (as discussed in [11] using the slowness space framework). A single proof on the completeness of the eigensolutions at the long wave like limit is found in [11]. Asymptotically the model and its solutions are also valid for the long wave approach. The value of the completeness was exploited to study the differences between the nonlinear wave interactions in the shallow water and the long wave like model approach. Model equations for the nonlinear interactions were developed. However, to simplify, the analysis is restricted to the reduced dynamics of single resonant triads. Two triads are studied: one of them formed by an slow mode and two fast modes, and the other formed by only slow modes. Both, the effective nonlinear coupling and interaction coefficient are computed for the general problem. It was verified that there is an effective reduction of the interaction frequency (independently of the triad studied) and consequently an increase of the energy exchange period of the triad components as δ decreases. This result has implications for the memory of the tropical region to wave disturbances. The larger the anisotropy (smaller δ), the larger the nonlinear energy exchange period. However, an asymptotic upper limit is attained as δ tend to zero. The magnitude of the energy exchange is also modified but, in this case, depends on the specific resonant triad studied, as well as on the initial energy partition. Numerical experiments confirm the above results. Within this asymptotic framework there is a potential to better understand some tropical large scale phenomena such as the El Niño and the MJO in connection with smaller spatial scales of variability, as for example those related with small scales Rossby waves. The value of asymptotic methods was also demonstrated in other contexts. For instance, [14] used asymptotic methods to overcome the difficulty associated with the degeneracy of the eigenmodes at the zero frequency limit.

Acknowledgments

We are grateful to Rosio Camayo and Leon Sinay, their light, help and comments were very useful. To FAPESP for the financial support of the first author through grant 2006/60488-3. Part of this work was done at CPTEC as part of PAE/USP and is a contribution to the SUPERCLIMA project. Carlos Raupp was supported by FAPESP through grants 06/53606-0 and 09/11643-4.

References

[1] Cane M A,Sarachik E S, "Forced baroclinic ocean motions I: The linear equatorial unbounded case," *Journal of Marine Research* **34**, No. 4, pp. 629-665, 1976.

[2] Majda A J,Stechmann S N, "The skeleton of tropical intraseasonal oscillations," *PNAS*, Vol. **106**, No. 21, pp. 8417-22, 2009.

[3] Matsuno T, "Quasi-Geostrophic motions in the equatorial area" *Journal of the Meteorological Society of Japan*, Vol. **44**, 25-42, 1966.

[4] Gill A E, "Some simple solutions for heat-induced tropical circulation" *Quarterly Journal of the Royal Meteorological Society*, **106**, 447-62, 1980.

[5] Jin F F, "An equatorial ocean recharge paradigm for ENSO. Part I: Conceptual model" *JAS*, **54**, 811-29, 1997.

[6] Jin F F, "An equatorial ocean recharge paradigm for ENSO. Part II: A stripped-down coupled model" *JAS*, **54**, 830-47, 1997.

[7] Dutrifoy A, Majda A, Schochet S, "A simple justification of the singular limit for equatorial shallow-water dynamics" , **00**, 1-11, 2000.

[8] Pedlosky J, "Geophysical Fluid Dynamics", Springer-Verlag, New York, 1997.

[9] Gill A E, "Atmosphere-Ocean Dynamics", Academic Press, 1982.

[10] Schubert W H, Silvers L G, Masarik M T, Gonzales A O, "A filtered model of Tropical wave motions", *JAMES*, Vol. **1**, No. 3, 11 pp., doi:10.3894/JAMES.2009.1.3, 2009.

[11] Ramírez E, Raupp C F, Silva Dias P L, "The family of anisotropically scaled equatorial waves," *in preparation* A, Vol. **X**, No. X, XXX-XX, 2010.

[12] Raupp C F, Silva Dias P L, "Dynamics of resonantly interacting equatorial waves," *Tellus Series* A, Vol. **58**, No. 2, 263-79, 2006.

[13] Ripa P, "Weak Interactions of Equatorial Waves in a One-Layer Model. Part I: General Properties," *JPO*, **13**, 1208-26, 1983.

[14] Ripa P, "Nonlinear wave-wave interactions in a one-layer reduced-gravity model on the equatorial β plane", *JPO*, **12**, 97-111, 1982.

[15] Domaracki A, Loesch A Z, "Nonlinear interactions among equatorial waves", *JAS*, **34**, 486-98, 1977.

[16] Loesch A Z, Deininger R C, "Dynamics of closed systems of resonantly interacting equatorial waves", *JAS*, **36**, 1490-97, 1977.

Journal of Physics: Conference Series **285** (2011) 012020 doi:10.1088/1742-6596/285/1/012020

Figure 1. Slow energy exchange for the [(5,1,R);(1,4,G^-);(6,4,G^+)] trio. For each wave, two values of δ are plotted δ = 1.0 and 0.5. These values correspond to the shallow water and to the long wave like approximation respectivelly. The initial energy partition is $[A_1^2 = 1.0; A_2^2 = 1.0; A_3^2 = 100.0]$. It is possible to note that nonlinear resonant triad interactions in the long wave like approach (thick lines) promotes an slower energy exchange than the same interactions in the shallow water approach (thin lines).

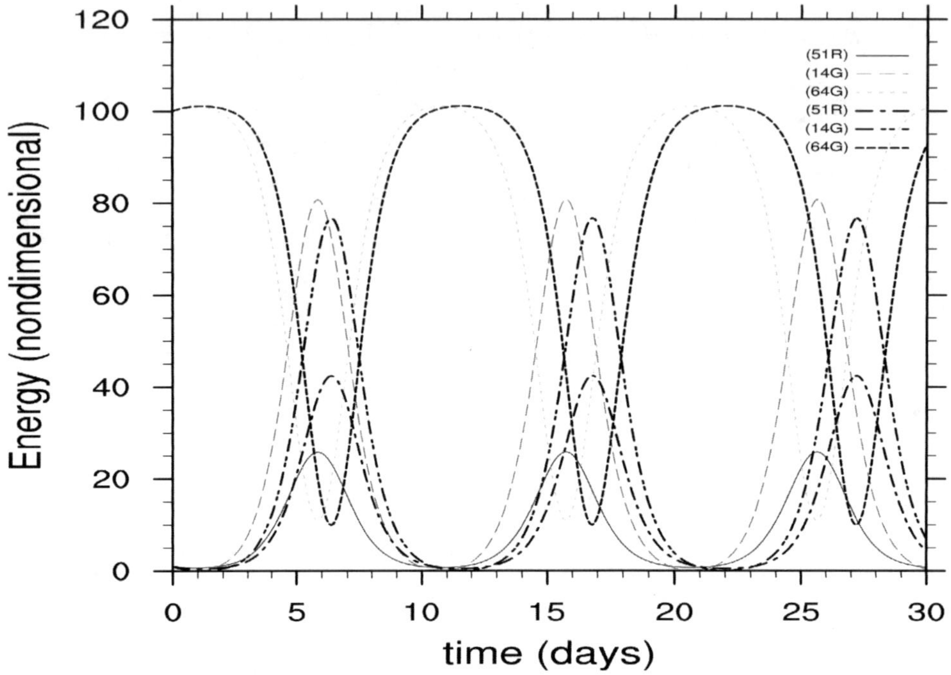

Figure 2. Same as Fig.(1) but for the [(0,1,R);(5,2,R);(5,0,M)] trio. The initial energy partion is also different $[A_1^2 = 1.0; A_2^2 = 6.25; A_3^2 = 100.0]$. Slower energy exchanges for the long wave like approach (thick lines) than for the shallow water (thin lines) are also noticeable.

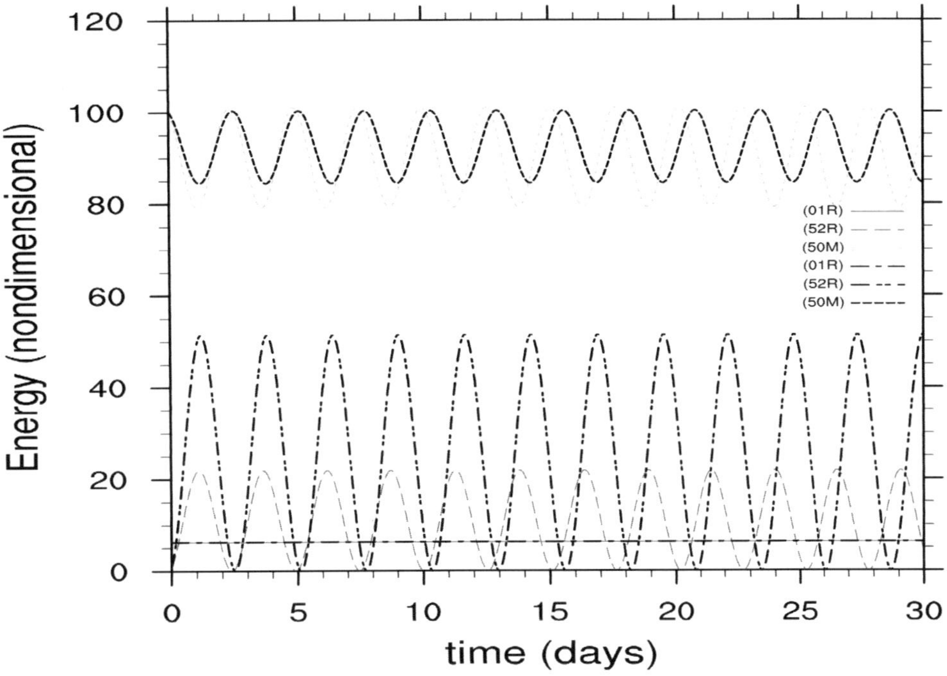

Dynamic Days South America 2010 IOP Publishing
Journal of Physics: Conference Series **285** (2011) 012021 doi:10.1088/1742-6596/285/1/012021

Chaotic fluctuations in mathematical economics

Hiroyuki Yoshida

College of Economics, Nihon University, Chiyoda-ku, Tokyo 101-8360, Japan

E-mail: yoshida.hiroyuki@nihon-u.ac.jp

Abstract. In this paper we examine a Cournot duopoly model, which expresses the strategic interaction between two firms. We formulate the dynamic adjustment process and investigate the dynamic properties of the stationary point. By introducing a memory mechanism characterized by distributed lag functions, we presuppose that each firm makes production decisions in a cautious manner. This implies that we have to deal with the system of integro-differential equations. By means of numerical simulations we show the occurrence of chaotic fluctuations in the case of fixed delays.

1. Introduction

The study of economic dynamics is one of the most important topics in the field of economics. Economic dynamics is concerned with changes and variations in the economic phenomena which take place in the course of time. In other words, economic dynamics seeks to explain the time paths of economic variables at the company, industry, or macroeconomic level. For this purpose, economists use extensively mathematical models, which are typically represented by difference equations or differential equations.

At an early stage of the development of economic dynamics from the 1950s to the 1970s, a significant portion of research was devoted to resolving the problems on the local stability of steady states and the existence of limit cycles in economic systems. It was not until the late 1970s that economists began to investigate chaotic behavior by using nonlinear dynamical system theory. Such a revolution was induced by the seminal contribution of Li and York [1]. The development of chaos theory in discrete-time dynamical systems has stimulated the publication of a large number of papers in economics (for example, see Rand [2], Benhabib and Day [3, 4], Day [5, 6], Grandmont [7], Puu [8], Nishimura and Yano [9]).

The main purpose of the present paper is to investigate the occurrence of chaotic fluctuations in the dynamic adjustment process by adopting the Cournot duopoly setting. Unlike Puu [8] who shows the emergence of chaotic trajectories in a discrete-time model, we use a continuous-time model of Cournot duopoly dynamics. By introducing a memory mechanism characterized by distributed lag functions, we presuppose that each firm makes production decisions in a cautious manner. This implies that we have to deal with the system of integro-differential equations.

2. The basic framework

As mentioned in Introduction, we consider a Cournot duopoly model, which formulates the strategic interactions between two firms in a market for a single homogeneous commodity. Each firm tries to maximize its profit by taking the output choice of the rival firm as given. The equilibrium point is defined as the intersection of the reaction functions.

Two firms X and Y produce a homogeneous product with output levels x and y. The market demand function is assumed to be

$$p = a - b(x + y), \ a > 0, b > 0, \tag{1}$$

where p is the price of output. Furthermore, the cost of each firm is assumed to be linear, $c_i x_i$ with $c_i > 0$ $(i = X, Y)$. Thus the profit of firm i is given by

$$\pi_X = [a - b(x + y)]x - c_X x, \tag{2}$$

$$\pi_Y = [a - b(x + y)]y - c_Y y. \tag{3}$$

As usual we assume that each firm maximizes its profit taking the quantity supplied by the rival firm as given. Consequently, the first order conditions for profit maximization are

$$-bx + a - b(x + y) - c_X = 0, \tag{4}$$

$$-by + a - b(x + y) - c_Y = 0, \tag{5}$$

which result in the following best-reaction functions:

$$x = R_X(y) = -\frac{1}{2}y + \frac{a - c_X}{2b}, \tag{6}$$

$$y = R_Y(x) = -\frac{1}{2}x + \frac{a - c_Y}{2b}. \tag{7}$$

According to (6), the optimal quantity for firm X depends on the output level of the rival firm. Equivalently, from (7) we obtain the result that the optimal quantity for firm Y depends on the output level of the rival firm. The intersection point of two lines (6)–(7) defines the Cournot-Nash equilibrium:

$$x_* = \frac{a - 2c_X + c_Y}{3b}, \ y_* = \frac{a + c_X - 2c_Y}{3b}. \tag{8}$$

In order to guarantee positive values of x_* and y_* we assume that $a - 2c_X + c_Y > 0$ and $a + c_X - 2c_Y > 0$.

Let us now consider the dynamic adjustment process. We adopt the following mechanism:

$$\frac{\dot{x}}{x} = \alpha_X(R_X(y) - x), \ \alpha_X > 0, \tag{9a}$$

$$\frac{\dot{y}}{y} = \alpha_Y(R_Y(x) - y), \ \alpha_Y > 0, \tag{9b}$$

where α_i indicates an adjustment speed parameter, which is a positive constant. The adjustment process shows that each firm controls the growth rate of its output according to the difference between its profit maximizing output and its actual output.

Substituting (6) and (7) into (9) leads to the following system:

$$\dot{x}(t) = \alpha_X \left(-\frac{1}{2}y(t) + \frac{a - c_X}{2b} - x(t) \right) x(t), \tag{10a}$$

$$\dot{y}(t) = \alpha_Y \left(-\frac{1}{2}x(t) + \frac{a - c_Y}{2b} - y(t) \right) y(t). \tag{10b}$$

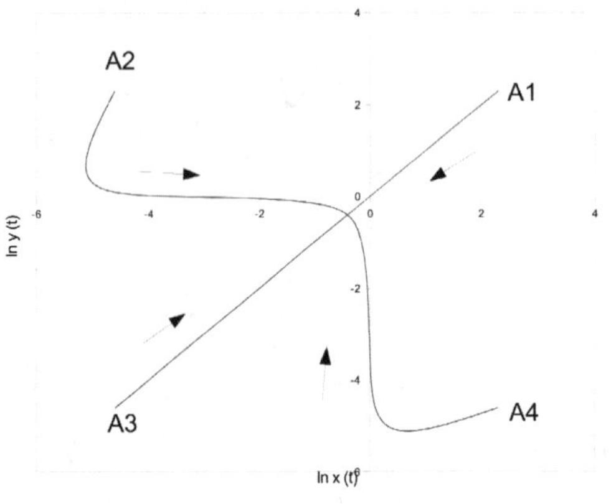

Figure 1. Convergence to the equilibrium point.

It is easy to check that the stationary point of system (10) coincides with the Cournot-Nash equilibrium. The Jacobian matrix at the Cournot-Nash equilibrium is

$$\begin{bmatrix} -\alpha_X x_* & -(1/2)\alpha_X x_* \\ -(1/2)\alpha_Y y_* & -\alpha_Y y_* \end{bmatrix}. \tag{11}$$

By evaluating the Jacobian matrix, we can conclude that the Cournot-Nash equilibrium is locally stable. This is because we can certify that the trace is negative and the determinant is positive. These are the conditions for the local stability of the stationary point.

Figure 1 shows the trajectories resulting from four different initial conditions: $A_1(\ln 10, \ln 10)$, $A_2(\ln 10, \ln 0.01)$, $A_3(\ln 0.01, \ln 0.01)$, $A_4(\ln 0.01, \ln 10)$. All trajectories of the system converge to the unique Cournot-Nash equilibrium. This result suggests that the equilibrium point is globally stable in the economically meaningful region. Note that this figure plots the natural logarithm of the output level of firm Y on the vertical axis versus the natural logarithm of the output level of firm X on the horizontal axis.

3. Extended model

In this section we shall extend the previous analysis by introducing a memory mechanism characterized by distributed lag functions. We assume that each firm adjusts its output by considering the past values as well as the current value of its own output. In this sense each firm makes production decisions in a cautious manner. We specify the following system of integro-differential equations:

$$\frac{\dot{x}}{x} = \alpha_X \left(R_X(y(t)) - \int_{-\infty}^{t} w_X(s)x(s)ds \right), \tag{12a}$$

$$\frac{\dot{y}}{y} = \alpha_Y \left(R_Y(x(t)) - \int_{-\infty}^{t} w_Y(s)y(s)ds \right). \tag{12b}$$

Here we assume that $w_i(s)$ indicates the following distributed lag function, i.e.,

$$w_i(s) = \left(\frac{n_i}{\tau_i}\right)^{n_i} \frac{(t-s)^{n_i-1}}{(n_i-1)!} e^{-(n_i/\tau_i)(t-s)}, \ \tau_i > 0,$$

where n_i is a positive integer ($i = X, Y$). On the account of the fact that $\int_{-\infty}^{t} w_i(s)ds = 1$, the function $w_i(s)$ can be interpreted as a weighting function, which is equivalent to a density function with the mean τ_i and the variance ${\tau_i}^2/n_i$.

It is important to note that we can classify the weighting function in three possible cases according to the values of n_i. First, we consider the case $n_i = 1$. In this case we obtain $w_i(s) = (1/\tau_i)e^{-(t-s)/\tau_i}$, which corresponds to the exponentially declining weighting function. This case implies that the firm evaluates the most recent event as being the most relevant, since the function $w_i(s)$ places more weight on more recent data. Secondly, we explain the case of $n_i \geq 2$. In this case the function $w_i(s)$ has a one-hump shape, which shows that a sharp peak emerges around time $t - \tau_i$ as n_i increases. Thirdly, we discuss the case $n_i \to +\infty$. When n_i becomes indefinitely large, $w_i(s)$ leads to a Dirac delta function that appears as a sharp peak at $t = \tau_i$. To put it simply, we consider a fixed delay $\tau_i (> 0)$:

$$\lim_{n_i \to \infty} \int_{-\infty}^{t} w(s)x(s)ds = x(t - \tau_i).$$

The important point to note is that MacDonald's linear chain trick transforms system (12) into an expanded system of ordinary differential equations. We can therefore obtain the analytical results about the Hopf bifurcation phenomena by using the linear chain trick. In this sense, the linear chain trick is a useful tool for investigating the systems of integro-differential equations. On this point, see MacDonald [10].

In the present paper we concentrate on the case of fixed time delays: $n_X \to \infty$, $\tau_X > 0$, $n_Y \to \infty$, and $\tau_Y > 0$. Since the weighting function $w_i(s)$ tends to a Dirac delta function as $n_i \to \infty$, we obtain the following system of delay differential equations:

$$\frac{\dot{x}}{x} = \alpha_X(R_X(y(t)) - x(t - \tau_1)), \tag{13a}$$

$$\frac{\dot{y}}{y} = \alpha_Y(R_Y(x(t)) - y(t - \tau_2)). \tag{13b}$$

The dynamic equations mean that at time t firm i adjusts its output at a rate proportional to the difference between its profit maximizing output at time t and its actual output at some preceding time $t - \tau_i$ ($i = X, Y$).

With (13) and the reaction functions, we obtain the following dynamical system:

$$\dot{x}(t) = \alpha_X \left(-\frac{1}{2}y(t) + \frac{a - c_X}{2b} - x(t - \tau_1) \right) x(t), \tag{14a}$$

$$\dot{y}(t) = \alpha_Y \left(-\frac{1}{2}x(t) + \frac{a - c_Y}{2b} - y(t - \tau_2) \right) y(t). \tag{14b}$$

Let us now start our analysis of system (14). The model to consider here is precisely identical to a model developed by Shibata and Saito [11]. Their model is an interesting contribution to the field of mathematical biology. They investigated the population dynamics of two competing species with fixed time lags and showed the appearance of chaotic dynamic trajectories by means of numerical simulations.

Following Shibata and Saito, we hereafter resort to numerical experiments to examine the dynamic properties of our system. We set the following parameter values: $a = 3, b = 1, c_X = c_Y = 1, \tau_X = 1.6, \tau_Y = 0.9, \alpha_X = \alpha_Y = 2$.

Figure 2 illustrates the emergence of a chaotic attractor in system (14). The trajectories never converge to the Cournot-Nash equilibrium, but exhibit perpetual fluctuations, while remaining forever in a bounded region of the state space. Although they are governed by the deterministic system, the behavior of phase trajectories looks random. This result implies that we can observe complex endogenous fluctuations in a particular market even if exogenous shocks are absent.

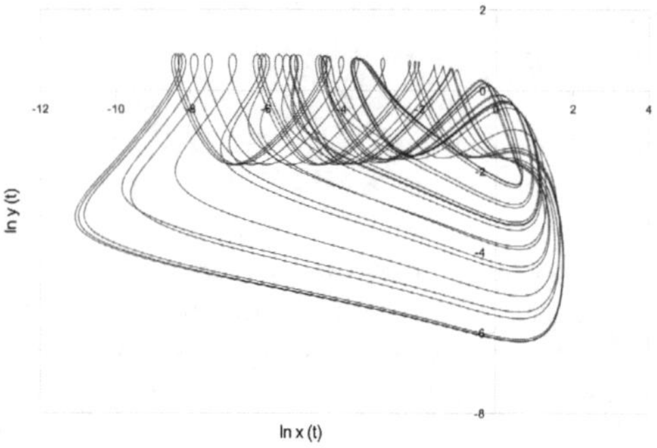

Figure 2. Chaotic attractor.

4. Conclusion

We have investigated the dynamic properties of Cournot duopoly models, which express the strategic interaction between two firms. By means of numerical simulations we have shown the occurrence of chaotic oscillations in the case of fixed delays.

References

[1] Li T and Yorke J A 1975 *Am. Math. Mon.* **82** 985–992.
[2] Rand D 1978 *J. Math. Econ.* **5** 173–184
[3] Benhabib J and Day R H 1980 *Econ. Lett.* **6** 113–117.
[4] Benhabib J and Day R H 1982 *J. Econ. Dyn. Control* **4** 437–55
[5] Day R H 1982 *Am. Econ. Rev.* **72** 406–414
[6] Day R H 1983 *Q. J. Econ.* **98** 201–213
[7] Grandmont J M 1985 *Econometrica* **53** 995–1045
[8] Puu T 1991 *Chaos Soliton. Fract.* **1** 573–581,
[9] Nishimura K and Yano M 1995 *Econometrica* **63** 981–1001
[10] MacDonald N 1978 *Time Lags in Biological Models* (Berlin: Springer)
[11] Shibata A and Saito N 1980 *Math. Biosci.* **51** 199–211

Study of the attractor structure of an agent-based sociological model

André M. Timpanaro[1], **Carmen P. C. Prado**[2]
Instituto de Física da Universidade de São Paulo, São Paulo, Brazil
E-mail: `timpa@if.usp.br`[1], `prado@if.usp.br`[2]

Abstract. The Sznajd model is a sociophysics model that is based in the Potts model, and used for describing opinion propagation in a society. It employs an agent-based approach and interaction rules favouring pairs of agreeing agents. It has been successfully employed in modeling some properties and scale features of both proportional and majority elections (see for instance the works of A. T. Bernardes and R. N. Costa Filho), but its stationary states are always consensus states. In order to explain more complicated behaviours, we have modified the *bounded confidence* idea (introduced before in other opinion models, like the Deffuant model), with the introduction of prejudices and biases (we called this modification *confidence rules*), and have adapted it to the discrete Sznajd model. This generalized Sznajd model is able to reproduce almost all of the previous versions of the Sznajd model, by using appropriate choices of parameters. We solved the attractor structure of the resulting model in a mean-field approach and made Monte Carlo simulations in a Barabási-Albert network. These simulations show great similarities with the mean-field, for the tested cases of 3 and 4 opinions. The dynamical systems approach that we devised allows for a deeper understanding of the potential of the Sznajd model as an opinion propagation model and can be easily extended to other models, like the voter model. Our modification of the *bounded confidence* rule can also be readily applied to other opinion propagation models.

1. Introduction and Model Description

In the last decade, econophysics and sociophysics models have attracted the attention of physicists, due to the current availability of great amounts of data and computing power. In this context, the Sznajd model simulates the propagation of opinions, using the idea that bigger groups of agreeing people have a stronger convincing power. The society is modeled as a network with N nodes, each node in the network is an agent (person), each edge is a social connection (friendship, marriage, acquaintances, etc.) and each agent has an integer between 1 and M, representing its opinion σ (*i.e.* $\sigma \in \{1, 2, \ldots, M\}$). At each time step, a pair of connected nodes, i and j, are chosen at random (first one of the nodes is chosen and then one of its neighbours). If they agree (that is, if $\sigma_i = \sigma_j$) a neighbour k of j is chosen and convinced to opinion $\sigma_j = \sigma_i$ with some probability. Otherwise ($\sigma_i \neq \sigma_j$), nothing happens. The idea is that the first step represents a conversation between two people that know each other and discuss some issue. If they disagree, none manages to convince the other. But, if they agree, they may set to convince another person, that one of them knows.

In the original model the conviction is always effective, meaning the pair always convinces the isolated site with probability 1. This leads to a dynamic where the only stationary states are consensus states (all the sites have the same opinion). Some modifications of the original model have been suggested in the literature in order to create stable coexistence states, like the *Bounded Confidence* rule [3] which says that all opinions are nuances between 2 extremes and opinion changes are not abrupt. More precisely, in a discrete opinion model, it means that agents with opinion σ can only interact with agents that have opinion $\sigma \pm 1$. Another modification that has been used is to add an undecided state to agents [1].

In a recent work [4], we generalized the idea of Bounded Confidence, introducing a modification to the dynamics of the Sznajd Model that we called *Confidence Rules.* In this generalized version of the model, the probability that a pair of agreeing sites with opinion σ convinces a site with opinion σ' is $p_{\sigma' \to \sigma}$, that is a parameter of the model. The reason why this probability should depend on both opinions is that, usualy an opinion includes prejudices about differing points of view. The set of probabilities $p_{\sigma \to \sigma'}$ will be refered to as the *Confidence Rule.*

We can recover the original Sznajd model and some of its modifications in the following limits:

- The usual Sznajd model is the case $p_{\sigma' \to \sigma} = 1$ for all $\sigma \neq \sigma'$.
- If an opinion σ has $p_{\sigma' \to \sigma} = 0$ and $p_{\sigma \to \sigma'} = p \neq 0$, for all $\sigma' \neq \sigma$, then σ behaves analogously to an undecided state.
- Usual bounded confidence is the case $p_{\sigma' \to \sigma} = 1$ if $\sigma' = \sigma \pm 1$ and $p_{\sigma' \to \sigma} = 0$ otherwise.

Note that $p_{\sigma \to \sigma}$ has no relevance whatsoever for any of the opinions σ.

2. Mean-Field Attractor Struture

The mean field version can be obtained from the master equation of the model defined in a complete network. When this network is large, this equation reduces to the flow

$$\dot{\eta}_\sigma = \sum_{\sigma'} \left(\eta_\sigma^2 \eta_{\sigma'} p_{\sigma' \to \sigma} - \eta_\sigma \eta_{\sigma'}^2 p_{\sigma \to \sigma'} \right), \tag{1}$$

where η_σ is the expected value of the fraction of sites holding opinion σ and one time unit corresponds to N simulation time steps, where N is the number of nodes. Since $\sum_{\sigma=1}^{M} \eta_\sigma = 1$, the phase space of this flow can be represented by an $(M-1)$-simplex (that is embedded in an M dimensional space for convenience), where the vertices correspond to consensus states and the other states are convex combinations of the vertices, with coeficients η_σ.

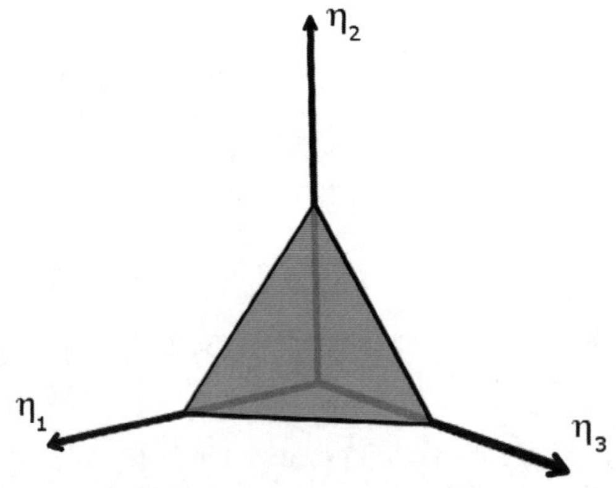

Figure 1: The 2 dimensional phase space (gray) for 3 opinions embedded in $\mathbb{R}^3$ ($\eta_1 + \eta_2 + \eta_3 = 1$; $\eta_1, \eta_2, \eta_3 \geq 0$)

The fixed points are given by

$$\eta_\sigma = 0 \text{ or } \sum_{\sigma'} \left(\eta_\sigma \eta_{\sigma'} p_{\sigma' \to \sigma} - \eta_{\sigma'}^2 p_{\sigma \to \sigma'}\right) = 0$$

for each opinion σ. Let $\vec{F}$ be the time derivative of $\vec{\eta} \equiv (\eta_1, \eta_2, \ldots, \eta_M)$, $\vec{F} = \dot{\vec{\eta}}$ and $\mathcal{J}$ be the jacobian of $\vec{F}$. If $\mathcal{J}^*$ is this jacobian, evaluated at a fixed point $\vec{\eta}^*$, then the analysis of the eigenvalues yields the following conclusions:

- If all opinions survive, $\mathcal{J}^*$ is positive semidefinite, but all eigenvalues with null real part are either due to embedding artifacts or to degeneracies (when the model can be split into simpler, non-interacting models), playing no role in the linear stability analysis. Hence, these fixed points have no stable manifolds. Also, with the exception of the case where the opinions are all non-interacting (when $\mathcal{J}^* = 0$), these fixed points have at least one unstable manifold.
- If there is a set Ω of opinions that don't survive, then there exists a permutation that leads $\mathcal{J}^*$ to
$$\begin{bmatrix} \mathcal{J}^*_\Delta & \mathcal{M} \\ 0 & \mathcal{J}^*_\Omega \end{bmatrix},$$
where $\mathcal{J}^*_\Delta$ ($\Delta \equiv \bar{\Omega}$, the set of opinions not in Ω) is the jacobian restricted to the surviving opinions and $\mathcal{J}^*_\Omega$ is the jacobian restricted to opinions in Ω (note that $\mathcal{M}$ is not relevant to the eigenvalues of $\mathcal{J}^*$). We can use for the block $\mathcal{J}^*_\Delta$ the former conclusion (for a coexistence fixed point in the model restricted to the surviving opinions). The block $\mathcal{J}^*_\Omega$ is a real, diagonal and negative semidefinite matrix. An analysis beyond linear stability shows that the null eigenvalues in $\mathcal{J}^*_\Omega$ correspond to unstable manifolds of the fixed point. Moreover, we can associate each of these eigenvalues to an opinion in Ω; the associated eigenvalue is negative iff the confidence rule is such that the opinion in Ω can be convinced by some of the surviving opinions.

One interesting aspect of these results is that the only property of the parameters $p_{\sigma\to\sigma'}$ relevant to the qualitative attractor structure is which of them are null and which are positive. The system has $M(M-1)$ of these probability parameters ($p_{\sigma\to\sigma} = 0$ in what follows), where M is the number of opinions and $p_{\sigma\to\sigma'} \in [0,1]$. Also, the parameters $p_{\sigma\to\sigma'}$ can be thought as the elements of the adjacency matrix of a weighted graph (where an arrow $(\sigma \to \sigma')$ means that the opinion σ can be convinced to opinion σ'), that will also be called confidence rule (as it is just a representation of the parameters that constitute the confidence rule) and denoted R. This graph helps to schematize the interactions among the different opinions. We can assign a skeleton $Skel(G)$ to a weighted graph G, which is the directed graph made of the same nodes as the weighted graph and all the arrows that have a non-zero weight (Figure 2). We can also assign, to each group of nodes Δ in the confidence rule, the manifold M_Δ, where only the opinions in Δ survive. In this way, the flow restricted to M_Δ is equivalent to the model with a confidence rule R_Δ, the subgraph of R induced by Δ.

Our results for the mean field model can then be summed up as

- The stability proprierties of a fixed point depend only on $Skel(R)$ and on which opinions survive in the fixed point. Particularly, if 2 interacting opinions survive, then the fixed point is always Lyapunov unstable.
- The manifold M_Δ is an attractor, where all of its points are fixed points iff the opinions in Δ are non-interacting, and every opinion not in Δ can be convinced by at least one opinion in it.

A phase portrait that helps to illustrate the linear analysis results for $M = 3$ can be found in figure 3.

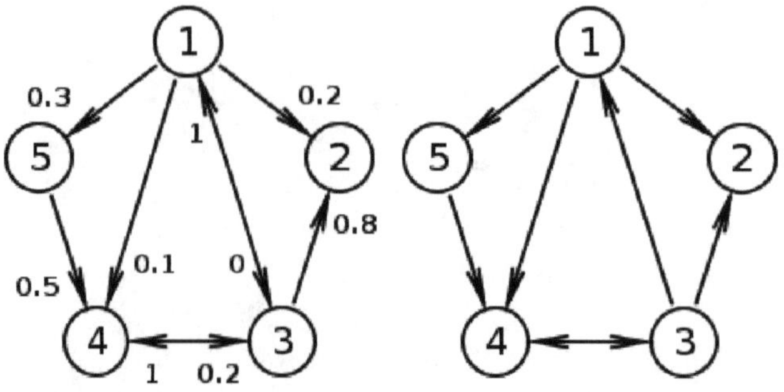

Figure 2: A confidence rule for 5 opinions and its skeleton. Here $p_{1\to2} = 0.2$, $p_{1\to4} = 0.1$, $p_{1\to5} = 0.3$, $p_{3\to1} = 1$, $p_{3\to2} = 0.8$, $p_{3\to4} = 1$, $p_{4\to3} = 0.2$, $p_{5\to4} = 0.5$ and $p_{\sigma\to\sigma'} = 0$ otherwise. The arrow $(1 \to 3)$ does not appear in the skeleton because it has weight 0, even though it was represented as part of the directed graph.

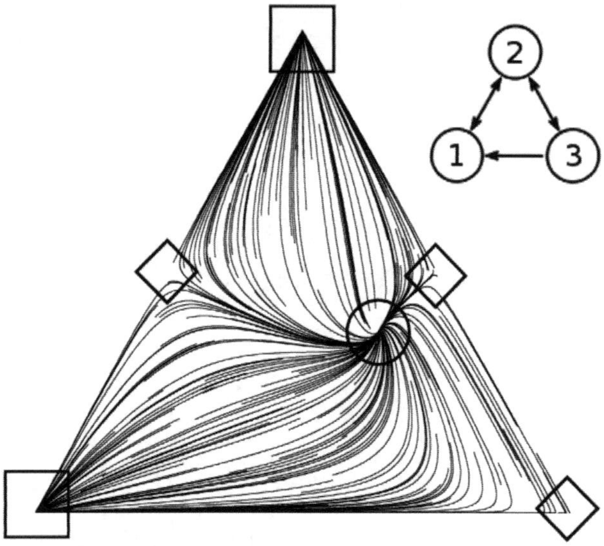

Figure 3: Phase portrait for a three opinions case. The arrows in the skeleton give the direction of the trajectories along the borders. The circles indicate unstable nodes, the diamonds indicate saddle points and the squares indicate stable nodes. The parameters $p_{\sigma\to\sigma'}$ are either 0 or 1.

3. Simulations

Simulations of the model in a Barabási-Albert (BA) network show a very similar phase portrait (Figure 4 shows a comparison between the two) for 3 opinions and the attractor structure agrees with what was found in the mean-field analysis. The "phase space" in this case is actually a projection to the $(M-1)$-simplex, where the states are described by η_σ. As such, some odd phenomena are possible, as the crossing of 'trajectories' seen in figure 4. Moreover, the derivation of equation 1 makes it clear that the variables in the flow are actually the expected values for the simulation (in a complete network). So, to draw these 'stochastic trajectories', we made an average over many simulations and the initial conditions were enforced only by changing the probabilities that a site would start with a given opinion.

Figure 4 shows that for 3 opinions the fixed points found in the mean-field approximation give the "fixed points" for the simulated model with good accuracy and the trajectories also show similar patterns. For 4 opinions, the qualitative attractor structure remains the same but the trajectories are different, especially in the bulk of the simplex. Figure 5 shows time series for a 4 opinions case.

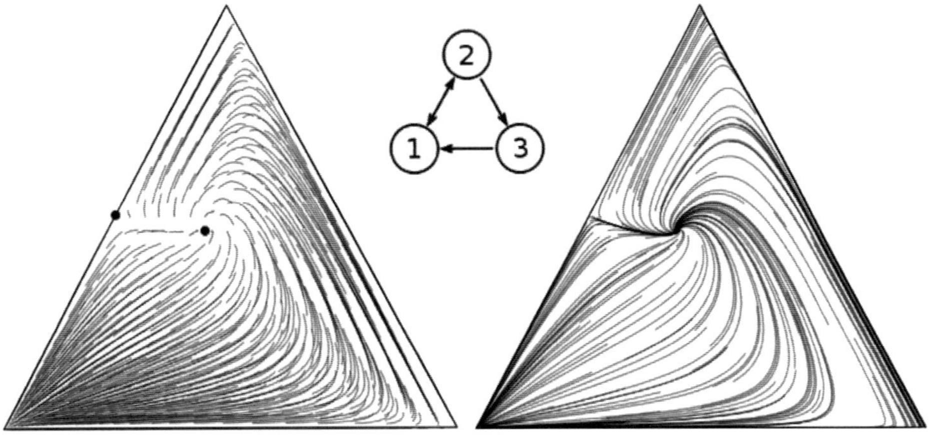

Figure 4: Comparison between a confidence rule simulated in a BA network (10^5 sites) (left) and its mean-field counterpart (right). The marks in the phase space to the left indicate the locations of the fixed points of the mean-field case. The parameters $p_{\sigma \to \sigma'}$ are either 0 or 1.

These correspondences are also network-dependent. For a square lattice and at least 4 opinions, the attractor structure (for the usual bounded confidence case) is completely different [5]. In a general network, the stationary states are characterized by contiguous domains of sites with the same opinion, arranged in such a way that domains with interacting opinions don't touch each other (isolated sites may violate this, depending on the confidence rule adopted).

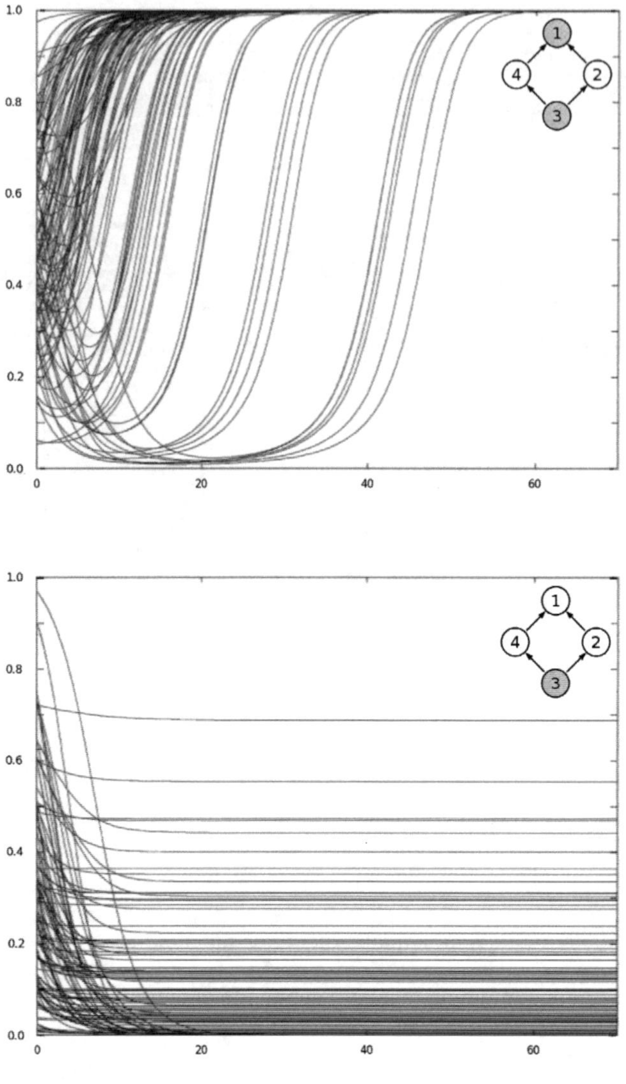

Figure 5: Time series for $\eta_1 + \eta_3$ (top) and η_3 (bottom). The mean field results imply that the only attractor is a coexistence of opinions 1 and 3, even though opinion 3 has no convincing power. This happens because opinion 1 always extinguishes opinions 2 and 4, before they can extinguish opinion 3. Each time step corresponds to 10^5 (network size) steps. The parameters $p_{\sigma \to \sigma'}$ are either 0 or 1.

4. Conclusions

We have shown that the Sznajd model can be extended to include more complex interactions among opinions, mimicking biases and prejudices. We have found out most of the relevant qualitative features for the mean-field case of this generalized version of the model, through a typical dynamical systems approach. The simulations in a Barabási-Albert network show great similarities with the mean-field case (even quantitative similarities for 3 opinions). We believe this connection may be related to the small-world properties of BA networks, as they make it more difficult the formation of isolated domains.

The dynamical systems approach we used for the mean-field analysis can be generalized to other similar models, like the voter model as well as more complex variants of the Sznajd model. Our generalization of the bounded confidence rule can be readily applied to other discrete opinion models or adapted to continuous opinion models.

Acknowledgments

The authors acknowledge FAPESP for financial support.

References

[1] Bernardes A T, Stauffer D and Kertész J 2002 *Eur. Phys. J. B* **25** 123–127
[2] Costa Filho R N, Almeida M P, Andrade Jr J S and Moreira J E 1999 *Phys. Rev. E* **60** 1067(2)
[3] Deffuant G, Neau D, Amblard F and Weisbuch G 2000 *Advances in Complex Systems* **3** 87–98
[4] Timpanaro A M and Prado C P C 2009 *Phys. Rev. E* **80** 021119(8)
[5] Stauffer D 2002 *Int. J. of Mod. Phys.* **13** 315–318

Solving the Levins' paradox in the logistic model to the population growth

Evaldo Araújo de Oliveira
Instituto de Matemática e Estatística - Universidade de São Paulo
R. do Matão 1010, 05508-090, São Paulo, Brazil

Vicente Pereira de Barros
Instituto Federal de Educação, Ciência e Tecnologia de São Paulo
Av. João Olímpio de Oliveira 1561, 18202-000, Itapetininga, Brazil

Roberto André Kraenkel
Instituto de Física Teórica - Universidade Estadual Paulista
R. Dr. Bento Teobaldo Ferraz 271, Bloco B, 01140-070, São Paulo, Brazil

Abstract. We introduce a new method to improve Markov maps by means of a Bayesian approach. The method starts from an initial map model, wherefrom a likelihood function is defined which is regulated by a temperature-like parameter. Then, the new constraints are added by the use of Bayes rule in the prior distribution. We applied the method to the logistic map of population growth of a single species. We show that the population size is limited for all ranges of parameters, allowing thus to overcome difficulties in interpretation of the concept of carrying capacity known as the Levins paradox.

1. Introduction

Since the very beginnings of the studies of what is now called *population biology*, by Thomas Malthus in 1798, the development of mathematical methods to adequately express the biological concepts has been a central point of the theory. Differential equations, maps and stochastic processes are a few of the modern methods employed, allowing the results to be put in analogy to systems in the physical sciences, a well as with the economical theory [1].

The Malthusian theory is based on a constant rate of compound interest, implying an exponential population growth. The underlying hypothesis is that the population is unchecked, that is, there are no limitations to its growth. In 1838, P. F. Verhulst proposed a mathematical model describing the self-limited growth of a biological population, where the saturation tends to halt exponential growth, driving the population size to an equilibrium. Verhulst proposed essentially to adjust the intrinsic rate of increase, defined as $R_t = (N_{t+1} - N_t)/N_t$ and constant in the Malthusian model, by a nonlinear factor:

$$R_t = r\left(1 - \frac{N_t}{K}\right), \tag{1}$$

with r being the *growth rate* and K *the carrying capacity*. This concept models the self-interaction of a species and accounts for the intra-species competition.

Equation (1) is called logistic model and recovers the Malthusian model by setting K to infinity, which means an unlimited capacity. Besides, $r > 0$ means a total growth rate greater than the total reduction rate and the opposite for $r < 0$.

Although the logistic model overcomes the problem of explosive growth, conceptual problems arise in the interpretation of the carrying capacity K. A criticism consists in what is known as *Levins' paradox* [2]. This paradox appears when we have a negative growth rate and an initial population greater than the carrying capacity ($N_t > K$). In this case, instead of reductions on the population size, since a negative r means a death rate greater then the birth, the logistic model presents unbounded growth.

Much effort has been made to overcome this problem, as for example, to model a population in the limit of the carrying capacity by analyzing the effect of processes such as culling or sterilization, but avoiding the problematic set of parameters r and K [3]. Another usual procedure is to define new constants in nondimensional quantities in the logistic equation [4] and be free of the paradox, but left with no clear interpretation of the constant in front of the nonlinear term.

In this direction, some authors work out a new definition of carrying capacity. Gabriel *et al.* [5] discussed this scenario using a restriction for region of validity of the equation. First, these authors showed that setting K in function of r, in such a way that K and r have the same sign, would be enough to solve the Levins' paradox. But, again, this gauge would bring difficulties to interpret K, since a negative carrying capacity makes no sense. Therefore, in a second step, the authors redefined the carrying capacity as

$$K_\infty = \lim_{t\to\infty} N_t, \text{ with } N_0 > 0 \tag{2}$$

having as consequence

$$K_\infty = \begin{cases} \text{constant}, & \text{for } r > 0 \\ 0, & \text{for } r < 0 \end{cases} \tag{3}$$

and kept K as a simple model parameter with $\text{sign}(K) = \text{sign}(r)$.

Although this procedure solves the mathematical paradox and in this way has been useful in practice, we are indeed changing the concept of the carrying capacity. It gives us the population of a species that is supported, given the mount food supply, habitat and other resources available within an environment. Of course, this capacity can be depredated by the population (*e.g.* when the population acts as plague) or improved by planed acts. But this changing (improvement or depredation) can also be developed by the environment independent of the population actions, *e.g.* the introduction of a lethal virus in the environment that interacts only with the specie observed or its resources. Therefore, it is interesting to search by another solution that takes account more realistic properties of the observed systems. In this sense, a interesting approach would be the probability theory.

Usually, the application of probability theory is recommended when we have to deal with assertions in the absence of complete information. But in the Bayesian Theory, probabilities can be interpreted as an extension of the logic [6, 7] and be can used to incorporate information on deterministic models.

2. Adding informations by the Bayes Theorem

Given a map $N_{t+1} = N_t + f(N_t)$, parameters from $f(...)$ and some constraints involving $\Delta N_t \equiv N_{t+1} - N_t$, the proposal is to incorporate these constraints in a prior probability distribution and obtain a new map by averaging the variable of interest over the probability of ΔN_t be given by $f(N_t)$ times the prior.

This procedure is different from usual application of Bayesian theory, which is applied to analysis of experimental or observational data. Here, the idea is not to fit some experimental data set, but to use the concept of probability in the Bayesian theory to improve a given map by adding new constraints. We thus discuss the model instead of the data. However, the main essence of the Bayesian Theory — the concepts and distinction between the prior, the likelihood and the posterior distributions — is required at all moment as in typical subjective probability models.

Here the likelihood is used to set the probability of observing a variation ΔN_t on the population given the actual state N_t and the chosen model defined by $\Delta N_t = f(N_t)$, *i.e.* $L \equiv P(\Delta N_t|N_t)$. On the other hand, the constraints which represents the prior informations, is embedded in the prior distribution over N_t. Thus, by using the Bayes theorem, the joint distribution $P(\Delta N_t, N_t) = P(\Delta N_t|N_t)P(N_t)$ is calculated and the model update by averaging the variable of interest, *e.g.* $R_t = g(\Delta N_t, N_{t+1})$, over $P(\Delta N_t, N_t)$.

The rest of this article is divided in two sections. In section (3) we describe the new method by applying it on the Verhulst model, wherefrom the main equations are developed and implemented. In section (4), we discuss the results as well as the method to improve map models, showing that our approach is free of the Levins' paradox and gives a bounded population in a natural way in all parameter ranges.

3. Improving the Verhulst model

The main point in the Verhulst model is to consider the rate of increase $R_t = \Delta N_t/N_t$ as a function of population size and some parameters. Is also in it, the rate of increase, that the Levins' paradox is found. Therefore, if we want to improve the Verhulst model, a good point to start is R_t.

As mentioned, the proposal is to consider that the map models the average behavior of population, which suggest to change R_t by $\langle R_t \rangle$, where $\langle ... \rangle$ means $\int ... P(\Delta N_t, N_t) \mathrm{d}\Delta N_t \mathrm{d}N_t$ and $P(\Delta N_t, N_t)$ is obtained by the Bayes' Theorem [8]:

$$P(\Delta N_t, N_t) = L(\Delta N_t|N_t)p(N_t), \tag{4}$$

where $P(\Delta N_t, N_t)$ is the probability to observe ΔN_t and N_t, $L(\Delta N_t|N_t)$ is the likelihood and $p(N_t)$ is the prior distribution.

The likelihood is defined as the Boltzmann factor

$$L \propto \exp\{-\beta V\},$$

with β been an inverse temperature parameter (related to noise) and V a function of $\Delta N_t - f(N_t)$. Therefore, using equation (1) as the initial model and considering a quadratic potential function, we obtain

$$V = \frac{1}{2}\left[\Delta N_t - \left(r - \frac{r}{K}N_t\right)N_t\right]^2$$

and

$$\begin{aligned} L(\Delta N_t|N_t) &= \sqrt{\frac{\beta}{2\pi}}\exp\left\{-\frac{\beta}{2}\right. \\ &\left.\left[\Delta N_t - \left(r - \frac{r}{K}N_t\right)N_t\right]^2\right\}. \end{aligned} \tag{5}$$

The prior distribution represents the knowledge from empirical source or what it is expected from the model. In our case, this information is about the decrease of the population size when

$r < 0$ and the convergence to K when $r > 0$. Besides, we can also add constraints to compensate some known limitations from the original map. In this context, we suggest to encode this prior information as

$$p(N_t) = \frac{[1 - \Theta(-r)\Theta(N_t - K)]N_t \exp\{-N_t/\kappa\}/\kappa^2}{1 - \Theta(-r)(1 + K/\kappa)\exp\{-K/\kappa\}} \tag{6}$$

where κ is directly connected with the population in the moment t, *i.e.* $\langle N_t \rangle = 2\kappa$, and $\Theta(-r)$ is the Heaviside function:

$$\Theta(-r) = \begin{cases} 1 & \text{for } r < 0 \\ 0 & \text{for } r > 0. \end{cases} \tag{7}$$

Given the equations (5) and (6), we obtain the posterior distribution by (4) and the next step is to calculate $\langle R_t \rangle$:

$$\langle R_t \rangle = r\frac{1 - 2\kappa/K + \Theta(-r)(1 + 2\kappa/K)\exp\{-K/\kappa\}}{1 - \Theta(-r)(1 + K/\kappa)\exp\{-K/\kappa\}}. \tag{8}$$

Note that $\langle R_t \rangle$ does not depend on β and this comes from its definition, that is linear on ΔN_t. But, when the parameter of interest is a nonlinear function of ΔN_t, β can play an important role to take account fluctuations on the map. This flexibility is required when we deal with experimental time-series, that have multiple sources of stochasticity [9]. Naturally, the biological meaning of β could be explored in future works.

Once given $\langle R_t \rangle$, not matter how it was calculated, the variation on population is estimated by the product of $\langle R_t \rangle$ by the population today: $\Delta N_t = \langle R_t \rangle N_t$. But, in biological problems there exist others events, not considered in the logistic model, which cause some fluctuation on the expected growth. Therefore, if we want to adjust the parameters of the model from experimental data, it is suitable to consider the averaged estimative of the last expression, *i.e.*

$$\langle \Delta N_t \rangle = \langle \langle R_t \rangle N_t \rangle = \langle R_t \rangle \langle N_t \rangle . \tag{9}$$

To compare the resulting model from (9) to the Verhulst model, we rewrite κ in terms of $\langle N_t \rangle$ in the equation (8):

$$\langle R_t \rangle = r\frac{1 - \langle N_t \rangle / K + \Theta(-r)(1 + \langle N_t \rangle / K)\exp\{-2K/\langle N_t \rangle\}}{1 - \Theta(-r)(1 + 2K/\langle N_t \rangle)\exp\{-2K/\langle N_t \rangle\}}. \tag{10}$$

For $r > 0$ the equation (10) becomes similar to (1) and the averaged population goes to the limit of the carrying capacity in finite time. As in the Verhulst model, this time is independent from where the population starts, whether above or below the carrying capacity. In Fig.(1) a numerical implementation with both $r > 0$ and $r < 0$ illustrates this point.

4. Discussion

Considering two trials developed from initial populations $N_0(1)$ and $N_0(2)$ for $r > 0$, the difference between the times at which they become equal K is given by

$$t_1 - t_2 = \frac{1}{r}\ln\left[\frac{N_0(2)}{N_0(1)}\right]. \tag{11}$$

where t_1 is the time where $N_t \to K$ from $N_0(1)$ and t_2 is the time where $N_t \to K$ from $N_0(2)$.

The initial time is the same in both situations and $N_0(1) < K$ and $N_0(2) > K$. The ratio in the logarithm is always greater than one, but the absolute value of logarithm may be smaller then values of t_1 and t_2. This means that the accuracy to distinguish when the population achieve the *carrying capacity* is not a good parameter for quantitative statements.

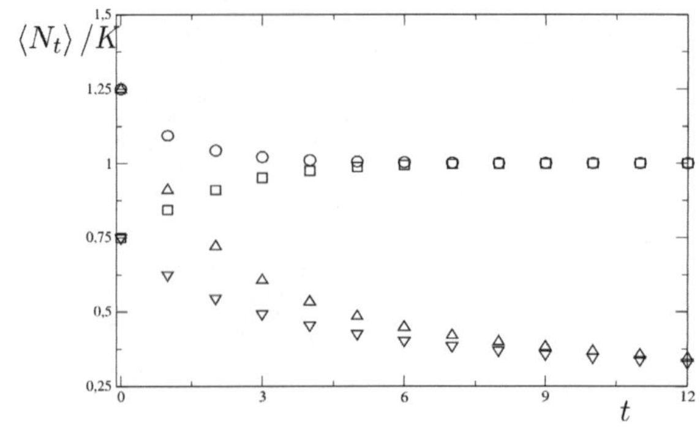

Figure 1. The evolution of the $\langle N_t \rangle / K$ given by the Eq. (10) with: $r = 0.5$ to circle and square symbols and $r = -0.5$ to triangles.

In the opposite scenario, with a negative growth rate, the population goes to extinction. Of course, the existence time of the population depends on the initial size $\langle N_0 \rangle$, on the growth rate and on the carrying capacity [10]. To illustrate these, consider a simple example: the arriving of a large emigrate population in an population subject to an epidemic.

The first thing to observe is the possibility of two growth rates, one of the native population and another of the foreign one. But, to avoid unnecessary complications it is assumed that the effective grown rate $\bar{r}$ is negative. If the sum of the native plus the foreign population is greater than the carrying capacity, then the total growth rate has to be less than $\bar{r}$ — representing the death due to the disease plus the limitations due to the lack of supplies. Besides, once the lack of supplies will affect the sick individuals more than it would affect the healthy ones, it is expected a total reduction of rate greater than the sum of the rate due the disease plus the rate due to the lack of supplies when $r > 0$. In the new model (10), if $\langle N_t \rangle \gg K$ and $r < 0$, $\langle R_t \rangle$ will be proportional to r, which implies a dynamics similar to the one of the Malthusian model. Besides, it can be interpreted as if, in the lack of supplies, the death due to diseases was greater than when there is abundance.

Naturally, the major contribution to the reduction rate should come from r when total population is far from the carrying capacity (less than K) — once it would have abundance. This can be observed in the limit $\langle N_t \rangle \ll K$, that produces $\langle R_t \rangle \to r(1 - \langle N_t \rangle / K)$, *i.e.* the old model is recovered.

While recent works [5] try to solve the Levins' paradox adjusting the sign of K with sign(r) or restricting the valid region to N_t, the proposed model keeps the meaning of the carrying capacity and allows for initial conditions either above or below the carrying capacity for both $r > 0$ and $r < 0$.

Another important detail is the Bayesian approach used in this paper to obtain a new map with the desired constraints. As the Lagrange multipliers, the prior distribution can be used to restrict the model to some region. In this way, the distribution is built from the geometrical information on the mathematical space induced by variables and parameters of the model.

This simple example indicates that the present method can be used as a promising tool to improve models, what can be verified in others dynamical population models [11]. Besides the application presented here, the method can be also applied to problems of other nature, as for example, is the case of mechanical systems.

Once redefining the Logistic model, the next step is to investigate models where r can change, *i.e.* with $r \equiv r_t$. The super abundance problems, competition systems or structured populations, maybe implemented by the method by using another tricks of the Bayesian approach developed in the study of learning systems [12, 13].

Acknowledgments

The authors wish to acknowledge M. Amaku for helpful discussion and *FAPESP* for financial support.

References

[1] Finnoff D and Tschirhart J 2008 *Res. Energy Economics* **30** 91–114
[2] Hutchinson G E 1978 *An Introduction to Population Ecology* (New Have: Yale Univ. Press)
[3] Barlow N D, Kean J M and Briggs C J 1997 *Wildlife Research* **24** 129–141
[4] Murray J D 1989 *Mathematical Biology* 2nd ed (Spring-Verlag)
[5] Gabriel J P, Saucy F and Bersier L F 2005 *Ecological Modelling* **185** 147–151
[6] Cox R T 1946 *Am. J. Phys.* **14** 1–13
[7] Jaynes E T and Bretthorst G L 2003 *Probability Theory: The Logic of Science* (Cambrigde University Press)
[8] Bernardo J and Smith A 2000 *Bayesian Theory* (John Wiley & Sons Ltd) wiley series in probability and statistics
[9] Calder C, Lavine M, Müller P and Clark J S 2003 *Ecology* **84** 1395–1402
[10] Mertz D B 1971 *The American Naturalist* **105** 437–453
[11] Buckland S T, Newman K B, Fernández C, Thomas L and Harwood J 2007 *Statist. Sci.* **22** 44–58
[12] de Oliveira E A and Alamino R C 2007 *IEEE Trans. on Neural Networks* **18** 902–905
[13] de Oliveira E A and Caticha N 2010 *IEEE Trans. on Neural Networks* **21** 1015–1020

Evolution of planet crossing asteroids in the inner Main Belt

A. O. Ribeiro and F. Roig

Observatório Nacional, Rio de Janeiro, 20921-400, RJ, Brazil

E-mail: anderson@on.br

Abstract. We studied the dynamical evolution of asteroids in terrestrial planet crossing orbits, located between 2.1 and 2.5 AU. The evolution is analyzed by direct numerical integration of massless particles under the gravitational influence of all planets from Venus to Neptune. The simulations include the Yarkovsky effect, introduced as a non conservative force that produces a slow variation of the average orbital semimajor axis. Our analysis focuses on the test particles that can reach the middle and outer regions of the Main Belt (semimajor axis $>$ 2.5 AU) during their evolution, since these may be relevant for understanding the transport mechanisms of asteroids from the inner Belt. These mechanisms could help to explain, for example, the existence of basaltic asteroids beyond 2.5 AU assuming that these bodies originate in the Vesta family, located at $\sim$ 2.3 AU. We found that, although some orbits that reach the middle and outer regions of the Belt can become temporarily detached from the planet crossing regime, and may have their orbital eccentricities damped due to capture at some mean motion resonances, such orbits survive for only a few hundred thousand years and, ultimately, the test particles return to the planet crossing regime being eventually discarded by close encounters with the planets. These results seem to indicate that a transport mechanism based only on planetary encounters and resonant capture might not be efficient enough to justify the presence of basaltic asteroids beyond 2.5 AU.

1. Introduction

One aspect of the dynamical evolution of small Solar System bodies that has called the attention of astronomers during the last decade, refers to the occurrence of mechanisms of dynamical transport that produce significant changes of the mean distances to the Sun. These mechanisms help to explain some phenomena observed in the Main Asteroid Belt, like the structure of asteroid families and the origin of small bodies populations as the Near-Earth Asteroids (NEAs).

The mechanisms responsible for these variations in orbital semimajor axis can be conservative or non conservative. Among the former, there is the gravitational swing-by effect that occurs when a small body approaches a planet and undergoes a significant change of its orbital energy and angular momentum relative to the Sun. Among the latter, there is the Yarkovsky effect which is a force caused by the thermal reemission at the surface of a small rotating body. These two effects could be combined with other dynamical processes, such as the chaotic evolution at mean motion and secular resonances, that favor the resonance capture through certain phenomena like the resonant stickiness.

Nesvorn et al. (2008) and Roig et al. (2008) presented results of numerical simulations showing that the variation of the semimajor axis due to the Yarkovsky effect and the interaction with mean motion and secular resonances helps to explain the presence of basaltic asteroids in the inner Belt outside the Vesta family. We must recall that the only well established source of basaltic asteroids in the Main Belt is an impact crater excavated on the surface of asteroid 4 Vesta, and that some basaltic asteroids are located too far from Vesta as to be explained only such collisional impact. Nesvorn et al. (2008) analyze the evolution of real and cloned objects of the known Vesta family over a time interval of 2×10^9 years. Their results show that a large fraction of the original family members may have evolved beyond the borders of the family, reaching a final distribution compatible with the presently observed one. Moreover, Roig et al. (2008) show that Vesta family members with diameters smaller than 3 km, can cross the 3:1 mean motion resonance with Jupiter reaching stable orbits in the middle Belt ($a > 2.5AU$).

A serendipitous result of the two works mentioned above was the identification of a dynamical path starting in the Vesta family and ending on a stable orbit in the middle Belt. This path is shown in Fig. 1 and involves four stages: i) inwards drift in semimajor axis due to the Yarkovsky orbital (red), ii) successive captures into some chaotic mean motion resonances, with the consequent increase of the orbital eccentricity (cyan), iii) capture into a planet crossing regime with Mars, and possibly also with the Earth (black), and vi) detach from the planet crossing regime due to resonance stickiness and temporary capture into the 8:3 mean motion resonance with Jupiter (magenta). Nevertheless, this behavior was observed in only 1 out of 6600 test orbits, which implies that it should be quite unlikely.

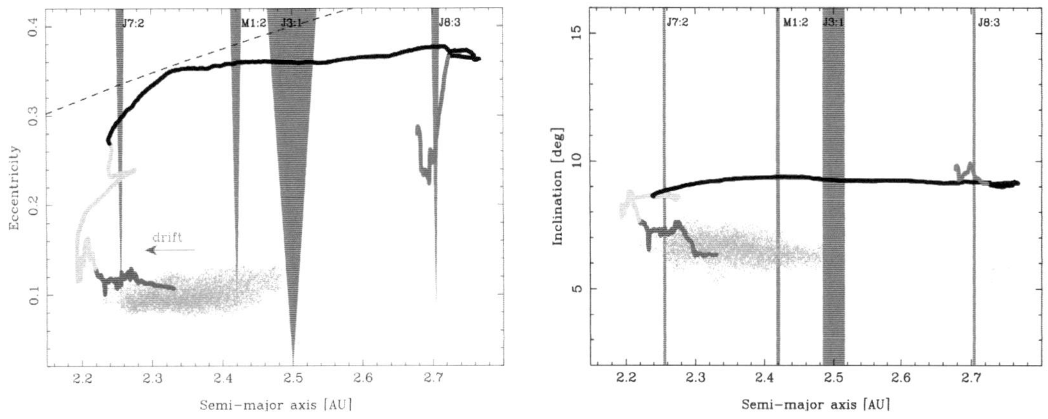

Figure 1. Evolutionary path of a test particle from the Vesta family (yellow dots) to the middle Main Belt. The different colors identify the different stages of the path. Left: eccentricity vs. semimajor axis. Right: inclination vs. semimajor axis. The plotted orbital parameters are averages over 10^6 years.

The purpose of this work is to analyze in more detail some stages of this dynamical path, aiming for a better estimate of its occurrence probability and its actual efficiency for transporting asteroids from the inner Main Belt to beyond 2.5 AU. We focus particularly on the evolution of orbits that already entered the terrestrial planet crossing regime, and follow their paths into the middle and outer Belt. Along these stages, the evolution times are relatively short, of the order of 10^7 years, which allows to perform a large number of numerical simulations at a low computational cost. Moreover, the analysis of these dynamical stages turns to be sufficient to address the efficiency of the whole transport mechanism from its initial stages.

In section 2 we introduce the applied methodology, and in section 3 we present our results. Section 4 is devoted to the conclusions.

2. Methodology

We studied the dynamical evolution of test particles through numerical integration of a Solar System model including perturbations of the planets from Venus to Neptune and a non conservative force mimicking the Yarkovsky effect. We considered orbits of real and fictitious asteroids initially crossing the orbit of Mars. The initial conditions for the planets and the real asteroids were obtained from the ephemeris of the Jet Propulsion Laboratory (JPL). The mass of Mercury was added to the mass of the Sun applying the corresponding barycentric correction, and the orbits were referred to the invariable plane of the Solar System.

The equations of motions were propagated using the symplectic scheme of Wisdom-Holman-Skeel, that consistently reduce the integration time step during close encounters. The time step was set to 0.03 years, and data output was recorded every 10^4 years.

The test particles were removed from the simulation if: i) they had a close encounter with any planet at a distance less than 1% of the corresponding Hill's sphere, ii) they reached a heliocentric distance larger than 10 AU, and iii) their heliocentric or pericentric distances were smaller than 0.01 AU.

Our simulations spanned a total time of 10^7 years, consistent with the times of the dynamical stages we are analyzing. It is worth noting, however, that this time span is only 10% the typical lifetime of a Mars crosser asteroid ($\sim 110 \times 10^6$ years; Migliorini et al. 1998). Therefore, our results must be considered with caution and regarded only as a first approach to the dynamics of the problem.

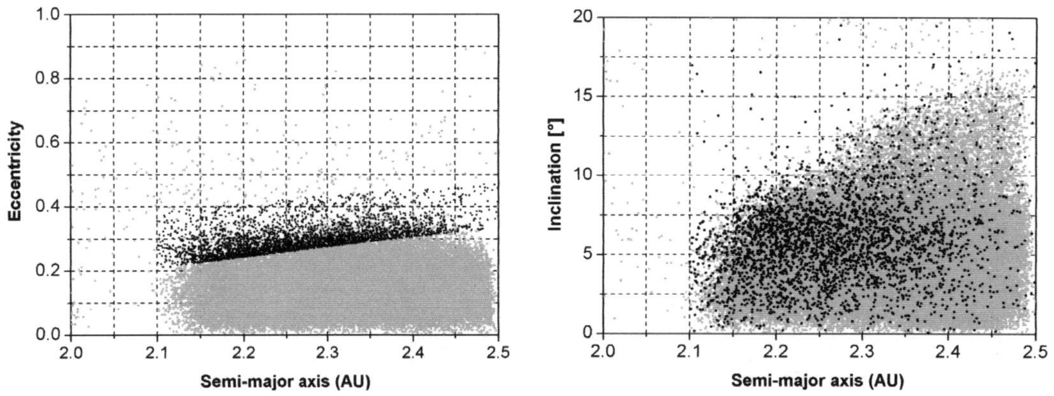

Figure 2. Distribution of the initial conditions of the sample $Real3263$ (blacks dots). Gray dots represent the asteroids in the inner Main Belt. Left: eccentricity vs. semimajor axis. Right: inclination vs. semimajor axis.

We selected initial conditions of 3263 real Mars crossers with multi oppositional orbits between 2.1 and 2.4 AU. This sample was named $Real3263$ and is shown in Fig. 2. We furthermore generated three grids of 1000 fictitious initial conditions each, initially crossing the orbit of Mars. These orbits were randomly sorted from a uniform deviate within the ranges 2.1 to 2.4 AU in semimajor axis, 0° to 19° in inclination, and 0° to 360° in λ, ϖ and Ω. The eccentricities were set accordingly among three different intervals of perihelion distance: $1.33 < q < 1.44$ AU, $1.44 < q < 1.55$ AU, and $1.55 < q < 1.66$ AU. These samples were named P_p1000, P_m1000 and P_g1000, respectively, and are shown in Fig. 3.

The evolution of each test particle was monitored until it eventually reached a semimajor axis > 2.5 AU. We assumed that a particle was effectively "transported" beyond the inner Main Belt if it continuously remained with $a > 2.5$ AU for a time interval $\Delta T \geq 10^5$ years. It may

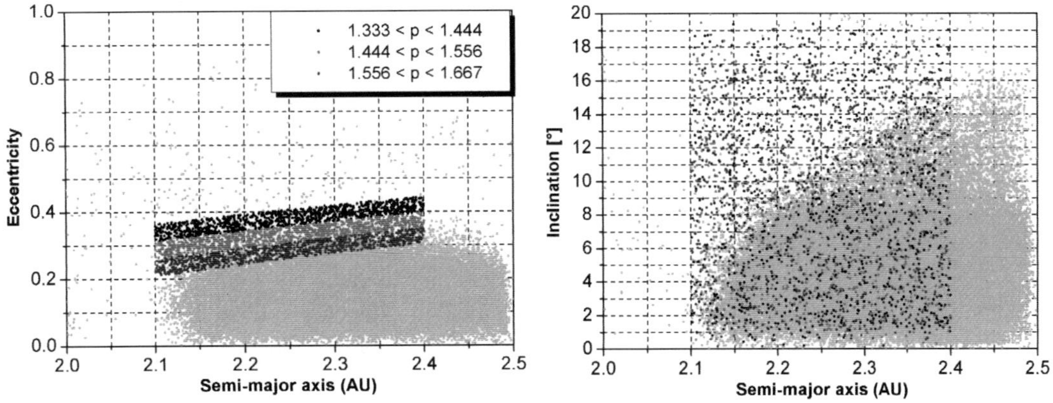

Figure 3. Same as Fig. 2 but for the samples P_p1000 (black dots), P_m1000 (red dots), and P_g1000 (blue dots).

happen that a given particle was transported once by a given time interval ΔT_1, then returned back to the inner Belt ($a < 2.5$ AU), then it was transported again by another time interval ΔT_2, and so on. In such case, we considered $\Delta T = \sum_i \Delta T_i$

In order to qualitatively analyze the behavior of the transported particles, we also built maps of residence times. We subdivide the space of eccentricity vs. semimajor axis in 20×70 cells, each of 0.05×0.01 AU in size, within the limits $0 < e < 1$ and $2.5 < a < 3.2$ AU. Then, we determined the corresponding time interval that each orbit continuously remained in each cell during the simulation, and for each cell we computed average time interval over all the orbits. We thus obtained for each cell the factor Γ, which represents the (average) residence time of any particle in that cell.

3. Results

Figure 4(left) shows the number of surviving particles in the sample *Real*3263 as a function of time. The black curve represents the total population, while the blue and red curves represent the populations with initially "low" and "high" inclination orbits, respectively. As expected, the half-life of this sample of real Mars crossers is actually much larger than the time span of our simulation. The distinction between orbits with initially "low" and "high" inclination is an rough indicator of evolution outside and inside the ν_6 secular resonance, respectively. It is worth recalling that the chaotic evolution in this secular resonance is responsible for the notorious depletion of real asteroids at high inclinations, as shown for example in Fig. 2.

Figure 4(right) shows the map of residence times beyond 2.5 AU for all the particles of the sample. The thin stalactite shape observed around 2.9-3.0 AU is related to temporary captures into resonance, possibly caused by resonant stickiness.

Figure 5(left) shows the distribution of the individual residence times, ΔT, of the "transported" particles during the simulation the sample P_g1000. The residence times are very short, in most cases of the order of the imposed minimum (10^5 years), and only in a few cases the particles stayed a few million years. Figure 5(right) shows the corresponding map of residence times for the whole sample. Similar portraits are presented in Figs. 6 and 7 for samples P_m1000 and P_p1000, respectively.

A comparative analysis between these three samples indicates that the smaller the initial perihelion distance, the larger the residence time beyond $2.5AU$. But this is likely to be related to the larger number of transported particles for smaller initial perihelion distances. The maps

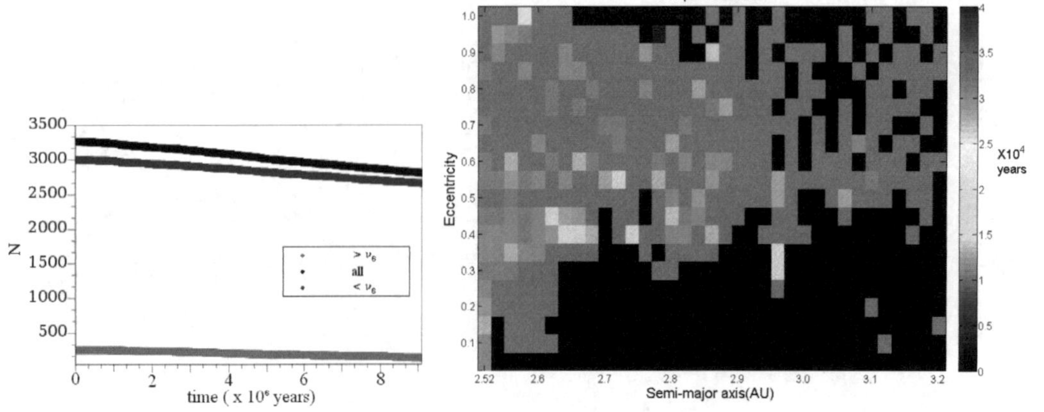

Figure 4. Results for the sample $Real3263$. Left: total number of surviving particles as a function of time (black curve), and the corresponding low (blue) and high (red) inclination populations. Right: map of residence times Γ.

of residence times show that the smaller the initial perihelia, the larger the area "visited" by the transported orbits. This favors the occurrence of temporary captures into resonances, as shown by the stalactite structures in Figs. 6 and 7.

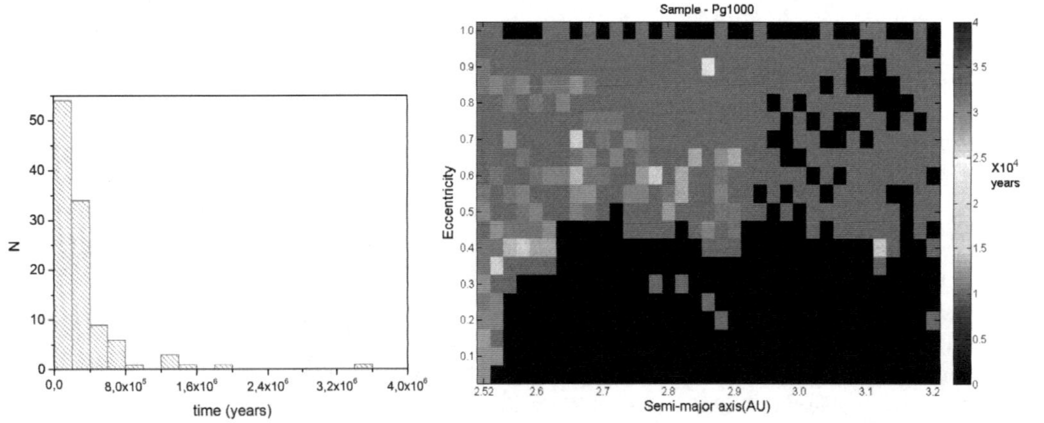

Figure 5. Results for the sample P_g1000. Left: distribution of ΔT for the "transported" particles. Right: map of residence times Γ.

Our results indicate that the larger the number of transported orbits beyond 2.5 AU, the greater the probability of having resonant captures Therefore, the results for the samples having lifetimes much longer than our simulation time span should be considered with caution, since they could change significantly if the simulations are extended over longer time spans.

4. Conclusions

We can conclude that, although some orbits in our simulations may reach the middle and outer regions of the Main Belt, becoming temporarily detached from the planet crossing regime and having their eccentricities damped due to captures into mean motion resonances (probably

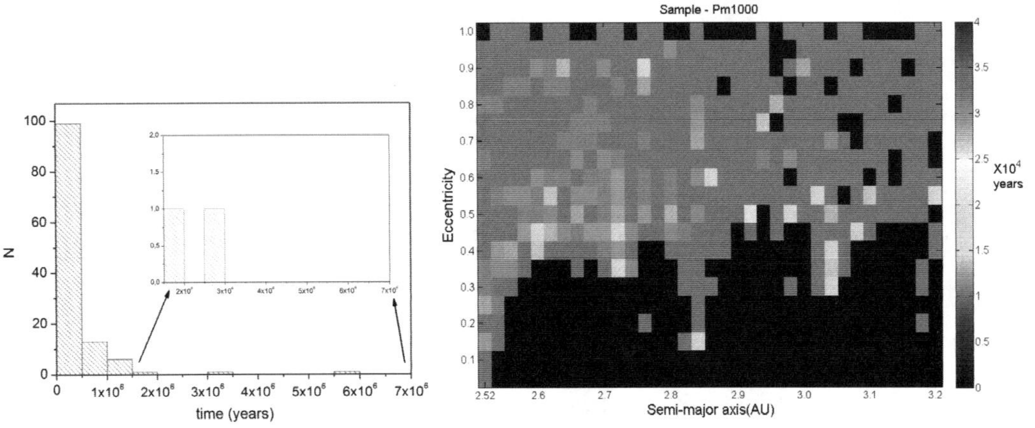

Figure 6. Same as Fig. 5 but for the sample P_m1000.

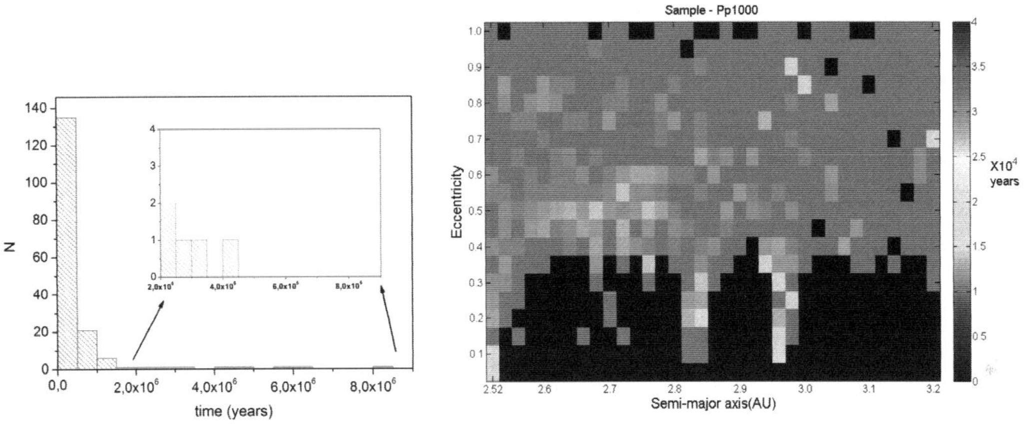

Figure 7. Same as Fig. 5 but for the sample P_p1000.

through resonant stickiness), such orbital configurations only survive for time intervals up to 10^5–10^6 years. These orbits eventually return back to the planet crossing regime, and the particles are discarded due to a close encounter with a planet.

These results seem to indicate that a transport mechanism similar to that shown in Fig. 1 might not be efficient enough to justify the presence of basaltic asteroids beyond 2.5 AU. We must note, however, that our simulations do not span sufficiently long times, and by extending them over longer intervals (e.g. 10^8 years) we might get more favorable results.

References

[1] F. Migliorini, P. Michel, A. Morbidelli, D. Nesvorny, and V. Zappala. Origin of Multikilometer Earth- and Mars-Crossing Asteroids: A Quantitative Simulation. *Science*, 281:2022–+, Sept. 1998.

[2] D. Nesvorný, F. Roig, B. Gladman, D. Lazzaro, V. Carruba, and T. Mothé-Diniz, *Fugitives from the Vesta family*, Icarus **193** (2008), 85–95.

[3] F. Roig, D. Nesvorný, R. Gil-Hutton, and D. Lazzaro, *V-type asteroids in the middle main belt*, Icarus **194** (2008), 125–136.

Navigation towards a goal position: from reactive to generalised learned control

Valdinei Freire da Silva

Laboratório de Técnicas Inteligentes – LTI, Escola Politécnica da Universidade de São Paulo, Av. Prof. Luciano Gualberto, trav.3, n.158, Cidade Universitária São Paulo - Brazil

E-mail: valdinei.freire@gmail.com

Antonio Henrique Selvatici

Universidade Nove de Julho, Rua Vergueiro, 235, São Paulo, Brazil

E-mail: antoniohps@uninove.br

Anna Helena Reali Costa

Laboratório de Técnicas Inteligentes – LTI, Escola Politécnica da Universidade de São Paulo, Av. Prof. Luciano Gualberto, trav.3, n.158, Cidade Universitária São Paulo - Brazil

E-mail: anna.reali@poli.usp.br

Abstract. The task of navigating to a target position in space is a fairly common task for a mobile robot. It is desirable that this task is performed even in previously unknown environments. One reactive architecture explored before addresses this challenge by defining a hand-coded coordination of primitive behaviours, encoded by the Potential Fields method. Our first approach to improve the performance of this architecture adds a learning step to autonomously find the best way to coordinate primitive behaviours with respect to an arbitrary performance criterion. Because of the limitations presented by the Potential Fields method, especially in relation to non-convex obstacles, we are investigating the use of Relational Reinforcement Learning as a method to not only learn to act in the current environment, but also to generalise prior knowledge to the current environment in order to achieve the goal more quickly in a non-convex structured environment. We show the results of our previous efforts in reaching goal positions along with our current research on generalised approaches.

1. Introduction

A key task in building mobile robots is navigation, i.e., moving from an initial position to the target one. When knowledge about constraints (obstacles) and dynamics are considered to be known, the navigation problem can be addressed analytically from the point of view of control theory. However, a different approach must be used if unknown environments are considered.

Robots are physical agents that must perform their tasks autonomously in the real world. Depending on the type of the robot, different sensors input are available that must be processed so that meaningful information can be extracted. The performance of a robot is the result of its perception, given by the sensory inputs, and its control system, all integrated in a *robot architecture*. The interaction between the robot and the environment occurs through alternating cycles of perception (acquiring sensory information) and action (choosing appropriate controls) [1].

If we consider mobile robots as autonomous agents, it is natural that the design of its architecture depends on the type of agents chosen to model the robot. Russel and Norvig [2] propose a classification of agents considering the level of abstraction of the information employed in the control of their actions. The information can represent knowledge about the environment, about the agent's objective, and about the agent itself.

Regarding knowledge about the environment, consider an agent whose actions are controlled by the current or previous short time inputs; we call it *pure reactive agent* [3, 4]. Such an agent can be programmed with condition-action rules, but due to its limited knowledge, the control can only be guaranteed to be suitable in a short period of time, although adequate long-term behaviour can be achieved, depending on task and the environment.

A more robust agent considers previous long-term inputs, constructing a model of the environment on most architectures; for this reason, we call it *model-based agent* [5, 6]. The model of the environment may rely on two classes of knowledge: knowledge about how the world evolves independently of the agent actions and knowledge about how its own actions affect the world.

Regarding knowledge about the agents's objective, instead of providing an agent with rules based only on sensor inputs (in the long or short previous time), it is possible to consider also an intrinsic objective state, which also interferes in the condition of condition-action rules. If such state models the agent's objective it can be used in order that the agent presents flexibility in dealing with changes in the world. If it is possible to model agent's objectives as a set of desired situations we call it a *goal-based agent* [7, 8]. In this case the agent can combine the knowledge about the results of its own actions in order to choose those that attain the goal (a desired situation). Planning techniques relies on models of the environment and on reasoning about future possible situations in order to provide an adequate control to attain its objective [9]. Goal-based agents divides control in two qualitative classes: controls that reach the goal and controls that do not reach the goal, however, depending on the task, it is also important to analyse each control quantitatively. This can be achieved by means of a more general performance measure that allows a comparison of the possible satisfaction level among different trajectories to a goal state. The agent can implement such measure through a utility function, which maps a sequence of states of the environment into a real number; that is why we call such agent *utility-based agent* [10, 11]. Utility-based agents are especially important in stochastic environments, where a goal cannot be guaranteed to be attained. In this case, the utility function provides a mean of pondering the probability of success with regard to the importance of the goals.

As for the agent himself, the agent can modify internal states autonomously in order to adapt to each environment; we call it *learning agent*. In addition to the model of the environment, the agent can also learn rules about how to control itself autonomously so that its objective can be attained. Reinforcement Learning [12, 13] is an interesting approach for learning how to control an agent in an unknown environment without knowledge about its dynamics, but requiring an appropriated state model of the environment, so that association can be made among states and actions.

In this paper we investigate the use of Relational Reinforcement Learning as a method to not only learn how to navigate in the current environment, but also to generalise the acquired knowledge in order to achieve the goal more quickly in an unknown non-convex but structured environment. In the next sections, we present our ongoing and planned efforts to combine all the previously presented architectures of agents in order to build a hybrid robot architecture that can reach goal positions fast in unknown environments.

Section 2 describes our first experiments with a pure reactive agent, where we combine simple motor schemas in order to pursue a goal position avoiding obstacles. Even if the robot cannot reach the goal position in structured environment, it is highly effective in environments with convex obstacles. Section 3 describes our architecture to combine goal-based, utility-based and learning agent in order to better coordinate motor schemas so that performance is improved. Section 4 describes our algorithm to combine low level control (motor schemas) with high level control (Reinforcement Learning), where we show that although finding a goal-position randomly can be time expensive, an adequate control can be learned fast. In the direction of learning generic rules which can be optimal among similar structured environments, in section 5 we present our current efforts in applying Relational Reinforcement Learning to the navigation task of reaching a goal position.

2. REACT: a pure reactive architecture

The behaviours integrating our reactive architecture REACT [14, 15] are based on Arkin's motor schemas [1]. A motor schema is a behaviour encoded by the potential fields method that translates sensor readings into a movement vector directly and in a continuous fashion.

Each motor schema is composed of two modules: the perceptual schema and the encoding module. The perceptual schema is responsible for the sensorial processing, determining relevant information for controlling the robot from sensors input. The encoding module calculates a movement vector based on the potential fields method using the information returned by the perceptual schema.

The movement vector is analogous to a force derived from some potential function, usually generated by associating repulsive charges to the obstacles and attractive charges to the target position. At each instant, each motor schema calculates the forces generated by the interaction of the robot with the virtual potential field, and returns the resultant force as a movement vector.

Before calculating the final action, the influence of each motor schema must be coordinated. The potential fields method suggests a very simple way to do it: the multiplication of each motor schema response by a pondering weight. Robot actions are determined by means of weighing the movement vectors returned by each motor schemas and then performing their vectorial summation.

Encoded as motor schemas, the behaviours that integrate the REACT architecture are: avoid collision, move to goal, and move ahead. For each behaviour a pondering weight must be defined in order to coordinate the final action (w_{AC}, w_{MTG} and w_{MA}).

The avoid-collision behaviour aims at avoiding the collision to the obstacles presented in the environment. The perceptual schema processes range readings in order to identify the location of the detectable obstacles. Then the movement vector is calculated analogously to the electrostatic force in Coulomb's law: repulsive charges are associated to each identified obstacle, generating move-away vectors with magnitudes that grow with the proximity to the obstacles. The behaviour response corresponds to the vectorial sum of the calculated vectors for all detected obstacles. The equations that determine the movement parameters for each obstacle are:

$$V(d) = \begin{cases} V_{AC}\, e^{\frac{S-d}{T}} & \text{for } d > S \\ V_{AC} & \text{for } d \leq S \end{cases} \quad \text{and} \quad \phi = \pi - \phi_{\text{rob-obst}}, \tag{1}$$

where V is the response magnitude (speed), d is the distance from the mass centre of the robot to the obstacle, V_{AC} is the maximum speed allowed for the behaviour, S is the stand off distance from the obstacle, T is the scale constant for the exponential function, ϕ is the motion direction, π is the mathematical number, and $\phi_{\text{rob-obst}}$ is the direction defined by the straight line that passes by the obstacle and the robot's centre of mass.

The move-to-goal behaviour aims at attracting the robot to a pre-determined location in the environment. The target position in the global coordinate system is informed by an external agent. The current robot position is determined by some localisation method. The resultant motion direction is equal to the target direction, given by the straight line that passes by the robot's centre and the target location ($\phi = \phi_{\text{rob-targ}}$). The magnitude is given by a constant ($V = V_{MTG}$).

The move-ahead behaviour provides a certain trend for the robot of not changing its heading direction. No sensory information is used and the motion parameters are determined in a very simple manner: the magnitude is a constant ($V = V_{MA}$) and the direction is equal to the current robot heading ($\phi = \phi_{\text{robot}}$).

REACT can reach a goal very fast in small convex-obstacle worlds. If there are large obstacles to be avoided the robot may choose the most expensive direction to go, what will result in a significantly worse path. However, it does not matter how large an obstacle is, as long as it is convex, the goal will be reached. On the other hand, if there are non-convex obstacles in the world, the agent may get stuck in one of them. This happens because of the activation and deactivation of the avoid collision behaviour when inside a non-convex obstacle whose concavity is opposite to the goal.

A less restrict but essential point of view is regarding the robot performance. Depending on the parameters used in the motor schemas, the robot may perform poorly as evaluated by a given performance measure. In the next section we show how to tune the pondering weights in order to obtained a more balanced behaviour.

3. AAREACT: tuning reactive behaviours

AAREACT is a robot architecture that combines learning and reactive behaviours in order to not only reach the goal position, but also to improve a performance measure [15, 16]. The architecture consists of a learning apparatus which modifies parameters of the embedded REACT architecture based on the robot performance. The architecture schema is outlined in figure 1.

The learning part of the architecture is called *coordination layer*. Its role is that of adapting the values of the influence parameters that define the weight of each behavioural response in the resultant action depending on the environment situation. Defined by a suitable interpretation of the robot sensors, each situation must be mapped into a weight vector. Thus, the coordination module has to learn the best way to coordinate the behaviours regarding the current sensory data.

The situation of the environment is determined by the supervisor module present in this layer, while the critic module observes the robot performance. One should notice that both modules may use raw sensor data and the perception information determined by robots perception. However, while the supervisor module considers only current information, the critic module has an internal state being updated during the interval of one situation. When the situation changes, the critic module defines a reinforcement value that summarises how well it behaved while that situation was observed.

The learning model adopted in AAREACT is Reinforcement Learning (RL). RL is a generic model towards autonomous agents. An agent must maximise a given performance measure through trial-and-error interactions with the environment. The performance measure is described by a reinforcement function [13]. The underlying model of several RL algorithms are Markovian Decision Processes (MDP). An MDP is described by a tuple $\langle \mathcal{S}, \mathcal{A}, P(s'|s,a), r(s,a) \rangle$, where $\mathcal{S}$ is a set of states (situations), $\mathcal{A}$ is a set of actions (vectors of weights), $P(s'|s,a)$ models the probability of transition to the state s' when the action a is chosen in the state s and $r(s,a)$ is the reinforcement incurred. The sets of states and actions are considered to be known, whereas the probabilities $P(s'|s,a)$ and reinforcements $r(s,a)$ must be inferred directly from the environment.

The RL module maintains a function $Q(s,a)$ of the estimated utility of choosing a certain vector of weights to be sent to the Merging module in each possible situation of the environment. In every change of situation, the RL module gets the new situation and the reinforcement value from the supervisor and critic modules, respectively, and updates its $Q(s,a)$ by using an RL algorithm. Then, in most cases, the parameters chosen to be passed to the reactive layer are those considered to have the greatest utility among all for the current situation. However, random actions must also be tried in order to learn improved control rules, when it is possible.

When the agent observes a situation s and executes an action a, it obtains a response from the environment in terms of the reinforcement r which can be a reward or a penalty. Then, a new situation s' is perceived. Once in this new situation, the agent has to choose a new action a'. Being the most popular algorithm, Q-learning [17] update is done through:

$$Q(s,a) \leftarrow Q(s,a) + \alpha(r + \gamma \max_{a' \in \mathcal{A}} Q(s',a') - Q(s,a)). \tag{2}$$

where $r \in \mathbb{R}$ is the reinforcement signal, $\alpha \in]0,1[$ is the learning rate, and γ is the discount rate. Both α and γ are project parameters previously defined.

If one decides to always trust in the current result of the learning process, the adopted strategy should be the greedy one which always chooses the action to which is associated the greatest value of utility Q for the situation. However, a less confident strategy is more suitable to explore the possibility of actions, what is useful mainly in the initial phases of learning, when the agent is unlike to have enough experience to decide the best action for the current situation.

Analogously to Kalmár's work [18], the environment situation space is defined by a set of on/off features, abstracted from the sensors' data. The environment situation is then defined by a vector indicating the activation or not of each feature, called features vector. The features defined for AAREACT are based on the presence of obstacles between the agent and the target position, relative direction of the target point, and relative direction and distance of obstacles, as depicted in figure 2. The environment situation is defined by the vector of features, signaling which features are on and which are off. A change of situation occurs when one or more inactive features are activated, or also when one or more activated features are deactivated.

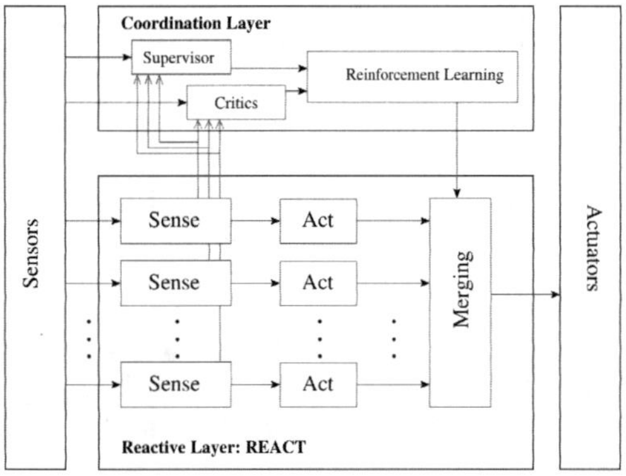

Figure 1. AAREACT outline.

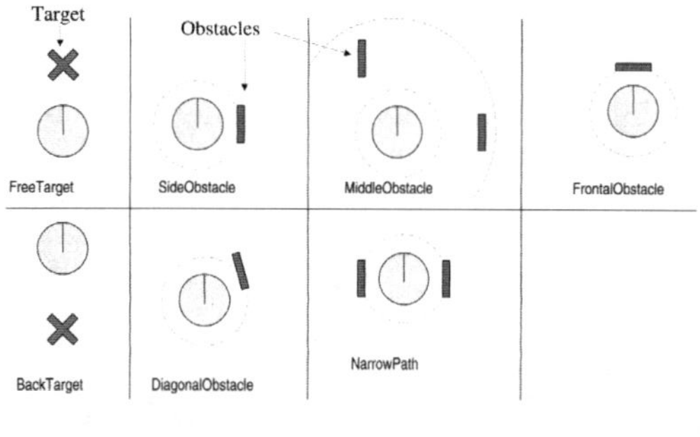

Figure 2. Illustration of the defined features. The robot is represented by the yellow circle, the red cross represents the target position, and the rectangles represent the detected obstacles.

Since the space of weight vectors is continuous, we proposed associating a weight vector for each feature, besides a generic weight vector which ponders all of the behaviour with non-null weights. Only weight vectors associated to features which are on and the generic weight vector are considered when choosing a coordinating action. Such consideration avoids the need of exploring clearly bad actions and allows the agent deciding autonomously which feature is more important.

In RL, positive reinforcements are used to reward desirable situations, while negative reinforcements also can be used to penalise undesirable situations. The robot's primary objective is to get to the target location. When it occurs, the critics module generates a large reward, defined by r_{goal}. It is also desirable that the robot presents a good performance during its actuation, whatever is desired from the robot. Whereas time to get to the target position is clearly desirable, other performance measures, such as energy consumption, may be chosen arbitrary by a programmer.

When a situation change is detected, a reward proportional to the chosen performance measures is calculated, characterising an *intermediate reinforcement*, received before reaching the target. Although such reinforcement can be positive, it must be small when compared to r_{goal}, so that a trajectory that reach the target position can always be a winner trajectory.

4. Accelerating Goal-driven Reinforcement Learning

Whereas the avoid-collision and go-ahead behaviours can be substituted by different suitable behaviours, the same is not true regarding the move-to-goal behaviour. This behaviour is essential in order to avoid random exploration and an accurate relative position between robot and target position must be obtained, so that the agent can define such a behaviour.

In this section we explore a global state space which may be provided by simultaneous localisation and mapping techniques (SLAM) [19, 20], which can map the environment while exploring it for the target position. Mapping the world is specially important in places where the robot can get stuck. If the target position is always reached, the location of both the robot and the target could be less accurate if the sensory information could guide the robot approach to the target position. In this case, the target position could be represented as an object to be approached and an object-based map could be used [21].

When dealing with continuous state space, it is necessary to discretise the state space into discrete macro states before traditional RL algorithms can be used. The final rules of control that are learned depend on the pattern and granularity of the discretisation. While high-definition discretisation allows near optimal control, the time spent in learning grows exponentially with the number of states. An alternative to deal with this problem is to discretise the space of states non-uniformly. In navigation tasks such high-definition discretisation is mainly required near obstacles, whereas it is unnecessary in free areas [22, 23].

Besides providing a better control near obstacles, it is important to provide an appropriate continuous control. Appropriate robot architectures use two levels of control: one low-resolution discretisation of the state space to control the robot globally (figure 3), and one high-resolution discretisation of the state space to control the robot locally. Since the position of the goal is used mainly to estimate a direction to the goal (move-to-goal behaviour), discretised directions are one interesting choice to model a set of actions to apply in the low-resolution discretisation. The AAREACT or REACT architectures can be used to control continuously the robot in the low-level of the hierarchy. An alternative approach is to define a behaviour that deviates the robot from obstacles.

We propose the CFQ-Learning algorithm (Compulsory Flow Q-Learning) [24] that can do this low level control, instead of using AAREACT at the lowest level of the hierarchy. In the CFQ-Learning algorithm the robots previously learn a local behaviour to follow around obstacles, and this behaviour is named compulsory flow (figure 4). At the highest level of the hierarchy the robot chooses a main direction to go and such action is executed in the lowest level until: 1) an obstacle is found, and the compulsory flow takes control; or 2) a transition among macro-states occurs and a new action is chosen in the highest level.

Another method proposed by us is the Heuristically Accelerated Q-learning (HAQL) [25]. Whereas in its traditional definition the RL algorithm learns optimal control from scratch, HAQL uses heuristics to accelerate the learning process. Such heuristics can be introduced in two different ways: 1) by choosing an appropriate initial Q-function; and/or 2) by choosing appropriate actions in the trial-and-error process. Control actions provided by the AAREACT architecture is a good choice to be used as heuristics when no knowledge about the environment

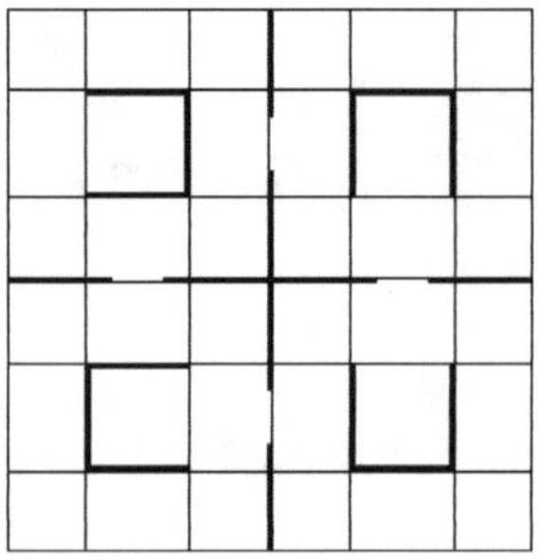

Figure 3. An example of discretisation with low definition and non-convex obstacles.

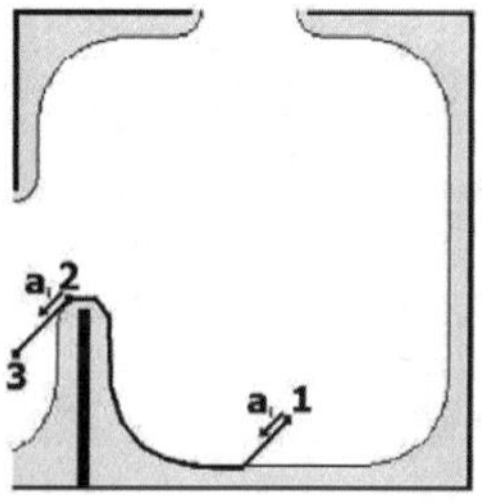

Figure 4. An agent's movement, which starts at point 1 and follows the compulsory flow until reaching the point 2, when it is released, reaching point 3.

is known. The distance in a straight line between two states can also be used as an optimistic heuristic function. The performance of the learning process using HAQL can be improved even using very simple heuristic functions. Also, the heuristic function can be modified or adapted online, as learning progresses and new information for enhancement of the heuristic becomes available. Another interesting feature of HAQL is that even if the heuristic is not completely adequate, performance enhancement can take place due to the partial correctness of the heuristic.

We are currently investigating such alternatives. First, by applying the CFQ-Learning algorithm in the Pioneer robot, using both the compulsory flow control coded by hand and the compulsory flow control learned directly from interactions with the environment. We are also implementing a higher level of control in the AAREACT, in the same way as CFQ-Learning, so that the compulsory flow control can be replaced by the AAREACT control. In both architectures we will also experiment RL algorithms based on heuristics so that the target position is reached more quickly.

Whereas robot navigation presents randomness in control dynamics and robot perception, if the constructed partial map is accurate and we use the assumption that connected neighbour position can be reached by the AAREACT architecture, planning algorithms can be used in order to avoid the robot getting stuck. However, RL algorithms can couple better with randomness in the control dynamics.

Since we are mainly interested in the robots reaching fast the target position in its first trial with the environment, we plan working with algorithms which keep track of previous experiences and constantly update the Q-function where and when it is necessary [26, 27]. In this case, the definition of the learning rate α should be non-stationary providing an appropriate use of initial experiences.

Prioritised Sweeping is an algorithm which keeps information about visited states-action-states triads (s, a, s') and its dependence on the Q-values of next states s'. If after an experience at time t the difference between the updated max value $\max_{a' \in \mathcal{A}} Q_t(s', a')$ and the previous max value $\max_{a' \in \mathcal{A}} Q_{t-1}(s', a')$ is beyond a threshold, the value $Q(s, a)$ is updated simulating the previous experiences related to the triad (s, a, s'). The order in which simulation occurs is prioritised on the difference regarding current update and the dependence of a triad on such difference.

5. Relational RL as a generic solution

Given the time needed to learn an optimal control, some works in the area of RL were devoted to the possibility of reusing learned information [28, 29]. This reuse can be done among different environments or among different states in the same task [30]. The first case is usually obtained through the use of locally controlled sub-tasks, whereas the second case is based on some state similarity measure. A global reuse of information can only occur when the representation of states is the same among environments.

Factored states are commonly used as a technique to allow reuse of information [31]. A factored state is represented by a vector of n factors. Similarity among states can be measured using a sub-set of factors. Reuse can be done in two ways: Q-function and actions. Sometimes the similarity is taken as equality of Q-function values, meaning that the states are at the same distances to the goal. In this case the value attributed to one state can be attributed to similar states. A more common approach is to consider that we can simulate a transition between two states as occurring in other similar states or that the optimal actions to be performed are the same for some set of similar states.

Whereas in the AAREACT architecture a locally factored state is considered in order to provide generalisation, the set of factors is chosen *a priori* and it is incomplete. If generalisation is possible because of such incompleteness, it is also the source of problems when such *a priori* factors are not enough. A relative new set of methods, Relational Reinforcement Learning (RRL) proposes starting with a complete set of factors and finding an adequate subset of them [32].

The RRL framework considers an MDP where the states and actions are structured into first-order propositions. First-order propositions are based on an object domain set $\mathcal{D}$ and a set of predicates $\mathbb{P}$. A ground state is a conjunction of predicates $p \in \mathbb{P}$ applied to objects $o \in \mathcal{D}$. An abstract state considers variables instead of objects when applying predicates, then representing a set of ground states. Generalisation is obtained choosing actions to abstract states.

RRL algorithms have been applied to block world problems, showing generalisation among tasks involving different numbers of blocks in the world. However, it strongly depends on the formulation of the factored states. The application of these methods to navigation problems have been less fruitful [33, 34]. Some works have taken advantage of structured buildings (rooms, halls and lifts) to infer general control rules like: entering room first needs to access a hall, accessing a hall first needs to access the lift, accessing the lift first needs to access any hall, etc. But, if buildings present connections among rooms, this kind of generalisation fails (the robot may found a target room navigating only among rooms). In this case a recover strategy is used based on a list of actions previously experienced. However, such solution does not couple with stochastic transitions in the world.

If generalisation cannot always be accurate, we believe that it can be a good source of heuristic so that we can have better guesses when exploring an unknown environment. For instance, when looking for a target point, if the robot is not far away from the target point, the AAREACT control should be a good generalisation, whereas when the robot is far away from the target point, the control rule "get access to a hall" should be a good generalisation. However, when the robot is in the hall, it can learn rules stating the disposal of visiting rooms in the opposite direction to the target and exploring rooms that are in the same direction and that had not yet been visited.

Our current research aims at formalising partial generalisation. If generalisation cannot be accomplish fully, partial generalisation may be a strong advice of how pondering appropriately the rate of exploration of each action. RRL algorithms results in the indication of an action for abstract states, i.e., for a set of ground states. However, some states may not agree about the optimal action, even if the full set of predicates is considered; in this case it should be considered the proportion of each winner action as an indication of exploration rates.

In order to apply RRL algorithms to the navigation towards a goal position, the world must be modelled regarding objects and predicates. We consider a mapping relating walls, doors and convex obstacles. Convex obstacles near the robot can immediately be observed and treated appropriately by the architecture AAREACT. Walls and doors must be identified in order to generate abstract states and abstract actions.

We use the following predicates in order to describe an abstract state: `hasDoor(X)`, `hasFreeSpace(X)`, `isDirection(X,Y)`, `targetDirection(Y)` and `isTargetVisible()`. The predicate `hasDoor(X)` indicates that a door `X` can be seen, whereas the predicate `hasFreeSpace(X)` indicates there is a free space `X`, i.e., without walls. The predicate `isDirection(X,Y)` indicates the relative direction `Y` (east, north, etc.) of a door or an open space `X` regarding the current robot position, the predicate `targetDirection(Y)` indicates the direction `Y` of the target and `isTargetVisible()` indicates if the robot can reach the target position.

We consider three abstract actions: `moveToDoor(X)`, `moveDirection(Y)` and `moveToTarget()`. All of this actions is executed by the AAREACT architecture, where in the case of the action `moveDirection(X)` the move-to-goal behaviour keeps constant the direction to the goal.

A second and more complex direction to take consists in generalising not only among optimal actions, but also generalising among alternative actions as long as they present similar performance. Although the Q-value can be used to analyse the quality of an action in one step, being an indicator of possibility of improvement, when analysing worse actions it may not perform well as an indicator of performance. Policy gradient analysis may help when evaluating the true value of changing one of the known optimal actions [35].

6. Conclusion

In this paper we present our advances in the task of navigating towards a goal position. We divide them in two parts: reactive solutions and learned solutions.

Despite being quite simple, reactive solutions are essential in many aspects. First, they serve as an interface between the continuous analog world and the discrete digital world. Whereas intelligent solutions backed by one of these two worlds are not satisfactory, the union between them can bring the best of both worlds. Second, hand-coded motor schemas are easy ways of providing safe control to the robot. Using motor schemas and appropriate sensors, it is very easy to do implement collision avoidance, falling edge avoidance, among other risky behaviours. Third, even with these hand-coded behaviours, arbitrary performances can be learned to adapt the motor schemas to a particular environment

Our approaches to accelerate Reinforcement Learning have also been promising. Although the approaches have been developed separately from our reactive approaches, we believe that they will unite naturally. The use of Relational RL seems to be very promising, and despite some discouraging examples in navigation tasks we believe our designed approach will be fruitful.

Although not explored in this paper, SLAM has been another focus of our research, mainly regarding semantic knowledge augmenting the map with label to objects and situations. Besides fitting well with our current necessities, our mapping solution is very useful to more complex tasks involving objects. Although the task of navigating to a target appears to be largely independent from objects, navigation is a basic task in a top layer of hierarchical architectures and most often is motivated by some objects in the environment. Navigating toward a target position often comes along with interactions with objects, so that the basic navigation task would be more like: "go to the red table". Then our map solution will be very helpful in such cases.

Acknowledgments

This work was conducted under project LogProb (FAPESP proc. 2008/03995-5). Valdinei F. Silva thanks FAPESP (proc. 09/14650-1) and Anna H. R. Costa thanks CNPq (proc. 305512/2008-0).

References

[1] Arkin R C 1998 *Behavior-Based Robotics* (Cambridge, MA: The MIT Press)
[2] Russel S and Norvig P 2003 *Artificial Intelligence: A Modern Approach* 2nd ed (Upper Saddle River, New Jersey: Prentice Hall) ISBN 0-13-790395-2
[3] Nolfi S 2002 *Neurocomputing* **42** 119 – 145 ISSN 0925-2312
[4] Woolley B and Peterson G 2009 *Journal of Intelligent & Robotic Systems* **55**(2) 155–176 ISSN 0921-0296
[5] Hester T and Stone P 2009 *AAMAS '09: Proceedings of The 8th International Conference on Autonomous Agents and Multiagent Systems* (Richland, SC: International Foundation for Autonomous Agents and Multiagent Systems) pp 717–724 ISBN 978-0-9817381-7-8

[6] Szita I and Szepesvári C 2010 *ICML* ed Fürnkranz J and Joachims T (Omnipress) pp 1031–1038
[7] Macal C M and North M J 2010 *Journal of Simulation* vol 4 (Operational Research Society) pp 151 – 162
[8] Kwiatkowska M, Norman G and Parker D 2007 *Formal Methods for Performance Evaluation* (*Lecture Notes in Computer Science* vol 4486) ed Bernardo M and Hillston J (Springer Berlin / Heidelberg) pp 220–270
[9] Bonet B and Geffner H 2001 *Applied Intelligence* **14** 237–252
[10] Haddawy P and Hanks S 1998 *Computational Intelligence* **14** 392–429
[11] Keeney R L and Raiffa H 1976 *Decisions with Multiple Objectives: Preferences and Value Tradeoffs* (New York: Wiley)
[12] Sutton R and Barto A 1998 *Reinforcement Learning: An Introduction* (The MIT press, Cambridge, MA)
[13] Kaelbling L P, Littman M L and Moore A 1996 *Journal of Artificial Intelligence Research* **4** 237–285
[14] Pacheco R N and Costa A H R 2002 *Workshop de Computação – WORKCOMP'2002* ed Sakude M T S and de A Castro Cesar C (ITA) pp 125–130
[15] Selvatici A H P and Costa A H R 2007 *Mobile Robots: The Evolutionary Approach* (The Netherlands: Springer-Verlag) chap 11, pp 161–184
[16] Selvatici A H P and Costa A H R 2007 *Revista Controle e Automação* **18** 173–186
[17] Sutton R S and Barto A G 1998 *Reinforcement Learning: An Introduction* (Massachussets, MA: MIT Press)
[18] Kalmár Z, Szepesvári C and Lörincz A 1998 *Machine Learning* **31** 55–85
[19] Bailey T and Durrant-Whyte H 2006 *Robotics and Automation Magazine* **13** 1–10
[20] Durrant-Whyte H and Bailey T 2006 *Robotics and Automation Magazine* **13** 1–9
[21] Selvatici A H P, Costa A H R and Dellaert F 2008 *IV Workshop de Visão Computacional* (Bauru, Brazil)
[22] Munos R and Moore A 2002 *Machine Learning* **49** 291–323
[23] Reynolds S I 2000 *Proceedings of the 17th International Conference on Machine Learning* pp 783–790
[24] Silva V F d and Costa A H R 2009 *Journal of the Brazilian Computer Society* **15** 65 – 75 ISSN 0104-6500
[25] Bianchi R, Ribeiro C and Costa A 2008 *Journal of Heuristics* **14**(2) 135–168 ISSN 1381-1231 10.1007/s10732-007-9031-5 URL `http://dx.doi.org/10.1007/s10732-007-9031-5`
[26] Moore A W and Atkeson C G 1993 *Machine Learning* **13** 237–285
[27] Sutton R S 1988 *Machine Learning* **3** 9–44
[28] Taylor M E, Kuhlmann G and Stone P 2008 *AAMAS '08: Proceedings of the 7th international joint conference on Autonomous agents and multiagent systems* (Richland, SC: International Foundation for Autonomous Agents and Multiagent Systems) pp 283–290 ISBN 978-0-9817381-0-9
[29] Taylor M E and Stone P 2009 *J. Mach. Learn. Res.* **10** 1633–1685 ISSN 1532-4435
[30] Ribeiro C H C, Pegoraro R and Costa A H R 2002 *International Joint Conference on Autonomous Agents and Multi-Agent Systems AAMAS´2002* ed Castelfranchi C and Johnson W L pp 1239–1245
[31] Boutilier C, Dearden R and Goldszmidt M 2000 *Artificial Intelligence* **121** 49–107
[32] van Otterlo M 2009 *The Logic of Adaptive Behavior - Knowledge Representation and Algorithms for Adaptive Sequential Decision Making under Uncertainty in First-Order and Relational Domains* (*Frontiers in Artificial Intelligence and Applications* vol 192) (IOS Press) ISBN 978-1-58603-969-1
[33] Cocora A, Kersting K, Plagemann C, Burgard W and De Raedt L 2006 *Proc. of the IEEE/RSJ International Conference on Intelligent Robots and Systems (IROS)* (Beijing, China)
[34] Kersting K, Plagemann C, Cocora A, Burgard W and De Raedt L 2007 *Advanced Robotics. Special Issue on Imitative Robots* **21**
[35] Peters J and Schaal S 2008 *Neural Netw.* **21** 682–697 ISSN 0893-6080

SCALING LAWS IN THE TRANSIENT DYNAMICS OF FIREFLY-LIKE OSCILLATORS

N Rubido[1], C Cabeza[1], G. M. Ramírez Ávila[2,3] and A. Martí[1]

[1] Iguá 4225, Instituto de Física, Facultad de Ciencias, Montevideo, Uruguay.
[2] Institut für Physik, Humboldt Universität zu Berlin, Germany.
[3] Instituto de Investigaciones Físicas, Universidad de San Andrés, La Paz, Bolivia.

E-mail: nrubido@fisica.edu.uy

Abstract. Fireflies constitute a paradigm of pulse-coupled oscillators. In order to tackle the problems related to synchronisation transients of pulse-coupled oscillators, a Light-Controlled Oscillator (LCO) model is presented. A single LCO constitutes a one-dimensional relaxation oscillator described by two distinct time-scales meant to mimic fireflies in the sense that: it is capable of emitting light in a pulse-like fashion and detect the emitted by others in order to adjust its oscillation. We present dynamical results for two interacting LCOs in the torus for all possible coupling configurations. Transient times to the synchronous limit cycle are obtained experimentally and numerically as a function of initial conditions and coupling strengths. Scaling laws are found based on dimensional analysis and critical exponents calculated, thus, global dynamic is restricted. Furthermore, an analytical orthogonal transformation that allows to calculate Floquet multipliers directly from the time series is presented. As a consequence, local dynamics is also fully characterized. This transformation can be easily extended to a system with an arbitrary number of interacting LCOs.

1. Introduction

Pulse coupling is extensively common in Biology —the chirp of crickets, pacemaker cells firing and fireflies flashing, among many—. The study of how pulse-coupled oscillators achieve synchrony is important due to experimental observations of synchronous neural firing patterns of various animal species [1, 2]. However, the scope of synchronous behaviours runs far beyond neural dynamics [3, 4]. Synchronisation exhibits itself in physical [5], chemical [6], biological [7] and electronic [8] systems. Furthermore, phenomena involving synchronisation in complex networks [9, 10], chaotic systems [11] or time-delayed systems [12], have been intensively studied in recent years.

While experimental works dealing with these problems are usually based in complex models, transient times are only elucidated for simpler models in some configurations of phase locked loops [13] and Light-Controlled Oscillators (LCOs) [14, 15, 16, 17]. These oscillators constitute simple electronic devices that allow to study synchronisation dynamics in a simple way. From a theoretical point of view, synchronisation can be characterized studying transient times towards phase-locking states. In the vicinity of these state, transient regimes are equivalent to the study of reaching a limit cycle. Nevertheless, despite huge efforts, a general theoretical framework for global dynamics is still missing.

In this sense, we present a dimensional analysis that allows to calculate scaling laws for two coupled oscillators that restrict the global dynamics transients. In particular, the technique is exhibited for LCOs and compared with results from the well-known Kuramoto oscillators. On the other hand, a general orthogonal transformation deduced from the model that allows to calculate the Floquet eigenvector directions directly from the time series is presented. Using this transformation, the limit cycle multipliers for the torus local dynamics of coupled LCOs are calculated in the two possible configurations. Furthermore, an analytical extension of these transformations to higher dimensions is direct.

This paper is organized as follows. In Sec. 2, the LCO experimental set-up and analytical model is presented. In Sec. 3 we explain the phase variables and dynamical quantities relevant for the representation of the evolution of two coupled oscillators in the toroidal state space. Experimental and numerical results for transients behaviours are shown in Sec. 4. Finally, conclusions and perspectives are presented in Sec. 5.

2. Light-controlled oscillators (LCOs)

2.1. Experimental set-up

The LCOs presented here constitute one dimensional dynamical systems that oscillate in a piecewise analytic form with two characteristic time-scales: a slow charging stage and a fast discharge. IR light is emitted in a pulse-like fashion during discharge modifying the receiver's stage. Experimentally, each LCO is composed of an LM555 chip operating in an astable oscillating mode and a dual *RC* circuit [16, 17] with a common capacitor. The chip acts as a logical gate between these two circuits by establishing well-defined thresholds for the capacitor voltage: $2V_{cc}/3$ and $V_{cc}/3$, being V_{cc} the source voltage. As a consequence, one *RC*-circuit defines the charging stage as long as the upper threshold is not achieved. Whenever the upper threshold is reached, the chip switches to the other *RC*-circuit, generating the discharging stage until the lower threshold is attained. During the discharge, the system emits an Infra-Red (IR) light pulse by means of LEDs connected in series with the resistors.

The associated charging and discharging times may be adjusted by modifying the resistors R_λ and R_γ connected in series with the LEDs. The duration of these stages are set in the order of milliseconds for experimental practicality. The period of each oscillator is determined by the sum of these time intervals. IR emissions can be detected due to the inclusion of photo-diodes set in parallel with the *RC* circuits. Consequently, pulse detection modifies at any instant the charge in the capacitor, increasing its voltage in either stage. Thus, shortening the charging stage and/or lengthening the discharge.

The time-scale parameters of LCOs are chosen in a way that the resulting oscillating pattern resembles an integrate-and-fire model (firing interval corresponds to $\sim 1\%$ of the total period), as this oscillating form results in smaller transients. Coupling strengths can be modified by two means: varying the distance between oscillators or by forcing smaller emissions due to the addition of an extra resistor connected in series with the LEDs. On the other hand, different coupling configurations can be attained if photo-sensors are masked differentially. An uncoupled LCO is named as, *in dark*. No time delays in the couplings are considered, in view of the characteristics of the light pulses.

The experimental measurements have been carried out using a data acquisition card (NI-DAQ) that allowed us to obtain all the underlying dynamics of the system with a 20kHz sampling rate. Each LCO has its own source voltage, all set approximately equal to $V_{cc} = 9$V. Therefore, different initial conditions are generated by switching on the sources in a differential way. However, this method results in a random initial charging state, thus, many measurements are done in order to have a uniform distribution of initial conditions throughout the voltage domain $[V_{cc}/3,\ 2V_{cc}/3]$.

2.2. Dynamical model
From a dynamical point of view, an LCO presents a one-dimensional flow, periodic and continuous, nevertheless, piecewise analytic. Charging and discharging time-scales are related to the RC characteristics. For the uncoupled dynamics, these scales can be expressed as a function of the circuit parameters by:

$$\begin{cases} T_\lambda^0 = \frac{\log(2)}{\lambda} = \log(2)\,(R_\lambda + R_\gamma)\,C\,, \\ T_\gamma^0 = \frac{\log(2)}{\gamma} = \log(2)\,R_\gamma\,C\,, \end{cases} \tag{1}$$

where λ (γ) is the inverse charging (discharging) characteristic time-scale, and the superscript ^{0} is used to denote the *in dark* value. Figure 1 shows a typical *in dark* LCO evolution.

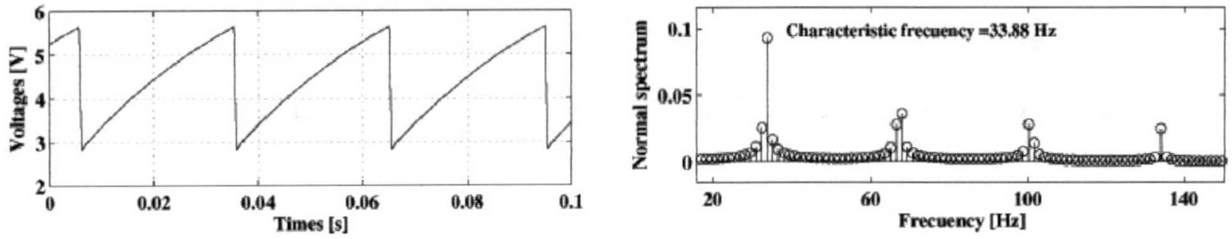

Figure 1. Time window for the *in dark* voltage evolution of an LCO (left panel) and frequency spectrum (right panel) for typical parameter values.

Based on the nature of the electronic devices that constitute an LCO and their experimental behaviour, the system of LCOs interacting is described by the following set of piecewise linear differential equations:

$$\dot{V}_i(t) = \lambda_i\,[V_{cc} - V_i(t)]\,\epsilon_i(t) - \gamma_i V_i(t)\,[1 - \epsilon_i(t)] + \sum_{j=1,\,j\neq i}^{N} \beta_{ij}\,[1 - \epsilon_j(t)],\ i = 1, \ldots, N, \tag{2}$$

being $V_i(t)$ the instantaneous voltage of the i-th oscillator of the set of N coupled LCOs and ϵ_i a discrete variable that can take the values 0 (discharge) or 1 (charge) according to the capacitor stage. This variable is created for each LCO in order to represent the change of stage and to simplify the notation of the set of differential equations of the voltage.

Equation (2) shows that the coupling is caused by the addition of a constant value: $\sum_k \beta_{ik}$, where k is an index that takes into account only the LCOs from the set that are discharging at that particular instant. We consider that all elements of the coupling matrix are non-negative and that there is no feedback, meaning that all diagonal elements are null. In particular, for two coupled LCOs, the coupling matrix can be characterized by two distinct forms: the master-slave (MS), where there is only one interaction in play, thus, one LCO acts as an external forcing to the other; and mutual-interaction (MI), where both LCOs interact. This coupling forms constitute the configurations dealt within this paper.

3. Evolution in the torus
LCOs constitute one-dimensional oscillators, thus, a pair of interacting LCOs can be described by phase angle variables that evolve in a toroidal states space. These phase variables are directly linked to the voltage evolution.

3.1. Voltage phase variables

The ith-oscillator of the set described by Eq.(2) can be described at all times by a pair of variables (V_i, ϵ_i), though, the only dinamical variable relevant of these is the voltage. The proof lies in the construction of the state space of the corresponding oscillator and its phase variable. While $\epsilon = 1$ the voltage increases monotonically, whereas, when $\epsilon = 0$ it decreases.Thus, a phase variable can be constructed directly from the voltage evolution taking these two stages into account by the use of a simple symmetry operation.

According to the pair of variables (voltage and stage), the evolution of a single LCO is always restricted to a state space composed of two intervals of equal length ($[V_{cc}/3, 2V_{cc}/3]$) that are connected in a discontinuous fashion by the discrete stage variable (ϵ). Coupling affects the time that the LCO spends in each stage. If these discontinuities are removed by connecting the two capacitor stages in a convenient way, the state space could be transformed into a circle. Assuming monotonicity, a phase variable can be constructed reflecting the discharging stage around its mean value and making a proper voltage translation. This transformation corresponds to passing from the initial oscillating state:

$$[V_{cc}/3,\, 2V_{cc}/3]\,,\ \epsilon = 1 \rightarrow [2V_{cc}/3,\, V_{cc}/3]\,,\ \epsilon = 0, \rightarrow [V_{cc}/3,\, 2V_{cc}/3]\,,\ \epsilon = 1 \rightarrow \cdots,$$

to a situation where the interval of voltages expands to $[V_{cc}/3, V_{cc}]$ in the following manner,

$$[V_{cc}/3,\, 2V_{cc}/3]\,,\ \epsilon = 1 \rightarrow [V_{cc}/3,\, 2V_{cc}/3] + V_{cc}/3,\ \epsilon = 0 \rightarrow [V_{cc}/3,\, 2V_{cc}/3]\,,\ \epsilon = 1 \rightarrow \cdots.$$

Therefore, a phase variable can be created directly from this evolution due to the fact that voltage now increases monotonically along a simple period for any coupling situation. Once the lower voltage threshold of the discharging situation is reached (which after the symmetry operations corresponds to a voltage of V_{cc}), then the oscillation retraces its evolution and the voltage is set to $V_{cc}/3$ in a charging stage ($\epsilon = 1$). As a consequence, a new voltage variable that increases monotonically can be expressed as a function of this new oscillating form, simply by adding V_{cc} after a period has passed.

Naming this new voltage variable for all LCOs as $\tilde{V}_i$, the phase angle variable is defined as:

$$\phi_i\,(t) = \pi\,\frac{\tilde{V}_i\,(t) - V_{cc}/3}{V_{cc}/3}\,,\quad \forall\, i = 1,\, \ldots,\, N\,. \tag{3}$$

For instance, starting from $V(0) = V_{cc}/3$, when the LCO reaches for the first time the upper threshold ($V\,(T_\lambda) = \tilde{V}V\,(T_\lambda) = 2V_{cc}/3$) the phase angle is π, whereas when it reaches the lower threshold ($V(T_\lambda + T_\gamma) = V_{cc}/3$ corresponding to $\tilde{V}(T) = V_{cc}$), the phase increases to take the value 2π.

3.2. State space

By means of the phase angle variables, a pair of LCOs can be described in a toroidal state space with a continuous evolution depicted in Fig. 2. It has been shown [18] analytically that the system exhibits two fixed points in the return map (one stable and the other unstable) for a wide range of frequency detuning between oscillators and coupling strengths. Thus, coupled dynamics tends to create two limit cycles, one attractive and other repelling.

The standard variance of two voltage time series is defined as the time average ($\langle\cdot\rangle$) of the voltage difference by:

$$\sigma^2 = \frac{1}{2}\left\langle\left(\frac{V_1 - V_2}{2}\right)^2\right\rangle. \tag{4}$$

Transients are then investigated using a Poincaré section of the flow as well as taking the standard variance σ^2 of the whole time series. The first method allows to calculate synchronisation times by adding the return times that the system needs to achieve the fixed point of the return map. On the other hand, σ^2 is generally used to measure the synchronicity of a system after transients have died out. If synchronised, $\sigma^2 \to 0$, nevertheless, it does not need to withdraw any portion of the time series if transient calculations are being made. In this cases, the value of σ^2 increases if the transient is longer, thus, σ^2 will measure the initial condition dependence of the system.

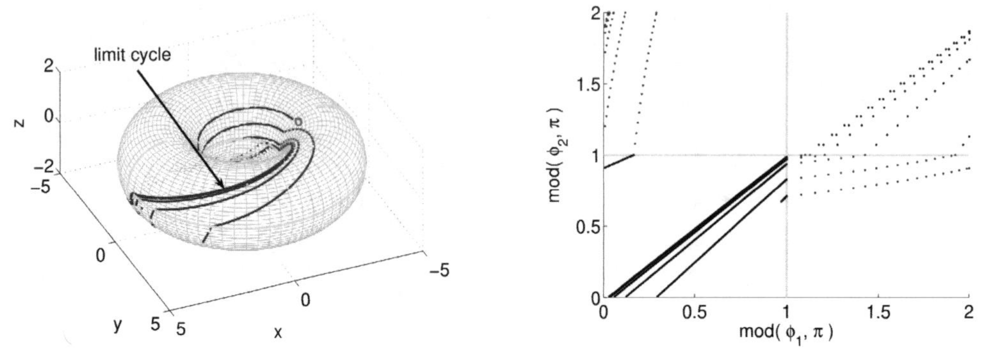

Figure 2. Numeric transient dynamics for two coupled LCOs in a MS configuration with $\beta = 200\ V/s$ and an initial voltage difference of $\Delta V(0) = 0.3\ V$ with both LCOs charging. In the right panel, the toroidal state space is represented. The flow is presented with blue lines, the initial state with a red circle and cyan lines represent a guide for the eye in order to recognize the four possible voltage stages. The same system is exposed in the right panel, where charging-charging stages and discharging-discharging stages are represented with black dots and mixed stages with blue dots (sampling rate: $10\,kHz$). The slow and fast evolution characteristics of each region are directly unveiled in this representation.

These methods are used to show that the global dynamic behaviour is restricted in a somehow universal fashion, leading to scaling laws of the system that make transient paths independent of coupling strengths (β) with the aid of proper dimensionless variables.

As the right panel in Fig. 2 shows, the attractive limit cycle corresponds to the diagonal of T^2. Thus, a simple orthogonal transformation of the state space would inevitably lead to create the Floquet eigenvectors that allow to calculate the local dynamic multipliers.

The coordinate transformation needed is a counter-clockwise rotation of 45° of the T^2 covering state space variables (right panel in Fig. 2) around the origin. This fact implies that the new coordinate system is:

$$\tilde{\phi}_1 = \frac{\phi_2 + \phi_1}{\sqrt{2}}, \quad \tilde{\phi}_2 = \frac{\phi_2 - \phi_1}{\sqrt{2}}, \tag{5}$$

where the $\tilde{\phi}_1$ new coordinate corresponds to the diagonal, thus a null multiplier is expected, and $\tilde{\phi}_2$ is the normal limit cycle coordinate. The same transformation can be applied to the voltage time series and the results are identical.

4. Results

4.1. Local dynamics

Equation (5) is an orthogonal transformation that can be easily generalized to higher dimension systems, due to the fact that the *quid* of the question lies in the fact that the synchronisation limit cycle of N coupled LCOs always lies in the covering state space diagonal. Thus, for higher dimensions, a composition of these transformations can generate all Floquet eigenvectors [19].

Therefore, a direct analysis from the transformed voltage time series can be done in order to calculate Floquet multipliers (see the right panel in Fig. 3).

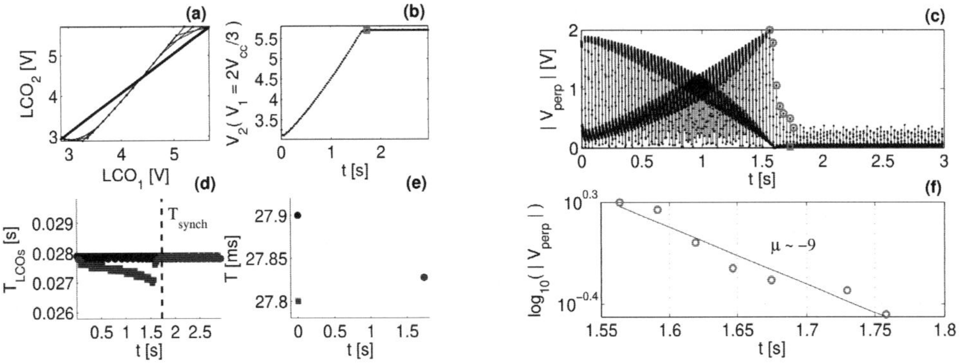

Figure 3. Experimental time series analysis of two coupled LCOs in MS configuration. In panel **(a)**, the Lissajous figure of the synchronised state is shown. In panel **(b)**, the Poincaré return map evolution with a red square that corresponds to the synchronisation time is plotted. The transformed voltage series for the cycle normal coordinate corresponds to panel **(c)**. Red circles correspond to the exponential decay in amplitude of the coordinate, and are represented in logarithmic fashion in panel **(f)**. Panel **(d)** displays the period evolution of the two coupled LCOs, and panel **(e)** the corresponding initial and final (synchronised) values.

Experimental results show that the Floquet multipliers μ for both directions in MS configuration and all initial conditions in average are:

$$\begin{cases} \bar{\mu}_2 = -10.45 \pm 0.29 \\ \bar{\mu}_1 = 0.012 \pm 0.02 \end{cases}, \tag{6}$$

where $\bar{\mu}_2$ ($\bar{\mu}_1$) corresponds to the normal (parallel) direction with respect to the limit cycle averaged over all initial conditions used. This shows that Floquet exponents should be independent from initial conditions as expected. Nevertheless, it is also found that exponents for two coupled LCOs are independent of the coupling strength and only depend on the coupling configuration, in other words, if they are coupled via an MS or MI configuration.

In order to improve the calculations of these exponents and reduce the noise coming from the experimental time series (caused by minor noise fluctuations of the source voltage), new calculations are being done over the time series. As the T^2 evolution in the right panel of Fig. 2 shows, the flow reaches the diagonal only on the $(\epsilon_1, \epsilon_2) = (1, 1)$ region (lower left section). Whereas in the (0, 0) (upper right section) is slightly translated. This is explained by analysing the return maps, where the fixed point is not located exactly in a corner [18]. Therefore, calculations of Floquet exponents introduce a systematic error that can be dismissed if the charging part of the time series is taken into account. This problem is increased when Floquet exponents are calculated for MI configuration, thus, μ_2 results in greater dispersion from $\bar{\mu}_2 \sim -7$.

4.2. Experimental and numerical synchronisation times

The experimental results obtained for MS and MI configurations for the synchronisation times as a function of initial coupling voltages are shown in Fig. 4. The initial time is taken to be the first time one of the two LCOs reaches the upper threshold for the first time due to the fact that former states can be treated as uncoupled dynamics. Thus, the name initial coupling voltage.

The LCO_1 and LCO_2 period *in dark* for the MS configuration are $T_1^{dark} = 28.6$ ms, $T_2^{dark} = 28.4$ ms; and for the MI configuration are $T_1^{dark} = 24.7$ ms, $T_2^{dark} = 24.3$ ms. Different coupling strengths are achieved by placing LCOs at different distances from each other. Distances used are: $d_1 = 5$ cm, $d_2 = 10$ cm, $d_3 = 15$ cm, $d_4 = 20$ cm and $d_5 = 25$ cm. Then, the connectivity matrix elements are calculated directly from the resulting time series by comparison with the solutions of the Eq.(2) for these parameters.

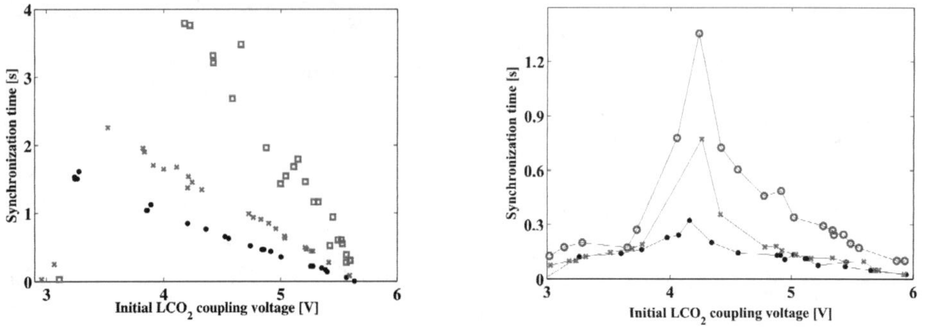

Figure 4. Synchronisation times as a function of the initial coupling voltage of LCO_2 in MS (left figure) and MI (right figure) configuration (LCO_1 initial voltage corresponds to the first time that it starts to discharge, meaning that $V_1(0) = 2V_{cc}/3$). Red circles (right panel) correspond to a distance d_4, squares (left panel) to d_3, grey crosses to d_2 and black points to d_1. On the right hand side a thin black line that conects the data is showed as a guide for the eye. LCO_1 is taken as the leader (or master) LCO.

Experimental synchronisation times for both configurations are resolved from the Poincaré section corresponding to registering voltages every time LCO_1 reaches the upper threshold (starts to discharge). As can be seen from these figures, there is a single initial condition value for LCO_2 voltage that gives a nearly asymptotic behaviour for all configuration schemes. This value reveals the repeling limit cycle that would correspond to a null synchronisation time. Experimental results cannot achieve this state due to the action of electronic noise, that presents itself as a kind of symmetry breaker. Moreover, these figures demonstrate that all initial conditions are attracted to the synchronisation manifold as is expected.

In an analogous fashion, numerical synchronisation times are resolved, both, by the Poincaré section and by Eq.(4). Figure 5 exhibits the resultant curves as a function of initial conditions for the MI configuration of coupled LCOs and Kuramoto oscillators. The Kuramoto model is represented by:

$$\begin{cases} \dot{\theta}_1 = \omega_1 + K \sin(\theta_1 - \theta_2) , \\ \dot{\theta}_2 = \omega_2 + K \sin(\theta_2 - \theta_1) , \end{cases} \tag{7}$$

where K is the coupling strength (it is comparable only with β/V_{cc}). Time-scale parameters for both systems are set in order to have identical individuals oscillating at a $\omega = 30$ rad/s frequency. The Kuramoto model is solved by means of a fourth order Runge-Kutta schemes, whilst LCOs are solved by using piecewise analytic solutions numerically implemented as in previous works [18].

As can be seen from these figures, transients behave in a self-similar way. Therefore, scaling laws should be possible to find in order to achieve a universal form. In Sec. 4.3 we present a method that uses the single initial condition that correspond to the nearest registered value to the repelling limit cycle to scale all quantities. The analysis made is based on dimensional analysis.

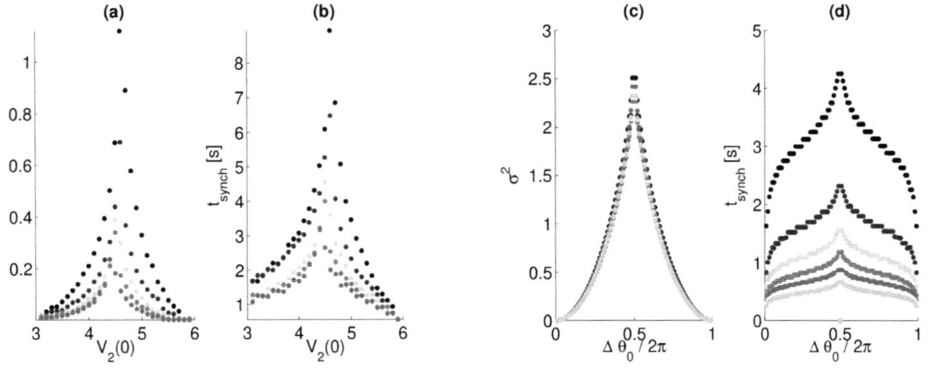

Figure 5. Panels **(a)** and **(b)** show the numerically simulated transients for two coupled identical LCOs in MI configuration. Voltage initial conditions can also be represented as phase angular variables using Eq.(3). Panels **(c)** and **(d)** exhibit the same dynamic for two coupled Kuramoto oscillators. For LCOs, coupling strengths correspond to $\beta = 100$ (black dots), 150 (blue), 200 (cyan), 250 (magenta) and 300 V/s (red). On the other hand, $K \simeq 1.21$ (black dots), 2.32 (blue), 3.43 (cyan), 4.54 (magenta) and 6.76 rad/s (red).

4.3. Global dynamics

As can be seen from both configurations plotted in Fig. 4, synchronisation times exhibit self-similar forms as functions of initial conditions for all coupling strengths. Thus, scaling laws are sought and critical exponents calculated. In particular, we consider that a MS configuration is achieved for a connectivity matrix given as $\beta_{12} = 0$ and $\beta_{21} = \beta$, while a MI configuration is formed by having $\beta_{12} = \beta_{21} = \beta$.

In order to achieve the scaling laws for these configurations we have to consider a dimensionless version of the model given in Eq.(2). Therefore, voltage signals are normalized using the source voltage ($V_n = V/V_{cc}$), times variables are made dimensionless by using LCO_1 *in dark* period ($t_n = t/T_1^0$), and as a result, β is changed to $\tilde{\beta} = T_1^0\beta/V_{cc}$. Then, the Pi theorem [20] assures that the relevant dimensionless parameters needed to describe the particular dynamic transient are: the initial voltage difference (in particular, the LCO_2 voltage, V_n^{init}), the synchronisation time (t_n^{synch}) and the coupling strength $\tilde{\beta}$, which obey a certain functional relation expressed as:

$$t_n^{synch} = f\left(V_n^{init}, \tilde{\beta}\right). \tag{8}$$

The self-similar behaviour found in both configurations for the experimental and numerical solutions of Eq.(2), allows to define two characteristic dynamical scales: the dimensionless maximum time, namely $t^* := \max\left(t_n^{synch}\right)$, and the corresponding dimensionless initial voltage, namely $V^* := V_n^{init}\left(t^*, \tilde{\beta}\right)$. These scales represent the nearest location to the repelling limit cycle that in T^2 the system can achieve. They are used to achieve the collapse of the transient curves and to find the scaling exponents that relates t^* with the ratio $V^*/\tilde{\beta}$. The exponents found correspond to the following linear relation

$$\log(t^*) = \alpha \log\left(\frac{V^*}{\tilde{\beta}}\right) + \kappa. \tag{9}$$

The exponents found for these linear relation in the configurations analysed are:

$$\left\{\begin{array}{ll} \alpha_{exp}^{MS} = 0.71\,, & \kappa_{exp}^{MS} = 1.76 \\ \alpha_{num}^{MS} = 1.04\,, & \kappa_{num}^{MS} = 1.96 \end{array}\right. , \quad \left\{\begin{array}{ll} \alpha_{exp}^{MI} = 0.74\,, & \kappa_{exp}^{MI} = 2.00 \\ \alpha_{num}^{MI} = 0.88\,, & \kappa_{num}^{MI} = 2.06 \end{array}\right. , \tag{10}$$

giving a relative difference between numeric and experimental values of

$$\begin{cases} \Delta\alpha/\alpha_{num} = 0.32 \\ \Delta\kappa/\kappa_{num} = 0.10 \end{cases}, \quad \begin{cases} \Delta\alpha/\alpha_{num} = 0.16 \\ \Delta\kappa/\kappa_{num} = 0.03 \end{cases}.$$

Once exponents and scaling laws are calculated, collapse curves are obtained by plotting

$$\frac{t_n^{synch}}{t^*} = \frac{f\left(V_n^{init}/V^*, \tilde{\beta}\right)}{10^\kappa \left(\frac{V^*}{\tilde{\beta}}\right)^\alpha} = g\left(\frac{V_n^{init}}{V^*}\right). \tag{11}$$

Results of this universal behaviour are presented in Fig. 6. It can be seen directly from the graphics, that numerical simulations reinforce the universality expectations of the transients as the collapsed curves exhibit great agreement.

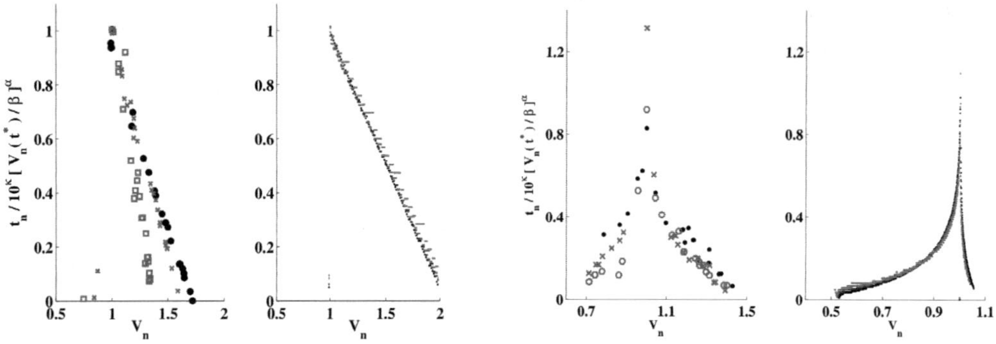

Figure 6. Synchronisation times collapsed as a function of the dimensionless LCO_2 initial conditions for the MS (two left panels) and MI (two right panels) configuration. For both configurations, the experimental data are plotted on the left and the numerical simulations on the right. Numerical simulations (dots) correspond to different coupling strengths: $\beta = 250$ (black), 200 (blue), 150 (red) and 50 (magenta) V/s and time-scale parameters given in Sec. 4.2.

5. Conclusions

The design of LCOs allowed us to modify on the spot their intrinsic period and pulse-like IR light emissions, enabling quantitative measurement of phase differences and period variations with fine precision, and produce electronically and physically different coupling strengths. This electronic set-up allowed us to obtain synchronisation times as well as stable states.

The evolution of the system in state space was used to construct phase variables and retrieve local dynamic behaviours by means of Floquet multipliers. The orthogonal transformation method that we present can be generalized to higher dimensions. Thus, systems with many interacting LCOs can also have their local dynamics quantified. Results in this matter corroborate that multipliers are independent of initial conditions. Nevertheless, for both configurations, exponents also seem to be independent of the coupling strength used. Current work is being done in order to improve the exponents calculations. The new method restricts the transformation to the charging-charging region of T^2.

The search for scaling exponents is done in order to analyse the transients global dynamics independently of the coupling strength, meaning that we can disregard the physical coupling magnitude used when studying the synchronisation times for coupled LCOs in MS and MI configurations. Also, these exponents constitute a classification of the LCOs transients within a certain universality class of relaxation oscillators yet to be found, regardless of their particular dynamics. With this spirit is that results from Kuramoto model are shown. Nevertheless, the self-similar behaviour that this system also shows when the oscillators are identical does not accept the similar scales than LCOs do in order to create the collapsed curves. This is due to the fact that the nearest initial condition to the repelling limit cycle does not change location if coupling strength is varied when Kuramoto model is investigated. However, recent results with non-identical Kuramoto oscillators retrieve the LCOs scale choice.

Scaling results for LCOs were contrasted with numerical ones showing good agreement in MI configuration. MS mismatch can be explained by observing Fig. 4, where experimental synchronisation times for smaller coupling strengths were not recorded for an interval around the asymptotic state initial condition. Thus, the maximum synchronisation time registered is not a reliable quantity. This recording problem is inherent in our initial condition experimental extracting method, due to the fact that initial conditions are randomly generated, and in practice the system seems to avoid these asymptotic state. Currently, these problems are addressed as part of future works.

Finally, the dimensional analysis that helps to introduce the scaling laws shows promising results and constitute a great encouragement for further studies and generalizations of this model. Paths to be followed, can be different characteristic time-scale ratios or LCOs coupled networks, where currently achieved simulations exhibit self-similar transient behaviours for three coupled LCOs in various configuration schemes. Results in these lines would show the versatility and experimental value of the model.

6. Acknowledgements

We acknowledge financial support from PEDECIBA, CSIC (U. de la República, Uruguay) and ANII (Uruguay).

References

[1] Lehnertz K, *et al* 2009, *J. Neuroscience Methods*, **183**, 42-48.
[2] Sandra R.F.S.M. Gois and Savi M A 2009, *Chaos, Solit. and Fract.*, **41**, 2553-2565.
[3] Hodgkin A L and Huxley A F 1952, *J. Physiol.*, **117**, 500-544.
[4] Fitzhugh R 1961, *Biophys. J.*, **1**, 445466.
[5] Argonov V Yu and Prants S V 2005, *Phys. Rev. A*, **71**, 053408-11.
[6] Fukuda H, Morimura H and Kai S 2005, *Physica D*, **205**, 80.
[7] Glass L 2001, *Nature* (London), **410**, 277.
[8] Pisarchik P A, Jaimes-Reátegui R and García-López J H 2008, *Int. J. Bif. Chaos*, **18**, 1801.
[9] Timme M, Wolf F and Geisel T 2002, *Phys. Rev. Lett.*, **89**, 258701.
[10] Zumdieck A, Timme M, Geisel T and Wolf F 2004, *Phys. Rev. Lett.* **93**, 244103.
[11] Mahmoud G M, Aly S A and Farghaly A A 2007, *Chaos, Sol. & Fract.*, **33**, 178-187.
[12] Masoller C and Martí A C 2005, *Phys. Rev. Lett.*, **94**, 134102.
[13] Goldsztein G and Strogatz S H 1995, *Int. J. Bif. Chaos.*, **5**, 983.
[14] Mirollo R E and Strogatz S H 1990, *SIAM J. App. Math.*, **50**, 1645.
[15] Bottani S 1995, *Phys. Rev. Lett.*, **74**, 4189.
[16] Ramírez Ávila G M, Guisset J L and Deneubourg J L 2003, *Physica D*, **182**, 254.
[17] Rubido N, Cabeza C, Martí A C and Ramírez Ávila G M 2009, *Phil. Trans. Royal Soc. A*, **367**, 3267-3280.
[18] Rubido N, Cabeza C, Kahan S, Ramírez Ávila G M and Martí A C 2010, *Eur. Phys. J. D*, DOI 10.1140/epjd/e2010-00215-4.
[19] Guckenheimer J and Holmes P, *Nonlinear Oscillations, Dynamical Dystems, and Bifurcations of Vector Fields* (Springer-Verlag, 1983).
[20] Buckingham E 1914, *Phys. Rev.*, **4**, 345376.

Dynamic Days South America 2010 IOP Publishing
Journal of Physics: Conference Series **285** (2011) 012027 doi:10.1088/1742-6596/285/1/012027

Dynamic valid models for the conservative Hénon-Heiles system

S. B. Bastos[1] **and E. M. A. M. Mendes**[1]

[1] Universidade Federal de Minas Gerais, Belo Horizonte, Brazil

E-mail: saulo@cpdee.ufmg.br, emmendes@cpdee.ufmg.br

Abstract. In this work the discretization of the Hénon-Heiles system obtained by applying the Monaco and Normand-Cyrot method is investigated. In order to obtain dynamically valid models, several approaches covering from the choice of terms in the difference equation originated from the discretization process to the increase of the discretization order are analyzed. As a conclusion it is shown that discretized models that preserve both the symmetry and the stability of their continuous counterpart can be obtained, even for large discretization steps.

1. Introduction

When dealing with the solution of conservative nonlinear differential equations, several problems such as energy loss and symmetry break can occur. In order to avoid such problems as much as possible, several numerical integration methods can be found in the literature [1]. Although the primary objective of these methods is the solution itself, a possible and interesting by-product is a difference equation (a discretized model) that hopefully reproduces the same behavior as the one generated by the original differential equations.

In [2] Mickens proposes a discretization scheme that it is claimed to be robust to numerical instabilities. Such instabilities may occur when difference equations do not meet restrictions imposed by the differential equations: conservation of energy, monotonicity, boundedness, and positivity, just to mention a few. A different discretization scheme was proposed by Monaco and Normand-Cyrot [3], which was studied in [4] in the context of dynamical systems that can exhibit chaotic behavior.

The authors of [5, 6] have successfully obtained discretized models for a dissipative system using the aforementioned method. The models found there are topologically equivalent to the original continuous system, except for a small displacement in the parameter space. However, for conservative systems, [7] shows that the symmetry is not conserved and numerical instabilities may occur when the discretization step increases, whether for the discretization scheme proposed by Mickens or for the one proposed by Monaco and Normand-Cyrot when only low order parameter is considered.

The purpose of this work is to find valid discretized models using the discretization scheme proposed by Monaco and Normand-Cyrot. Here the discrezation order is no longer as low as in the previous cases. The example used throughout the paper is the Hénon-Heiles system and the main objective is to define high-order discretized models that exhibit the same behavior as the original counterpart and therefore conserve the energy and symmetry of the solution even for large discretization steps.

The work is organized as follows. Section 2 describes the Hénon-Heiles system. Section 3 briefly introduces the discretization scheme proposed by Monaco and Normand-Cyrot. Then, a few remarks in section 4 about energy conservation in discrete systems are given. Section 5 shows results of the discretization scheme proposed by Monaco and Normand-Cyrot applied to the Hénon-Heiles system. Finally, the conclusions are presented in section 6.

2. The Hénon-Heiles system

The Hénon-Heiles system was firstly devised as a model to describe planetary motion, during large time intervals, inside a gravitational potential U_0, in a galaxy with cylindrical symmetry [1, 8, 9]. Nowadays, this well known nonlinear conservative system is widely studied [1, 7] in the realm of dynamical systems. After an order reduction, [1, 7] shows that the Hamiltonian function with two degrees of freedom is:

$$H(x, y, u, v) = \frac{1}{2}\left(u^2 + v^2\right) + U(x, y) \tag{1}$$

where $U(x, y)$, the potential function, is given by:

$$U(x, y) = \frac{1}{2}\left(x^2 + y^2 - \frac{2}{3}y^3\right) + x^2\, y \tag{2}$$

After the derivation of the equations of motion, the Hénon-Heiles system can be described by the following four ordinary differential equations:

$$\begin{cases} \dot{x} &= u \\ \dot{y} &= v \\ \dot{u} &= -x - 2\, x\, y \\ \dot{v} &= -x^2 + y^2 - y \end{cases} \tag{3}$$

For the initial condition ($x_0 = 0.000, y_0 = 0.670, u_0 = 0.093, v_0 = 0.000$) used in [7, 9] the energy is equal to $H = 0.128546999$, which means that the system is behaving chaotically.

The Hénon-Heiles system was simulated with the aforementioned condition using the implicit Runge-Kutta method with coefficients given by the 6th order Gauss-Legendre method, a symplectic integrator with area preserving [1] - and integration step $\tau = 0.001s$. The Poincaré section is shown in Figure 1. In this chaotic tangle, there are several islands, which are associated to orbits of periods 1 to 5. Symmetry, a characteristic of conservative systems, occurs for $v_n = 0$.

When the sampling frequency f_s is higher than twice the highest significative frequency f_{max} in the power spectrum, the Nyquist criterion states that sufficient information is available for a trustworthy description of the dynamics. This value works as an upper limit for the discretization time step of any discretization scheme, so $f_s \geq 2\, f_{max} \Rightarrow \frac{1}{f_s} = T_s = h_{max} \leq \frac{1}{2\, f_{max}}$. As long as the Nyquist criterium is satisfied, numerical instabilities are avoided and discretized models generate solutions which are equivalent to the one of the original system except for a small displacement in the parameter space [5, 6, 7]. For the variable x, Fourier analysis (Figure 2) shows that the maximum frequency is $f_{max} \approx 0.75Hz$. Therefore, the maximum discretization step must be less than $h_{max} = 0.67s$ to satisfy the Nyquist criterium.

3. Monaco and Normand-Cyrot discretization scheme

Consider the dynamical system:

$$\dot{\mathbf{x}} = \mathbf{f}\,(\mathbf{x}) \tag{4}$$

where $\mathbf{x} = (x_1, \ldots, x_m) \in \mathbb{R}^m$ are the state variables, $\mathbf{f}\,(\mathbf{x}) = (f_1(\mathbf{x}), \ldots, f_m(\mathbf{x}))$ are analytic functions. The derivative of $\mathbf{x}$ with respect to time is denoted by $\dot{\mathbf{x}}$.

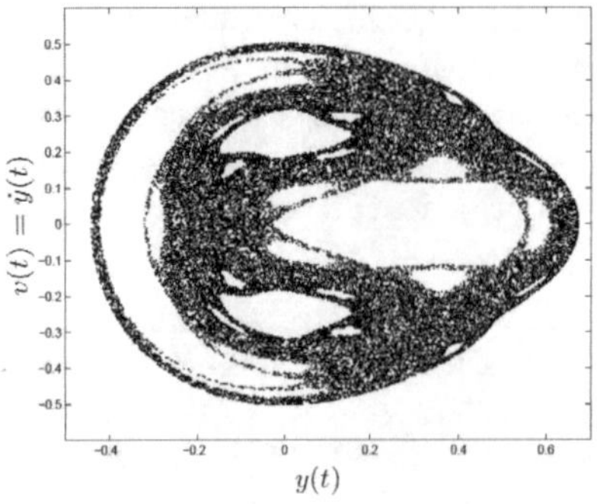

Figure 1: Poincaré section of the Hénon-Heiles system defined by $P \equiv \{(y_n, u_n, v_n) \in \mathbb{R}^3 \mid x_n = 0\}$ with initial condition $(x_0 = 0.000, y_0 = 0.670, u_0 = 0.093, v_0 = 0.000)$. The continuous system was simulated using the implicit Runge-Kutta method with 6th order Gauss-Legendre coefficients [1], and integration step $\tau = 0.001s$.

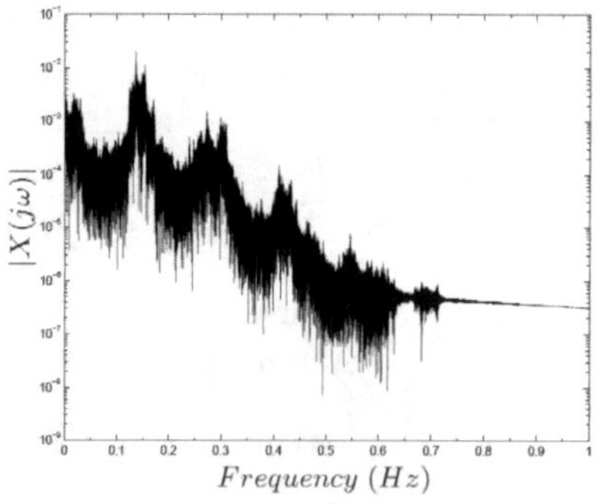

Figure 2: Frequency spectrum of the variable x of the Hénon-Heiles system, with initial conditions $(x_0 = 0.000, y_0 = 0.670, u_0 = 0.093, v_0 = 0.000)$.

The discrete model of eq (4) is given by:

$$\mathbf{x}_{k+1} = \mathbf{g}\left(\mathbf{x}_k, h\right) \tag{5}$$

where $\mathbf{x}_k \in \mathbb{R}^m$ are discrete state variables at time $t = t_0 + kh$, and h is the discretization step. Since discrete models derived from a set of differential equations have an extra parameter, the discretization step h, the parameter space to be considered is larger the continuous counterpart. The sole choice of h may lead to numerical instabilities [7].

In [4, 7] it is shown that the discretization, originally proposed by Monaco e Normand-Cyrot [3], can be accomplished by the Lie exponential expansion of Eq. (4), as follows:

$$\mathbf{x}_{k+1} = \exp\left(h\,\mathcal{L}_f\left(\mathbf{x}_k\right)\right) = \mathbf{x}_k + \sum_{n=1}^{\eta} \frac{h^n}{n!} \mathcal{L}_f^n\left(\mathbf{x}_k\right) \tag{6}$$

where η is the expansion order. The Lie derivative is given by:

$$\mathcal{L}_f\left(\mathbf{x}_k\right) = \sum_{j=1}^{m} f_j \frac{\partial \mathbf{x}}{\partial x_j} \tag{7}$$

where f_j represents the $j - th$ component in the vector field. Higher derivative orders can be calculated recursively by:

$$\mathcal{L}_f^n\left(\mathbf{x}_k\right) = \mathcal{L}_f\left(\mathcal{L}_f^{n-1}\left(\mathbf{x}_k\right)\right) \tag{8}$$

Equation (6) can be rewritten as follows:

$$\mathbf{x}_{k+1} = \mathbf{x}_k + h\,\mathbf{f} + \frac{h^2}{2} J(\mathbf{f})\,\mathbf{f} + \frac{h^3}{6} J(J(\mathbf{f})\,\mathbf{f})\,\mathbf{f} + \cdots \tag{9}$$

where $J\left(\cdot\right)$ is the Jacobian matrix of the argument. Note that the above equation should be truncated to avoid an excessive amount of terms and thus making computational simulations feasible. Besides, when $\eta \to \infty$, terms tend to zero which means that no information is added to the model output due to the lack of computational precision.

$\eta = 1$ results in the well-known explicit Euler method that preserves the number and location of the fixed-points of the original differential equations as shown in [4]. For $\eta \geq 2$, the location of the original system fixed-points is also preserved, but spurious fixed-points are added. The number and location of the spurious fixed points depend on the discretization order η and the discretization step h, respectively. It is worth noticing that spurious fixed points are a characteristic of the discretization scheme.

4. Energy analysis of discrete models

4.1. Hamiltonian Function

The energy is given by the hamiltonian function. The energy value of the discrete model is related to the model structure and the discretization step h. Energy mean and standard deviation of the discrete model simulation can be used to measure if the energy is conserved along the simulation trajectory for different discretization steps and are defined as follows

$$\mu_H = \frac{1}{N+1} \times \sum_{k=0}^{N} H(x_k) \tag{10}$$

$$\sigma_H = \sqrt{\frac{1}{N} \times \sum_{k=0}^{N} \left(H(x_k) - \mu_H\right)^2} \tag{11}$$

In order to check whether the discretized model can reproduce the dynamic motion of the continuous system, μ_H must be equal to the energy defined for the continuous system, and σ_H must be equal to 0 in order to guarantee that μ_H is constant $\forall k$.

As observed in [7], but also inferred from [10], the energy conservation for discretized models can be checked using the following relation

$$H(x_{k+1}) - H(x_k) = 0 \tag{12}$$

Thus, considering the simulation given by a discretized model at instants $k = (0, 1, ..., N)$, the following additional conditions must be derived in order to check the energy conservation along the trajectory

$$\mathcal{H}_1 = \frac{1}{N\,h} \times \sum_{k=0}^{N-1} \left(H(x_{k+1}) - H(x_k)\right) = \frac{1}{N\,h} \times \left(H(x_N) - H(x_0)\right) \to 0 \tag{13}$$

and also:

$$\mathcal{H}_2 = \frac{1}{N\,h} \times \sum_{k=0}^{N-1} \left|H(x_{k+1}) - H(x_k)\right| \to 0 \tag{14}$$

where $\mathcal{H}_1$ is the mean of the total energy variation and $\mathcal{H}_2$ is the mean of the absolute energy variation, both normalized in a simulation period of $1s$.

4.2. Jacobian Matrix

For the discrete system of Eq. (6), the jacobian matrix is given by:

$$J_\eta(\mathbf{x_k}, h) = \frac{\partial \mathbf{g}}{\partial \mathbf{x_k}} \tag{15}$$

where, for each element (i,j), one have:

$$\begin{array}{rcll} (J_\eta)_{ij} & = & \mathcal{O}(h^\eta) & i \neq j \\ (J_\eta)_{ij} & = & 1+\mathcal{O}(h^\eta) & i = j \end{array} \tag{16}$$

and $\mathcal{O}(h^\eta)$ is a polynomial of h of ηth order, without the constant term. Therefore, when $h \to 0$, J_η tends to an identity matrix and $\mathbf{x}_{k+1} \to \mathbf{x}_k$.

The volume conservation in the context of discrete systems is accomplished by [11]:

$$\mathcal{L}(\mathbf{x_k}, h) = \prod_{s=1}^{m} \lambda_s = 1 \tag{17}$$

where λ_s is an eigenvalue of the jacobian matrix J_η, Eq. (15).

The eigenvalue product mean and standard deviation are another measurement of interest and are defined as follows

$$\mu_{\mathcal{L}} = \frac{1}{N+1} \times \sum_{k=0}^{N} \mathcal{L}(\mathbf{x_k}, h) \tag{18}$$

$$\sigma_{\mathcal{L}} = \sqrt{\frac{1}{N} \times \sum_{k=0}^{N} \left(\mathcal{L}(x_k, h) - \mu_{\mathcal{L}}\right)^2} \tag{19}$$

If the above product is very close to 1 for a discretized model then the energy is said to be conserved.

An equivalent analysis, based on the jacobian matrix trace, was proposed in [7]:

$$tr\{J_\eta\} = m + \mathcal{O}(h^\eta) \approx m \tag{20}$$

Similarly to the case of eigenvalue product, if the trace is practically equal to the system dimension, the energy is conserved and, therefore, approximately constant to small h values.

Another interesting measure, which will be used latter on, is the trace increment when discretization order increases:

$$\Delta\, tr\{J_\eta\} = tr\{J_\eta\} - tr\{J_{\eta-1}\} \tag{21}$$

5. Results: Hénon-Heiles discretization

5.1. Mickens discretization scheme

Using Mickens discretization scheme [2] and the following tranformation

$$\left\{ \begin{array}{lcl} 1^{st}\ Eq. & : & (x_k, y_k, u_k, v_k) \mapsto (x_k, y_k, u_{k+1}, v_{k+1}) \\ 2^{nd}\ Eq. & : & (x_k, y_k, u_k, v_k) \mapsto (x_k, y_k, u_{k+1}, v_{k+1}) \\ 3^{rd}\ Eq. & : & (x_k, y_k, u_k, v_k) \mapsto (x_k, y_k, u_k, v_k) \\ 4^{th}\ Eq. & : & (x_k, y_k, u_k, v_k) \mapsto (x_k, y_k, u_k, v_k) \end{array} \right. \tag{22}$$

results in the discrete model (also used in [7])

$$\left\{ \begin{array}{rcl} x_{k+1} & = & x_k + h\, u_{k+1} \\ y_{k+1} & = & y_k + h\, v_{k+1} \\ u_{k+1} & = & u_k + h\,(-x_k - 2\, x_k\, y_k) \\ v_{k+1} & = & v_k + h\,\left(-y_k + y_k^2 - x_k^2\right) \end{array} \right. \tag{23}$$

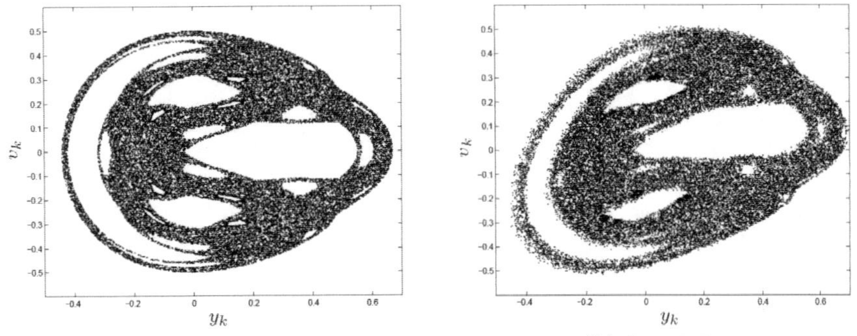

(a) $h = 0.02s$ (b) $h = 0.67s$

Figure 3: Poincaré section of the Hénon-Heiles system, defined by $P \equiv \{(y_n, u_n, v_n) \in \mathbb{R}^3 \mid x_n = 0\}$, using the method proposed by Mickens, with initial condition ($x_0 = 0.000, y_0 = 0.670, u_0 = 0.093, v_0 = 0.000$) and last simulation instant $N \approx \frac{120000s}{h}$.

Simulating Eq. (23) with $h < 0.3s$, therefore $h < h_{max}$, generates valid solutions as shown in [7]. Poincaré section of the Hénon-Heiles discretized model with $h = 0.02s$, shown in Figure 3*a*, is practically the same as the one generated by the original continuous system (Figure 1).

When the discretization step increases, the symmetry is broken due to the impossibility to estimate the derivatives and the variables at the same time and thus breaking the time invariance in the discretized form. The Poincaré section of the discretized model with $h = 0.67s$, the maximum step size suggested by the Nyquist criteria, is shown in Figure 3*b*. Note that the symmetry is not conserved, as already observed in [7].

The energy indexes calculated under the simulation conditions given above are presented at Table 1. Energy mean is closer to the original energy value for small *h*s, but, as *h* increases, the energy increases as well which suggests that the model starts to behave like it would somehow stretch indefinitely. $\mathcal{H}_1$ shows that the change in total energy slowly increases as *h* increases. In addition to that, it can be noticed the variance of this change also increases with *h*.

5.2. Monaco and Normand-Cyrot discretization scheme

In this section several discretized models were obtained using the Monaco e Normand-Cyrot method. Each one was simulated with different discretization steps.

The simulation of the 3rd order discretized model, eq (6) with $\eta = 3$, generates a Poincaré section that is similar to the one generated by the original continuous system for small discretization steps ($h \leq 0.02s$). The symmetry and the chaotic islands of the original system are preserved. For larger step sizes ($h > 0.02s$), numerical instabilities start to occur, resulting in Poincaré sections with points converging to the fixed-point located at the origin. This fact, observed in [7], is explained by analyzing the energy. As the step size increases, the energy decays, as shown in Figure 4, resulting in points closer to the origin. It can be also observed that as the step size increases, the eigenvalue product of the jacobian matrix is not preserved, as shown in Figure 5. This also indicates that the energy is not conserved.

The energy indexes calculated for the 3rd order discretized model with $h = 0.01s$ and $N \approx 6.0 \times 10^6$ are listed at Table 1. The energy mean, $\mu_H = 0.12811895$, differs the original energy only on the 4th decimal digit. The energy mean for higher *h* values is smaller, thus, the original energy is not conserved. The total energy variation and the absolute energy variation within a 1*s* period are small, but a large variation is seen for higher *h* values, demonstrating that the model structure is not sufficient to conserve the energy for higher *h* values.

Authors of [7] added/deleted terms in Eq. (6) in order to force the jacobian matrix trace to be equal to 4, no matter the choice of the step size value. In [7] it was also checked that the

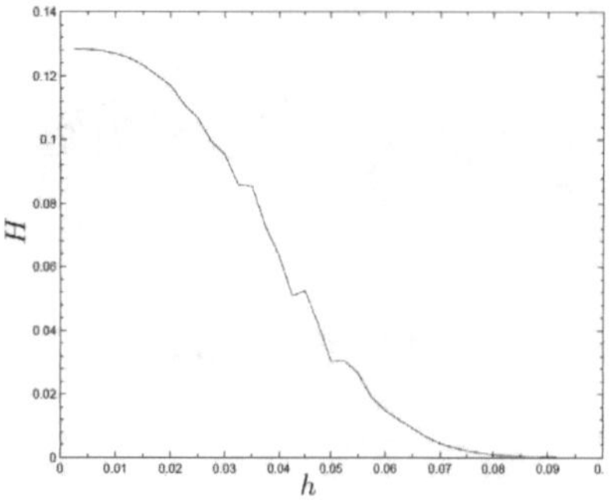

Figure 4: Stationary energy of 3rd order discretized model of the Hénon-Heiles system versus the step size h. Stationary energy was calculated considering the average of the last 1000 simulation values.

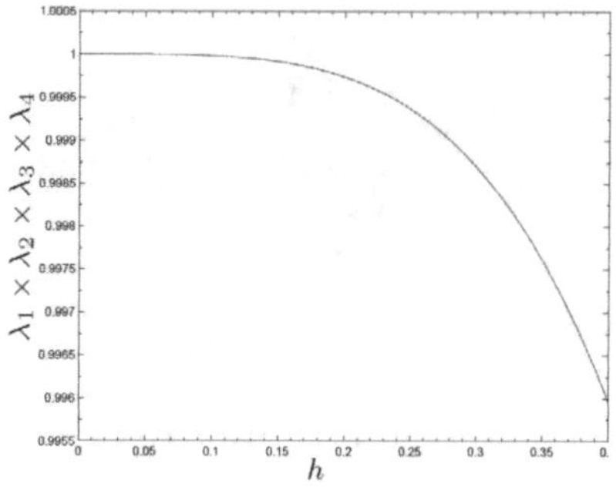

Figure 5: Eigenvalue product of the jacobian matrix of the 3rd order discretized model of the Hénon-Heiles system versus the step size h. $(-)$ $\mu_{\mathcal{L}}$; $(\cdots)$ $\mu_{\mathcal{L}} \pm \sigma_{\mathcal{L}}$

jacobian matrix trace of the 2nd order discretized model was equal to $4 - 2\,h^2$ and the one of the 3rd order was equal to $4 - 2\,h^2$. To accomplish this, 2nd order term from the 3rd order discretized model was eliminated. Unfortunately simulations show that these results are not satisfactory, because they lead the model to instability.

The addition of a single 3rd order term to the Euler approximation is not enough to produce a valid model. The same reasoning to avoid the stability problem and symmetry loss was applied to the discretized models.The jacobian matrix trace variation was studied for each increment in the discretization order, using Eq. (21). A further attempt to solve the problem was to add higher order terms considering only the jacobian matrix trace. The 3rd order discretized model was chosen because it can successfully reproduce the original Hénon-Heiles system for $h < 0.02s$. 6th and 7th order terms were added to this model since their low contribution to the trace. The resulting model with the chosen terms was simulated and presented better results than the plain 3rd order model. The model can successfully reproduce the Poincaré section with step sizes up to $h \approx 0.4s$.

The Normand-Cyrot method was again applied, without the choice of terms, and the 9th order discretized model was simulated. The 9th order model proved to be much more robust than the ones shown previously. For discretization steps less then $h \approx 0.5s$, the Poincaré section is very close to the section of the original continuous system.

With the purpose of checking the robustness of the 9th order discretization, new sections in the Poincaré section were generated as a function of the step h. New sections at $v_n = 0$ were achieved by getting points of the Poincaré section in which $-0.01 \leq v_k \leq 0.01$ was verified. Same reasoning was made for the Poincaré section at $y_n = 0$. Such graphics for the 9th order discretized model, without chosen terms, as well as the stationary energy and eigenvalue product, are shown in Figures 6, 7 and 8.

Region I of Figure 6 shows that for $h < 0.5s$, the islands of the Poincaré section are the same as the original continuous system. Region I of the energy plot (Figure 7) shows that the energy is practically constant, resulting in an attractor similar to the original one. The Poincaré section for $h = 0.5s$ is shown at Figure $9a$. For $h > 0.5s$, islands begin to cease to exist (region II of Figure 6). Region II of the energy plot (Figure 7) points to the beginning of a more considerable energy loss, resulting in slightly smaller attractors. Even though, at $h = 0.67s$, when the energy is equal to 0.122 (5% of the initial energy), the model still has a solution topologically equivalent to the original system one. For region III of the energy plot (Figure 7), $h > 0.67s$, spurious behavior and numerical instabilities begin to occur, as expected, because the discretization step is above the maximum allowed by the Nyquist criteria. For $0.67s < h \leq 0.87s$, some details of

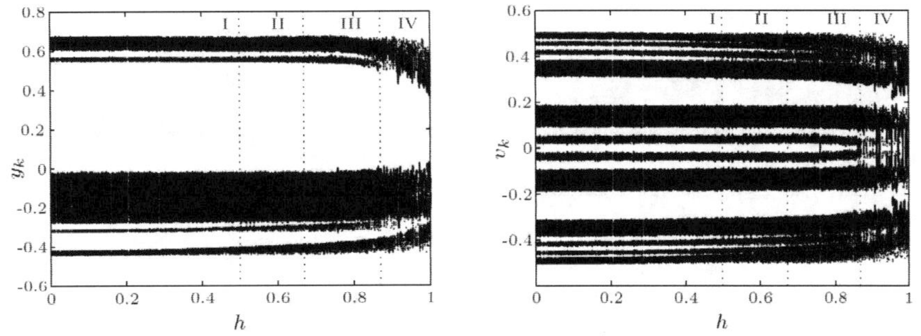

(a) Cut at $v_n = 0$ (b) Cut at $y_n = 0$

Figure 6: Cuts in the Poincaré section of the Hénon-Heiles system for the 9th order discretized model, using the method proposed by Monaco e Normand-Cyrot, versus the discretization step h. The attractor of the model at region (I) is similar to the attractor of the continuous system $(0s < h \leq 0.5s)$; (II) slightly smaller than the attractor of the continuous system $(0.5s < h \leq 0.67s)$; (III) without a few details and with some numerical instabilities, due to the simulation with step sizes higher than the one limited by the Nyquist Criteria $(0.67s < h \leq 0.87s)$; and (IV) predominancy of numerical instabilities $(h > 0.87s)$.

the Poincaré section are lost, but the symmetry still remains, as seen in the Poincaré section at $h = 0.87s$ (Figure 9b). Only after $h \approx 0.87s$ that numerical instabilities take over. This can be inferred by several abrupt changes in the Poincaré section (region IV of Figure 6).

Eigenvalue product of the jacobian matrix slightly decreases as the discretization step increases, as shown in Figure 8. But this characteristic is much smoother than the one of the 3rd order model.

The energy indexes calculated for the 9th order discretized model are listed in Table 1. The energy mean calculated in the simulation with this model and $h = 0.50s$ is closer to the original energy than the simulation with the 3rd order model and $h = 0.01s$. The total energy variation and the absolute energy variation within a period of 1 *sec* are of the same order of magnitude. This reassures that the 9th order model structure is better for energy conservation for higher h values than the 3rd order one. As the step size h increases, the energy variation increases as well.

A few remarks must be made for the simulation with $h = 0.87s$ (Figure 9b): i) points are agglomerated in well defined regions, ii) the Poincaré section lacks details, and iii) the presence of aliasing phenomenon. This is explained by the frequency analysis of variable x. Figure 10a

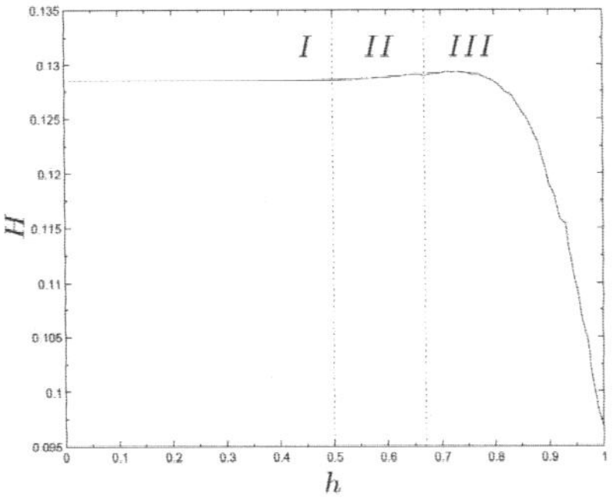

Figure 7: Stationary energy of 9th order discretized model of the Hénon-Heiles system versus the step size h. Stationary energy was calculated considering the average of the last 1000 simulation values.

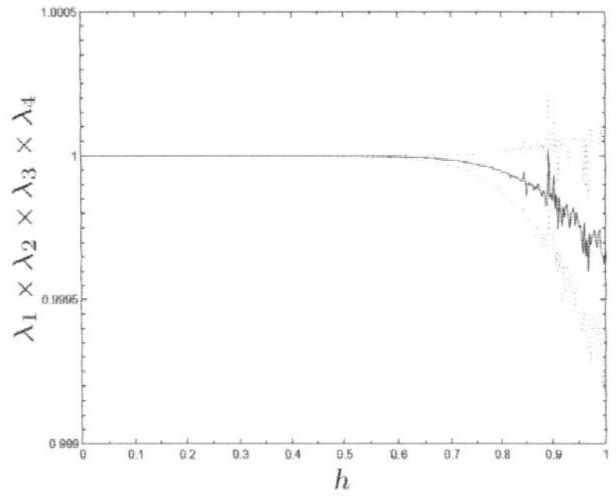

Figure 8: Eigenvalues product of the jacobian matrix of the 9th order discretized model of the Hénon-Heiles system versus the step size h. $(-)$ $\mu_{\mathcal{L}}$; $(\cdots)$ $\mu_{\mathcal{L}} \pm \sigma_{\mathcal{L}}$

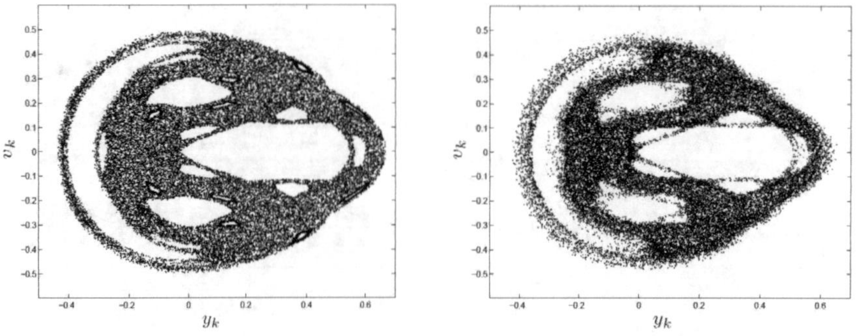

(a) $h = 0.50s$ (b) $h = 0.87s$

Figure 9: Poincaré section of the Hénon-Heiles system, defined by $P \equiv \{(y_n, u_n, v_n) \in \mathbb{R}^3 \mid x_n = 0\}$, using the method proposed by Monaco and Normand-Cyrot, $\eta = 9$, initial condition ($x_0 = 0.000, y_0 = 0.670, u_0 = 0.093, v_0 = 0.000$) and last simulation instant $N \approx \frac{120000s}{h}$.

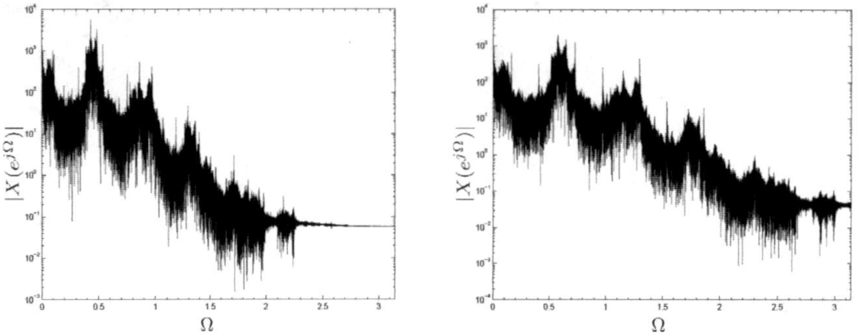

(a) $h = 0.50s$ (b) $h = 0.67s$

Figure 10: Frequencies of the variable x of the Hénon-Heiles system, with initial conditions ($x_0 = 0.000, y_0 = 0.670, u_0 = 0.093, v_0 = 0.000$), simulation of the 9th order discretized model.

Table 1: Energy indexes calculated for the discrete models.

Model	h (s)	μ_H	σ_H	$\mathcal{H}_1$	$\mathcal{H}_2$
Mickens	0.01	0.12852172	3.656872×10^{-4}	5.588389×10^{-9}	6.436041×10^{-4}
Mickens	0.02	0.12852644	7.284305×10^{-4}	-8.847700×10^{-9}	1.297851×10^{-3}
Mickens	0.05	0.13267830	1.894329×10^{-2}	-2.536240×10^{-8}	3.208169×10^{-2}
Mickens	0.67	0.13651811	2.639399×10^{-2}	4.064695×10^{-8}	4.268835×10^{-2}
Nor.-Cyr. $\eta = 3$	0.01	0.12811895	2.322869×10^{-4}	-1.333164×10^{-8}	1.360052×10^{-8}
Nor.-Cyr. $\eta = 9$	0.50	0.12869485	1.019699×10^{-4}	2.895236×10^{-9}	4.990294×10^{-8}
Nor.-Cyr. $\eta = 9$	0.67	0.12932747	5.262554×10^{-4}	1.503759×10^{-8}	7.072343×10^{-7}
Nor.-Cyr. $\eta = 9$	0.87	0.12264505	3.211799×10^{-3}	-9.508887×10^{-8}	7.150463×10^{-6}

shows that there is no border effect for frequencies near π at $h = 0.5s$. $h = 0.67s$ (Figure 10b) is the upper bound of the discretization step. Thus, higher frequencies begin to overlap for $h > 0.67s$.

6. Conclusion

In the context of the analysis of nonlinear conservative systems, the discretization scheme proposed by Monaco and Normand-Cyrot achieved better results and shown to be more robust than the one proposed by Mickens as the discretization step increases.

The maximum discretization step for the simulations using the discretization scheme proposed by Mickens is $h \approx 0.3s$. For such a range of values the Poincaré section generated from the iteration of the discretized models is close to the original continuous counterpart. $h \approx 0.85s$ was achieved before the trajectories were ejected to infinity [7]. The symmetry is not kept as the step size increases.

As for the discretization scheme proposed by Monaco e Normand-Cyrot the maximum discretization step was found to be $h \approx 0.5s$. For the 9th order model the Poincaré section is close to the original. The maximum step size provided by the Nyquist criteria, $h \approx 0.67s$, produced a Poincaré section that resembles the original one but with an displacement in the parameter space. For $0.67s < h \leq 0.87s$, the Poincaré sections lacked details due to border effect - overlapping of higher frequencies. For $h > 0.87s$ the discretized models no longer generate Poincaré sections that resemble the original continuous Poincaré section. The symmetry of the conservative system was preserved in all simulations, even for discretization steps higher than the limit imposed by Nyquist criterium.

Simulations with different model structures showed that low order terms are necessary for the model stability. The 3rd order model managed to reproduce the Poincaré section for $h < 0.02s$. Even if the simulation was performed with a higher order model, the higher order terms would not be representative, because h is too small. But removing the 2nd order term makes the 3rd order model to unstable. As h increases, higher order terms become more representative and cannot be disregarded, and therefore they are crucial for long simulations. These higher order terms introduce tiny increments that are responsible for the success of long simulations for such non-linear systems.

Sensitivity to initial conditions is a well known problem in non-linear systems [12]. An analogous reasoning can be applied to the discretization method proposed by Monaco and Normand-Cyrot. The initial conditions are the same, but the model structure is slightly different as the discretization order increases. Numerical instabilities appear when the model structure cannot reproduce the original continuous system for long simulations at a certain h. In this case, it is necessary to add higher order terms to improve the model approximation.

In order to verify whether the discretized system can reproduce the continuous system behavior, the eigenvalue product of the jacobian matrix was used as a measure of energy conservation. If it is closer to one, the system is closer to conserve energy.

Finally, the discretization scheme proposed by Monaco and Normand-Cyrot has an advantage to be a straightforward and simple method that provides robust discretized models. The only necessary step is to determine the desired order to obtain the difference equations; no further study of the system is required. The disadvantage is that the increase in the discretized model order means more computational effort. For conservative systems, it was shown that method can generate valid models that preserves the symmetry for any value of the discretization step.

References

[1] Hairer E, Lubich C and Wanner G 2006 *Geometric Numerical Integration: Structure-Preserving Algorithms for Ordinary Differential Equations* 2nd ed (Netherlands: Springer)
[2] Mickens R E 2002 *Journal of Difference Equations and Applications* **8** 823–847
[3] Monaco S and Normand-Cyrot D 1990 *Lecture Notes in Control and Information Sciences* vol 144 ed Thoma M and Wymer A (New York: Springer-Verlag) pp 788–797
[4] Mendes E M A and Billings S A 2002 *Chaos* **12** 66–71
[5] Letellier C and Mendes E M A M 2004 *Journal of Physics A* **37** 1203–1218
[6] Letellier C and Mendes E M A M 2005 *Chaos* **15**
[7] Letellier C, Mendes E M A M and Mickens R E 2007 *International Journal of Bifurcation and Chaos* **17** 891–902
[8] Hénon M and Heiles C 1964 *Astron* **69** 73–79
[9] Letellier C, Aguirre L A and Maquet J 2005 *Phys. Rev. E* **71** 066213
[10] Monaco S, Normand-Cyrot D and Tiefensee F 2009 *Joint 48th IEEE Conference on Decision and Control and 28th Chinese Control Conference* 1782–1787
[11] Egger J 1996 *Monthly Weather Review* **124** 1955–1964
[12] Strogatz S H 2000 *Nonlinear Dynamics and Chaos: With Applications to Physics, Biology, Chemistry, and Engineering* 1st ed (USA: Westview Press)

Dynamic Days South America 2010 IOP Publishing
Journal of Physics: Conference Series **285** (2011) 012028 doi:10.1088/1742-6596/285/1/012028

The method of separation of variables for the Frobenius-Perron operator associated to a class of two dimensional chaotic maps

José-Rubén Luévano

Departamento de Ciencias Básicas, Universidad Autónoma Metropolitana, Unidad Azcapotzalco, CP 02200, México, D.F., México.

E-mail: jrle@correo.azc.uam.mx

Abstract. Analytical expressions for the invariant densities for a class of discrete two dimensional chaotic systems are given. The method of separation of variables for the associated Frobenius-Perron equation is introduced. These systems are related to nonlinear difference equations which are of the type $x_{k+2} = T(x_k)$. The function T is a chaotic map of an interval whose chaotic behaviour is inherited to the two dimensional one. We work out in detail some examples, with T an expansive or intermittent map, in order to expose the method. Finally, we discuss how to generalize the method to higher dimensional maps.

1. Introduction

Explicit computations of invariant densities for higher dimensional maps are seldom found. In contrast, in one dimensional dynamics there are lots of exactly solvable chaotic maps, which can be generated e.g., using the conjugation property [1, 2, 3, 4, 5, 6, 7] or by means of the Schröder method, see [8, 9] and references therein. Recently the problem of continuous iteration has become an interesting topic which can be associated to our treatment [10].

From the statistical point of view, the knowledge of the invariant density of a map is of particular importance because that enables us to obtain with detail some of its statistical properties [6, 8, 11, 12, 13, 14, 15].

The invariant density of a chaotic map is an eigenfunction of the Frobenius Perron operator induced by it [6, 11, 12, 14, 16]. The associated eigenvalue problem is given by a functional equation but there are no known general methods to solve it. The methods used in the field of functional equations have not yet provided a general way to solve them, instead one can find a lot of operational methos adapted to each class of equations, see for example [17, 18]. Therefore, to find reliable methods to obtain these densities becomes a very important task.

In this work a methodology is developed to solve the Frobenius-Perron functional equation, which roughly consists in introducing the method of separation of variables for it. We use several examples to show how it works.

The paper is organized as follows: in section 2 we discuss the Lauwerier map; section 3 introduce the Frobenius-Perron operator and the notion of invariant measure; sections 4 and 5 develops the method of separation of variables for the Frobenius-Perron equation and presents some examples of it; section 6 extend that treatment to some more general discrete systems;

section 7 briefly discuss the relation with the theory of nonlinear difference equations, and finally we give some discussion and conclusions.

2. Lauwerier's map

We start considering a map introduced by Lauwerier in [19]. For any $(x, y) \in \mathbb{R}^2$, the first order nonlinear system of difference equations:

$$\begin{aligned} x_{n+1} &= y_n, \\ y_{n+1} &= 4x_n(1 - x_n), \end{aligned} \tag{1}$$

defines a discrete dynamical system for which the unit square $\Omega = [0, 1] \times [0, 1]$ is invariant, and the dynamics in Ω is chaotic. As remarked in that paper, the invariant density for this example can be exactly computed and is given by:

$$\rho(x, y) = \frac{1}{\pi^2 \sqrt{xy(1-x)(1-y)}}. \tag{2}$$

However, no comments about its computation are given. We point out that $\rho(x, y) = A(x)A(y)$, where $A(x) = 1/(\pi\sqrt{x(1-x)})$. In other words, ρ is the product of two identical functions A of a single variable, each one depending on x and y independently. The invariant measure $\mu(x, y) = \int_{(0,0)}^{(x,y)} \rho(u, v)dudv$, is given by

$$\mu(x, y) = \frac{1}{\pi^2} \arcsin(\sqrt{x}) \arcsin(\sqrt{y}). \tag{3}$$

The main purpose of the present paper is to generalize this result to more general two dimensional maps. To this end we are implementing the method of separation of variables for the Frobenius-Perron operator associated to the corresponding transformation.

3. The Frobenius-Perron equation

Let us consider a higher dimensional nonlinear transformations $F : U \subseteq \mathbb{R}^d \to \mathbb{R}^d$, and denote $F(X) = (f_1(X), f_2(X), \ldots, f_d(X))$ for $X = (x_1, x_2, \ldots, x_d) \in U \subseteq \mathbb{R}^d$. We are interested in the study of the discrete dynamical system $X_{n+1} = F(X_n)$. We assume that F is, at least, a C^1 function. Also, that there exists a set of functions $\phi_j : U \subseteq \mathbb{R}^d \to \mathbb{R}^d$, such that $F(\phi_j(X)) = X$, for $j = 1, 2, \ldots, d$. In other words, $\phi_j(X)$ is the $j-th$ branch of the inverse function of F. The Frobenius Perron operator associated to F is defined by [6, 11, 12, 14, 16]:

$$\mathcal{L}_F[\rho(X)] = \sum_{F(Y)=X} \frac{\rho(Y)}{|\det D_Y F|}, \tag{4}$$

where the Jacobian matrix $D_Y F = [\partial F_i/\partial x_j]$ is evaluated at Y. The scalar function supported on a given $U \subseteq \mathbb{R}^d$, denoted by ρ, which solves the functional equation

$$\sum_{F(Y)=X} \frac{\rho(Y)}{|\det D_Y F|} = \rho(X), \tag{5}$$

is called the invariant density for F.

4. The method of separation of variables

The method presented in this section is called separation of variables, because of its resemblance with the traditional one used in differential equations. The method is exposed in detail in all the examples given in the following sections. For convenience, in the sections below, we use the standard representation of the Frobenius-Perron equation by means of the Dirac delta function.

4.1. One dimensional maps

We need a brief exposition of the one-dimensional case. Let us start considering a piecewise continuous transformation $T : I \to I$ where the interval $I = [a, b] \subset \mathbb{R}$. Moreover, for a partition of I, i.e. $I = \cup_j C_j = \cup_j [a_j, a_{j+1}]$, where $a = a_0, \ldots, a_k = b$, we assume the map T is monotonous, continuous and onto each $C_j = [a_j, a_{j+1}]$, and also that T is differentiable on each (a_j, a_{j+1}).

If T has an invariant density $\rho(x)$, then it must satisfy the Frobenius-Perron equation:

$$\mathcal{L}_T[\rho(x)] = \int_I dy \delta(x - T(y))\rho(y) = \rho(x), \tag{6}$$

$$= \sum_{j=1}^{r} \frac{\rho(T_j^{-1}(x))}{\left|T_j' \circ T_j^{-1}(x)\right|} = \rho(x). \tag{7}$$

In particular, we study some examples defined in terms of the Chebyshev polynomials because these functions considered as maps share nice ergodic properties, not only ergodicity but also stronger ones: mixing and exactness [6]. Notice moreover that these polynomials satisfy the desired properties mentioned above [2, 6, 8].

4.2. Two dimensional maps

For the sake of clarity we begin our exposition with a straightforward generalization of the equation (1). Now, for any $(x, y) \subset \mathbb{R}^2$ we consider the system:

$$x_{n+1} = y_n,$$

$$y_{n+1} = T(x_n), \tag{8}$$

where $T : I \to I$ with $I \subseteq \mathbb{R}$, is a map satisfying conditions stated in the preceding section. Also, we assume that the set $\Omega = I \times I$ is invariant. Now, the Frobenius-Perron equation associated to equation (8) is given by

$$\mathcal{L}[\rho(x, y)] = \sum_{j=1}^{r} \frac{\rho(T_j^{-1}(y), x)}{\left|T_j' \circ T_j^{-1}(y)\right|} = \rho(x, y). \tag{9}$$

We can see that this functional equation can be solved by the *ansatz* $\rho(x, y) = A(x)B(y)$, whereby

$$\mathcal{L}[\rho(x, y)] = \mathcal{L}_T[A(y)]B(x) = A(x)B(y). \tag{10}$$

The method of separation of variables for the equation (9) works as follows:

$$\frac{\mathcal{L}_T[A(y)]}{B(y)} = \frac{A(x)}{B(x)} = c, \tag{11}$$

where we denote by c the constant of separation, such that

$$\mathcal{L}_T[A(y)] = cB(y),$$

$$A(x) = cB(x). \tag{12}$$

Note now that this is a system of two coupled functional equations. To proceed to decoupling them we make the substitution of the second equation in the first one. Hence, we obtain $\mathcal{L}_T[A(y)] = A(y)$. In other words, if the map T has an invariant density A, we must finally have

$$\rho(x, y) = \frac{1}{c}A(x)A(y). \tag{13}$$

The constant c is obtained by normalization:

$$\int_{\Omega} \rho(x,y)dxdy = 1$$

hence,

$$c = \left[\int_I A(x)dx\right]^2. \tag{14}$$

4.3. *Analytical examples*

In this section we provide three exactly solvable examples of systems like equation (8).

4.3.1. Chebyshev maps I If the transformation T in equation (8) is given by the k-th Chebyshev polynomial, $P_k(x)$, then as it is well known, the invariant density for this map is given by $\rho(x) = 1/(\pi\sqrt{1-x^2})$. Hence, by the equation (12) the invariant density for the corresponding two dimensional system is given by:

$$\rho(x,y) = \frac{1}{\pi^2\sqrt{(1-x^2)(1-y^2)}}, \tag{15}$$

where the support for the invariant measure is the set $\Omega = [-1,1] \times [-1,1]$.

4.3.2. Intermittency An example of a map with a marginally stable fixed point is the cusp-shaped map $T(x) = 1 - 2\sqrt{|x|}$. It is well known that this map has an invariant density $\rho(x) = (1-x)/2$, $x \in [-1,1]$, see [4]. Therefore, the corresponding system, given by equation (8) has an invariant density

$$\rho(x,y) = (1-x)(1-y)/4. \tag{16}$$

for any $(x,y) \in \Omega = [-1,1] \times [-1,1]$.

4.3.3. Transformations of the plane onto itself From Newton's method to search the roots of the equation $x^2+1=0$, [20, 21, 22], a chaotic iteration of the real line is found $T(x) = (x - 1/x)/2$. Also, its invariant density is known to be $\rho(x) = 1/(\pi(1+x^2))$. Therefore the corresponding system, equation (8), has an invariant density

$$\rho(x,y) = \frac{1}{\pi^2\sqrt{(1+x^2)(1+y^2)}}, \tag{17}$$

where $(x,y) \in \Omega = \overline{\mathbb{R}} \times \overline{\mathbb{R}}$.

5. A further generalization

Now we consider two different transformations, $S: J \to J$ and $T: I \to I$, both of them satisfying conditions stated in section (4.1), but each one having r and s monotone pieces respectively. Therefore, T^{-1} and S^{-1} have r and s monotone branches on their respective intervals. Then, defining a discrete dynamical system on $(x,y) \in \Omega = I \times J$ by

$$\begin{aligned} x_{n+1} &= S(y_n), \\ y_{n+1} &= T(x_n), \end{aligned} \tag{18}$$

the associated Frobenius Perron operator is given by

$$\mathcal{L}[\rho(x,y)] = \int_{\Omega} dudv\delta(x - S(v))\delta(y - T(u))\rho(u,v). \tag{19}$$

Now, using the definition of a higher dimensional δ function [?], we have

$$\mathcal{L}[\rho(x,y)] = \sum_{j=1}^{r}\sum_{l=1}^{s} \frac{\rho(T_j^{-1}(y), S_l^{-1}(x))}{\left|T' \circ T_j^{-1}(y)\right| \left|S' \circ S_l^{-1}(x)\right|}. \tag{20}$$

By analogy with to the previous section we introduce the *ansatz* $\rho(x,y) = A(x)B(y)$, which allows us to have

$$\mathcal{L}[\rho(x,y)] = \mathcal{L}_T[A(y)]\mathcal{L}_S[B(x)] = A(x)B(y). \tag{21}$$

This time the method of separation of variables gives us

$$\frac{\mathcal{L}_S[B(x)]}{A(x)}\frac{\mathcal{L}_T[A(y)]}{B(y)} = 1, \tag{22}$$

then, introducing a constant c, it provides the following system

$$\begin{aligned} \mathcal{L}_T[A(y)] &= cB(y), \\ \mathcal{L}_S[B(x)] &= \frac{1}{c}A(x). \end{aligned} \tag{23}$$

Now, this system can be decoupled giving us $\mathcal{L}_{S\circ T}[A(x)] = A(x)$ and $\mathcal{L}_{T\circ S}[B(y)] = B(y)$. In other words, the system (23) is decoupled in two functional equations of a single variable. More important, this time the functions $A(x)$ and $B(x)$ are (if they exists) the invariant densities of $S \circ T$ and $T \circ S$ respectively. We finally obtain

$$\rho(x,y) = A_{S\circ T}(x)B_{S\circ T}(y). \tag{24}$$

5.1. *Examples*

The following two examples are related to the last discussed case.

5.1.1. Chebyshev maps II As stated before, the Chebyshev maps are a very important examples among the class of integrable chaotic maps. Also, it is well known that all the Chebyshev maps share the same invariant density $A(x) = 1/(\pi\sqrt{1-x^2})$. Now we consider the following system

$$\begin{aligned} x_{n+1} &= P_l(y_n), \\ y_{n+1} &= P_k(x_n), \end{aligned} \tag{25}$$

where P_k and P_l are the Chebyshev polynomials of k and l degrees respectively. From the semigroup property $P_l \circ P_k = P_k \circ P_l = P_{kl}$, we have that $\mathcal{L}_{P_{kl}}[A(x)] = A(x)$. Therefore

$$\rho(x,y) = \frac{1}{\pi^2\sqrt{(1-x^2)(1-y^2)}}, \tag{26}$$

for any $(x,y) \in \Omega = [-1,1] \times [-1,1]$. We show a plot of this density in figure (1).

5.1.2. Cournot maps Some mathematical models in economy are known as Cournot duopoly games [23, 24]. They are described by on a system consisting of two coupled logistic maps $F_r(x) = rx(1-x)$:

$$\begin{aligned} x_{n+1} &= ry_n(1-y_n), \\ y_{n+1} &= sx_n(1-x_n), \end{aligned} \tag{27}$$

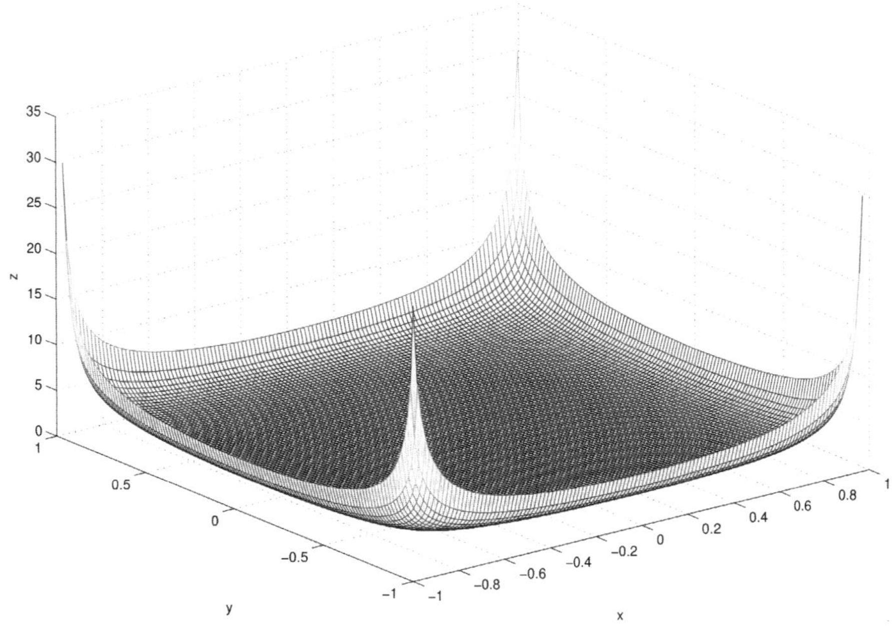

Figure 1. The invariant density for the map given by equation (25), $\rho(x,y) = \frac{1}{\pi^2\sqrt{(1-x^2)(1-y^2)}}$.

where $r, s > 0$ and $(x, y) \in [0,1] \times [0,1]$. We restrict our analysis to the case $r = s = 4$. It is well known that $F_4(x) = 4x(1-x)$ is conjugated to $N_2(x) = (-1)^{[2x]}\{2x\}$. Here $[rx]$ denotes the integer part of rx and $\{rx\}$ means $rx \bmod 1 = rx - n$ where n is the largest integer such that $rx - n \geq 0$ for any $r \in \mathbb{N}$. Hence, the map $F_4 \circ F_4(x) = F_4^{(2)}(x)$ is conjugated to $N_4(x) = (-1)^{[4x]}\{4x\}$, so that both maps share the same invariant density $A(x) = 1/\pi\sqrt{x(1-x)}$. To be more specific, $F_4(\sin^2(\theta)) = \sin^2(2\theta)$ then $F_4^2(\sin^2(\theta)) = \sin^2(4\theta)$. Because the invariant measure for F_4 is given by $\mu(x) = (2/\pi)\arcsin(\sqrt{x})$, therefore $\mathcal{L}_{F_4 \circ F_4}[A(y)] = \mathcal{L}_{F_4^{(2)}}[A(y)] = A(y)$, and finally

$$\rho(x,y) = \frac{1}{\pi^2\sqrt{xy(1-x)(1-y)}}. \tag{28}$$

6. General maps

Our method is not restricted to the class of systems strictly given in the form of equations (8) and (18). Of course, it is possible to find new exactly solvable maps which are equivalent to these systems by means of a change of variables. In the following we give an example showing how this works. A nontrivial example which is studied in [2] is provided by

$$\begin{aligned} x_{n+1} &= (y_n^2 - x_n - 1)^2 - (y_n^2 + x_n - 1)^2, \\ y_{n+1} &= \sqrt{(y_n^2 - x_n - 1)^2 + (y_n^2 + x_n - 1)^2}. \end{aligned} \tag{29}$$

In that reference the authors shows that this system is exactly solvable one. After making the change of variables (this one is different from that employed in [2]) $x_n = (\eta_n - \xi_n)/2$ and

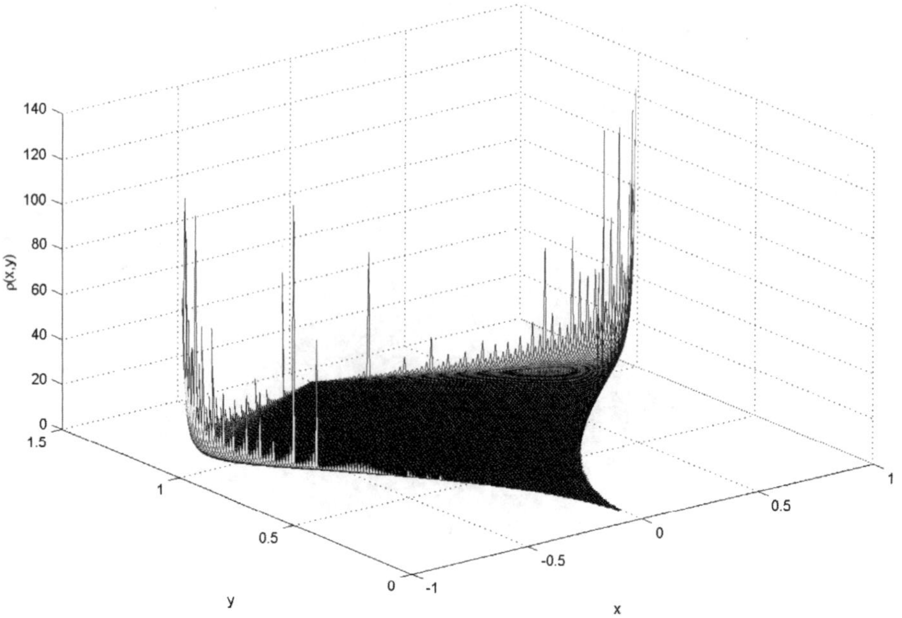

Figure 2. The invariant density for the map given by equation (29). It is noted the sharp singularities at $(\pm 1, \pm 1)$. $\rho(x,y) = \frac{4y}{\sqrt{(2-(y^2-x))(y^2-x)(2-(y^2+x))(y^2+x)}}$.

$y_n^2 = 1 + (\eta_n + \xi_n)/2$ we obtain the system

$$\xi_{n+1} = 2\eta_n^2 - 1, \tag{30}$$
$$\eta_{n+1} = 2\xi_n^2 - 1, \tag{31}$$

which is clearly of the form of the equation (18). Hence, the (unnormalized) density is $\rho(\xi,\eta) = 1/\sqrt{(1-\xi^2)(1-\eta^2)}$. Now, the $|\det(J)| = |4y|$, then the invariant density in the variables x, y is computed by $\rho(\xi,\eta)d\xi d\eta = \rho(x,y)\,|4y|\,dxdy$, such that

$$\rho(x,y) \propto \frac{4y}{\sqrt{(1-(y^2-x-1)^2)(1-(y^2+x-1)^2)}}. \tag{32}$$

The argument in its denominator can be rewritten as

$$\left(2-(y^2-x)\right)\left(y^2-x\right)\left(2-(y^2+x)\right)\left(y^2+x\right). \tag{33}$$

After some algebra, we find that the domain of that density is given by the set $\Omega = \{(x,y) \in \mathbb{R}^2\}$ such that the following inequalities (also given in [2])

$$y^2 - x < 2,\ y^2 - x > 0,\ y^2 + x < 2,\ y^2 + x > 0, \tag{34}$$

are valid. Hence, this is the invariant set for the system. Since these inequalities are satisfied for $y > 0$, the absolute value in equation (32) becomes irrelevant. A plot of the associated invariant density is given in figure (2).

7. Delayed nonlinear difference equations

In recent times the study of nonlinear difference equations has become an intensive area of study from the point of view of dynamical systems [24, 25]. It is worth mentioning that the system (8) is equivalent to the nonlinear second-order difference equation of the form:

$$x_{n+2} = T(x_n). \tag{35}$$

We point out that the change of variables $x_{n+1} = y_n$ give us $y_{n+1} = T(x_n)$, i.e. the two dimensional system given by equation (8). A straightforward generalization is given by the delayed difference equation:

$$x_{n+k} = T\left(x_n\right) \tag{36}$$

where $k \in \mathbb{N}$. We can see that equation (36) becomes equivalent to a k-dimensional first order system which is a generalization of the equation (8).

8. Discussion and Conclusions

We have given a generalization of a discrete two dimensional chaotic system studied by Lauwerier some years ago. For these systems the corresponding Frobenius Perron equation is solved by a method of separation of variables. In this way we can transform a two dimensional system into 2 one-dimensional systems. Moreover, if the one-dimensional systems are chaotic, then this property is inherited to the two dimensional one. Several examples were worked out in detail to describe the method. The Chebyshev polynomials and logistic maps play an important role in our study, but others maps like the Jacobi sine function and rational functions could be considered, see [3, 5, 7].

Also, we related these systems to some class of nonlinear delayed difference equations. This relation can be exploited to build higher dimensional maps which can be solved by our method. Moreover, it can be interesting to apply our method to coupled maps lattices.

In this work we were only concerned with the computation of invariant densities. Some important topics in ergodic theory were not addressed, for example mixing and exactness properties. In particular, it will be interesting to explore the spectral theory of the Frobenius-Perron operator for the systems worked out here. Moreover, there is still the question about the possibility to give explicit expressions for the generalized spectra as studied in [29]. Some progress in such direction will be reported elsewhere.

Acknowledgments

I wish to thank Ismael Muñoz M. for valuable comments and Sudeshna Sinha to bring to my attention the problem of Coupled Maps Lattices. This work was partially supported by CONACYT (SNI), México.

References

[1] Fukuda, W and Katsura, S 1986 *Physica* A **136** 588-99
[2] Gardini, L, Mira, C, Barugola, A and Cathala, J C 1996 *Chaotic Dynamcis in Two Dimensional Nonivertible Maps* World Scientific Series in Nonlinear Science (Series A, vol 20)
[3] Grosjean, C C 1987 *J. Math. Phys.* **28** 1265-74
[4] Hemmer 1984 The exact invariant density for a cusp-shaped return map *J. Phys.* A: Math. Gen. **17**, L247-49
[5] Jafarizadeh, M A and Behnia, S 2003 Hierarchy of one and many parameter families of elliptic chaotic maps of **cn** and **sn** types *Phys. Lett.* A **310** 168-76
[6] Lasota A and Mackay M C 1994 *Chaos, Fractals and Noise* (Appl. Math. Sc. vol 97) (New York: Springer)
[7] Umeno, K 1997 Method of constructing exactly solvable chaos *Phys. Rev. E* **55** 5280-84
Umeno, K 1999 Exactly solvable chaos and addition theorems of elliptic functions *RIMS Kokyuroku* **1098** 104-17
[8] Luévano, J R and Piña, E 2008 The Schröder functional equation and its relation to the invariant measures of chaotic maps *J. Phys. A: Math. Theor.* **41** 265101. Also in arXiv:0907.3765v1 [math-ph] 22 July 2009

[9] Schröder, E 1871 *Math. Ann.* **3** 296-322. Available through http://gdz.sub.uni-goettingen.de/en (search by title)
[10] Curtright, T and Zachos, C 2009 Evolution Profiles and Functionals Equations *J. Phys. A: Math. Theor.* **42** 485208
[11] Collet, P and Eckmann, J P 2006 *Concepts and Results in Chaotic Dynamics* (Berlin Heidelberg: Springer-Verlag)
[12] Dorfman J R 1999 *An introduction to chaos in nonequilibrium statistical mechanics*, Cambdrige Lect. Notes in Phys. (Cambridge: Cambridge University Press)
[13] Hunt, B R, Kennedy, J A, Li, T Y and Nusse, H E 2004 *The Theory of Chaotic Attractors* (New York: Springer-Verlag Inc.)
[14] Mackay M C 2003 *Time's Arrow* (New York: Dover Publ. Inc.)
[15] Wang, K *et al* 2008 Statistical independence in nonlinear maps coupled to non-invertible transformations *Phys. Lett.* A **372** 6593-6601
[16] Katok A and Hasselblatt B 1995 *Introduction to Modern Theory of Dynamical Systems* (Cambridge: Cambridge Univ. Press)
[17] Castillo, E and Reyes Ruiz-Cabo, M 1992 *Functional Equations and Modeling in Science and Engineering*, Pure and Applied Math. (New York:Marcel Dekker, Inc.)
[18] Small, C 2007 *Functional Equations and How to Solve Them* (Berlin: Springer)
[19] Lauwerier, H A 1986 Two-dimensional iterative maps, in *CHAOS* ed Holden, A V (New Jersey: Princeton Univ. Press) chapter 4 pp 58-95
[20] Bender, C M and Orszag, S A 1978 *Advanced Mathematical Methods for Scientists and Engineers* (New York: McGraw-Hill) 233-39
[21] Beardon, A F 1991 *Iterations of Rational Functions* Graduate Texts in Mathematics, vol 132 (N.Y.: Springer-Verlag Inc.)
[22] Skwroner, L and Gora, P 2007 Chaos in Newtonian iterations: searching for zeros which are not there *Acta Phys. Pol.* B **38** 1909-24
[23] Canovas, J S and Linero, A 2001 Topological dynamics classification of duopoly games *Chaos, Solitons and Fractals* **12** 1259-66
[24] Kulenović, M R S and Merino, O 2002 *Discrete Dynamical Systems and Differential Equations with Mathematica* (Boca Raton Florida: Chapman & Hall/CRC)
[25] Elaydi, S N 1996 *An Introduction to Difference Equations* Second Edition (New York: Springer-Verlag)
[26] Rasband, N 1990 *Chaotic Dynamics of Nonlinear Systems* (N.Y.: John Wiley & Sons, Inc.)
[27] Tél, T and Gruiz, M 2006 *Chaotic Dynamics* (Cambrigde: Cambrigde Univ. Press)
[28] Sprott, J C 2000 *Strange Attractors: Creating Patterns in Chaos* M&T Books (New York: 1993) out of print but available at http://sprott.physics.wisc.edu
[29] Beck C and Schlögl F 1993 *Thermodynamic of chaotic systems* (Cambridge: Cambridge University Press)

Dynamic Days South America 2010 IOP Publishing
Journal of Physics: Conference Series **285** (2011) 012029 doi:10.1088/1742-6596/285/1/012029

Measuring Asymmetry in Insect-Plant Networks

Cláudia P. T. Cruz[1], Adriana M. de Almeida[2] and Gilberto Corso[3]

[1] Programa de Pós-Graduação em Física, Universidade Federal do Rio Grande do Norte, UFRN - Campus Universitário, Lagoa Nova, CEP 59078 972, Natal, RN, Brazil.
[2] Departamento de Botânica, Ecologia e Zoologia, Centro de Biociências, Universidade Federal do Rio Grande do Norte, UFRN - Campus Universitário, Lagoa Nova, CEP 59078 972, Natal, RN, Brazil.
[3] Departamento de Biofísica e Farmacologia, Centro de Biociências, Universidade Federal do Rio Grande do Norte, UFRN - Campus Universitário, Lagoa Nova, CEP 59078 972, Natal, RN, Brazil.

E-mail: claudia@dfte.ufrn.br, adrianam@ufrn.br and corso@cb.ufrn.br

Abstract. In this work we focus on interaction networks between insects and plants and in the characterization of insect plant asymmetry, an important issue in coevolution and evolutionary biology. We analyze in particular the asymmetry in the interaction matrix of animals (herbivorous insects) and plants (food resource for the insects). Instead of driving our attention to the interaction matrix itself we derive two networks associated to the bipartite network: the animal network, D_1, and the plant network, D_2. These networks are constructed according to the following recipe: two animal species are linked once if they interact with the same plant. In a similar way, in the plant network, two plants are linked if they interact with the same animal. To explore the asymmetry between D_2 and D_1 we test for a set of 23 networks from the ecologic literature networks: the difference in size, ΔL, clustering coefficient difference, ΔC, and mean connectivity difference, $\Delta{<}k{>}$. We used a nonparametric statistical test to check the differences in ΔL, ΔC and $\Delta{<}k{>}$. Our results indicate that ΔL and $\Delta{<}k{>}$ show a significative asymmetry.

1. INTRODUCTION

In ecology, networks are mainly used to visualize and describe food webs [1]. In the last two decades, scientists show a growing interest in networks in order to study other ecological interactions such as pollination, parasitism [2], seed dispersion or detrivory [3]. Plants and their pollinators or animals and their parasites are examples of interaction networks in community ecology. All these community ecology systems can be described by interaction matrices, or bipartite graphs. In pollinator networks, for example, the two functional groups are pollinators and flowering plants. The pollinators and flowering plants are the vertices of the bipartite network and observed interactions are drawn as links. In this context a species (pollinator or plant) that interacts with many species is called a generalist, while a species that have exclusive interactions or with few species is called specialist. In Ecology, the field data corresponding to the digraph is composed by two sets of species and the corresponding links (interactions) between them.

Some metrics over the interaction matrizes have being used to characterize its order or structure, for instance, the modularity, the connectivity, the nestedness and the assymetry [3].

The main objective of this manuscript is to explore the asymmetry of the interaction matrix using a new tool. Instead of driving our attention to the interaction matrix itself we focus the analysis on the two networks derived from the bipartite network: the animal network, D_1, and the plant network, D_2. These networks are built in the standard way in network analysis [4]. This work is organized as follows: in section 2 we outline the interaction matrix, where we show the mathematical background and presented the biological data set; in section 3 we present our results; finally in section 4 we discuss the results of the work.

2. INTERACTION MATRIX

2.1. MATHEMATICAL BACKGROUND

In order to fix the notation we call digraph an object D formed by two sets of vertices V_1 and V_2 and a set of links between these two sets. The digraph is completely described by the adjacency matrix, M, of order $L_1 \times L_2$, where L_1 and L_2 are the number of elements of V_1 and V_2, respectively. By definition:

$$M_{i,j} = \begin{cases} 0, & \text{if } i \text{ and } j \text{ are not linked.} \\ 1, & \text{if there is a link between vertices } i \text{ of } V_1 \text{ and } j \text{ of } V_2 \end{cases} \tag{1}$$

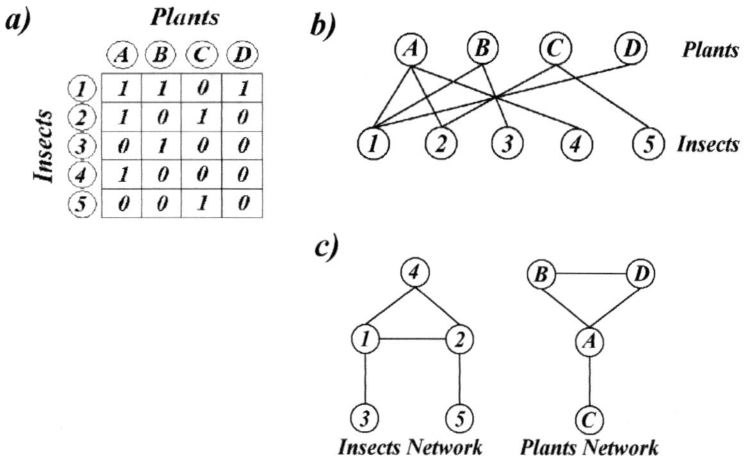

Figure 1. In (a) we have a example of an adjacency matrix, M, with zeros and ones sites; In (b) we have the representation of bipartite graph of the adjacency matrix, M, and, in (c) we have the two networks, D_1 and D_2, derived from the interaction network.

Moreover, the number of links of a vertex l is k_l and the distribution of links of D_1 and D_2 is P_{l_1} and P_{l_2} respectively. We define the occupancy number ρ as the fraction of ones in the adjacency matrix. For N, the total number of ones in M, we have

$$\rho = \frac{N}{L_1 L_2} \tag{2}$$

We study the asymmetry of a bipartite graph M using the two associated networks of M: the animal network, D_1, and the plant network, D_2. These networks are constructed according to the following recipe: two animal species are linked once if they interact with the same plant. In a similar way, in the plant network, two plants are linked if they interact with the same animal. In figure 2.1 we show the formation of two associated networks from a source bipartite graph.

2.2. ECOLOGICAL DATA SET

In the present paper we select a set of 23 plant-insect matrices of herbivorous insects and its host plants, which is characterized as antagonist interactions [4]. All the analyzed matrices consist of insects observed feeding on vegetal tissues of host plants. We used this data set by convenience, but the same study conducted here could be performed with mutualistic interactions, as well as with other antagonist interactions. In table 1 we show the main characteristics of the 23 interaction matrices used in this work: the number of plants (flowering plants) D_2, the number of animals (insects) in the matrix D_1, the occupancy of the matrix ρ and the reference of the data in the literature.

Table 1. Ecological networks analyzed for asymmetry, where L_2 is the number of plants (flowering plants) in the matrix, L_1 is the number of animals (insects) in the matrix, ρ is the occupancy of the matrix and in the last column is shown the reference of the data in the literature.

Matrix	L_2	L_1	ρ	Reference
CJ	33	21	0.13	[5]
PO	13	12	0.44	[6]
AN	33	29	0.05	[7]
MG	53	92	0.02	[8]
AP	27	22	0.09	[9]
SR	15	8	0.49	[10]
CW	33	55	0.09	[11]
JM	54	24	0.13	[12]
RA	13	9	0.23	[13]
CP	18	15	0.16	[14]
BF	43	14	0.14	[15]
PD	55	43	0.04	[16]
FJ	107	104	0.01	[17]
FG	18	57	0.43	[18]
NE	46	22	0.06	[19]
BL	63	25	0.08	[15]
DW	10	18	0.11	[20]
CE	21	32	0.20	[21]
JS	51	27	0.05	[22]
PP	8	11	0.15	[23]
PN	11	11	0.35	[6]
BJ	30	34	0.08	[9]
JA	52	22	0.16	[24]

3. RESULTS

In this section we analyze asymmetry between plant and animal interaction network. To explore the asymmetry between D_2 and D_1 we test for the set of 23 antagonist networks the following quantities: the difference in size, ΔL, clustering coefficient difference, ΔC, and mean connectivity difference, $\Delta{<}k{>}$.

We start our analysis with the size difference to test for size bias. The clustering coefficient C quantifies how much the vertices connected with a given vertex are connected among them. Indeed, C counts the number of triangles in the network [4]. Finally $<k>$ measures the average number of connections of the vertices, $<k>= \frac{2L}{N}$.

Figures 2, 3 and 4 summarize the differences between animal and plant networks. In this set of figures we show in the horizontal axis the 23 matrices. In figure 2 we show size difference, $\Delta L = L_2 - L_1$. In figure 3 we show the clustering coefficient difference, $\Delta C = C_2 - C_1$ and, finally, in figure 4 we have the mean connectivity difference, $\Delta{<}k{>}={<}k{>}_2 - {<}k{>}_1$.

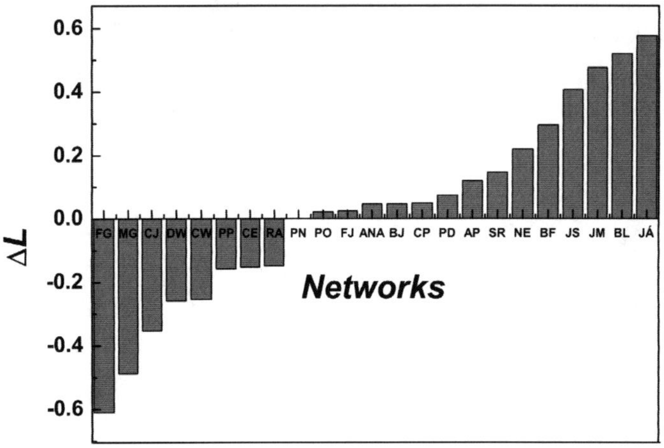

Figure 2. Normalized size difference, $\Delta L = \frac{L_2 - L_1}{L_2 + L_1}$, for the 23 networks depicted in table 1. This figure reveals an asymmetry between the two sets, the size of the insect set is statistically larger than the plants. This visual result is supported by statistical inference in the text.

In figure 2 we ranked the networks according to size difference to better visualize the results. The figures 2, 3 and 4 have the studied quantities for the set of the 23 networks, in all these figures we plot the networks in the same order to compare the results. In the inset of figures 3 and 4 the same result ranked. The figure in the inset allows direct visual check for asymmetry while the main figure is useful to compare the differences among specific networks. A simple visual inspection of the figures reveals that the ΔL is not responsible for the behavior of ΔC or $\Delta{<}k{>}$. That means, a potential asymmetry in ΔC or $\Delta{<}k{>}$ is not due to the most obvious asymmetry in ΔL.

In figure 2 is possible to see that the size distribution is asymmetric. This result is not surprising considering that the number (abundance and diversity) of insect exceed the number of plants in nature, and that interaction networks usually present more animals than plants. Otherwise, there is a paper in the literature has found, for a set of mutualist interaction networks, that there is an asymmetry between animal and plant sizes [2]. In figure 3 is plotted the clustering coefficient difference, ΔC. A simple inspection in the data show an almost symmetric

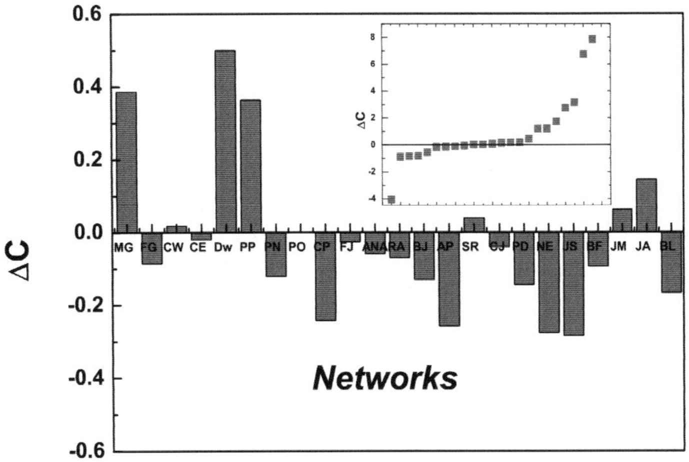

Figure 3. Clustering coefficient difference, $\Delta C = L_2 - L_1$, for the 23 networks depicted in table 1. We use in this picture the same network order of figure 2. In the inset we show the same data, but ranked to visualize the asymmetry. Indeed, the statistical test reveals no significative difference between the sets, ΔC do not reveals asymmetry.

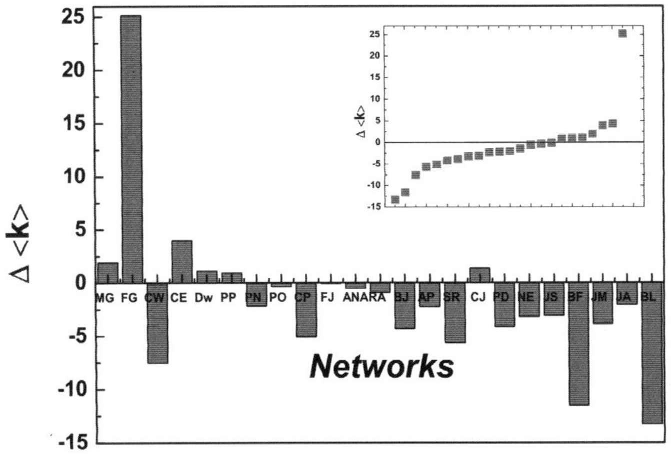

Figure 4. Mean connectivity difference, $\Delta{<}k{>}={<}k{>}_2 - {<}k{>}_1$, for the 23 networks depicted in table 1. We use in this picture the same network order of figure 2. In the inset we show the same data set, but ranked to highligth the asymmetry between the two sets. In the text we confirm the assymetry of the figure using a statistical test.

distribution of ΔC. This result indicates that ΔC is not an index appropriate to measure asymmetry in this class of bipartite networks. In the next section we discuss the possible use of the normalized clustering coefficient. In figure 4 is plotted the mean connectivity difference, $\Delta{<}k{>}$. This result seems to indicate that the $\Delta{<}k{>}$ is a good candidate to be used as an asymmetry index in matrices networks.

We use the Wilcoxon test and the signal test to compare the difference between the properties of the two sets (plants and insects). The two tests produce the same result. For a significance level $p < 0.05$, we verified asymmetry for $L(L_1 > L_2)$, $<k>$ $(<k>_1<<k>_2)$ but not for $C(C_1 < C_2)$.

4. CONCLUSION

In this work we developed a simple and effective way to measure the differences between two groups of ecological actors in a community context. We project the bipartite interaction network into two sets of vertices and construct two new networks. We use the difference between the indices of these new networks to characterize the asymmetry between these groups. We tested our technique in a set of 23 antagonist interaction networks insect and flowers. We observed for a significant level ($p < 0.05$) difference between the groups for the network size ΔL and the mean connectivity $\Delta<k>$. The clustering coefficient C have not shown difference between the sets.

We have two perspectives in mind. The first is to use other network indices to evaluate the asymmetry. The second point is not related to asymmetry, but to apply the same technique of this work to characterize the difference between nested and modular interaction networks.

We have explored the clustering coefficient which shows a distribution almost symmetric for the studied group of matrices. In the network literature it is common to use the normalized clustering coefficient, $C_{norm} = \frac{C}{C_{rand}}$, where $C_{rand} = \frac{<k>}{N}$ is the clustering coefficient of an associated random network with number of vertices N. This quantity is used when we compare a network with its random counterpart. This normalized index expresses directly how far a network is from the random model and it is used to measure the complexity of a network [4]. The quantity C_{norm} is as asymmetric as $<k>$ (we have not shown this result in this paper), but its asymmetry comes from $<k>$. Nevertheless, we point that our interaction matrix problem is asymmetric for C, a common measure of network science.

In a future work we will explore other network indices in the search for asymmetry in community networks. Indeed, there is a large set of indices to test, we have not explored, for instance, neither any centrality or betweeness index, nor the degree distribution $P(k)$ of the networks. There is an important study on the subject that claims that $P(k)$ of animals and plants in mutualistic interaction networks follow power-law distributions, or truncated power-law [29]. This result was criticized by Okuyama, T. [30], that remark that $P(k)$ of interaction networks are not easily classified into a single distribution class. The question we pose, otherwise, is other, we are not interested in $P(k)$ of the interaction network, but of animal and plant networks, which is not the same. By construction the networks D_2 and D_1 present much higher $<k>$ values than the interaction matrices, a fact that has deep consequences on $P(k)$. Finally, the road to achieve a good index to quantify asymmetry in interaction networks is only at the beginning. What is clear in this project is that the understanding of asymmetry in mutualistic and antagonistic networks is a major challenge in communities ecology and evolutionary biology.

ACKNOWLEDGEMENTS

The authors gratefully acknowledge the financial support of CNPq (Conselho Nacional de Pesquisa, Brazil). We also thanks to Umberto Kubota, Graciela Valladares and Thomas Lewinsohn made unpublished data available. Most tables were gathered by A.M. Almeida during a postdoctoral fellowship with grant from FAPESP.

REFERENCES

[1] Odum E P 1983 *Basic Ecology* (CBS College Publishing)
[2] Vásquez D P, Bluthgen N, Cagnolo L and Chacoff N P 2009 Uniting pattern and process in plat-animal mutualistic networks: a review *Ann. Bot.* **103** pp 1445-57
[3] Bascompte J and Jordano P 2007 Plant - animal mutualistic networks: the architecture of biodiversity *Annu. Rev. Ecol. Evol. Syst.* **38** pp 567-93
[4] Réka Albert and Albert-László Barabási 2002 Statistical mechanics of complex networks *Rev. Mod. Phys.* **74** pp 47-97
[5] Almeid, A M 2001 Biogeografia de interações entre Eupatorieae (Asteraceae) e insetos endófagos de capítulos na Serra da Mantiqueira *PhD Thesis*
[6] Pielou E C 1974 Biogeographic range comparisons and evidence of geographic variation in host-parasite relations *Ecology* **55** pp 1359-67
[7] Valladares G Unpublished data
[8] Memmott J, Godfray H C J and Gauld I D 1994 The structure of a tropical host-parasitoid community *J. An. Ecol.* **63** pp 521-40
[9] Lewinsohn, T M Unpublished data
[10] Sheldon J K and Rogers L E 1978 Grasshopper food habitats within a shrub-steppe community *Oecologia* **32** pp 85-92
[11] Claridge M F and Wilson M R 1981 Host plant associations, diversity and species-area relationships of mesophyll-feeding leafhoppers of trees and shrubs in Britain *Ecol. Ent.* **6** pp 217-38
[12] Joern A 1979 Feeding patterns in grasshoppers (Orthoptera: Acrididae): factors influencing diet specialization *Oecologia* **38** pp 325-47
[13] Ratchke B J 1976 Competition nd coexistence within a guild of herbivorous insects *Ecology* **57** pp 76-87
[14] Prado P I, Almeida A M, Lewinsohn T M, Norrbom A, Buys B D, Macedo A C and Lopes M B 2002 The Fauna of Tephritidae (Diptera) from Capitula of Asteraceae in Brazil *Proc. Entomol. Soc. Wash.* Washington **104**(4) pp 1006-27
[15] Buruga J H and Olembo R J 1971 Plant food preferences of some sympatric drosophilids of Tropical Africa *Biotropica* **3** pp 151-58
[16] Podlussány A, Jermy T and Szentesesi À 2001 On the leguminous host plants of seed predator weevils (Coleoptera: Apionidae, Curculionidae) in Hungary *Acta Zool. Acad. Sci. Hung.* **47** (4) pp 285-99
[17] Flowers R W and Janzen D 1997 Feeding records of Costa Rican leaf beetles (Coleoptera: Chrysomelidae) *Florida Entomol.* **80** pp 334-64
[18] Futuyma D J and Gould F 1979 Associations of plants and insects in a deciduous forest *Ecol. Monog.* pp 33-50
[19] Neck R W 1976 Lepidopteran foodplant records from Texas *J. Res. Lep.* **15** (2) pp 75-82
[20] Dawah H A, Hawkins B A and Claridge M F 1995 Structure of the parasitoid communities of grass-feeding chalcid wasps *J. An. Ecol.* **64** pp 708-20
[21] Kubota U 2003 Fenologia da comunidade de Asteraceae, variação temporal e determinantes locais de riqueza de insetos endófagos de capítulos *M S Thesis* Universidade de Campinas SP Brazil
[22] Jermy T and Szentesi A 2003 Evolutionary aspects of host plant specialisation - a study on bruchids (Coleoptera: Brichidae) *Oikos* **101** pp 196-204
[23] Pipkin S B, Rodriguez R L and Leon J 1966 Plant host specificity among flower-feeding neotropical drosophila (Diptera: Drosophilidae) *Am. Nat.* **100** pp 135-56
[24] Joern A 1979 Feeding patterns in grasshoppers (Orthoptera: Acrididae): Factors influencing diet specialization *Oecologia* **38** pp 325-47
[25] Vásquez D P and Aizen M A 2004 Asymmetric specialization: a pervasive feature of plant-pollinator interactions *Ecology* **85** pp 1251-57
[26] Guimarães Jr, P R, Rico-Gray V, Reis S F and Thompson J N 2006 Asymmetries in specialization in ant-plant mutualistic networks *Proc. R. Soc.* B **273** pp 2041-47
[27] Bascompte J, Jordano P and Olesen J M 2006 Asymmetric coevolutionary networks facilitate biodiversity maintenance *Science* **312** pp 431-33
[28] Vásquez D P and Aizen M A 2003 Null model analyses of specialization in plant-pollinator interactions *Ecology* **84** pp 2493-2501
[29] Jordano P, Bascompte J and Olesen J M 2003 Invariant properties in coevolutionary networks of plant-animal interactions *Ecol. Lett.* **6** pp 69-81
[30] Okuyama T 2008 Do mutualistic networks follow power distributions? *Ecological Complexity* **5** pp 59-65

The influence of energy changes in breathers

G G Slade, E Drigo Filho and J R Ruggiero

UNESP – Universidade Estadual Paulista, Instituto de Biociências, Letras e Ciências Exatas, Departamento de Física, Rua Cristóvão Colombo 2265, 15054-000, São José do Rio Preto, Brazil.

Email: ggslade@hotmail.com

Abstract. This work studies the dynamical behavior of breathers in a single nonlinear lattice under the influence of energy changes. To create the breather we used the anti-continuous limit and studied its stability through the Floquet theory. Using the information entropy we calculated the effective number of oscillators with significant energy and determined if there is or not the formation of a spatially localized structure.

1. Introduction

Breathers are time periodic and spatially localized solutions [1]. This kind of solution is generated by the combination of two factors: the discreteness and nonlinearity [2]. The first one makes the dispersion relation (linear spectrum) limited and discrete, while the second one makes the frequency of vibration dependent on the amplitude motion. Mackay and Aubry [3] proved the existence of breathers in the one-dimensional lattice when the system is composed by weakly coupled nonlinear solutions. Their proof was based on the anti-continuous limit and showed that the solution persists in a space of periodic solutions with a fixed period and decays exponentially in space.

In the last two decades many studies have been done to study this kind of solution [4,5]. They analyze both numeric and analytic ways to create the breather and study its stability for numerous discrete systems. Besides, there are some works [6, 7] that study the behavior of breathers interacting one to another and with some impurity of the lattice, like collisions and trapping. Another works try to apply this knowledge in some phenomenon of nature. In particular, some of them are intended to understand some biological aspects, like DNA transcription [8].

Here, we follow the numeric proceedings suggested by Marin and Aubry [9] to create the breather and use the Floquet theory to analyze its stability [10]. This study gives us solutions with different coupling parameter that can be stable or not. However, the energies involved are fixed and dependent to the breather frequency and coupling. In this work, we verify the influence of energy changes in the dynamic behavior of a breather. In order to do this, we use a stable solution and vary its energy by increasing or decreasing its amplitude. Based on the information entropy [11] we calculate the number of oscillators with significant energy and use it like a criteria to energy localization. The main idea is to know if despite of the energy variation we get a localized solution or not.

2. Prescription to create the breather

The Lagrangian of the system is:

$$L_y = \sum_{j=1}^{N} \left\{ \frac{m}{2} \dot{y}_j^2 - \frac{k}{2}(y_{j+1} - y_j)^2 - \frac{1}{2}V(y_j) \right\}, \tag{1}$$

and represents a chain of harmonic oscillators with an additional nonlinear on site potential $V(y_j)$. Here $V(y_j)$ is the Morse potential written, in a simplified way, as:

$$V(y_j) = \left(e^{-y_j} - 1\right)^2. \tag{2}$$

The equations of motion obtained from Lagrangian (1) are:

$$m\ddot{y}_j + k(2y_j - y_{j+1} - y_{j-1}) + e^{-y_j}(1 - e^{-y_j}) = 0 \tag{3}$$

The breather can be created using the principle of the anti-continuous limit suggested by MacKay and Aubry [3] and implemented by Marin and Aubry [9] and Cuevas [12]. So, at the beginning we considered the uncoupled system, i. e. $k = 0$. With this limit the system is composed of isolated oscillators under the influence of the nonlinear on site potential.

Because we are looking for periodic solutions, they can be written in terms of a Fourier cosine series:

$$y_j(t) = z_j^0 + 2\sum_{l=1}^{lm} z_j^l \cos(l\omega_b t), \tag{4}$$

where ω_b is the breather frequency. Thus, the equation of motion can be changed through substitutions of a set of algebraic equations to obtain the $lm + 1$ coefficients of z_j^l parameters (4), these equations can be solved by the Newton method.

Once we found the breather profile for the uncoupled system, $k = 0$, we introduced the coupling into the system by small increments on k, δk. The next step is to find the numerical solution for $k = \delta k$ by using the Newton method and the original profile obtained from $k = 0$. This new solution is used like a seed for the case $k = 2\delta k$ and so on until we reach the desirable value of k.

3. Breather stability

Here we use the Floquet theory to study the stability of the breathers, where the Floquet operator (F) determines the evolution of the system in a period T and it can be expressed by:

$$\Omega(T) = F\Omega(0) \tag{5}$$

with $\Omega(t)$ equal to the matrix form of the Hill´s equation.

Through this theory the system is linearly stable if none of the eigenvalues of F have a module greater than 1. Hence a necessary condition to linear stability is that all Floquet multipliers are on the unit circle in the complex plane. In other words, θ must be real. In order to have instability, one or more pairs of eigenvalues must go out of the unit circle. If the eigenvalues go out of the unit circle in the complex plane in $\theta = 0$, we obtain an harmonic bifurcation. If they go out in $\theta = \pi$, we get a sub-harmonic bifurcation. Finally, if the eigenvalues go out of the unit circle in $\theta \neq 0, \pi$, then the bifurcation is called oscillatory.

4. Results and discussion

The results reported below were obtained using chains with 21 oscillators $(N = 21)$ and the number of coefficients in the Fourier series was $lm = 17$. For numerical simulations we used periodic boundary conditions.

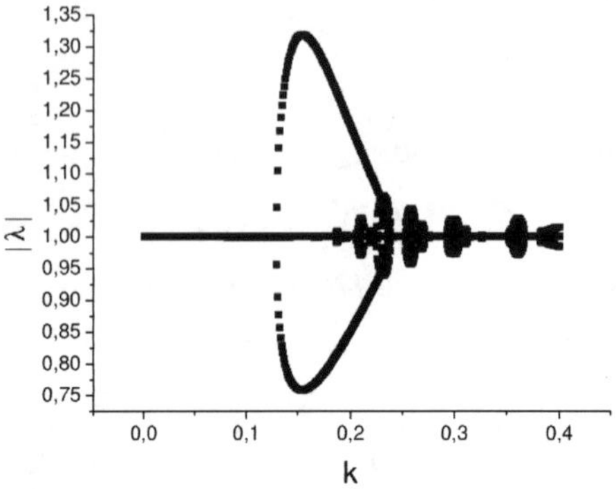

Figure 1. The absolute value of the Floquet multiplier versus coupling parameter k.

In figure 1 we present the results for the absolute value of the Floquet multipliers in function of the coupling parameter k with breather frequency $\omega_b = 0.8$. For small values of k all the Floquet multipliers have module equal one, so the stability is guarantee. After a certain value of k ($k = 0.1297$), some bifurcations starts to occur and the stability is lost.

The dynamical behavior of a stable solution is the formation of a spatially localized structure in the chain, and it persists for all time. The Floquet analysis allows knowing the behavior of the system as function of the coupling parameter, but does not give directly any information about the influence of energy changes in the breather. To test this we used a stable solution, $k = 0.1$, and changed its energy by decreasing or increasing all the initial positions of this solution. So, we got one of these new solutions and used it like an initial condition in a dynamic simulation, where we used a Runge-Kutta Nystron of tenth order method [13] to integrate the equations of motion (3). The criterion to analyze the behavior of the system relative to the energy localization is based on the information entropy and measures the number of oscillators with significant energy of the system, n_{osc}. This parameter is used as criteria for energy localization [14, 15] and it is given by the following expression:

$$n_{osc} = \frac{e^S}{N}, \tag{6}$$

where $S = -\sum_{j=1}^{N} e_j \ln e_j$ is the information entropy and $e_j = \dfrac{E_j}{\sum_{j=1}^{N} E_j}$ represents the normalized instant energies.

The figure 2 shows the average of the n_{osc} values in function of the energy. We note the existence of a minimum value of n_{osc} that is equivalent to the energy of the created stable breather $(E = 0.645)$. If we look carefully the values of energy near the breather one, we note that the n_{osc} values are lower than 0.2 in the range of energy equivalent to 0.291 – 1.203. It means that for the observable time ($t = 10000$ a.u.) the energy did not spread out through the lattice and remained localized in a small number of oscillators. This fact could be important for the use of breathers in the study of biological systems, like the DNA [14, 16], where a localized structure is observed [17, 18] and despite of the variation of the energy it remains localized.

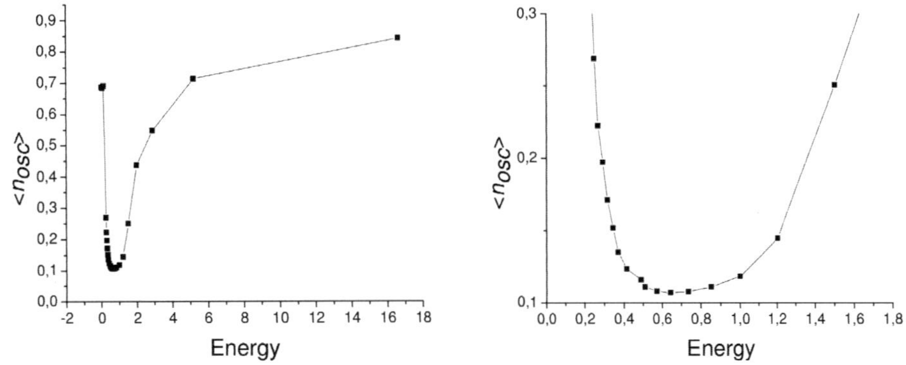

Figure 2. a) The number of oscillators with significant energy value in function of the energy. b) Detail of the figure 2a.

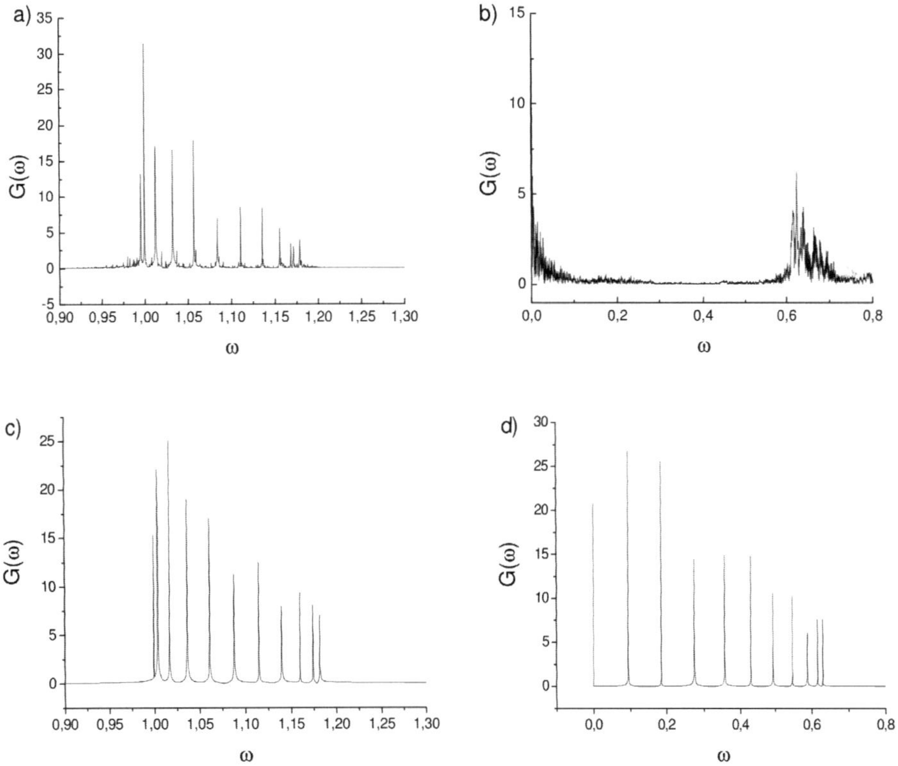

Figure 3. Modulus $G(\omega)$ of the complex Fourier transform of the eleventh oscillator. a) $E = 0.0556$ and $n_{osc} = 0.684$, b) $E = 1.5$ and $n_{osc} = 0.251$, c) The representation of the optical branch and d) The representation of the acoustic branch.

However, for energies outside this range of energy (0.291 – 1.203), the n_{osc} value starts to grow rapidly and the energy localization is lost. In these situations the system presents the behavior of harmonic chains, where the energy diffuses in the lattice. The Fourier analysis allows us to rationalize an explanation for this fact. Figure 3 shows the modulus $G(\omega)$ of the complex Fourier transform of the eleventh oscillator for two energies of the system: (a) $E = 0.0556$ and (b) $E = 1.5$. In the case of low energies, the Fourier analysis (figure 3a) shows that the system presents oscillations inside the optical branch, $1 \leq \omega \leq \sqrt{1+4k}$ (figure 3c). So the localization is lost because the nonlinear modes are not excited. When we deal with high energies, (figure 3b), all frequencies with significant intensity are inside the acoustic branch, $0 \leq \omega \leq \sqrt{4k}$ (figure 3d), so the increase in the n_{osc} value occurs. This happens because some oscillators have sufficient energy to overcome the Morse barrier and the system behaves like a purely harmonic one. These results agree with the conjecture that for the existence of localized modes, the system must oscillate outside the linear spectrum.

5. Conclusion

In this work we study the influence of energy changes in the dynamical behavior of a stable breather solution using a parameter based on the information entropy like a criteria. The variation of energy permits, in some range, the localization of the energy in a small number of oscillators (less than 20%) for the observable time and this fact can be important in the use of breather solution in biological system, where localized phenomena are commonly found. However when the energy of the system decreases from this range of energy, the localization is not found because the nonlinear modes are not excited and the system oscillates in the optical branch. For energy above this range, the system acquires energy sufficient to overcome the barrier of the Morse potential and some oscillators of the lattice lost their connection with the on site potential. So the system acquires a behavior similar to purely harmonic network. These results show the importance of the energy in localized phenomena that involves nonlinearity and discreteness.

Acknowledges

The authors are grateful to Capes and CNPq for financial support.

References

[1] Aubry S 1997 *Physica D* **103** 201-50
[2] Flach S and Willis C R 1998 *Physics Reports* **295** 181-264
[3] MacKay R and Aubry S 1994 *Nonlinearity* **7** 1623-43
[4] Aubry S 2006 *Physica D* **216** 1-30
[5] Flach S and Gorbach A V 2008 *Physics Reports* **467** 1-116
[6] Bena I, Saxena A, Tsironis G, Ibanes M and Sancho J M 2003 *Physical Review E* **67** 037601
[7] Alvarez A, Romero F R, Cuevas J and Archilla J F R 2008 *Physics Letters A* **372** 1256-64
[8] Ting J and Peyrard M 1996 *Physical Review E* **53**
[9] Marin J L and Aubry S 1996 *Nonlinearity* **9** 1501-28
[10] Hoppensteadt F C 2000 *Analysis and Simulation of Chaotic Systems* (New York: Springer) chapter 1 pp 1-19
[11] Luca J, Lichtenberg A J and Libermon 1995 *Chaos* **5** 283-97
[12] Cuevas J 2003 *Localizacion y Transferencia de Energia em Redes Anarmónicas No Homogénea* Phd thesis Universidad de Sevilla
[13] Tsitouras C 1999 *Celest. Mech.Dun. Astron.* **74** 223-30
[14] De Luca J, Drigo Filho E, Ponno A and Ruggiero J R 2004 *Phys. Rev. E* **70** 026213
[15] Silva J M, Drigo Filho E and Ruggiero J R 2009 *Eur. Phys. J. E* **29** 245-51
[16] Slade G G, Drigo Filho E and Ruggiero J R 2010 *J. Phys.: Conf. Ser.* **246** 012039

[17] Englander S W, Kallenbach N R, Heeger A J, Krumhansl J A and Litwin A 1980 *Proc. Natl. Acad. Sci.* **77** 7222-6
[18] Leroy J L, Kochoyan M, Huynh-Dinh T and Guéron M 1988 *J. Mol. Biol.* **200** 223-38

Particular solution for anomalous diffusion equation with source term

M T Araujo[1], E D Filho[2]
[1]Physics Department, Unesp, São José do Rio Preto, SP, Brazil
[2]Physics Department, Unesp, São José do Rio Preto, SP, Brazil

E-mail: [1]marcelot-araujo@hotmail.com and [2]edrigof@gmail.com

Abstract: In literature the phenomenon of diffusion has been widely studied, however for nonextensive systems which are governed by a nonlinear stochastic dynamic, there are a few soluble models. The purpose of this study is to present the solution of the nonlinear Fokker-Planck equation for a model of potential with barrier considering a term of absorption. Systems of this nature can be observed in various chemical or biological processes and their solution enriches the studies of existing nonextensive systems.

1. Introduction

A large number of phenomena can be related to kinetic processes, this is the case, for example, of electrochemical reactions in metals or in membranes. In these cases, the atoms or particles that interact are in the interface between the environments, the medium and the interior of the material. In many cases, this difference between the environments leads to a diffusion process in which particles of the medium are absorbed into the structure. This absorption can occur at a constant rate or, in some cases it can present temporal dependence [1].

The Fokker-Planck equation is widely used in literature to study stochastic processes of diffusion, where it is possible to consider the presence of forces acting on the interaction between the bodies [2]. Its representation is:

$$\frac{\partial P(x,t)}{\partial t} = \frac{\partial}{\partial x}\left[f(x)\cdot P(x,t)\right] + Q\frac{\partial^2 P(x,t)}{\partial x^2}, \tag{1}$$

where $P(x,t)$ is the distribution of probabilities or a density of particles depending on the problem concerned. Q is the coefficient of diffusion and $f(x)$ is the external force that acts in the environment.

The distribution of probability obtained from equation (1) maximizes the entropy of Gibbs-Boltzmann $S = \sum_i P_i \ln(P_i)$ and offers a very good description of the behavior of systems characterized by fluctuations which are proportional to the time variation, where there are short-range interactions among particles [3].

When considering anomalous systems, where the fluctuations are disproportionate to the time and where there is the presence of long-range interactions among particles, the common Fokker-Planck equation (1) is not able to describe well the model [4,5]. In order to solve this problem, the Fokker-Planck equation must be rewritten as:

$$\frac{\partial P(x,t)^{\mu}}{\partial t} = -\frac{\partial}{\partial x}\left[f(x)\cdot P(x,t)^{\mu}\right] + Q\frac{\partial^2 P(x,t)^{\nu}}{\partial x^2} \tag{2}$$

As it can be observed, the difference between (1) and (2) is the presence of the indexes μ and ν in $P(x,t)$ [6]. These indexes are real constants that must be adjusted according to the system analyzed and are constantly related to the process of anomalous diffusion in biological systems [7 – 10]. When $\mu = 1$, the probability provided by (2) is intimately associated with nonextensive systems, because it maximizes the entropy proposed by Tsallis [11, 12],

$$\frac{\partial P(x,t)}{\partial t} = -\frac{\partial}{\partial x}\left[f(x)\cdot P(x,t)\right] + Q\frac{\partial^2 P(x,t)^{\nu}}{\partial x^2}. \tag{3}$$

In the problem addressed, we will make a brief review of nonlinear diffusion in a simple system consisting of two regions separated by a barrier of potential. In section 2 we present the solution of the nonlinear Fokker-Planck equation (NLFP) with an additional term, which is interpreted as a source term or absorption term that favors the passage of particles through the potential barrier. In section 3 we discuss about the results obtained and, finally, we present the conclusions.

2. PURPOSE

In this paper, the function *P(x,t)* is interpreted as a density of particles of a medium. An additional term, proportional to the function $\mu(t)$ is included in equation (3) and is related to the absorption rate of the model. This way, we obtain:

$$\frac{\partial P(x,t)}{\partial t} = -\frac{\partial}{\partial x}\left[P(x,t)f(x)\right] + Q\frac{\partial^2 P(x,t)^{\upsilon}}{\partial x^2} - \mu(t)P(x,t) \tag{4}$$

The nonlinearity of the Fokker-Planck equation is due to the exponent ν present in the second derivative. Equation (1) is employed in the study of nonextensive systems, where there are long-range interactions and the thermodynamic relations meet the proposal introduced by Tsallis [5, 13].

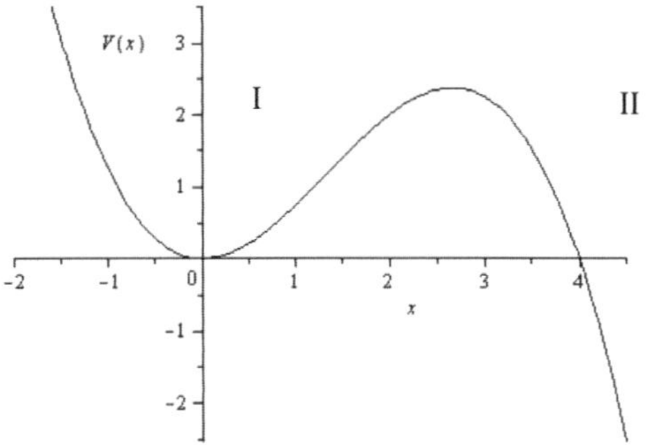

Figure 1: Representation of the potential considered with $k_1 = 1$ and $k_2 = 0.25$.

In this study, the function *f(x)* is given by the derivative of the potential:

$$V(x) = k_1 x^2 - k_2 x^3, \tag{5}$$

whose characteristic graph is shown in Figure 1. The solution for this system is the form:

$$P(x,t) = D(t)\left\{1-(\upsilon-1)\beta(t)\left[x^2 - M(x,t)\right]\right\}^{1/(\upsilon-1)}, \tag{6}$$

where *D(t)*, $\beta(t)$ and *M(x,t)* are functions dependent on the time to be defined later. For a long time, the *ansatz* above indicates the distribution of the stationary probability of the Fokker-Planck equation. As we are considering a nonextensive system, the solution needs to be given this way, because this function maximizes the Tsallis' entropy and allows us to obtain a good description of the environment through relevant thermodynamic information as, for example, the free energy.

Applying (6) in (4) we obtain a system of equations,

$$\begin{cases} i) \quad \dfrac{dD(t)}{dt} - 2D(t)k_1 + 2\nu Q D(t)^{\nu}\beta(t) + \mu(t)D(t) = 0 \\ ii) \quad -6k_2 D(t)x - \nu Q D(t)^{\nu}\beta(t)\dfrac{\partial^2 M}{\partial x^2} = 0 \\ iii) \quad x^2\left[\dfrac{1}{\beta(t)}\dfrac{d\beta(t)}{dt} - 4k_1 + 4\nu Q D(t)^{\nu-1}\beta(t)\right] = 0 \\ iv) \quad -\dfrac{\partial M}{\partial t} - \dfrac{M}{\beta(t)}\dfrac{d\beta(t)}{dt} + \left(2k_1 x - 3k_2 x^2\right)\dfrac{\partial M}{\partial x} - 6k_2 x^3 - Q\nu\beta(t)D(t)^{\nu-1}\left\{4x\dfrac{\partial M}{\partial x} + \left(\dfrac{\partial M}{\partial x}\right)^2\right\} = 0 \end{cases}$$

Disregarding the relations *ii)* and *iv)*, we go back to the system presented in [14], when the force is proportionate to *x*, only less than the source term $\mu(t)$. By isolating the term $2\nu Q\beta(t)D(t)^{\nu}$ from *i)* and substituting in *iii)*, we find the following expression:

$$\frac{1}{\beta(t)}\frac{d\beta(t)}{dt}-\frac{2}{D(t)}\frac{dD(t)}{dt}-2\mu(t)=0. \tag{7}$$

By integrating (7) with respect to t, we observe that $\beta(t)$ is defined as:

$$\beta(t)=\frac{\beta_0}{D_0^2}D(t)^2 e^{2T(t)}, \tag{8}$$

where the function $T(t)$ is equal to the integration of $\mu(t)$ in a determined time interval.

The equation *ii)* provides the function $M(x,t)$, which satisfies the problem. To that effect, integrating twice over *x* is enough. To facilitate, we will replace the function $\beta(t)$ by definition (8). Therefore, the expression is:

$$M(x,t)=\frac{-k_2 x^3 e^{-2T(t)}}{\gamma D(t)}, \tag{9}$$

where γ is a constant equal to $Q\nu\beta_0/D_0$, adopted to facilitate the notation.

Thus, after some manipulations we obtain the following expression:

$$-\frac{k_2(\nu+1)x^3 e^{-2T(t)}}{\gamma D(t)^{\nu+2}}\frac{dD}{dt}+\frac{k_2 x^3 e^{-2T(t)}}{\gamma D(t)^{\nu+1}}\frac{2}{D(t)}\frac{dD(t)}{dt}-\frac{6k_1k_2x^3e^{-2T(t)}}{\gamma D(t)^{\nu+1}}-6k_2x^3+12k_2x^3=0, \tag{10}$$

or:

$$-\frac{(\nu-1)}{3}\frac{dD}{dt}-2k_1D(t)+2\gamma D(t)^{\nu+2}e^{2T(t)}=0. \tag{11}$$

Until now, from a system of four equations (*i, ii, iii* and *iv*), we have defined $\beta(t)$ and $M(x,t)$ in terms of $D(t)$ and $\mu(t)$, which were not found yet. In order to obtain these expressions, it is necessary to compare equations (11) and *i)*:

$$\begin{cases}\dfrac{dD(t)}{dt}-2D(t)k_1+2\gamma D(t)^{\nu+2}e^{2T(t)}+\mu(t)D(t)=0\\[2ex] -\dfrac{(\nu-1)}{3}\dfrac{dD}{dt}-2k_1D(t)+2\gamma D(t)^{\nu+2}e^{2T(t)}=0\end{cases} \tag{12}$$

As we can observe, both expressions are similar except for the derivative in relation to $D(t)$ and the product $\mu(t)\cdot D(t)$. By subtracting one from the other, we have:

$$\frac{dD(t)}{dt}+\frac{(\nu-1)}{3}\frac{dD}{dt}+\mu(t)D(t)=0 \tag{13}$$

and by rearranging the terms of (13), we obtain a relation for $\mu(t)$ in function of $D(t)$:

$$\mu(t)=-\frac{(\nu+2)}{3D(t)}\frac{dD}{dt}. \tag{14}$$

Through the relation found above, we can rewrite $T(t)$ as

$$T(t)=\ln\left(D(t)\right)^{-(\nu+2)/3}. \tag{15}$$

By using (15), the exponential term is replaced by,

$$e^{2T(t)}=e^{\ln\left(D(t)\right)^{-2(\nu+2)/3}}=D(t)^{-2(\nu+2)/3}. \tag{16}$$

and by substituting the relation above in (11) we obtain an equation which is only dependent on the function $D(t)$, which, being solved provides the other functions: $\beta(t)$, $M(x,t)$ and $\mu(t)$. Due to the simplicity, we will work with the second equation of system (12) and, for this reason, we have rewritten it as:

$$\frac{dD^{-(\nu-1)/3}}{dt} - 2k_1 D^{-(\nu-1)/3} + 2\gamma = 0. \tag{17}$$

The expression of *D(t)* that satisfies the above is

$$D(t) = \left\{ \frac{Q\nu\beta_0}{k_1 D_0^2} + \left(1 - \frac{Q\nu\beta_0}{k_1 D_0^2}\right) e^{2k_1 \cdot t} \right\}^{-3/(\nu-1)}. \tag{18}$$

Accordingly with the relations (8), (9) and (14), the functions $\beta(t)$, *M(x,t)* e $\mu(t)$ are represented as:

$$\beta(t) = \frac{\beta_0}{D_0^2} \left\{ \frac{Q\nu\beta_0}{k_1 D_0^2} + \left(1 - \frac{Q\nu\beta_0}{k_1 D_0^2}\right) e^{2k_1 \cdot t} \right\}^2, \tag{19}$$

$$\mu(t) = -2\frac{(\nu+2)}{(\nu-1)} \left(k_1 - \frac{Q\nu\beta_0}{D_0^2}\right) \left\{1 - \left(e^{-2k_1 \cdot t} - 1\right) \frac{Q\nu\beta_0}{k_1 D_0^2}\right\}^{-1}, \tag{20}$$

and

$$M(x,t) = \frac{-k_2 k_1 D_0^2}{Q\nu\beta_0} \left\{ \frac{Q\nu\beta_0}{k_1 D_0^2} + \left(1 - \frac{Q\nu\beta_0}{k_1 D_0^2}\right) e^{2k_1 \cdot t} \right\} x^3. \tag{21}$$

Having defined all the expressions of *ansatz* (6) the probability is expressed as:

$$P(x,t) = \left\{ \frac{Q\nu\beta_0}{k_1 D_0^2} + \left(1 - \frac{Q\nu\beta_0}{k_1 D_0^2}\right) e^{2k_1 t} \right\}^{-3/\nu-1} \cdot \left\{ 1 - (\nu-1)\frac{\beta_0}{D_0^2} \left[\frac{Q\nu\beta_0}{k_1 D_0^2} + \left(1 - \frac{Q\nu\beta_0}{k_1 D_0^2}\right) \cdot e^{2k_1 \cdot t} \right]^2 \cdot \right.$$
$$\left. \cdot \left[x^2 - \frac{k_2 k_1 D_0^2}{Q\nu\beta_0} \left\{ \frac{Q\nu\beta_0}{k_1 D_0^2} + \left(1 - \frac{Q\nu\beta_0}{k_1 D_0^2}\right) \cdot e^{2k_1 \cdot t} \right\} x^3 \right] \right\}^{1/\nu-1}. \tag{22}$$

3. RESULTS AND DISCUSSION

Figures 2 and 3present the graphs of the distribution *P(x,t)* obtained in different time intervals, with *x* ranging from -1.5 to 2, which corresponds to the region of stable potential minimum (region I). The values assigned to constants were Q = 0,5 and $k_1 = 1$, $k_2 = 0.25$, ν = 1.5 for Figure 2 and $k_1 = 0.2$, $k_2 = 0.05$, ν = 0.5 for Figure 3.

The time taken by the particles density to reach zero, considering these values of the constants, was of approximately 1 unit, when ν is 1.5 and 2 units for ν equal 0,5.

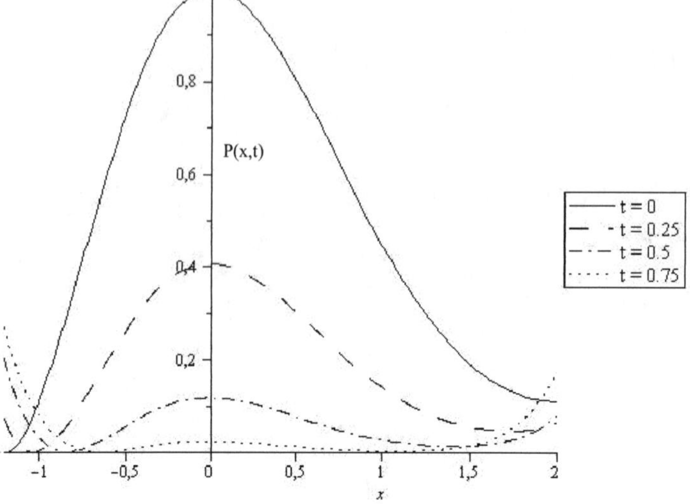

Figure 2: Density of particles vs. space and time for $k_1 = 1$, $k_2 = 0.25$ and $\nu = 1.5$.

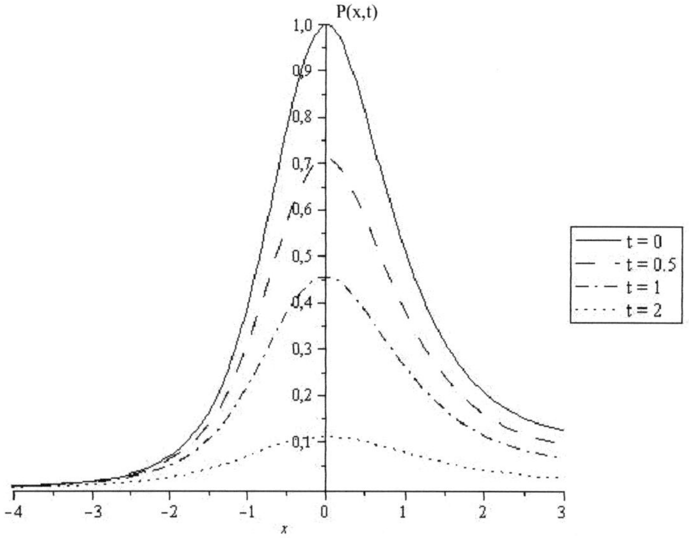

Figure 3: Density of particles vs. space and time for $k_1 = 0.2$, $k_2 = 0.05$ and $\nu = 0.5$.

The inclusion of a term referring to an absorption rate has been already discussed in some papers (see references [15] and [16]). However, in these cases, the function *f(x)* corresponds to a harmonic potential. In this model, a disturbance term was inserted in the harmonic potential, creating a barrier and two state levels. This change has generated a specific temporal dependence for this case in the absorption rate (20).

Temporal dependences of the kind presented here (20) can be observed in many biochemical systems [17, 18] when analyzing the absorption rates, creating the possibility of approximating this theoretical model with data of reactions or interactions obtained in the laboratory. In these cases, the limitation consists of the necessity of adjusting the nonlinear parameter (ν) to the system being treated, as it can be seen when comparing Figures 2 and 3.

The values of k_1 and k_2 in Figure 3 are smaller than in Figure 2, reducing the size of the barrier and facilitating the diffusion between regions I and II of Figure 1. The value of nonlinear parameter is altered due to the change in the size of the barrier, guaranteeing a physically acceptable behavior of the solution.

4. CONCLUSION

The graphs made from *ansatz* have suggested an expected behavior of the system, i.e., as time passes there is a tendency for particles to go from a region of higher potential to a region of lower potential.

This diffusion process is favored not only by the nonlinear factor of the Fokker-Planck equation, but also by the addition of the source term $\mu(t)$, which helps the particles to transpose the barrier. The incorporation of this temporal function in the Fokker-Planck nonlinear equation led to the solution of a simple system of potential barrier.

In literature there are a few soluble models for NLFP, which hinders the application of the nonextensive formalism in the study of chemical or biological processes.

ACKNOWLEDGMENTS

The authors thank Capes (Coordination for the Improvement of Higher Education Personnel) for the financial support.

References

[1] Vemulapalli G K 1993 *Physical Chemistry* ed Prentice-Hall (New Jersey: Inc. A Simon & Schuster Company Englewood Cliffs).
[2] Risken H 1989 *The Fokker-Planck Equation* (Berlim: Springer).
[3] Tomé T 2006 *Braz. Jour. Phys.* **36** - 4A 1285 - 89.
[4] Cohen E G D 2002 *Physica* **A 305** 19 – 26.
[5] Tsallis C 1999 *Braz. Jour. Phys* **29** 1-35.
[6] Frank T D 2005 *Nonlinear Fokker-Planck Equations – Fundamental and Applications* (Berlim: Springer).
[7] Pedron I T and Mendes R S 2005 *Rev. Bras. Ens. Fís.* **27** 251 – 258.
[8] Adler R S 1978 *Jour. Chem. Phys.* **69** 2849 – 63.
[9] Ott A, Bouchaud J P, Langevin D and Urbac W 1990 *Phys. Rev. Lett.* **65** 2201 – 04.
[10] Borland L 1998 *Phys. Rev.* E **57** 6634 – 42.
[11] Tatsuaki W, Takeshi S 2001 *Phys.* A **301** 284 – 90.
[12] Tsallis C 2009 *Introduction to Nonextensive Statistical Mechanics* Rio de Janeiro (Rio de Janeiro: Springer).
[13] Shiino M 2001 *J. Math. Phys.* **42** 2540-53.
[14] Plastino A R and Plastino A 1995 *Physica* A **222** 347 – 54.
[15] Drazer G, Wio H S and Tsallis C 2000 *Phys. Rev.* E **61** 1417-22.
[16] Pedron I T, Mendes R S, Malacarne L C and Lenzi E K 2002 *Phys. Rev.* E **65** 0411081-6.
[17] Crank J 1956 *The Mathematics of Diffusion* ed Oxford University Press (London).
[18] Hansen C M 2010 *Eur. Polym. J.* **46** 651-662.

Maximum entropy, fractal dimension and lacunarity in quantification of cellular rejection in myocardial biopsy of patients submitted to heart transplantation

L A Neves [1,5], F R Oliveira [2], F A Peres [2], R D Moreira [3], A R Moriel [3] and M F de Godoy [3,5]

[1]Universidade Estadual Paulista, IGCE, DEMAC, Rio Claro, SP, Brazil.
[2]Faculdade de Tecnologia de São José do Rio Preto, São José do Rio Preto, SP, Brazil.
[3]Faculdade de Medicina de São José do Rio Preto, FAMERP, São José do Rio Preto, SP, Brazil.
[4]Instituto de Anatomia Patológica e Citopatologia, São José do Rio Preto, SP, Brazil.
[5]Núcleo Transdisciplinar para Estudo do Caos e da Complexidade – NUTECC - São José do Rio Preto, SP, Brazil.

E-mail: laneves@rc.unesp.br

Abstract. This paper presents a method for the quantification of cellular rejection in endomyocardial biopsies of patients submitted to heart transplant. The model is based on automatic multilevel thresholding, which employs histogram quantification techniques, histogram slope percentage analysis and the calculation of maximum entropy. The structures were quantified with the aid of the multi-scale fractal dimension and lacunarity for the identification of behavior patterns in myocardial cellular rejection in order to determine the most adequate treatment for each case.

1. Introduction

The biological vision system is one of the main means humans use to explore the world and performs complex functions in a considerably natural fashion, such as the analysis, interpretation, recognition and classification of patterns. A large number of studies have attempted to produce models with the same efficiency as the biological system. However, this task is highly complex, especially with regard to transposing the quantification and qualification of information represented in different domains, such as intensities of gray scales, edges, contours and textures. These attributes are produced naturally by the human vision system when the signal being assessed is an image [1].

The exemplification of such difficulties is perceived in the process of assessing cardiac rejection, which the International Society for Heart and Lung Transplantation [2] standardizes in four grades (0R; 1R; 2R; 3R), among which grade 2R merits particular attention. For pathologists, there is a spectrum ranging from small points of inflammation involving one or two injured cells to a large degree of inflammatory infiltrate, with the injury of muscle cells. Even after the aforementioned standardization, the difficulty in diagnosing grade 2R involves the presence of muscle cell injury. Consequently, some diagnostic centers do not treat 2R patients and recommend monitoring with repeated biopsies at one-to-two-week intervals. Thus, cardiac rejection remains an issue of uncertainty and studies should be carried out with the aim of understanding and minimizing the difficulties involved.

Models employing methods that behave similarly to natural phenomena can assist in the quantification, understanding and diagnosis of cardiac rejection. For such, the fractal dimension and lacunarity are potential alternatives as natural descriptors of shapes and textures [3-5]. The efficiency of the fractal dimension and lacunarity is enhanced when combined with one of the most interesting properties of objects in nature – the presentation of different characteristics when observed on different scales. This association is known as the multi-scale fractal signature. However, the quantification and understanding of cardiac rejection using fractal methods depends on the appropriate segmentation of the region of interest.

Segmentation is one of the first steps in the digital processing of images, as it allows the separation of an object into parts based on a uniformity criterion [6]. In order to obtain high-quality segmentation, digital image processing systems primarily use thresholding, which consists of determining an intensity value at which an object is best distinguished from the background. In the literature, there are different automatic thresholding methods based on different criteria [7-12]. In many cases, however, a single threshold is not sufficient in furnishing adequate segmentation for the entire image. In such situations, variable and multilevel thresholding methods are administered. Aboud Neta [13] proposes one such method, but its sensitivity leads to high rates of non-significant thresholds, which can compromise the assessment of the image.

This paper defines a model for quantifying cardiac rejection made up of an automatic multilevel thresholding method based on maximum entropy and multi-scale fractal techniques. The method allows determining appropriate thresholds for the segmentation of regions of interest in myocardial biopsies and the quantification of these regions using the fractal dimension and lacunarity. This important model furnishes significant results, enabling a better understanding of behavior on each rejection level.

2. Methods

To facilitate the understanding of the proposed model, Subsection 2.1 describes the automatic multilevel thresholding method for the separation of the regions of interest and Subsection 2.2 describes the multi-scale fractal methods used to measure the complexity of the segmented regions.

2.1. Automatic multilevel thresholding method

The automatic multilevel thresholding method determines the threshold based on the identification of maximum entropy by comparing the entropies in different regions of the histogram of the image in question. This method is an improvement over the approach proposed by Aboud Neta [13], allowing better control over the thresholds identified. For such, properties were added to the method cited, such as the division of the histogram into classes, analysis of the slope percentages and the determination of the threshold by the calculation of maximum entropy.

2.1.1. Quantification of Histogram. The first step in segmentation is to determine the histogram of the image to be studied. The histogram is then divided into classes. A class is one or more gray scale values that make up the histogram. For this division, it is necessary to predefine the class size. If the value informed is 1, the process analyzes each of the intensities of the histogram. For values greater than 1, the increment considered in the iteration process of the method is the predefined value. For instance, consider a class made up of 10 intensities and a histogram with 256 gray levels. The first class encompasses levels 0 to 9, the second encompasses 10 to 19, and so forth, totaling 25 classes will 10 levels of gray and one class with six levels (figure 1).

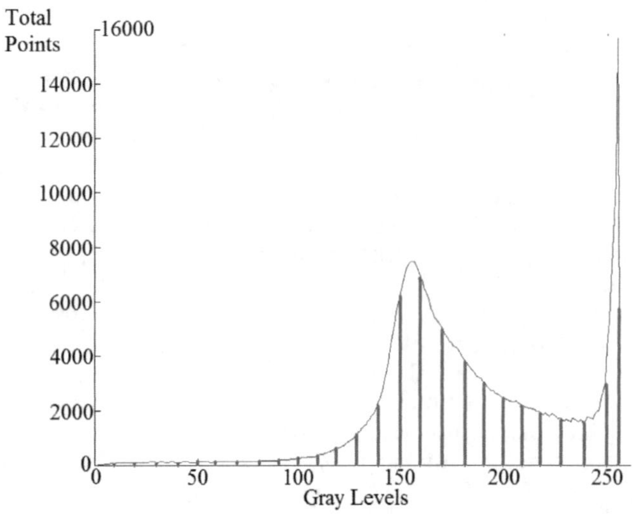

Figure 1. Example of histogram divided into 26 classes – 25 containing 10 levels and one containing six levels; vertical lines demark classes.

2.1.2. Valleys Analysis. This step consists of identifying what classes have valleys that are relevant to the specification of the thresholds. The method automatically identifies these classes by means of sign transition, analyzing the following characteristics:

- The mean of the points contained in the first half of the class analyzed is compared with the mean of the points in the second half of the same class. If the value of the first half is smaller, the values of the histogram are increasing toward a peak and the sign attributed to the class is positive. Otherwise, a negative sign is attributed to the class, indicating movement toward a valley.
- The sign of the next class is determined. If there is a transition from a negative to a positive sign, the valley is relevant to the specification of a threshold (figure 2).

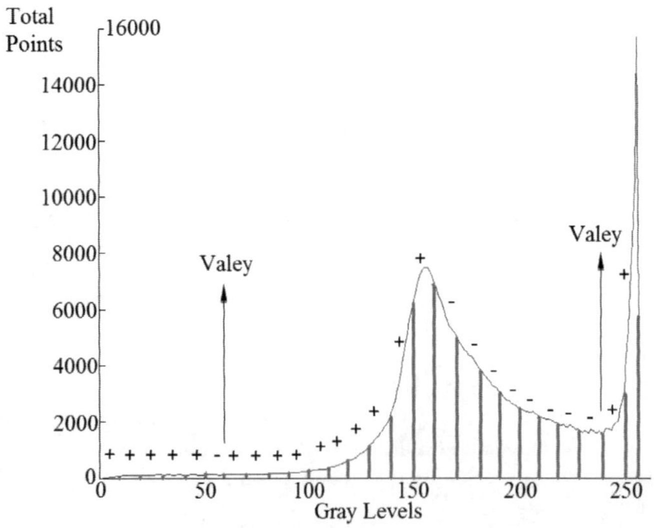

Figure 2. Example of histogram divided into 26 classes with indication of two relevant valleys to the determination of thresholds – one at level 50-59 and another at level 230-239.

The determination of thresholds based on valley analysis may be influenced by homogeneous regions in the histogram (with no valleys or with non-significant valleys). To solve this problem, the proposed method determines that a class is relevant if it has a slope percentage greater than a predefined value. This value can be adjusted to the type of image being studied. A slope percentage is the percentage difference between the means of the points in the first and second halves of the class being analyzed. Two or more classes are grouped together when the slope percentage is lower than the predefined value. This resource allows controlling the sensitivity of the method. A threshold is established when the slope percentage of the class is greater than the predefined value.

In the study of images of myocardial biopsies of transplant patients, the values used for the input parameters for segmentation were classes of size 10 and the slope was defined at 35%. These values proved sufficient for the appropriate segmentation of the regions of interest.

2.1.3. Threshold identification using maximum entropy. For each relevant class identified to define a valley, a threshold is calculated based on entropy, considering a probability of an intensity correctly segmenting a given group of objects. For such, the image is taken as the result of a random process in which probability p corresponds to the probability of a pixel in the image taking on an intensity value i (i=1,..,n) [8], as shown in equations (1) and (2). The intensity or gray level of the class with the greatest entropy is identified as a threshold.

$$H = -\sum_{i=1}^{n} p_i \log(p_i) \tag{1}$$

$$p_i = \frac{n_i}{N} \tag{2}$$

in which H is the entropy of the image; n is the total number of outputs (number of gray levels in the image); p_i is the probability of gray level i being found in the image; n_i is the number of pixels with intensity i; and N is the total number of pixels in the image.

2.2. Quantification of regions of interest

The texture in an image is characterized by the repetition of a model over a region; the model may be repeated exactly or with slight variations [6]. The analysis of texture allows distinguishing regions with the same reflectance characteristics, which is a process used for the recognition, analysis, description and classification of digital images. There are different approaches to the study of texture described in the literature [6]. The approach presented here is the quantification of textures in images of myocardial biopsies from heart-transplant patients using more natural methods, such as the fractal dimension and lacunarity.

2.2.1. Fractal dimension. The fractal dimension is defined as a measure of the complexity of objects. When applied to textures, this measure allows quantifying the complexity of the organization of its pixels (how much the space is filled in). Box counting is one of the best known and employed methods for estimating the fractal dimension of objects and images [14]. For the images studied, this method was applied to regions segmented as "cell nuclei", with the understanding that, in the case of an inflammatory process, such as cellular rejection, the number of cell nuclei is progressively greater and more diffusely distributed with the increase in the degree of rejection. The multi-scale approach was achieved by overlaying a grid on the image analyzed. The aim was to obtain the number of squares necessary to cover the image [15 -16]. For an myocardial biopsy image I as the input, the number of squares that contain part of the shape I, $N_r(I)$ depends on the size of the box r, equation (3). The relation defined in equation (3) allows estimating the fractal dimension D, based on equation (4).

$$N_r(I) = \mu r^{-D} \tag{3}$$

$$D = -\lim_{r \to 0} \frac{\ln(N_r(I))}{\ln(r)} \tag{4}$$

However, equation (4) imposes the calculation of a limit. One solution is to adapt this calculation to a discreet space, in which box counting $N_r(I)$ is performed for different values of r. As a consequence, it is necessary to define a set S with the possible sizes of sides r used in the diverse iterations of the method, based on the dimensions of the shape under analysis, equation (5).

$$\forall r_i \in S \begin{cases} r_0 = \max(\text{height}, \text{width}) \\ r_{i+1} = r_i / 2 \end{cases} \tag{5}$$

The approximation of a straight line is obtained from the regression of the log-log graph of $N_r(I)$ (number of boxes occupied) by r (size of side of this box), in which the angle coefficient is α. Thus, it is possible to define $D=-\alpha$ as the fractal dimension of I [15-16]. In the present paper, box counting was used as the texture descriptor. For such, the luminance of the pixel was considered the height of that point of the image. This modification implies dividing the image into cubes with side r rather than boxes with side r, thereby producing a new $N_r(I)$, depending on the number of cubes that contain part of the image.

2.2.2. Lacunarity. Lacunarity characterizes the way the pixels are distributed and organized in a given region of the image (quantifying how the space is filled in). Lacunarity is complementary to the fractal dimension, as shapes with the same fractal dimension may have different lacunarity values. One of the most popular methods for estimating lacunarity is the gliding-box method [17], which was used in the present study.

The process begins with a box with side r positioned in the upper left corner of the image and a count of the number of points in the image. This process is repeated for all the lines and columns of the image, producing a frequency distribution of the mass of the image. The number of boxes with side r containing a mass S of the image is designated by $n(S,r)$, with the total number of boxes counted designated by $N(r)$. This frequency distribution is then converted into a probability distribution $Q(S,r)$, equation (6). The first (Z^1) and second (Z^2) moments of this distribution are determined in equation (7). Lacunarity (L) for a box of size r is defined in equation (8).

$$Q(S,r) = n(S,r) / N(r) \tag{6}$$

$$\begin{aligned} Z^{(1)} &= \sum SQ(S,r) \\ Z^{(2)} &= \sum S^2 Q(S,r) \end{aligned} \tag{7}$$

$$L(r) = Z^{(2)} / \left(Z^{(1)}\right)^2 \tag{8}$$

It should be reiterated that lacunarity was used here as a quantifier of texture [3-5]. For such, the luminance of the pixel was considered the height of that point of the image. Consequently, rather than divide the image into boxes with side r, the image was divided into cubes with side r. This produced a new frequency distribution of the mass of the image [$n(S,r)$] and a new $N(r)$, with this new $N(r)$ representing the total number of cubes counted.

2.3. Image processing and comparison of parameters

The quantifications with fractal dimension and lacunarity were performed processing part of the image database of myocardial biopsies from heart-transplant patients provided by the *Núcleo Transdisciplinar para Estudo do Caos e da Complexidade of the Faculdade de Medicina de São José do Rio Preto*, Brazil. A total of 118 images with different grades of rejection were analyzed: 0R (22 images), 1R (24 images), 2R (29 images) and 3R (43 images). For the image processing, each biopsy was segmented using the automatic multilevel method proposed in Subsection 2.1 and the quantifications were performed using the methods described in Subsection 2.2.

To validate the proposed segmentation method, mean values were calculated of the thresholds, gray levels and total number of pixels contained in each class of the segmented structure. Three classes were considered in the images: cell nuclei or cellular rejection (Class 1), heart muscle tissue (Class 2) and interstitial space or absence of structures (Class 3). The values were determined in each class and compared with those provided by Otsu's method [7]. The quantification with fractal methods was for structures segmented as Class 1.

Means, standard deviations and confidence intervals were calculated (with a 95% confidence level) for the fractal dimension and lacunarity attributes. This procedure was performed for each rejection level (0R to 3R) in order to identify and understand the behavior in this type of image in light of the main attributes of fractal geometry.

3. Results

3.1. Segmentation

The segmentation method proposed here and Otsu's method [7] were both applied to all images. The comparative results are displayed in tables 1, 2 and 3.

Otsu's method provided significantly higher mean values of the thresholds, gray level and total number of pixels contained in each structure class. One possible explanation for this difference is that Otsu's method segmented muscle structures as being cell nuclei or cellular rejection, which had a negative influence over a possible pattern recognition or quantification step. Figure 3 illustrates the segmentation achieved with the proposed method and figure 4 allows comparing the results of the two segmentation methods.

Table 1. Mean threshold values obtained with proposed segmentation method and Otsu's method [7]. Two thresholds were identified for the segmentation of classes of interest.

	Mean threshold values	
	Classes 1 and 2	**Classes 2 and 3**
Otsu's method [7]	139	203
Maximum Entropy	70	240

Table 2. Mean intensity values in each class obtained with proposed segmentation method and Otsu's method [7].

	Mean gray level		
	Class 1	**Class 2**	**Class 3**
Otsu's method [7]	104.91	172.44	232.38
Maximum entropy	41.12	165.99	243.55

Table 3. Mean pixel total in each class obtained with proposed segmentation method and Otsu's method [7].

	Mean pixel total		
	Class 1	**Class 2**	**Class 3**
Otsu's method [7]	52160	277376	150464
Maximum entropy	13823	408421	57756

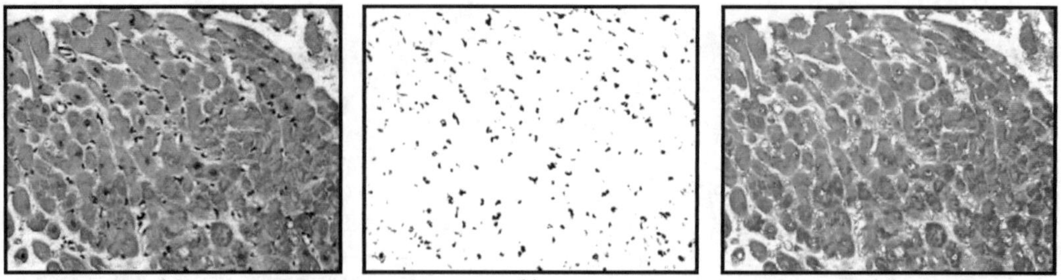

(a) (b) (c)

Figure 3. Illustrative case of image obtained from endomyocardial biopsy of heart-transplant patient (a) prior to processing; (b) result obtained with proposed segmentation method revealing cell nuclei and (c) heart muscle (gray) and interstitial spaces (white).

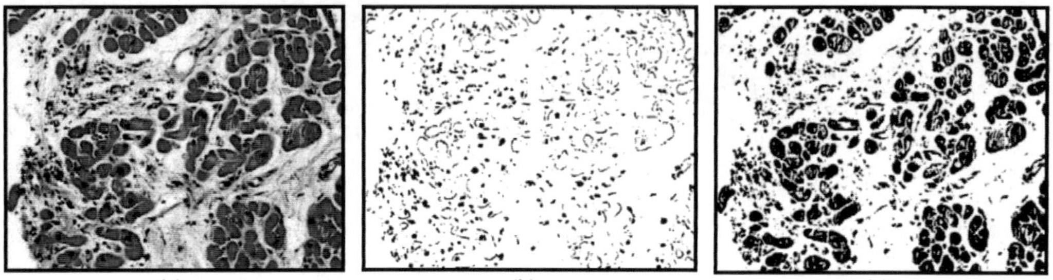

(a) (b) (c)

Figure 4. (a) Endomyocardial biopsy classified as 2R used to illustrate visual differences obtained with different segmentation methods; (b) cell nuclei segmented using proposed method and (c) using Otsu's method [7].

3.2. Quantification

The nuclei detected (both normal myocardial cell nuclei and nuclei of inflammatory cells found in states of rejection) in each group (0r to 3R) were quantified using the fractal dimension and lacunarity. The results are displayed in tables 4 and 5. Figure 5 allows viewing the behavior identified, with overlapping between rejection levels.

Table 4. Mean, standard deviation and 95% confidence interval calculated for fractal dimensions obtained on each rejection level.

Parameter	0R	1R	2R	3R
Mean	1.021	1.135	1.262	1.323
Standard deviation	0.140	0.088	0.143	0.170
		Confidence interval		
Lower value	0.944	1.089	1.194	1.257
Upper value	1.097	1.181	1.330	1.389

Table 5. Mean, standard deviation and 95% confidence interval calculated for lacunarity obtained on each rejection level.

Parameter	0R	1R	2R	3R
Mean	1.979	1.865	1.738	1.677
Standard deviation	0.14	0.088	0.14	0.17
		Confidence interval		
Lower value	1.903	1.819	1.670	1.611
Upper value	2.056	1.911	1.806	1.743

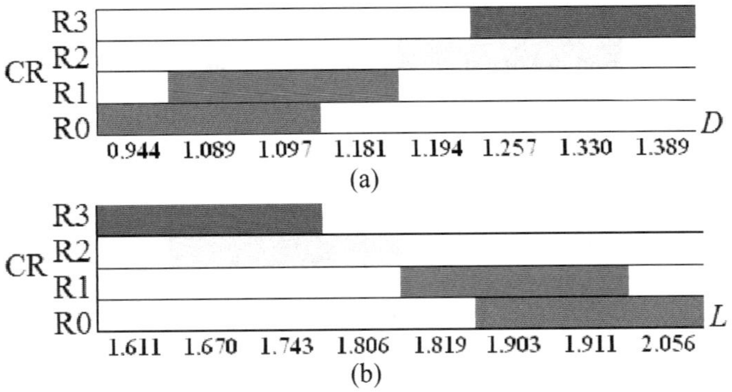

Figure 5. Visualization of behavior on each cardiac rejection (CR) level; (a) fractal dimension (*D*); (b) lacunarity (*L*); representation defined based on confidence intervals in tables 4 and 5.

4. Discussion and Conclusions

Based on the model used, the proposed thresholding method allowed the appropriate segmentation of the myocardial biopsy images and more precise quantification of the regions of interest. In comparison to Otsu's method [7], the proposed thresholding method achieved better results, as it allowed adjusting the parameters, such as class size and histogram slope percentage, thereby ensuring more appropriate thresholds. Otsu's method proved less flexible, with a tendency toward including heart muscle and cell nuclei (of both myocytes and inflammatory cells) in the same segmentation and higher thresholds. This behavior obviously implies an erroneous quantification, which did not occur with the thresholding methods proposed here.

The purpose of the proposed model was the quantification of regions of interest in images of myocardial biopsies from heart-transplant patients. For such, fractal dimension and lacunarity attributes were used. Cardiac rejection can be correctly categorized if the values are close to the means of the levels (tables 4 and 5). This is no trivial task for cases in which the dimension or lacunarity is closer to the lowest or highest values of the sets. This observation is based on the existing overlapping between rejection levels (figure 5), which may be explained as intermediate states in regression or progression. Another characteristic was the fact that, although overlapping occurred between levels 0R or 1R, no overlapping was found between these levels and the more advance levels of cellular rejection (2R and 3R) (figure 5), suggesting diagnostic security and a good prognosis. Patients categorized as 2R or 3R merit special attention due to the possibility of a mistaken classification and the possibility of progression or regression in cardiac rejection. The quantification of these characteristics furnishes data that allow medical services to pay greater attention to heart-transplant patients with grade 2R, considering the evident possibility of confusion with grade 3R.

This paper presented a study on cardiac rejection levels based on multilevel thresholding with maximum entropy, which ensured an appropriate quantification of each level in light of the fractal dimension and lacunarity. This method allowed the successful identification of the behavior of each cellular rejection level. These results are useful to the determination of earlier, more effective treatments, which could ensure a greater survival rate among patients submitted to heart transplantation.

Acknowledgments
This study was supported in part by UNESP/PROP (Pró-Reitoria de Pesquisa).

References

[1] Jain A K and Duin R 2000 Statistical pattern recognition: A review. *IEEE Transactions on Pattern Analysis and Machine Intelligence* **22** 14–37.
[2] Stewart et al 2005 Revision of the 1990 Working Formulation for the Standardization of Nomenclature in the Diagnosis of Heart Rejection *The Journal of Heart and Lung Transplantation* **24** 1710–1720.
[3] Manousaki A G, Manios A, Tsompanaki E I and Tosca A D 2006 Use of color texture in determining the nature of melanocytic skin lesions – a qualitative and quantitative approach, *Computer in Biology and Medicine* **36** 419-427.
[4] Dougherty G and Henebry G M 2001 Fractal signature and lacunarity in the measurement of the texture of trabecular bone in clinical CT images *Medical Engineering & Physics* **23** 369–380.
[5] Borys P, Krasowska M, Grzywna Z J, Djamgoz M B A and Mycielska M E 2008 Lacunarity as a novel measure of cancer cells behavior *BioSystems* **94** 276–281.
[6] Gonzalez R C and Woods R E 2008 *Digital Image Processing* ed 3 Prentice-Hall.
[7] Otsu N A 1979 Threshold selection method from gray-level histogram *IEEE Transactions on Systems, Man, and Cybernetics* **9** 1.
[8] Pun T 1980 A New Method for Gray-Level Picture Thresholding Using the Entropy of the Histogram *Signal Processing* **2** 223-237.
[9] Kapur J N, Sahoo P K and Wong K C 1985 A new method for gray-level Picture thresholding using the entropy of the histogram *Computer Vision, Graphics, and Image Processing* **29** 273-285.
[10] Abutaleb A H 1989 Automatic Thresholding of Gray-Level Pictures Using Two-Dimensional Entropy *Computer Vision, Graphics, and Image Processing* **47** 22-32.
[11] Beghdadi A, Négrate A L and De Lesegno P V 1995 Entropic Thresholding Using a Block Source Model *CVGIP: Graphical Models and Image Processing* **57** 197-205.
[12] Brink A D 1996 Using spatial information as an aid to maximum entropy image threshold selection *Pattern Recognition Letters* **17** 29-36.
[13] Aboud Neta S R, Dutra L V and Erthal1 G J 2008 Limiarização Automática em Histogramas Multimodais *Proceeedings of the 7th Brazilian Conference on Dynamics, Control and Applications* (FCT – Unesp at Presidente Prudente, Brazil 7-9 May 2008).
[14] Normant F and Tricot C 1991 Method for evaluating the fractal dimension of curves using convex hulls *Physical Review A* **43** 6518-6525.
[15] Chaudhuri B B and Sarkar N 1995 Texture Segmentation Using Fractal Dimension. *IEEE Transactions on Pattern Analysis and Machine Intelligence* **17** 72-76.
[16] Coelho R C and Costa L da F 1995 The Box-Counting Fractal Dimension: Does it provide an Accurate Subsidy for Experimental Shape Characterization? If So, How to Use It? *Symp. on Comp. Graphics and Image Processing* (São Carlos: Brasil, 24-27 October) 183-191.
[17] Plotnick R E, Gradner R H, Hargrove W W, Prestegaard K and Perlmutter M 1996 Lacunarity analysis: a general technique for the analysis of spatial patterns *Physical Review E* **53** 5461-5468.

Dynamic Days South America 2010 IOP Publishing
Journal of Physics: Conference Series **285** (2011) 012033 doi:10.1088/1742-6596/285/1/012033

Self-organization and pattern formation in coupled Lorenz oscillators under a discrete symmetric transformation

A Carrillo[1], **B A Rodríguez**[1]

[1] Instituto de Física, Universidad de Antioquia, AA1226 Medellín, Colombia

E-mail: acarril@pegasus.udea.edu.co

Abstract. We present a spatial array of Lorenz oscillators, with each cell lattice in the chaotic regime. This system shows spatial ordering due to self-organization of chaos synchronization after a bifurcation. It is shown that an array of such oscillators transformed under a discrete symmetry group, does not maintain the global dynamics, although each transformed unit cell is locally identical to its precursor. Alternatively, it is shown that in a 1-dimensional lattice, the coupling destroy the chaotic behavior but there are similar global behaviors between both coupled arrays, suggesting that is the local equivariance which controls the dynamics.

1. Introduction

A covering dynamical system is one with a discrete symmetry, and the equations of motion that describe it are left unchanged under the action of a discrete symmetry group. By means of a local diffeomorphism, is possible to map it to a locally equivalent dynamical system without symmetry [1]. The dynamical properties of this *new* system, such as Lyapunov exponents and fractal dimensions are left unchanged [2], although the resulting system (*image*) is topologically different from the former (*cover*). In this order of ideas, a system without symmetry is more easy to study, since there is a reduction of the branched manifolds in the *asymmetry* space, and this allow us to study a transformed asymmetric attractor, gather conclusions about its dynamics, and then make the inverse transformation without loss of information. For a more extense discusion about lost topological information in transformed spaces see [3].

Recently, the pattern formation in systems of chaotic coupled attractors has become of main importance [4, 5, 6]. However, in general, is difficult to anticipate that chaotic units can construct patterns with ordered space structures, mainly due to the sensitivity of chaotic trajectories to their initial conditions and the unpredictability of long term evolution of the orbits. However the possibility to study a simpler asymmetric image system should simplify, at least, the computational effort. In this work, we show how lattices of locally similar dynamical systems, although topologically inequivalent, can exhibit different pattern formations, and that is possible to perform global bifurcations by means of a parametric change in the coupling strength. This bifurcations are not the same for different topologies of the unit cells. We also investigate how this discrete equivariant transformation affect the dynamical behavior of 1-dimensional arrays of chaotic oscillators.

2. Model System

In order to study global differences between a cover and its image, we choose two different configurations, one and two dimensional lattices, and explore their spatiotemporal behavior with fixed control parameters.

2.1. Two-dimensional lattices

We take a two-dimensional lattice of coupled Lorenz oscillators, with fixed boundary conditions $x_{0,j} = x_{N,j}$ and $x_{i,0} = x_{i,N}$ for each variable:

$$\begin{aligned} \dot{x}_{ij} =& -\sigma(x_{ij} - y_{ij}) + \epsilon \bigtriangledown^2 x; \\ \dot{y}_{ij} =& -x_{ij}\, z_{ij} + r\, x_{ij} - y_{ij} + \epsilon \bigtriangledown^2 y; \\ \dot{z}_{ij} =& x_{ij}\, y_{ij} - \beta z_{ij} + \epsilon \bigtriangledown^2 z, \end{aligned} \tag{1}$$

where

$$\bigtriangledown^2 w = w_{i+1,j} + w_{i-1,j} + w_{i,j+1} + w_{i,j-1} - 4w_{i,j}, \tag{2}$$

is a two-dimensional diffusive coupling, and $i, j = 1, 2, \cdots, N$, with N the dimension of the $N \times N$ lattice. The parameter settings are: $\sigma = 10$, $\beta = 8/3$ and $r = 27$, so that each single independent Lorenz oscillator is chaotic, and are left unchanged for the following. The spatial orderings of the system are obtained varying homogeneously the control parameter ϵ, and in order to study this transition, we use the following error functions [7]:

$$\begin{aligned} E_{min} =& min\big(E_1(m), E_2(m)\big); \\ E_{max} =& max\big(E_1(m), E_2(m)\big), \end{aligned} \tag{3}$$

where

$$\begin{aligned} E_1(m) =& \frac{1}{N}\sum_{i}^{N} \left\| x(i, \frac{N}{2} + m) - x(i, \frac{N}{2} - m) \right\|; \\ E_2(m) =& \frac{1}{N}\sum_{j}^{N} \left\| x(\frac{N}{2} + m, j) - x(\frac{N}{2} - m, j) \right\|, \end{aligned} \tag{4}$$

and m is a spatial delay, and its value is chosen in the range $0 < m < N/2$. We use the notation $x(i,j) = x_{ij}$ writing eq. (3) for the sake of simplicity. With this definitions, we can identify three types of behavior: *i)* $E_{min} > 0$, $E_{max} > 0$ there is no pattern in the array; *ii)* $E_{min} = 0$, $E_{max} > 0$ there is a mirror symmetry respect to one of the axis: $i = N/2$, $j = N/2$ or the diagonal axis; *iii)* $E_{min} = 0$, $E_{max} = 0$, the symmetry is central, *i.e.*, there is a mirror symmetry for each axis.

Using the fact that eq. (1) is equivariant under z-axis rotations, we have the diffeomorphism

$$\begin{aligned} u &= x^2 - y^2; \\ v &= 2xy; \\ w &= z, \end{aligned} \tag{5}$$

and the mapping of the system eq. (1) into its locally-equivariant *image* system [8] is

$$\begin{aligned} \dot{u}_{ij} =& (-\sigma - 1)u_{ij} + (\sigma - r)v_{ij} + v_{ij}w_{ij} + (1 - \sigma)\rho + \epsilon \bigtriangledown^2 u; \\ \dot{v}_{ij} =& (r - \sigma)u - (\sigma + 1)v_{ij} - u_{ij}w_{ij} + (r + \sigma)\rho - w_{ij}\rho + \epsilon \bigtriangledown^2 v; \\ \dot{w}_{ij} =& -\beta w_{ij} + 0.5v_{ij} + \epsilon \bigtriangledown^2 w, \end{aligned} \tag{6}$$

where $\rho = \sqrt{u^2 + v^2}$. The system eq. (6) is topologically similar to a Rössler oscillator. We have not make the transformation on the coupling, in order to assure that the two systems are coupled in the same diffusive way.

2.2. One-dimensional lattices

Using both eq. (1) and eq. (6), we built one-dimensional lattices, with fixed and free borders, i.e., chain and ring configurations. The diffusive terms used are

$$\nabla w = w_{i+1} + w_{i-1} - 2w_i \quad \text{chain and in the bidirectional ring;} \tag{7}$$

$$\nabla w = w_{i-1} \quad \text{unidirectional ring,} \tag{8}$$

with $i = 1, 2, \cdots, N$, where N is the number of cells in the lattice, and w stands for x, y or z, and u, v or w. The parameter settings are the same as in the two-dimensional case, so each single independent Lorenz oscillator and his *image* are chaotic. The coupling parameter was set as $\epsilon = 2.5$.

3. Results

In Figures (1) and (2) the two regimes of the system eq. (1) are plotted for times greater than the transient. The only dynamical transition present when ϵ is changed, is from disorder to a central symmetry. By means of the error functions eq. (3), we observe that the dependency

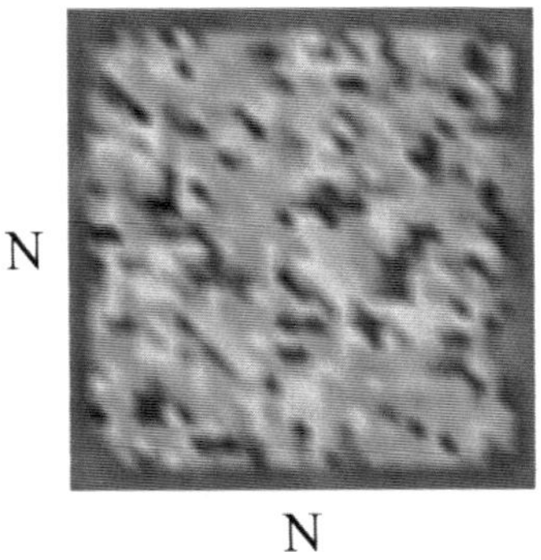

Figure 1. Space pattern of a Lorenz oscillator lattice with $\epsilon = 0.5$ and $N = 31$. There is an evident lack of synchronization.

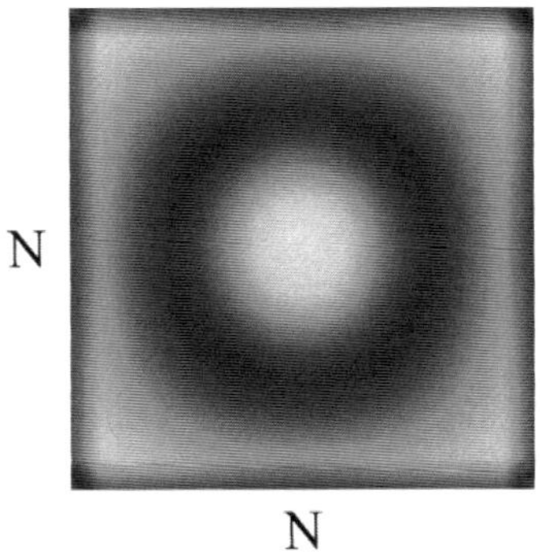

Figure 2. Space pattern of a Lorenz oscillator lattice with $\epsilon = 0.9$ and $N = 31$. The lattice behaves like a breathing system.

of the parameter ϵ_b where the bifurcation occur with the size of the array is roughly parabolic, different from the exponential dependence as was proposed in [7] for a configuration of two-lobe attractors. Although each central configuration at which the system arrives after the bifurcation depends on the initial conditions, the same qualitative behavior is observed for long enough times.

The results for the image system are plotted in Figures (3) and (4). We can see a very different behavior: in this system, made up of locally equivalent units of eq. (1), the array present two bifurcations: one that goes from disorder to a mirror symmetry, and from there to a central symmetry. We found that the most common mirror symmetry present in this array is about the diagonal axis, so a redefinition of the error functions was done in order to quantify it. This behavior is observed for systems without discrete symmetries, like Rössler oscillator, and

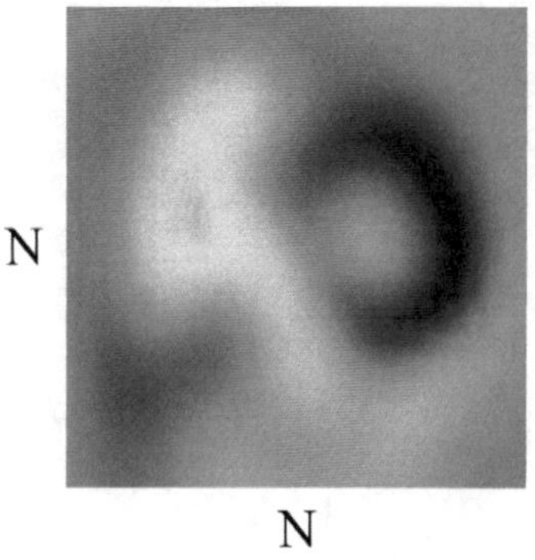

Figure 3. Space pattern of a Lorenz oscillator lattice with $\epsilon = 0.4$ and $N = 31$. The lattice behaves like a breathing system with two lobes

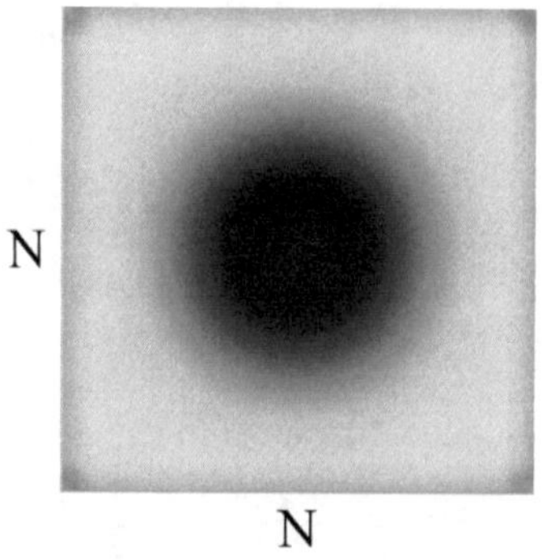

Figure 4. Space pattern of a Lorenz oscillator lattice with $\epsilon = 2.3$ and $N = 31$. The lattice behaves like a breathing system.

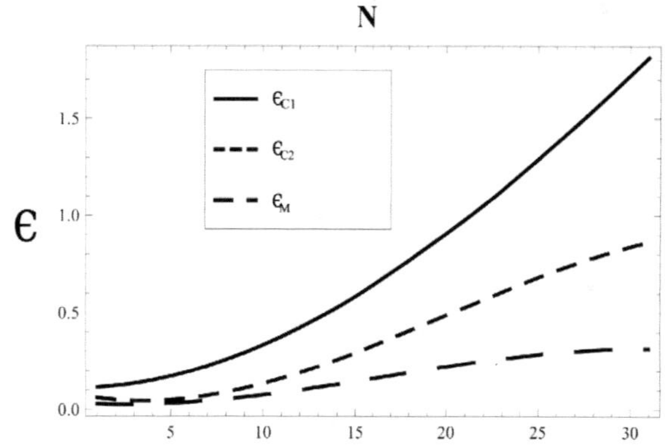

Figure 5. Variation of the control parameter ϵ with the lattice dimension. Solid line shows the random-to-central symmetry bifurcation in eq. (1); the other two show the bifurcation points in the transformed system eq. (6): dashed line shows the mirror to central bifurcation parameter ϵ_M and the long dashed line shows the *none to mirror* bifurcation parameter ϵ_{C2}.

the change of the control parameter with the dimension of the lattice is almost linear for both bifurcation points; the change of ϵ for the two systems considered are shown in the Figure (5).

Is important to note from this results that with the tools available to study spatiotemporal chaos, is impossible to claim for the origin of the double-bifurcated system: an observer can not say if the array is made with Rössler oscillators, or with Lorenz *images* oscillators. This means that there are unrecoverable information lost in the transformation, and that in coupled $2D$ lattices of oscillators is more important the topology (one or two lobes) than the local equivariance.

In Figures (6),(7) and (8), we plot the spatiotemporal dynamics of one-dimensional arrays of diffusively coupled Lorenz attractors and their *images*. In order to differentiate global behaviors we used their power spectrum, because patterns are most difficult to identify in this case. In the three different configurations we can see similar global dynamics: although in each case the way in which the unit cells synchronize are different, between a system and its image there are more complex correlations. This correlations are well described because the main frequencies in both (cover and image) configurations are the same in the first two lattices couplings (fixed borders, using eq. (8) and free borders, using eq. (7)). We saw cluster synchronization, correlated states and total synchronization in Figures (6),(7) and (8) respectively, but no spatio-temporal chaos.

The way in which the borders of the arrays are set, fixed the spatiotemporal dynamic, in both the cover and the image system. The fact that the phase space state of each cell is not chaotic, tell

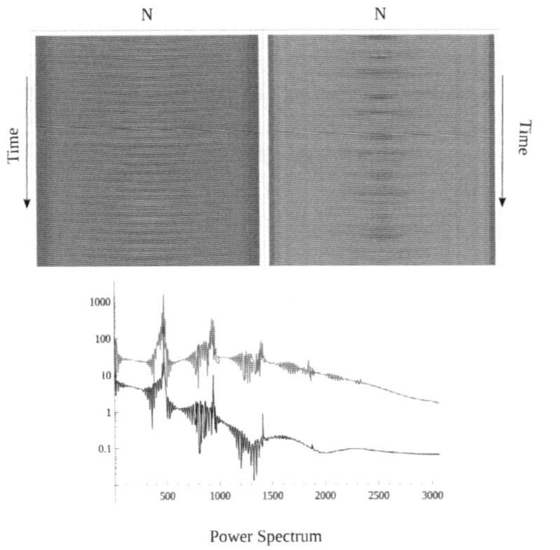

Figure 6. Evolution of $N = 40$ attractors coupled in a diffusive way, with fixed borders, using eq. (8). Upper left: Lorenz lattice; upper right: Lorenz Image lattice; bottom: Power Spectrum of (red) Lorenz and (black) its image.

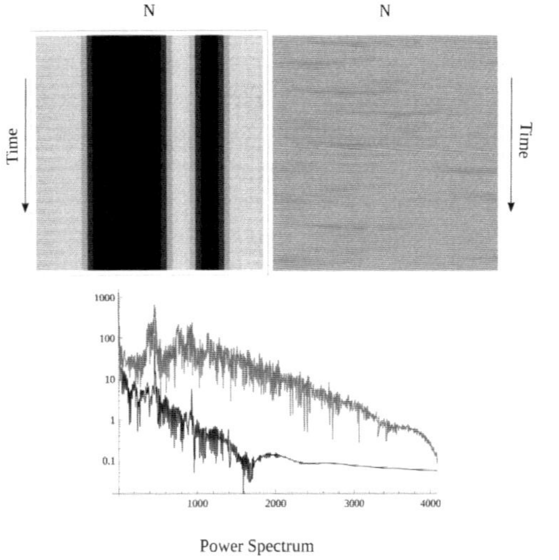

Figure 7. Evolution of $N = 40$ attractors coupled in a diffusive way, with variable borders, using eq. (7). Upper left: Lorenz lattice; upper right: Lorenz Image lattice; bottom: Power Spectrum of (red) Lorenz and (black) its image.

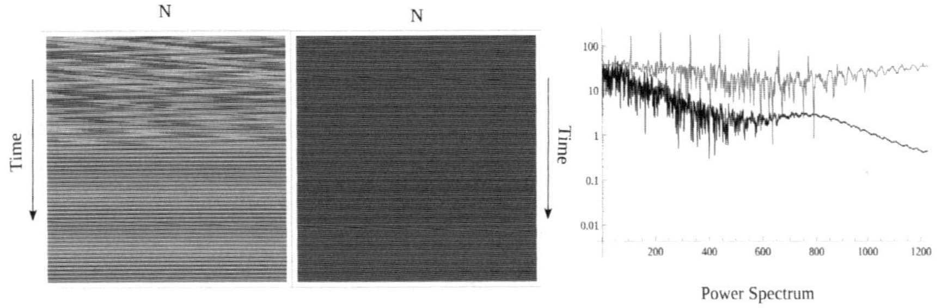

Figure 8. Evolution of $N = 40$ attractors coupled in a diffusive way, with variable borders, using eq. (8). Left: Lorenz lattice; center: Lorenz Image lattice; right: Power Spectrum of (red) Lorenz and (black) its image.

us that is the local nature which determine the global dynamic, and hence is the equivalent local nature between the cover and its image which makes the spatiotemporal dynamic to behave in a similar fashion, i.e., to oscillate around the same fundamental frequencies. However is important to note that when both system shows total synchronization, as in Figure (8), their frequencies are not correlated.

4. Conclusion

Although identical dynamic information can be gathered from a cover system and his image (Lyapunov exponent, fractal dimensions), their behavior in phase space is topologically different. The system without symmetry is more easy to study, since there is a reduction of the manifolds in the *asymmetry* space. However we observed that the information that can be recovered from the mapped lattice is different in an essential way from the former. More important is the fact that although the phase space dynamics of a individual cell image is *simpler*, because in the case of a Lorenz oscillator it is a one-lobe attractor, the dynamic of *images* lattice is more complicated: it seems to show globally a pattern that recalls the untransformed individual cell phase space dynamics, a *two-lobe* global attractor. We found that for large enough values of ϵ, the systems reach a stable state with no chaotic unit-cell behavior as expected. We also show numerically that the dependence of the control parameter with the dimension of an array of one-lobe coupled systems is almost linear and that of two-lobe coupled systems is parabolic, in clear difference with the results found in [7], making possible to study finite but high dimensional lattices without high computational efforts.

In one-dimensional lattices the way in which the borders of the arrays are set, fixed the spatiotemporal dynamics: in the cases shown by Figures (6) and (7) there is the same fundamental frequency, as it is shown by the Power Spectrum. When the chaotic nature of each cell is lost, the shape (topology) of the attractors is also lost. This suggest that is the local nature which determine the global dynamic, and hence is the equivariance between the cover and its image which makes the spatiotemporal dynamic to behave in a similar fashion, i.e., to oscillate around the same fundamental frequencies.

This imply that although in a 2-D array is not possible to relate an image to its cover based on the global lattice behavior, in a $1D$ array it is possible, even when there are not chaotic information (Lyapunov exponent, fractal dimensions, etc.), given that they will oscillate with the same frequencies. The fact that when both systems (image and cover) shows total synchronization their frequencies are not related, suggests that there is a global mixing destroying all the information given to the system by the way in which the cells are coupled, or by the way in which the borders are fixed.

Although more work on theorical differentiation of lattice dynamics is needed, a small insight has been gained through this work.

References

[1] Gilmore R 2007 *Chaos* **17** 013104
[2] Anastassiou S, Bountis T and Petalas Y G 2008 *J. Phys. A: Math. Theor. Vol.* **41** 485101
[3] Cvitanović P, Artuso R, Mainieri R, Tanner G and Vattay G 2010 *Chaos: Classical and Quantum* (Niels Bohr Institute, Copenhagen) chap 9
[4] Pecora L M 1998 *Phys. Rev. E* **58** 347–60
[5] Hu G, Zhang Y, Cerdeira H A and Chen S 2000 *Phys. Rev. Lett.* **85** 3377–80
[6] Liu Z and Lai Y C 2001 *Phys. Rev. Lett.* **86** 4737–40
[7] Zhang X, Fu M, Xiao J and Hu G 2006 *Phys. Rev. E* **74** 015202
[8] Letellier C and Gilmore R 2000 *Phys. Rev. E* **63** 016206

Dynamic Days South America 2010 IOP Publishing
Journal of Physics: Conference Series **285** (2011) 012034 doi:10.1088/1742-6596/285/1/012034

Two upwinding schemes for nonlinear problems in fluid dynamics

Giseli A. B. Lima, Laís Corrêa, Valdemir G. Ferreira

Instituto de Ciências Matemáticas e de Computação - USP: Av. Trabalhador são-carlense 400, Zip code: 13560-970, São Carlos, SP, Brazil

E-mail: giabl,lacorrea,pvgf@icmc.usp.br

Abstract. The correct modeling for processes involving convection, without introducing excessive artificial damping while retaining high accuracy, stability, boundedness and simplicity of implementation continues being nowadays a challenging task for CFD practitioners. The objective of this study is to present and evaluate the performance of two new upwinding schemes, namely SDPUS-C1 and EPUS, for nonlinear convection term discretization. Both SDPUS-C1 and EPUS schemes satisfy the TVD principle of Harten and are based on the NVD formulation of Leonard. Firstly, a description of the schemes is presented and then the numerical results are provided for one- and two-dimensional hyperbolic conservation laws. Finally, as an application, the SDPUS-C1 and EPUS schemes are employed for the simulation of two-dimensional incompressible fluid flows involving moving free surfaces. The numerical experiments show that the proposed upwinding schemes perform very well.

1. Introduction

The numerical solution of complex fluid dynamics problems has emerged as a viable alternative to both experimental and analytical studies. To make the simulations of these problems more acceptable and reliable, there is an increasing demand for development, analysis and implementation of a upwinding convective scheme (in general nonlinear) that offers simplicity, accuracy, robustness and versatility. Such a scheme is particularly important when one wants to solve full transient balance laws that model incompressible viscous flows involving moving free surfaces.

In the published literature there is a variety of techniques to approximate convective terms. For instance, first order upwind schemes, such as FOU [6], are unconditionally stable, but they have a diffusive character that, in general, smooths the solution. On the order hand, classical high resolution schemes, such as central differences and QUICK [11] scheme (and its related QUICK with estimated stream terms - QUICKEST) can often produce unphysical oscillations which, in the most of the time, can lead to numerical instability. Indeed, all of the linear schemes with accuracy above first order are oscillatory, according to Godunov's theorem.

To overcome these defects produced by the aforementioned schemes, a number of monotonic high-order upwind schemes have appeared such as, for example, the sharp and monotonic algorithm for realistic transport (SMART) [8], the variable-order non-oscillatory scheme (VONOS) [20] and an adaptative bounded version of the QUICKEST (ADBQUICKEST) [7]. In addition to this list, one may cite the monotone upstream scheme for conservation laws (MUSCL) originally pioneered by van Leer [19] and the associated limiters developed for the past twenty years, for example the van Albada [18] (and its variants). Thus, in this work, we presented the development of two new monotonic techniques, namely Six-Degree Polynomial Upwind Scheme of C^1 Class (SDPUS-C1) and Eight-degree Polynomial Upwind Scheme (EPUS), for the discretization of the convection flux terms, yielding stable and accurate upwinding schemes. These schemes are nonlinear and based on the NVD formulation of Leonard [12] and satisfy the TVD [9] principle. In the development of the schemes, we have also included the ideas of Lin and Chieng [15], which say if a scheme is of C^1 class then it provides good results on coarse meshes. In this context, the main focus of this study is to evaluate the performance of the SDPUS-C1 (a scheme of C^1 class) and EPUS (a scheme of C^2 class) for the numerical solution of hyperbolic system of conservation laws and related fluid mechanics equations. The primary motivation for the development of simple, stable, and accurate upwinding schemes, for hyperbolic conservation equations and related fluid dynamics problems, is that there are few flux-based upwinding schemes that are relatively simple to understand and easy to implement into a new or existent computer code.

The structure of the article is as follows. In section 2, the theoretical foundations are described. Subsequently, in section 3, a description of the SDPUS-C1 and EPUS schemes is presented. In section 4, numerical results are provided for inviscid Euler and viscous Navier-Stokes equations. Finally, conclusions and future works are presented in section 5.

2. Theoretical foundations

We now concentrate on the theoretical foundations of high resolution total variation diminishing (TVD) techniques for homogeneous scalar conservation laws. In these laws, the numerical fluxes are calculated by using the modified flux approach of Harten [9], a well established technology to construct high resolution shock capturing upwinding schemes. This is accomplished by interpolating a numerical flux, say ϕ_f, at a boundary face f between two control volumes, using three neighboring grid points, namely D (Downstream), U (Upstream) and R (Remote-upstream), labeled according to the local wind direction (upwinding). For multidimensional systems, this strategy is applied in the same fashion, with each convective derivative approximated along of the relevant variable (direction-by-direction). In summary, the numerical flux is calculated according to $\phi_f = \phi_f(\phi_D, \phi_U, \phi_R)$. To simplify this functional relationship, the original variables are transformed in NV as

$$\hat{\phi}_{[\ ]} = \frac{\phi_{[\ ]} - \phi_R}{\phi_D - \phi_R}. \tag{1}$$

The advantage of this new formulation is that the interface value $\hat{\phi}_f$ depends of $\hat{\phi}_U$ only, since $\hat{\phi}_D = 1$ and $\hat{\phi}_R = 0$. Thus, the functional relationship can be rewritten as

$$\hat{\phi}_f = \hat{\phi}_f(\hat{\phi}_U). \tag{2}$$

The normalized variable diagram (NVD), proposed by Leonard [12], is based on the definition (1) and functional relationship (2). According to Leonard [12], for $0 \leq \hat{\phi}_U \leq 1$, it is possible to derive a nonlinear monotonic third order NV scheme by imposing the following conditions: $\hat{\phi}_f(1) = 1$ and $\hat{\phi}_f(0) = 0$ (necessary conditions); $\hat{\phi}_f(1/2) = 3/4$ (a necessary and sufficient condition to reach second order of accuracy); and $\hat{\phi}_f'(1/2) = 3/4$ (a necessary and sufficient condition to reach third order of accuracy). In addition, Leonard recommends that, for values of $\hat{\phi}_U < 0$ or $\hat{\phi}_U > 1$, the scheme must be extended in a continuous manner using the FOU [6] scheme, namely

$$\hat{\phi}_f = \hat{\phi}_U. \tag{3}$$

Another important convective stability criterion (a form of nonlinear stability) is the TVD constraint of Harten [9], which is a purely scalar property and ensures that spurious oscillations are removed from the numerical solution. In summary, by considering a sequence of discrete approximations $\phi(t) = \phi_i(t)_{i\in\mathbb{Z}}$ to a scalar, the total variation (TV) at time t of this sequence is defined by $TV(\phi(t)) = \sum_{i\in\mathbb{Z}} |\phi_{i+1}(t) - \phi_i(t)|$. Then, an upwinding scheme is TVD if the following condition is satisfied $TV(\phi^{n+1}) \leq TV(\phi^n), \quad \forall n \in \mathbb{N}$.

After one has developed a NVD/TVD-based upwinding scheme, one may derive the associated flux-limiter by rewriting the original scheme in the following way (see, for example, Sweby [17] and Waterson and Deconinck [21]):

$$\hat{\phi}_f = \hat{\phi}_U + \frac{1}{2}\psi(r)(1 - \hat{\phi}_U), \tag{4}$$

where $\psi(r)$ is the flux-limiter that determines the level of antidifusividade, r being a local shock sensor given by ratio of consecutive gradients as $r = \left(\frac{\partial\phi}{\partial x}\right)_g \Big/ \left(\frac{\partial\phi}{\partial x}\right)_f$. On uniform meshes, it is given by $r = (\phi_U - \phi_R)/(\phi_D - \phi_U)$ and in NV by

$$r = (\hat{\phi}_U)/(1 - \hat{\phi}_U). \tag{5}$$

With the concepts introduced by Harten [9], Sweby [17] proposed, in the NVD context, restrictions for a TVD scheme. These define the TVD regions shown in figures 1 and 2. So a scheme is TVD, if the its functional relationship $\phi_f(\phi_U)$ and flux limiter $\psi(r)$ are contained in the regions 1 and 2, respectively.

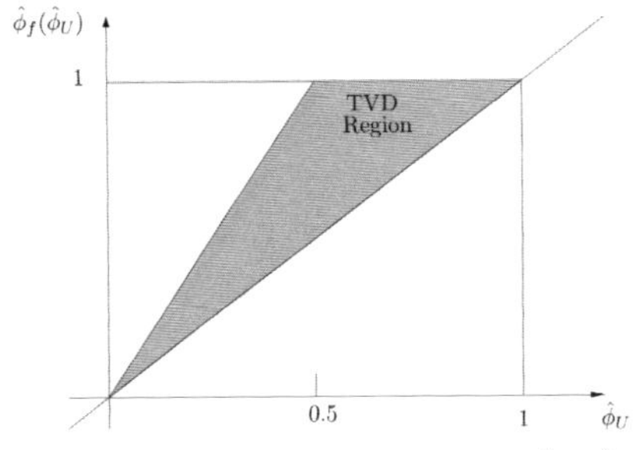

Figure 1. TVD region for $\hat{\phi}_f(\hat{\phi}_U)$.

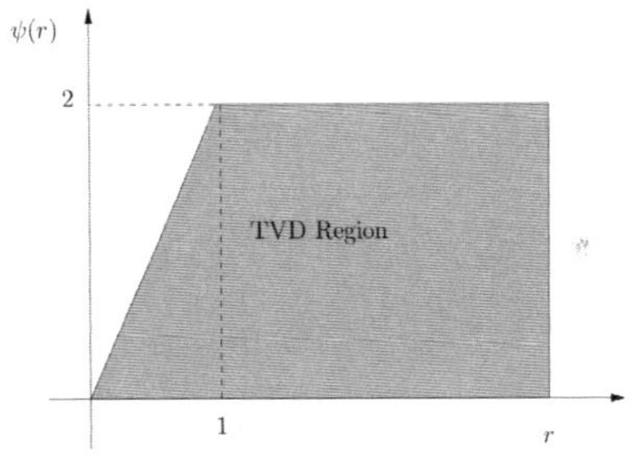

Figure 2. TVD region for $\psi(r)$.

3. The SDPUS-C1 and EPUS schemes

The SDPUS-C1 and EPUS schemes are derived, in the context of NV of Leonard [12] and TVD constraint of Harten [9], by following the procedure described above. The numerical solutions obtained with these schemes can lead to accurate results (up to 3th order) away from discontinuities, as well as well defined, very steep, monotone profiles at those locations where discontinuities (or large gradients) occur in the solution. The SDPUS-C1 scheme is developed by assuming that the NV at the cell interface f, $\hat{\phi}_f$, are related to $\hat{\phi}_U$ as a six-degree polynomial function of the form

$$\hat{\phi}_f = \sum_{k=0}^{6} b_k \hat{\phi}_U^k, \tag{6}$$

for $0 \leq \hat{\phi}_U \leq 1$, and a linear function (the FOU scheme), given by (3), for $\hat{\phi}_U < 0$ or $\hat{\phi}_U > 1$. We set a free parameter (say $b_2 = \gamma$) in equation (6) and determine the other coefficients by imposing the four conditions of Leonard outlined above, plus the continuous differentiability condition for $\hat{\phi}_f = \hat{\phi}_f(\hat{\phi}_U)$; in another words, the equations (6) and (3) are linked on the points $(0,0)$ and $(1,1)$ with the same values to the first derivatives. So, the SDUPS-C1 scheme becomes a continuously differentiable function. It is important to observe here that, according to Lin and Chieng [15], if this property is not satisfied then convergence problems may be found in unsteady calculations when large time steps are employed. Similarly, the EPUS scheme is derived by considering a eight-degree polynomial function of the form

$$\hat{\phi}_f = \sum_{k=0}^{8} c_k \hat{\phi}_U^k, \tag{7}$$

for $0 \le \hat{\phi}_U \le 1$, and the FOU scheme for $\hat{\phi}_U < 0$ or $\hat{\phi}_U > 1$. We set a free parameter (say $c_3 = \gamma$) and determine the other coefficients by imposing the conditions of Leonard and continuous differentiability property for $\hat{\phi}_f = \hat{\phi}_f(\hat{\phi}_U)$, plus the condition $\hat{\phi}''(0) = \hat{\phi}''(1) = 0$ for obtaining C^2 class.

In summary, the SDPUS-C1 and EPUS upwinding schemes in NV are given by

- **SDPUS-C1:**

$$\hat{\phi}_f = \begin{cases} (-24+4\gamma)\hat{\phi}_U^6 + (68-12\gamma)\hat{\phi}_U^5 + (-64+13\gamma)\hat{\phi}_U^4 + \\ \qquad (20-6\gamma)\hat{\phi}_U^3 + \gamma\hat{\phi}_U^2 + \hat{\phi}_U, & \hat{\phi}_U \in [0,1], \\ \hat{\phi}_U, & \hat{\phi}_U \notin [0,1], \end{cases} \tag{8}$$

- **EPUS:**

$$\hat{\phi}_f = \begin{cases} -4(\lambda-24)\hat{\phi}_U^8 + 16(\lambda-23)\hat{\phi}_U^7 + (528-25\lambda)\hat{\phi}_U^6 \\ +(19\lambda-336)\hat{\phi}_U^5 + (80-7\lambda)\hat{\phi}_U^4 + \lambda\hat{\phi}_U^3 + \hat{\phi}_U, & \hat{\phi}_U \in [0,1], \\ \hat{\phi}_U, & \hat{\phi}_U \notin [0,1], \end{cases} \tag{9}$$

where γ and λ are free parameters.

In the same way as was done by Ferreira et al. [7], the corresponding flux-limiter function for SDPUS-C1 is derived by combining equations (4), (5) and (8) to obtain

$$\psi(r) = \max\left\{0, \quad \frac{0.5(|r|+r)(4r^2+12r)}{(1+|r|)^4}\right\}. \tag{10}$$

Analogously, by combining equations (4), (5) and (9) one derives the flux-limiter for EPUS scheme, namely

$$\psi(r) = \max\left\{0, \quad \frac{0.5(|r|+r)[(2\lambda-32)r^4+(160-4\lambda)r^3+2\lambda r^2]}{(1+|r|)^7}\right\}. \tag{11}$$

It can be seen that both SDPUS-C1 (a scheme of C^1 class) and EPUS (a scheme of C^2 class) are nonlinear and combine boundedness with second/third order of accuracy. In particular, it can be shown that SDPUS-C1 and EPUS schemes are TVD if $\gamma \in [4,12]$ and $\lambda \in [16,95]$, respectively. In figures 3 and 4, it is depicted the flux-limiter functions for the cases $\gamma = 12$ and $\lambda = 95$. Lima [14] and Corrêa [5] demonstrate, from numerical experimentation, that these values have provided good results for problems possessing discontinuous solutions.

4. Numerical results

In order to demonstrate the behavior, validity, flexibility and robustness of the SDPUS-C1 and EPUS schemes, some numerical tests will be carried out. Comparisons with reference solutions are initially assessed. The SDPUS-C1 and EPUS schemes are then used to solve incompressible Navier-Stokes equations involving arbitrarily moving free surfaces.

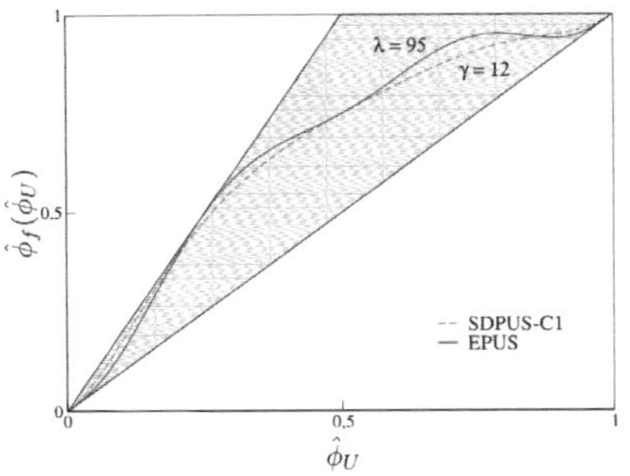

Figure 3. SDPUS-C1 and EPUS schemes in TVD region for $\hat{\phi}_f(\hat{\phi}_U)$.

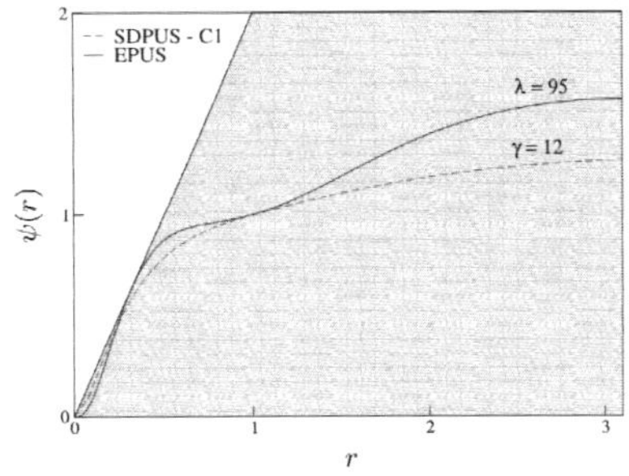

Figure 4. SDPUS-C1 and EPUS schemes in TVD region for $\psi(r)$.

4.1. Hyperbolic conservation laws

Many problems in science and engineering involve quantities that are preserved and that lead to certain types of nonlinear PDEs called hyperbolic conservation laws. In the two-dimensional (2D) case, they are defined by

$$\frac{\partial \phi}{\partial t} + \frac{\partial F(\phi)}{\partial x} + \frac{\partial G(\phi)}{\partial y} = 0, \tag{12}$$

where $\phi = \phi(x, y, t) : \mathbb{R}^2 \times \mathbb{R} \to \mathbb{R}^m$ is the m-dimensional vector of the conserved variables. $F(\phi) = F(\phi(x, y, t))$ and $G(\phi) = G(\phi(x, y, t)) : \mathbb{R}^m \to \mathbb{R}^m$ are convective flux functions. In this study, we consider two particular cases namely, 1D and 2D Euler equations. In order to resolve these hyperbolic systems, we have been used the well established CLAWPACK (Conservation LAW PACKage) code of LeVeque [13]. This code employs the finite volume methodology and has been equipped with the SDPUS-C1 and EPUS flux-limiters.

4.1.1. Test case 1 (1D Euler equations): We consider an inviscid compressible gas modeled by 1D Euler equations (12) with $\phi = [\rho, \rho u, E]^T$, being ρ, u, ρu, E and p the density, velocity, momentum, total energy and the pressure, respectively. In order to close this system, the ideal gas equation of state $p = (\gamma - 1)(E - \frac{1}{2}\rho u^2)$ is assumed, where $\gamma = 1.4$ is the ratio of specific heats. Here, we solve the problem suggested in [23], which involves multiple interactions of strong shocks. The initial condition is

$$[\rho_0, \ v_0, \ p_0]^T = \begin{cases} [1, \ 0, \ 1000]^T, & \text{if} \quad 0 \le x \le 0.1, \\ [1, \ 0, \ 0.01]^T, & \text{if} \quad 0.1 < x \le 0.9, \\ [1, \ 0, \ 100]^T, & \text{if} \quad 0.9 < x \le 1.0. \end{cases} \tag{13}$$

The implemented boundary condition is the zero-order extrapolation (see [13]).

The reference solution was calculated on a mesh size of 2000 computational cells using the Godunov method with term correction with monotinized centered (MC) flux-limiter (see [13]). The numerical solutions were obtained on a mesh size of 1000 computational cells using the SDPUS-C1 and EPUS flux-limiters. Figure 5 depicts the reference and numerical solutions for ρ at time $t = 0.038$ and CFL=0.5. It can be seen that the shock is well resolved with both SDPUS-C1 and EPUS upwinding schemes, without apparition of spurious oscillations. It is also observed from this figure that the EPUS scheme provides more accurate results.

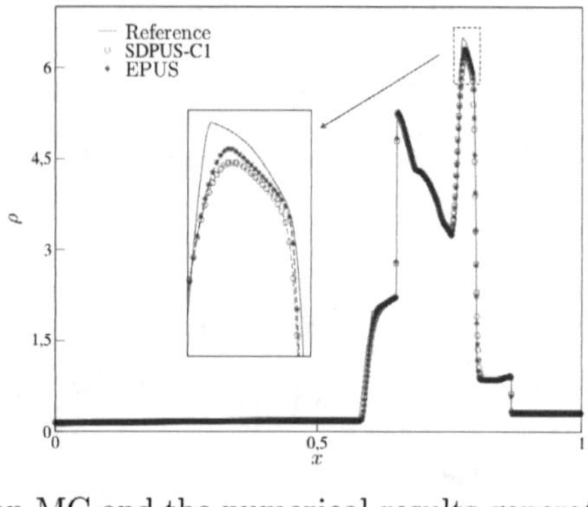

Figure 5. Reference solution MC and the numerical results generated with the SDPUS-C1 and EPUS schemes, for the 1D Euler equations.

4.1.2. Test case 2 (2D Euler equations): This case considers the 2D Euler equations (12), defined on $[0,1] \times [0,1]$, with $\phi = [\rho, \rho u, \rho v, E]^T$, $F(\phi) = [\rho, \rho u^2 + p, \rho uv, (E+p)u]^T$ and $G(\phi) = [\rho v, \rho uv, \rho v^2 + p, (E+p)v]^T$. The variables have been previously defined. In order to close the system, the ideal gas equation of state $p = (\gamma - 1)(E - \frac{1}{2}\rho(u^2 + v^2))$ with $\gamma = 1.4$, was considered. In particular, we considered the shock-shock interaction problem, which consists of the interaction of two oblique shocks (states ⓐ and ⓒ) with two normal shocks (states ⓑ and ⓓ). The initial conditions are

$$[\rho_0,\ u_0,\ v_0,\ p_0]^T = \begin{cases} [1.5,\ 0,\ 0,\ 1.5]^T & \text{state ⓐ}, \\ [0.13799,\ 1.2060454,\ 1.2060454,\ 0.0290323]^T & \text{state ⓑ}, \\ [0.5322581,\ 1.2060454,\ 0,\ 0.3]^T & \text{state ⓒ}, \\ [0.5322581,\ 0,\ 1.2060454,\ 0.3]^T & \text{state ⓓ}. \end{cases} \tag{14}$$

Zero-order extrapolation on the boundary was used.

The reference solution was calculated using the Godunov method with a correction term containing the MC flux-limiter [13]. Both the numerical and reference solutions were calculated using the CLAWPACK code on a mesh size of 200×200 computational cells at CFL=0.8. Figure 6 depicts the solutions for contours of the density profile at time $t = 0.8$. It is also report in figure 7 the distribution of the density along of the diagonal ($y = x$). It is seen that the SDPUS-C1 and EPUS schemes solved well the complex structure in the solution, with the EPUS scheme providing the best results.

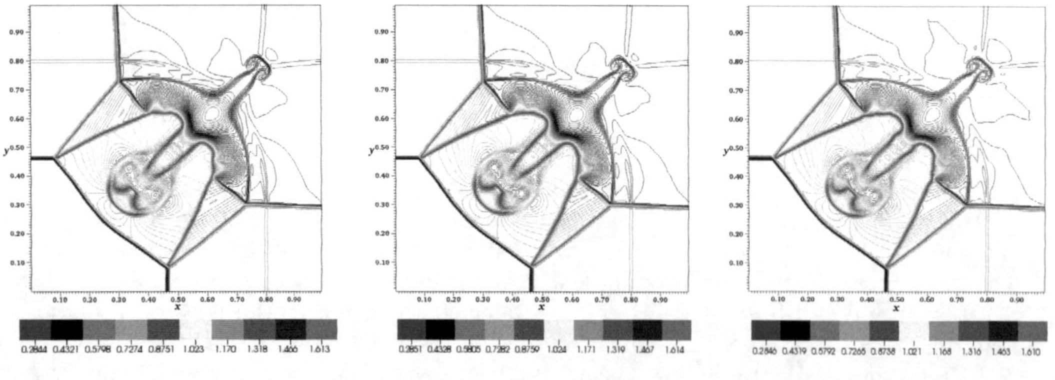

Figure 6. Reference solution and numerical results for Euler equations in terms of contours of the density ρ: reference solution (left), SDPUS-C1 (center) and EPUS (right) schemes.

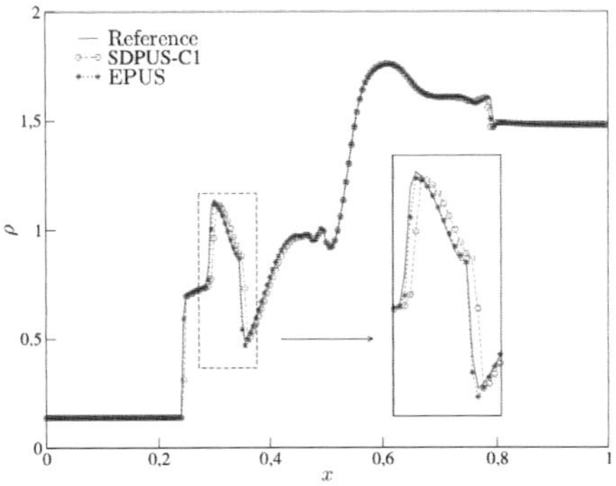

Figure 7. Reference solution and numerical results for Euler equations in terms of ρ on the line $y = x$.

4.2. Navier-Stokes equations

The governing PDEs considered for the simulation of incompressible laminar flows involving moving free surfaces are the Navier-Stokes and mass conservation equations. In the conservative form, these equations are given by

$$\frac{\partial u}{\partial t}\frac{1}{r^\alpha}\frac{\partial(r^\alpha uu)}{\partial r}+\frac{\partial(uv)}{\partial z} = \frac{\partial p}{\partial r}+\frac{1}{Re}\frac{\partial}{\partial z}\left(\frac{\partial u}{\partial z}-\frac{\partial v}{\partial r}\right)+\frac{g_r}{Fr^2}, \tag{15}$$

$$\frac{\partial v}{\partial t}\frac{1}{r^\alpha}\frac{\partial(r^\alpha vu)}{\partial r}+\frac{\partial(vv)}{\partial z} = \frac{\partial p}{\partial z}+\frac{1}{Re}\frac{1}{r^\alpha}\frac{\partial}{\partial r}\left(r^\alpha\left(\frac{\partial u}{\partial z}-\frac{\partial v}{\partial r}\right)\right)+\frac{g_z}{Fr^2}, \tag{16}$$

$$\frac{1}{r^\alpha}\frac{\partial(r^\alpha u)}{\partial r}+\frac{\partial v}{\partial z} = 0, \tag{17}$$

where $u = u(r, z, t)$ e $v = v(r, z, t)$ are, respectively, the components of velocity vector in r-direction and z-direction; $g = (g_r, g_z)^T$ is gravitational acceleration vector and p is pressure. The dimensionless parameters $Re = U_0L_0/\nu$ and $Fr = U_0/\sqrt{L_0/g}$ are, respectively, the Reynolds and Froud numbers, where ν is coefficient of molecular kinematic viscosity. U_0 and L_0 are velocity and length scales, respectively. The parameter α in equations (15)–(16) defines the coordinates system. When $\alpha = 0$, 2D Cartesian coordinates are considered with r interpreted as x and z interpreted as y, and when $\alpha = 1$ cylindrical coordinates is assumed.

4.2.1. Test case 3 (Broken dam problem): Results are presented now for the collapse of a column of water onto a horizontal wall for 2D case. Equations (15)-(17) are solved with $\alpha = 0$. By using the SDPUS-C1 scheme implemented into the 2D version of the Freeflow code of Castelo et al. [3] (a finite difference methodology on a structured grid), we performed a simulation of this unsteady moving free surface flow. The geometry used is a fluid column ($a = b = 0.057m$) in hydrostatic equilibrium and confined between walls. Initially, a wall is instantaneously removed and the fluid is subject to gravity and free to flow out along a rigid horizontal wall. The simulation was performed by using a mesh size of 1000 × 200 computational cells, and with the following input data: domain of $0.5m \times 0.1m$; $g = 9.81m/s^2$; $L_0 = a = 0.057m$; $U_0 = \sqrt{g \cdot L_0} = 0.74778m/s$; $\nu = 10^{-6}m^2/s$; and $Fr = 1$. Figure 8 shows the numerical results and the experimental data for the position of the fluid front $x_{\max}$ versus non-dimensional time. In order to provide a stiffer test for the performance of SDPUS-C1, we compared the calculated surge front position (as a function of non-dimensional time) against other sophisticated techniques presented by Colagrossi and Landrini [4], namely the smoothed particle hydrodynamics (SPH), boundary element method (BEM), level set, and an approach by Ritter. As shown in figure 8, the calculations with SDPUS-C1 agree fairly well with the data, giving us confidence in the numerical solution.

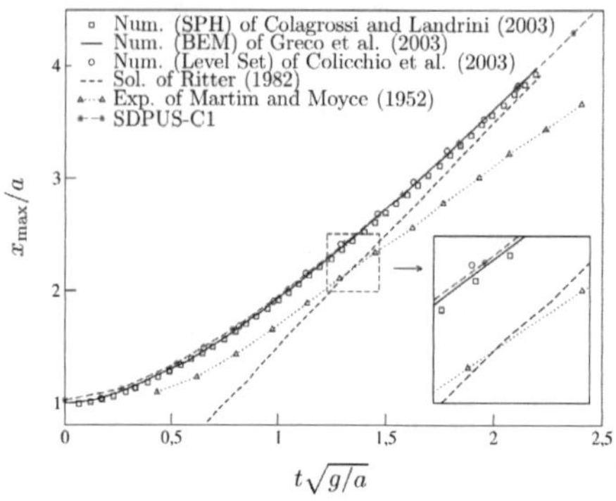

Figure 8. Computation with SDPUS-C1 scheme and experimental data for the fluid front position $x_{\max}$ versus non dimensional time.

4.2.2. Test case 4 (Circular hydraulic jump problem): The equations (15)-(17) are solved with $\alpha = 1$. When an vertical free jet of liquid impinges perpendicularly onto a flat (horizontal) plate it can, for certain moderate values of the Reynolds number, create a circular hydraulic jump. The phenomenon occurs at a critical radius, where there is a sudden transition from shallow rapidly flowing fluid to deep, much slower flowing fluid. A better understanding of this phenomenon and the instabilities when it is turbulent is of commercial interest, since jet impingement is often used in cooling systems and the flow of the fluid beyond the jump can degrade the efficiency of the system.

The axisymmetric version of the Freeflow code [3] equipped with the EPUS scheme was run on this problem using three meshes, namely 200 $\times$ 126 ($\delta_x = \delta_y = 0.00025$ m); 400 $\times$ 252 ($\delta_x = \delta_y = 0.000125$ m) and 800 $\times$ 504 ($\delta_x = \delta_y = 0.000625$ m) computational cells (known hereafter as Mesh I, Mesh II and Mesh III, respectively). The radius of the inlet $a = 0.008$ m and the velocity of the fluid at this boundary $U_0 = 3.75 \times 10^{-1}$ ms^{-1} have been used as the scaling parameters. The jet flow rate $Q = \pi U_0 a^2 = \nu Rea = 0.75 \times 10^{-5}$ m^{3}s^{-1}, producing a Reynolds number of 250, and a constant inflow-to-plate distance of $H = 0.03$ m were employed in the simulations. We begin by verifying that the EPUS scheme provides good estimates for the position of the jump (the results with the SDPUS-C1 scheme have been similar - not shown). For this, the scaling relations for the radius of the jump $r_{jump} = \left((27g^{-1/4})/(2^{-1/4}35\pi)\right)^{2/3} Q^{2/3} H^{-1/6} \nu^{-1/3}$ of Brechet and Néda [2], and $r_{jump} = Q^{5/8}\nu^{-3/8}g^{-1/8}$ of Bohr et al. [1] were used for comparison. Table 1 shows the jump radius obtained from the simulation results and the theoretical scaling laws. One can see that the calculated estimates for the jump with the EPUS scheme on the three meshes, particularly the one on the fine mesh (Mesh III), are in reasonable agreement with the theoretical scaling law of Brechet and Néda [2].

Table 1. Jump radii obtained of the scaling relations and simulation results.

Scaling relations		**EPUS estimate**		
Brechet and Néda [2]	Bohr et al. [1]	Mesh I	Mesh II	Mesh III
1.3 e-2	5.9 e-2	1.8 e-2	1.6 e-2	1.4 e-2

A comparison was then performed between the total thickness of the fluid layer h, obtained from the numerical results and the viscous solution of Watson [22]; this is displayed in figure

9. The numerical solutions were calculated by using Meshes I, II and III at a time step of 1.3×10^{-4}s. We restricted the analysis to the region $0.2 < (r/a)Re^{-\frac{1}{3}} < 0.8$ because Watson's analysis is only valid under the restriction $r >> a$ and the presence of the outflow-boundary (see [22]). One can clearly see from figure 9 that the numerical solutions do not show oscillatory behavior, and as the mesh size is decreased, the solution converges, indicating the convergence of the method for this complex nonlinear free surface flow problem. Watson's approximate solution is only valid over a restricted range of r and the results presented in figure 9 are extremely good over that range.

Finally, in order to check the effect of the time step on the numerical solution we compute, on the Mesh I (200×126), the fluid layer h using different time steps (from 10^{-4}s to 10^{-6}s). In figure 10, the numerical results of three simulations using the time steps 1.3×10^{-4}s, 6.5×10^{-5}s and 2.7×10^{-6}s are presented. It can be seen that no significant effect was detected in the numerical solutions by reducing or increasing the time step.

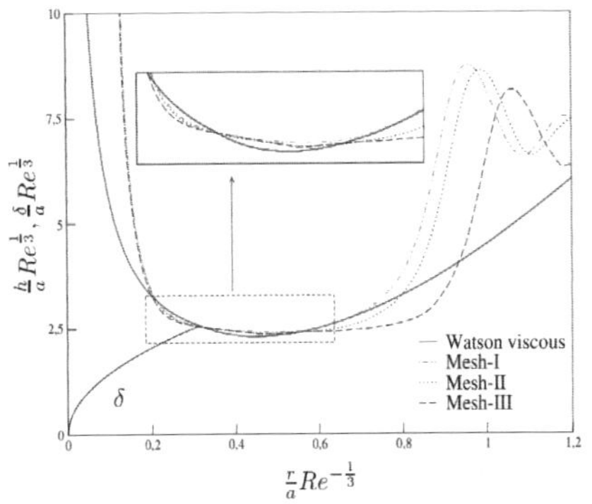

Figure 9. Numerical results (EPUS) and analytical solutions (viscous) of Watson, at $t = 2.5s$.

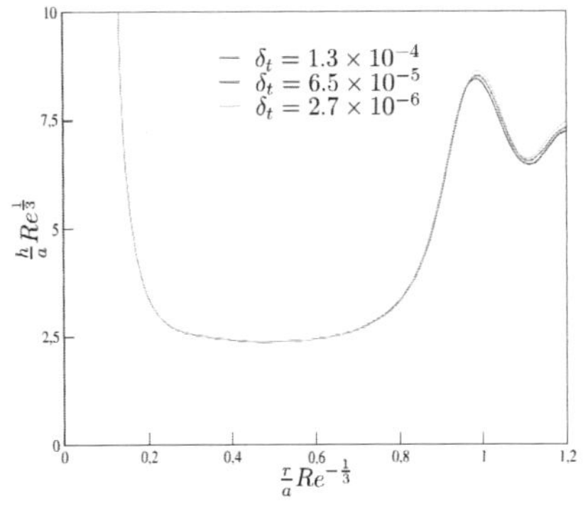

Figure 10. Calculated fluid layer h on the Mesh I (200×126) using three different time steps.

5. Conclusions and future work

We have presented two new TVD-based upwinding schemes (SDPUS-C1 and EPUS) for the numerical solution of nonlinear hyperbolic conservation laws and related incompressible moving free surface flows. The results of these computations clearly show that these upwinding schemes can be used to simulate hyperbolic systems and complex fluid flow problems without difficulty.

The results also clearly demonstrate that the main advantage of these schemes are their abilities to simultaneously handle calculations of slowly varying flows as well as rapidly varying flows containing shocks or discontinuities such as those commonly occurring in hydraulic jumps or breaking dams. In particular, the spurious overshoot in the vicinity of the shock may be damped by adding diffusion, i.e., simply decreasing the Courant parameter CFL.

It is important to be aware that there does exist another very successful class of high resolution shock-capturing schemes, namely the essentially non-oscillatory schemes (ENO) [10] (and its related weighted ENO). However, ENO schemes are generally thought of as applying to hyperbolic systems. The main point of this paper was to demonstrate that the both SDPUS-C1 and EPUS schemes can be employed to solve a wide range of difficult problems.

Encouraged by the results in this paper, the authors are planning for the future to apply SDPUS-C1 and EPUS schemes to the solution of 3D hyperbolic conservation laws and even more challenging 3D turbulent and viscoelastic flows involving moving free surfaces.

Acknowledgements
We thank the financial support given by FAPESP (Grants 2008/07367-9 and 2009/16954-8) and CNPq (Grants 300479/2008-5). This work was also carried out in the framework of the Instituto Nacional de Ciência e Tecnologia em Medicina Assistida por Computação Científica (CNPq, Brazil).

References

[1] Bohr T, Dimon P and Putkaradze V 1993 Shallow-water approach to the circular hydraulic jumps *Journal Fluid Mechanics* **254** 635-648
[2] Brechet Y and Z Néda 1999 On the circular hydraulic jump *Am. Journal Physics* **67** **(8)** 723-731
[3] Castelo A F, Mangiavacchi, Tomé M F, Cuminato J A, Ferreira V G and Mckee S 2001 Surface tension implementation for GENSMAC 2D *Journal of the Brazilian Society of Mechanical Sciences* **XXIII** 532-532
[4] Colagrossi A and Landrini 2003 Numerical simulation of inter facial flows by smoothed particle hydrodynamics *Journal Computational Physics* **191** 448-475
[5] Corrêa L 2010 Solving fluid dynamics problems using a new polynomial upwind convection scheme *Dincon'10 9th Brazilian Conference on Dynamics Control and their Applications*
[6] Courant R, Isaacson E and Rees M 1952 On the solution of nonlinear hyperbolic differential equations by finite differences *Community Pure Applied Mathematics* **5** 243-255
[7] Ferreira V G, Kurokawa F A, Queiroz R A B, Kaibara M K, Oishi C M, Cuminato J A, Castelo A, Tom M F and Mckee S 2009 Assessment of a high-order finite difference upwind scheme for the simulation of convection-diffusion problems *International Journal for Numerical Methods in Fluids* **60** 1-26
[8] Gaskell P H and Lau A K 1988 Curvature-compensated convective transport: SMART, a new boundedness-preserving transport algorithm *International Journal for Numerical Methods in Fluids* **8** 617-641
[9] Harten A 1983 High resolution schemes for hyperbolic conservation laws *Journal of Computational Physics* **49** 357-393
[10] Jiang G S and Shu C W 1996 Efficient implementation of weighted ENO scheme *Journal of Computational Physics* **126** 202-228
[11] Leonard B P 1979 The QUICK algorithm: a uniformly third-order finite difference method for highly convective flows *computational Methods Applicate Mechanic Engineering* **19** 59-98
[12] Leonard B P 1988 Simple high-accuracy resolution program for convective modeling of discontinuities *International journal for Numerical Methods in Fluids* **8** 1291-1318
[13] LeVeque R J 2004 Finite volumes methods for hyperbolic problems *Press Syndicate of the University of Cambridge*
[14] Lima G A B 2010 Desenvolvimento de estratégia de captura de descontinuidades para leis de conservação e problemas relacionados em dinâmica dos fluidos *Universidade de São Paulo - Instituto de Ciências Matemáticas e de Computação* master thesis
[15] Lin H and Chieng C C 1991 Characteristic-based flux limiters of an essentially third-order flus-splitting method for hyperbolic conservation laws *International Journal for Numerical Methods in Fluids* **13** 287-307
[16] Rai A , Dandapat B S and Poria S 2008 Circular hydraulic jump in generalized-Newtonian fluids *arXiv:0809.2231v3 [physics.flu-dyn]*
[17] Sweby O K 1984 High resolution schemes using flux-limiters for hyperbolic conservation laws *SIAM Journal on Numerical Analysis* **21** 995-1011
[18] van Albada G D, van Leer B and Roberts W W A comparative study of computational methods in cosmic gas dynamics *Astronomy and Astrophysics* **108** 76-84
[19] van Leer B 1977 Towards the ultimate conservative difference scheme IV a new approach to numerical convection *Journal of Computation Physics* **23** 263-275
[20] Varonos A and Bergeles G 1988 Development and assessment of a variable-order non-oscillatory scheme for convection term discretization *International Journal for Numerical Methods in Fluids***26** 1-16
[21] Waterson N P and Deconinck H 2007 Design principles for bounded higher-order convection schemes - a unified approach *Journal of Computational Physics* **224** 182-207
[22] Watson E J 1964 The radial spread of a liquid jet over a horizontal plane *Journal of Fluid Mechanics* **20** 481-499
[23] Woodward P and Collela P 2002 The numerical simulation of two-dimensional fluid flow with strong shocks *Journal Computational Physics* **149** 439-456

Properties of an Arithmetic Code for Geodesic Flows

Daniel P. B. Chaves[1], **Reginaldo Palazzo Jr.**[1] **and José R. Rios Leite**[2]

[1] Department of Telematics, School of Electrical and Computer Engineering, State University of Campinas, 13081-970 Campinas-SP, Brazil

[2] Department of Physics, Federal University of Pernambuco, 50670-901 Recife-PE, Brazil

E-mail: dpb.chaves@gmail.com, palazzo@dt.fee.unicamp.br, rios@df.ufpe.br

Abstract. Topological analysis of chaotic dynamical systems emerged in the nineties as a powerful tool in the study of strange attractors in low-dimensional dynamical systems. It is based on identifying the stretching and squeezing mechanisms responsible for creating a strange attractor and organize all the unstable periodic orbits in this attractor. This method is concerned with the manifold generated by the chaotic system. Furthermore, as a mathematical object, the manifolds have a well studied geometric and algebraic structure, particularly for the case of compact surfaces. Intending to use this structure in the analysis and application of chaotic systems through their topological characteristics, we determine properties of geodesic codes for compact surfaces necessary for the construction of encoders from the symbolic sequences of experimental data generated by the unstable periodic orbits of the strange attractor (related to the behavior changes of the system with the variation of control parameters) to the geodesic code sequences, which permits to use the surface structure to study the system orbits.

1. Introduction

In the works of Gilmore, Lefranc, Plumecoq and others [1–6], topological methods have been applied to analyze the chaos behavior of dissipative dynamical systems. The topological approach is appropriate to analyze low-dimension dynamical systems, that is, those with Lyapunov dimension less than three or, in other words, whose flow rapidly relaxes to a three-dimensional subspace of the phase space. It intends to complement other methods of analysis based on values of metrics and dynamical invariants. However, it goes beyond by describing how to model the dynamics, allowing the determination and validation of models for dynamical mechanisms that generate a specific chaotic data, and being robust under changes in control-parameter values. The topological analysis relies on representing the mechanisms that generate the strange attractor and organize its unstable periodic orbits (UPO's) through a branched manifold, that is, a simpler two-dimensional structure that preserves the topological properties of the strange attractor. The branched manifold is determined by extracting topological-invariant properties from the chaotic time series which need not be exceptionally clean nor long. Thus we may consider, up to some extent, the topological approach tolerates noise and inaccurate sampling for a fairly practical amount of data to set up the conclusions about the system.

The topological characterization of a strange attractor is divided into a gross and a fine structure. Both are determined by computing the linking numbers and relative rotation rates of the UPO's [1, 2], which demonstrate the central matter of the UPO's for topological analysis. The gross structure is provided by a set of integers derived from the underlying branched manifold, and it is robust under control-parameter variation. The fine structure relies on *unfolding* the

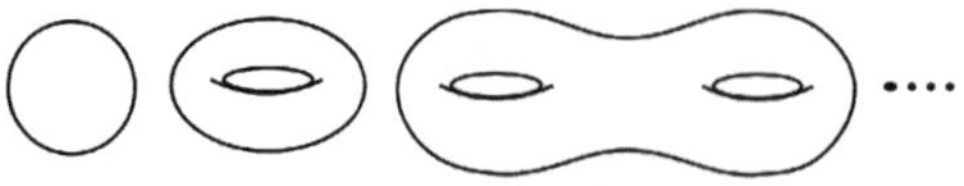

Figure 1. Canonical compact orientable surfaces.

branched manifold by changing the control parameters in such a way that some subsets of orbits are removed and the remaining orbits can be determined by a basis set of orbits that forces the presence of all remaining ones. These elements may become clearer if compared with a typical communication process. In this case, the gross structure can be compared with the subset of the Euclidean space that contains the signal constellation in a typical modulation scheme; and the fine structure specifies the constraints that have to be satisfied and how information is conveyed when using the branched manifold in a communication process.

The study of physical systems by applying geometric and topological concepts have already been proved to be successful in studying standard Hamiltonian systems [7, 8]. The trajectories of these systems can be seen as geodesics of the phase-space manifold equipped with the standard Jacobi metric, and therefore questions about the stability/instability properties of the dynamics can be tackled by studying the corresponding properties of the geodesics. Furthermore, this approach has provided an *alternative point of view* about Hamiltonian chaos with respect to previous explanations. In this work we follow a similar approach by proposing the application of the well known algebraic and geometric structure of the canonical surfaces to study streams of data obtained from experiments or simulations of low-dimensional dissipative dynamical systems. Any compact orientable surface can be deformed in a canonical surface by a homotopy deformation, that is, preserving its topological properties. The canonical surfaces are informally distinguished by the number of *holes* or *handles* as shown in Figure 1. This number is called the genus of the surface and can be rigorously specified [9, 10]. Thus, we can specify an embedding from the orbit flow in a branched manifold to one of the canonical surfaces, so that the topological properties of the flow in the branched manifold are matched to the topological properties of the geodesic flow in the canonical surface. Since we are working with symbol sequences, a natural way to achieve this task is to code the symbol sequences of the orbits, obtained from experiments or simulations, to symbol sequences coding the geodesics of the canonical surfaces. In this work we consider the geodesic code proposed by Adler and Flatto [11], since it has a well defined direct-graph representation of the code sequences and defines an easy-to-calculate hyperbolic measure determined from the code symbols. As additional information to specify an encoder, we should determine the entropy of the code sequences (that form a shift-space) and the intrinsic properties of the code sequences that can be helpful to the construction of an encoder. In Section 4 we determine this additional information, by specifying the maximum full-shift (set of all possible bi-infinite symbol sequences over a finite alphabet) embedded in a geodesic code and its topological entropy, as a function of the genus of the canonical surface.

Additionally, in Section 2 we give an overview of how topological analysis can be applied to study the chaotic behavior of low-dimensional dissipative dynamical systems. In Section 3 basic concepts in hyperbolic geometry, geodesic flow, and the considered geodesic code are presented. Finally, in Section 5 we draw the conclusions.

2. Topological elements

In spite of the simplicity of some nonlinear dynamical systems, exact analytical solutions are not known for the differential equations describing their behaviors. On the other hand, approximate solutions do not grasp the general behavior of the system. A breakthrough to this problem was given by the French mathematician Henry Poincaré, based on the study of how a set of nearby initial conditions (an entire neighborhood in phase space) evolves, which gave rise to the now called topological theory. His intent was to study conservative dynamical systems, the Hamiltonian ones. However, the Birman-Williams Theorem generalizes the results considered

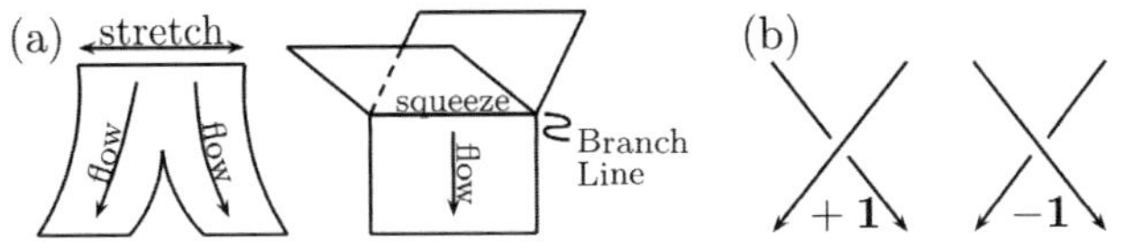

Figure 2. Stretching and squeezing processes (a), right handed "**+1**" and left handed "**−1**" crossings (b).

from the topological approach for low-dimensional dissipative dynamical systems whenever the Lyapunov dimension is less than three [1, 2].

2.1. Chaos mechanisms

A chaotic behavior is in essence recurrent, non-periodic and sensitive to initial conditions. By sensitive to initial conditions we mean the distance between any two nearby points in the phase space tends to increase exponentially, at least for a sufficiently small time interval, in other words, the points are stretched apart. Considering a motion in a bounded phase space, these points should eventually reach a maximum separation and then begin to approach each other again, in other words, the initial conditions are squeezed together. These behaviors are represented in Figure 2 (a), where we emphasize the relation between the stretching and squeezing mechanisms and the flow direction. The branched line determines the one-dimensional region of the branched manifold where the branch flows get so squeezed together that become indistinguishable from each other, however still different. This is the essential feature to justify irreversibility.

Furthermore, in a dissipative dynamical system all volumes in the phase space shrink asymptotically to zero overtime. In this case, the stretching and squeezing mechanisms imply that almost all initial conditions will asymptotically gravitate to a strange attractor, and that the flow takes place in a two-dimensional manifold almost everywhere but at the *tear point* of the stretching process and at squeezed regions meeting at the branch line. These are the two possible singularities that keep the region determined by the flow from being a two-dimensional manifold. The stretching and squeezing mechanisms form the fundamental blocks that built up the resulting two-dimensional structure (the branched manifold), by connecting every outflow to some inflow and vice-versa. The Birman-Williams Theorem makes rigorous the existence of such a branched manifold whenever the flow rapidly relaxes to a three-dimensional subspace of the phase space, in other words, the associated attractor has Lyapunov dimension less than three. From a finite length of discretely sampled scalar time-series data, the topological approach permits to determine the stretching and squeezing mechanisms that built up the attractor, and so to specify a dynamical system model that reproduces an experimental data set to an acceptable level. This is obtained by determining the organization of the UPO's within the strange attractor through the topological invariants: Linking numbers and relative rotation rates. As an advantage of the topological approach, such invariants are unaltered under smooth coordinate transformations, control-parameters changes, and do not depend on the orbits' stability provided that they exist. Thus, the topological approach is robust in identifying the chaos mechanisms, that is, the branched manifold. All necessary information to proceed with a topological analysis is represented in a template, a third topological invariant, that also exhibits the previously stated robustness properties. Details about its construction and properties can be found in [1, 2, 6].

2.2. Linking numbers and templates

The linking number amounts to the signed numbers of crossings between orbits, from an orbit with itself, or between branches of the branched manifold or template. The signs +1 and −1 are given for each cross as specified in Figure 2 (b). The linking numbers can be extracted from the template, that are a standard form of representing the branched manifold for purpose of computing the topological invariants. The template is a projection of the two-dimensional

$$\mathbf{T} = \begin{bmatrix} 0 & 0 & 0 & 0 \\ 0 & 2 & 2 & -1 \\ 0 & 2 & 1 & -1 \\ 0 & -1 & -1 & -1 \end{bmatrix}$$

$$\mathbf{A} = \begin{bmatrix} 2 & 0 & 1 & 3 \end{bmatrix}$$

$$\mathbf{I} = \begin{bmatrix} 1 & 1 & 0 & 0 \\ 0 & 0 & 1 & 1 \\ 0 & 0 & 1 & 1 \\ 1 & 1 & 0 & 0 \end{bmatrix}$$

(rows and columns labeled a, b, c, d)

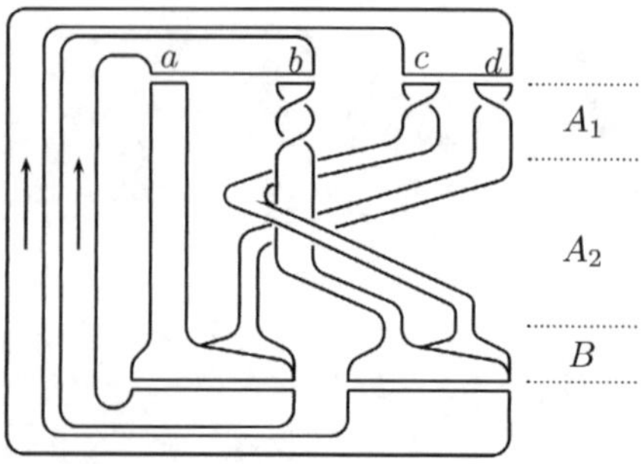

Figure 3. Standard template and its algebraic description. The one represented here corresponds to four branches.

branched manifold, which is embedded in $\mathbb{R}^3$, into a two-dimensional subspace, in such a way that the crossing information is preserved. Since the Birman-Williams Theorem guarantees that the strange attractor projects to a two-dimensional branched manifold, maintaining its topological organization invariant under the projection, thus the template is a standard form of summarizing the stretching and squeezing mechanisms responsible for generating the strange attractor. These mechanisms are represented in the segments A_1, B and A_2 in the template, as shown in Figure 3. The segment A_1 summarizes the signed number of twists in each branch of the flow. The segment A_2 describes the branch crossings, such that, each branch crossing is assigned an integer in the same way as shown in Figure 2 (b). In the region labeled B the branches are squeezed together, and the order in which they are squeeze is represented in the template. The algebraic counterpart of this description has three components, for a branched manifold with n branches: (**Topological Matrix - T**) An $n \times n$ symmetric matrix that describes the topological organization of the branches, where the diagonal element T_{ii} describes the branch twist of branch i, and the off-diagonal elements $T_{ij} = T_{ji}$ describe how branches cross each other; (**Joining Array - A**) A $1 \times n$ array that describes the order in which branches are joined at branch lines, such that, the closer a branch is to the observer, the lower the branch number; (**Transition Matrix - I**) An $n \times n$ matrix that describes which branches flow into which branches, following the convention that the element I_{ij} is one if branch i flows into branch j, and zero otherwise. Thus, matrix **I** gives a description of the symbolic dynamics of allowed periodic orbits of the associated Markov shift. The integers that appear in the algebraic description are extracted from the experimental data, and form the sufficient information to identify the stretching and squeezing mechanisms that generate the chaotic behavior.

Every usual chaotic system studied is non-hyperbolic. As a consequence of this the experimental symbol sequences that encodes the UPO's of its attractor will generate forbidden finite symbol sequences that vary with the control parameters. However, since the template is determined for a considered range of control parameters its structure remains invariant under the variation of them. Thus, once a template satisfying the attractor topology is identified, it remains invariant in the considered range of control parameters while a fixed setting of the control parameters restricts the flow to a subset of the template (sub-template). A physical interpretation of this phenomenon is that orbits are created, annihilated or even both as the control parameters vary, and their topological organization does not change with these variations. Thus, the template corresponds to the gross structure and the flow is pushed around on the template as the control parameters vary. Accordingly, the template establishes a communication framework that is determined only once and remains invariant as long as the orbits remain embedded in the strange attractor as the control parameters vary. Matrix **I** determines the restrictions that should be satisfied by every orbit-code sequence, and while the control parameters vary the set of forbidden sequences changes. This can be used to specify a relationship between the basis sets of orbits and the control parameters, and so encode information to be transmitted.

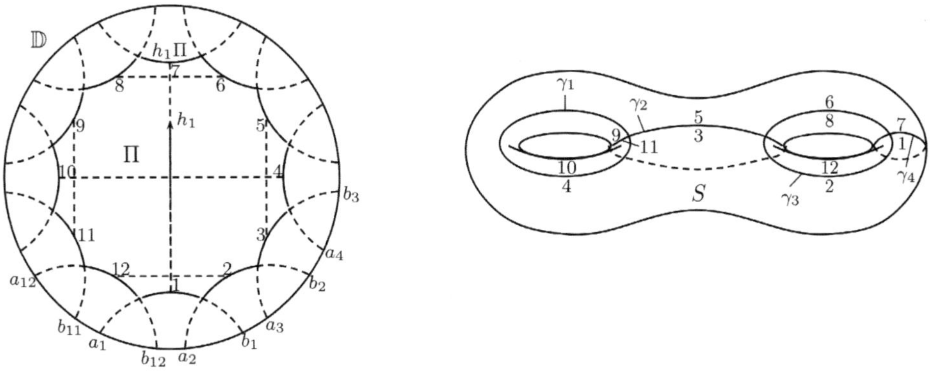

Figure 4. Fundamental region Π and hyperbolic surface S obtained by gluing the edges of Π.

3. Geodesic code

A canonical geometric surface originates from a regular tessellation of one of the three possible geometric planes: Euclidean ($\mathbb{R}^2$), spherical ($\mathbb{S}^2$), and hyperbolic ($\mathbb{H}^2$). Regarding the objective of this paper, a regular tessellation of $\mathbb{H}^2$ is a covering of the entire plane by non-overlapping regular $\mathbb{H}^2$-polygons meeting only along complete edges, or at vertices. We also consider the existence of a group of isometries acting on polygon Π. Furthermore, since the hyperbolic isometries are conformal mappings, they preserve area and angle. Thus, a tessellation can be seen as a mosaic, of equal polygons, of $\mathbb{H}^2$.

The set of all isometries that map Π to any other polygon of the tessellation forms a group Γ. The group Γ is called a Fuchsian group. Thus, the tessellation is generated by the action of Γ on Π called fundamental region of Γ. Γ is infinity, but it has a finite number of generators given by a subset $\Gamma_0 \subset \Gamma$ of isometries that maps Π to its adjacent images in the tessellation which share an edge with Π. In practice, Γ_0 is specified by a set of isometries that map each edge of Π to a non-adjacent edge. Thus, if $h \in \Gamma_0$ maps the edge e to the edge f of Π, we say that e and f are "glued" by h. Note that h^{-1} maps f to e by the inverse mapping. As a consequence, the action of the maps in Γ_0, are called pairing maps or transformations. By gluing together the edges of Π by the elements of Γ_0 we obtain a geometric surface that inherit the local metric structure of the covering plane, that is, the differential metric of the surface is equal to the differential metric of the hyperbolic plane. There exist more than one model for the hyperbolic plane, in this work we consider the disk model or the Poincaré disk model, represented by $\mathbb{D}^2$. Let Π^o be the interior of a polygon Π, then Π is a fundamental region if it satisfies : (i) $h_1\Pi^o \cap h_2\Pi^o = \emptyset$ for $h_1 \neq h_2$, $h_i \in \Gamma$; (ii) $\bigcup_{h\in\Gamma} h\Pi = \mathbb{D}^2$; (iii) $\lambda(\Pi - \Pi^o) = 0$, where λ denotes the hyperbolic area. These concepts are illustrated in Figure 4: The hyperbolic polygon Π is the closed region formed by the solid lines; the straight dashed lines indicate the edges of Π that are glued by side-pairing maps in Γ_0; h_1 is the side-pairing map that glues edge 1 to the edge 7; the dashed lines extending from the edges of Π to the boundary circle of $\mathbb{D}^2$ are geodesics whose segments form the edges of Π (a hyperbolic geodesic in $\mathbb{D}^2$ is a circular arc orthogonal to the boundary circle of $\mathbb{D}^2$); the closed curves γ_i in S are obtained by gluing the edges of Π together, the numbers above and below the curve indicate the edges that form it.

The process of determining the geometric surfaces is similar to the three possible planar geometries. In addition to the metric, they can also differ in their topological properties. It is well known that the sphere is the only orientable spherical surface, and that all compact orientable Euclidean surfaces are tori, hence a surface of genus one. All compact orientable surfaces of genus greater than one are hyperbolic surfaces. Furthermore, there exists a hyperbolic surface for any genus greater than one. This property justifies the choice of hyperbolic surfaces to embed the flow dynamics associated with a branched manifold, since they provide the necessary topological structure to embed the great amount of possible branched-manifold topologies.

3.1. Coding geodesics

The geodesic code in consideration is the one proposed by Adler and Flatto in [11] which is a special case of the more general coding process proposed by Series in [12]. It is based on the relationship identified by Bowen and Series between the geodesic flow on a compact surface of constant negative curvature and the interval map defined by the Markov partition of the boundary $\partial\mathbb{D}^2$ of the Poincaré disk, providing the relation between their symbolic dynamics [13]. The previous works explore the properties and existing interrelationships between these systems as for instance the use of ergodicity of the interval maps to prove ergodicity of geodesic flows. Conversely, explicit formulas for invariant measures of interval maps from invariant measures for flows were also developed in these works.

The coding process is similar to the binary expansion of the points in the unit interval $[0,1)$ by the piecewise linear expansion map $f(x) = (2x)$, where (α) denotes the fractional part of α. For this simple case, the subintervals $[0, \frac{1}{2})$ and $[\frac{1}{2}, 1)$ form a Markov partition of $[0,1)$. The n-th digit of the binary expansion of $x \in [0,1)$ is equal to 0 if $f^n(x) \in [0, \frac{1}{2})$, and 1 otherwise. The idea of the geodesic coding is similar, but the function f is substituted by the side-pairing transformations of a hyperbolic polygon Π with $(8g-4)$ edges that is the fundamental region of a Fuchsian group Γ, and the Markov partition is formed by the arcs in $\partial\mathbb{D}^2$ with endpoints determined by the geodesics whose segments form the boundary of Π. This is shown in Figure 4 where the geodesics are formed by the extension of the edges of Π as illustrated by the dashed arcs, and the endpoints specified by the geodesic obtained from the extension of the i-th edge of Π are a_i, b_i. By considering hyperbolic polygons with $8g-4$ edges as fundamental regions the proposed analytical method is not restricted, since it is well known that for any surface of genus $g \geq 2$ there exists a fundamental polygon Π whose edges are the $8g-4$ consecutive geodesics segments $s_1, s_2, \ldots s_{8g-4}$ with counterclockwise orientation. The side-pairing maps Γ_0 are given by the $\sigma(i)$ permutation of order 2 of the set $\{1, \ldots, 8g-4\}$, defined by

$$\sigma(i) = \begin{cases} 4g - i \mod (8g-4), & i \ \text{odd}, \\ 2 - i \mod (8g-4), & i \ \text{even}, \end{cases} \tag{1}$$

such that, there exists a unique element $h_i \in \Gamma_0$ that maps s_i to $s_{\sigma(i)}$. This is shown in Figure 4 for $g = 2$. Thus, the Markov partition is given by the intervals $i_1 = [a_i, b_{i-1})$ and $i_2 = [b_{i-1}, a_{i+1})$, and the expansion map is defined as $f(\xi) = h_i(\xi)$ whenever $\xi \in [a_i, a_{i+1})$, for $1 \leq i \leq 8g-4$. All operations are modulo $8g-4$, then $i+1 = 1$ if $i = 8g-4$ and $i-1 = 8g-4$ if $i = 1$.

The coding process is similar to the binary expansion of points in the unit interval. An example is shown in Figure 5, where part of Π is shown, and adjacent to Π the image of Π by the application of an isometry h_i on Π, which is indicated with the label $h_i^{-1}\Pi$. The regions Π and $h_i^{-1}\Pi$ have the same hyperbolic metric properties, however under a Euclidean perspective an exponential reduction of lengths and areas are observed. This fact is used to coding the extreme points ξ (forward) and η (backward) of a geodesic γ in $\mathbb{D}^2$. The coding process is recursive, let the region Π be the reference after the n-th application of the expansion map f, $n \in \mathbb{Z}$. Since the partition interval where the forward point ξ of the geodesic γ defined by ξ, η lies is i_2, it follows that if $w = \ldots w_{-1}w_0w_1w_2\ldots$ is the bi-infinite codeword of γ, then $w_n = i_2$. For the next step consider Figure 5, the region of the tessellation of $\mathbb{D}^2$ defining the partition of the interval i_2 in $\mathbb{D}^2$ is $h_i^{-1}\Pi$, due to the fact that this is the hyperbolic polygon adjacent to Π at the edge i determining the partition intervals i_1 and i_2. Now, either ξ belongs to the interval j_1 or j_2 or $(j+1)_1$ of the partition defined by $h_i^{-1}\Pi$ in the previous defined region i_2, thus w_{n+1} is equal to j_1, j_2 or $(j+1)_1$ depending on the interval that contains ξ. The same idea follows for the

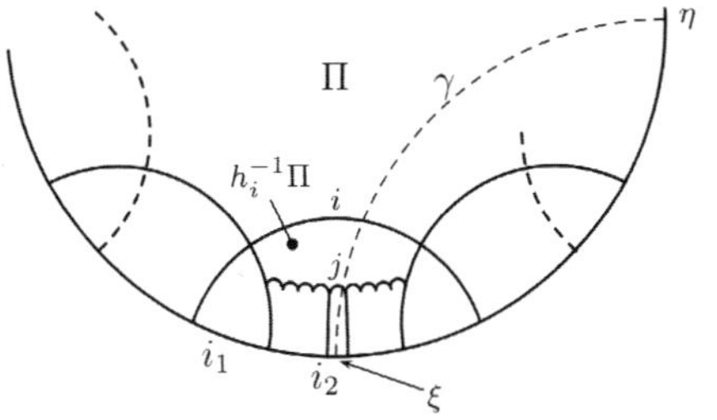

Figure 5. Diagram of coding method. See explanation in the text.

expansion of both extremes of the geodesics ξ and η, and the concatenation of these expansions determines the bi-infinite code sequence of the geodesic.

The space of sequences coding the geodesics forms a shift space X [11]. In fact, X is a Markov shift with admissible transitions given by (2), where σ is the permutation specified in (1).

$$\mathcal{T}: \begin{cases} i_1 \to (\sigma(i)+1)_2,\ (\sigma(i)+2)_1. \\ i_2 \to (\sigma(i)+2)_2,\ (\sigma(i)+3)_1, \ldots, (\sigma(i)-2)_1,\ (\sigma(i)-2)_2. \end{cases} \tag{2}$$

From the admissible transitions given in (2) we may construct a directed-graph presentation of the geodesic flow, where bi-infinite sequences obtained by following bi-infinite paths in the graph generate the bi-infinite codewords of the geodesics. Thus, we may encode the symbolic sequences generated by the orbits of a branched manifold obtained from a dynamical system to geodesics over a hyperbolic surface of any genus $g \geq 2$ [14]. The symbolic dynamics associated with the orbits code is determined by the transition matrix $\mathbf{I}$ obtained from the algebraic description of the template.

4. Maximum full-shift and topological entropy

In this section we present the main results of the paper. Theorem 2 is concerned with a structural property of the geodesic code, and it establishes the maximum full-shift embedded in the geodesic-code sequences as a function of the genus of the surface associated with the hyperbolic polygon Π with $(8g-4)$-edges applied in the geodesic-coding process. Theorem 6 specifies the topological entropy of the geodesic-code sequences as a function of the genus of the surface, this sets the capacity of the set of sequences for conveying information [14].

These results are related with concepts of symbolic dynamics, so we begin with some definitions and concepts about it. Thus, the set of all possible finite sequences over an alphabet $\mathcal{A}$ is denoted by $\mathcal{A}^*$, this include the empty sequence ε, such that, $w\varepsilon = \varepsilon w = w$ for every $w \in \mathcal{A}^*$. The set $\mathcal{A}^*$ is called the free monoid of $\mathcal{A}$, with concatenation of words as the considered operation. Let $\mathfrak{F} \subseteq \mathcal{A}^*$ and $\mathcal{A}^{\mathbb{Z}}$ be the full-shift formed by the set of all bi-infinite sequences over $\mathcal{A}$. A shift-space $X_{\mathfrak{F}}$ is the set of all elements in $\mathcal{A}^{\mathbb{Z}}$ having no factors in the forbidden set $\mathfrak{F}$, equivalently, $X_{\mathfrak{F}} = \{x \in \mathcal{A}^{\mathbb{Z}} |\ x_{[i,j]} \neq w \text{ for every } w \in \mathfrak{F} \text{ and } i,j \in \mathbb{Z},\ i \leq j\}$ where $x_{[i,j]} = x_i \ldots x_j$ for $x = \ldots x_{-1}x_0x_1 \ldots \in \mathcal{A}^{\mathbb{Z}}$. The language of a shift-space $X_{\mathfrak{F}}$ is the set $\mathcal{B}(X_{\mathfrak{F}}) = \{w \in \mathcal{A}^* |\ w = x_{[i,j]} \text{ for some } x \in X_{\mathfrak{F}}\}$ and includes ε.

4.1. Maximum embedded full-shift

Lemma 1. *Let $\mathcal{L} = \{1, \ldots, k\}$, where $k \geq 4$ and k even, and $\sigma : \mathcal{L} \to \mathcal{L}$ be an arbitrary permutation. Suppose a partition $\{\mathcal{C}, \mathcal{P}\}$ of $\mathcal{L}$, such that, $\mathcal{P} = \bigcup_{i \in \mathcal{C}} \mathcal{P}(i)$ where $\mathcal{P}(i) = \{(\sigma(i)-1), \sigma(i), (\sigma(i)+1)\}$. Then, the greatest value of $|\mathcal{C}|$ is reached for $|\mathcal{P}| = |\mathcal{C}| + 2$, and it only happens when the elements $\{\sigma(i)\}_{i \in \mathcal{C}}$ are consecutive.*

Proof: Initially, $i, j \in \mathcal{L}$ are consecutive if $|i - j \mod k| = 1$, observing that $(k-1) \equiv -1 \mod k$ and thus $|(k-1) \mod k| = 1$. We have to show there exist no $\sigma(\cdot)$ satisfying $|\mathcal{P}| = |\mathcal{C}|$

and $|\mathcal{P}| = |\mathcal{C}| + 1$, and that $|\mathcal{P}| = |\mathcal{C}| + 2$ is only satisfied when $\{\sigma(i)\}_{i\in\mathcal{C}}$ is a consecutive set of elements.

Suppose $|\mathcal{P}| = |\mathcal{C}|$. Thus, there exists $i, j_1, l \in \mathcal{C}$ such that $\sigma(j_1) = (\sigma(i) - 1)$ and $\sigma(l) = (\sigma(i) + 1)$, otherwise there might exist $t_1, t_2 \in \mathcal{C}$ such that $\sigma(t_1) = \sigma(t_2)$ and $t_1 \neq t_2$, a contradiction. Since $j_1 \in \mathcal{C}$ and $\sigma(j_1) = (\sigma(i) - 1)$ it implies that there exists $j_2 \in \mathcal{C}$ such that $\sigma(j_2) = (\sigma(i) - 2)$, then there exists $j_3 \in \mathcal{C}$ such that $\sigma(j_3) = (\sigma(i) - 3)$, etc. Hence, $\mathcal{L} = \mathcal{P}$.

Suppose $|\mathcal{P}| = |\mathcal{C}| + 1$, thus either $\sigma^{-1}(\sigma(i) - 1) \notin \mathcal{C}$ or $\sigma^{-1}(\sigma(i) + 1) \notin \mathcal{C}$ for some $i \in \mathcal{C}$. We consider just one case, the remaining ones follow *pari passu.* Let $\sigma(j_1) = (\sigma(i) + 1)$ for some $j_1 \in \mathcal{C}$, then there exists $j_2 \in \mathcal{C}$ such that $\sigma(j_2) = (\sigma(i) + 2)$, and so on. Again, $\mathcal{L} = \mathcal{P}$.

If $|\mathcal{P}| = |\mathcal{C}| + 2$ and $\{\sigma(i)\}_{i\in\mathcal{C}}$ is not a set of consecutive elements, then $\mathcal{C}$ has at least three elements j_1, j_2, j_3 such that $(\sigma(j_l) + 1) \notin \{\sigma(i)\}_{i\in\mathcal{C}}$ or $(\sigma(j_l) - 1) \notin \{\sigma(i)\}_{i\in\mathcal{C}}$, $1 \leq l \leq 3$. Thus, $|\mathcal{P}| = |\{\mathcal{P}(i)\}_{i\in\mathcal{C}}| \geq |\mathcal{C}| + 3$. Hence, $|\mathcal{P}| = |\mathcal{C}| + 2$ occurs only if the elements in $\{\mathcal{P}(i)\}_{i\in\mathcal{C}}$ are consecutive, it is easy to see that there exists only two elements in $\mathcal{P}$ not in $\{\mathcal{P}(i)\}_{i\in\mathcal{C}}$. In this case we obtain $|\mathcal{C}| = k/2 - 1$ and $|\mathcal{P}| = k/2 + 1$. □

Theorem 2. *The maximum* full-shift *in the geodesic-code shift space* X *has alphabet cardinality* $|\mathcal{A}|$ *equal to* $4(g-1)$.

Proof: For the specific permutation $\sigma(\cdot)$ given in (1), we show that does not exist a partition $\{\mathcal{C}, \mathcal{P}\}$ of $\mathcal{L} = \{1, \ldots, 8g-4\}$ such that $|\mathcal{P}| = |\mathcal{C}| + 2$, and also we have to specify a partition that satisfies $|\mathcal{P}| = |\mathcal{C}| + 4$.

From Lemma 1, there exists a partition such that $|\mathcal{P}| = |\mathcal{C}| + 2$ if and only if the elements in $\{\sigma(i)\}_{i\in\mathcal{C}}$ are consecutive. Then, the elements in $\mathcal{C}$ are also consecutive, and $|\mathcal{L}| = |\mathcal{C}| + |\mathcal{P}| = 2|\mathcal{C}| + 2 = 8g - 4$, and so $|\mathcal{C}| = 4g - 3$. Furthermore, $|\mathcal{L}|/4 = (8g-4)/4 = 2g - 1$ and so $|\mathcal{C}| - |\mathcal{L}|/4 = 2g - 2$, which for $g \geq 2$ is at least two. Thus, the set $\mathcal{C}$ has at least one of the cardinal elements $\{1, 2g, 4g-1, 6g-2\}$ and its nearest neighbor. From (1) the neighbors of a cardinal element is the image of each other. Thus, for $|\mathcal{P} = |\mathcal{C}| + 2$ there exists $i \in \mathcal{L}$ such that $i, \sigma(i) \in \mathcal{C}$, a contradiction.

Since does not exist a partition with $|\mathcal{C}| = (4g - 3)$, we proceed with the analysis for $|\mathcal{C}| = (4g - 4)$. Consider the partition $\{C_1, C_2\}$ of $\mathcal{C}$ where $C_1 = \{2, 3, \ldots, 2g-1\}$ and $C_2 = \{4g, 4g+1, \ldots, 6g-3\}$. Thus, $|\mathcal{C}| = |C_1| + |C_2| = 4(g-1)$. From (1), it follows that for

$$i \in C_1 : \begin{cases} 2g+1 \leq \sigma(i) \leq 4g-3, & i \text{ odd;} \\ 6g \leq \sigma(i) \leq 8g-4, & i \text{ even.} \end{cases} \qquad i \in C_2 : \begin{cases} 6g-1 \leq \sigma(i) \leq 8g-5, & i \text{ odd;} \\ 2g+2 \leq \sigma(i) \leq 4g-2, & i \text{ even.} \end{cases}$$

Thus, $\mathcal{P} = \{2g, 2g+1, \ldots, 4g-1\} \cup \{6g-2, 6g-1, \ldots, 1\}$, where $\sigma^{-1}(i) \notin \mathcal{C}$ for $i \in \{1, 2g, 4g-1, 6g-2\}$. It is easy to see that $\mathcal{C} \cap \mathcal{P} = \emptyset$ and $\mathcal{L} = \mathcal{C} \cup \mathcal{P}$. □

4.2. Topological entropy

A graph consists of a pair of finite sets $G = (\mathcal{V}, \mathcal{E})$ where $\mathcal{V}$ is the set of states and $\mathcal{E}$ is the set of edges, such that, to each edge $e \in \mathcal{E}$ is associated with an initial state $\mathsf{i}(e)$ and a terminal state $\mathsf{t}(e)$. A path π in G is a set of consecutive edges, that is, if $\pi = e_1 \ldots e_n$ then $\mathbf{t}(e_i) = \mathbf{i}(e_{i+1})$ for $1 \leq i < n$. A labeled graph is a pair $\mathcal{G} = (G, \mathcal{L})$ such that G is an underlying graph and $\mathcal{L} : \mathcal{E} \to \mathcal{A}$ is the label function where $\mathcal{A}$ is a finite set called alphabet. A labeled graph $\mathcal{G}$ is right-resolving if for any $e_1, e_2 \in \mathcal{E}$ such that $\mathsf{i}(e_1) = \mathsf{i}(e_1)$ then $\mathcal{L}(e_1) \neq \mathcal{L}(e_2)$. Observe that G can be considered a labeled graph by taking $\mathcal{L}$ as the identity function. Let A_{ij} denote the number of edges in G with initial state i and terminal state j, for every $i, j \in \mathcal{V}$. The adjacency matrix of G is $A = [A_{ij}]$, and its construction from G is denoted by A_G. For a labeled graph $\mathcal{G} = (G, \mathcal{L})$, we denote its adjacency matrix by $A_{\mathcal{G}} = A_G$. A sofic-shift $X_{\mathcal{G}}$ is a shift-space obtained by reading off the labels of consecutive edges in a labeled graph $\mathcal{G}$. In this case $\mathcal{G}$ is a presentation of $X_{\mathcal{G}}$. Our interest is on sofic-shifts known as topological Markov shifts (TMS), characterized by a forbidden set whose length of every element is two. It is straightforward to determine a presentation $\mathcal{G} = (G, \mathcal{L})$ for a TMS $X_{\mathcal{F}}$. The underlying graph $G = (\mathcal{V}, \mathcal{E})$ is given by $\mathcal{V} = \mathcal{A}$ and $\mathcal{E} = \{ij \in \mathcal{A} \times \mathcal{A} | \ ij \notin \mathcal{F}\}$, such that $\mathsf{i}(ij) = i$, $\mathsf{t}(ij) = j$ and $\mathcal{L}(ij) = j$ for every $ij \in \mathcal{E}$ [14].

Thus, a presentation $\mathcal{G}_X = (G, \mathcal{L}_X)$ of the geodesic code has the underlying graph G as given by (3), where $\mathcal{P}$ is the set of admissible sequences of length two specified by (2), and the label function is given by $\mathcal{L}_X(i_k i_l) = i_l$ for every $i_k i_l \in \mathcal{E}$.

$$G : \begin{cases} \mathcal{V} = \{i_k | \ 1 \le i \le 8g-4 \text{ and } k \in \{1,2\}\}; \\ \mathcal{E} = \{i_k i_l | \ i_k i_l \in \mathcal{P} \text{ for } 1 \le i,j \le 8g-4 \text{ and } k,l \in \{1,2\}\}. \end{cases} \tag{3}$$

We apply the symmetry of $\mathcal{G}_X$ to determine two other graphs with the same topological entropy of X. First, $\mathcal{G}_Y = (G, \mathcal{L}_Y)$ has the same underlying graph as $\mathcal{G}_X$ and label function given by (4).

$$\begin{cases} \mathcal{L}_Y(i_2(\sigma(i)+2)_2) = b_1, \ \mathcal{L}_Y(i_2(\sigma(i)+3)_2) = b_2, \dots, \\ \mathcal{L}_Y(i_2(\sigma(i)-2)_2) = b_{(k-3)}; \\ \mathcal{L}_Y(i_2(\sigma(i)+3)_1) = c_1, \dots, \mathcal{L}_Y(i_2(\sigma(i)-2)_1) = c_{(k-4)}; \\ \mathcal{L}_Y(i_1(\sigma(i)+1)_2) = d_1, \ \mathcal{L}_Y(i_1(\sigma(i)+2)_1) = a_1. \end{cases} \tag{4}$$

Graph $\mathcal{G}_Z = (G_Z, \mathcal{L}_Z)$ has underlying graph specified in (5) where $k = 8g-4$, and $\mathcal{L}_Z$ is the identity.

$$G_Z : \begin{cases} \mathcal{V}_Z = \{1,2\}; \\ \mathcal{E}_Z = \{a_1\} \cup \{d_1\} \cup \{c_1, \dots, c_{(k-4)}\} \cup \{b_1, \dots b_{(k-3)}\}, \end{cases} \tag{5}$$

such that

$$\begin{cases} \mathsf{t}(a_1) = \mathsf{i}(a_1) = 1, \text{ and } \mathsf{t}(b_l) = \mathsf{i}(b_l) = 2 \text{ for } 1 \le l \le (k-3); \\ \mathsf{i}(d_1) = 1, \ \mathsf{t}(d_1) = 2, \text{ and } \mathsf{i}(c_t) = 2, \mathsf{t}(c_t) = 1 \text{ for } 1 \le t \le (k-4). \end{cases} \tag{6}$$

Lemma 3. *The labeled graphs $\mathcal{G}_Y$ and $\mathcal{G}_Z$ are presentations of the same sofic-shift, that is, $\mathcal{B}(Y) = \mathcal{B}(Z)$.*

Proof: From the definitions of $\mathcal{G}_Y$ and $\mathcal{G}_X$, for any $i \in \{1, \dots, 8g-4\}$ and $n, m \in \{1,2\}$, there exists a path π in $\mathcal{G}_Y$ such that $\mathcal{L}_Y(\pi) = w$, $\mathsf{i}(\pi) = i_n$ and $\mathsf{t}(\pi) = i_m$ if and only if there exists a path ϖ is $\mathcal{G}_Z$ such that $\mathcal{L}_Z(\varpi) = w$, $\mathsf{i}(\varpi) = n$ and $\mathsf{t}(\varpi) = m$, whenever $|\pi| = |\varpi| = 1$.

By hypothesis, suppose the last statement holds for $|\pi| = |\varpi| = n$. Next we show it also holds for $|\pi| = |\varpi| = n+1$.

Let πe be a path in $\mathcal{G}_Y$ such that $\mathcal{L}_Y(\pi e) = \mathcal{L}_Y(\pi)\mathcal{L}_Y(e) = wa$ for $e \in \mathcal{E}_Y$, $\mathsf{i}(\pi) = i_n$ and $\mathsf{t}(\pi) = i_m$. By induction, there exists a path $\varpi \in \mathcal{G}_Z$ such that $|\varpi| = |\pi|$, $\mathcal{L}_Z(\varpi) = w$, $\mathsf{i}(\varpi) = n$ and $\mathsf{t}(\varpi) = m$. Furthermore, for some path f in $\mathcal{G}_Z$ such that $|f| = 1$, it follows that $\mathcal{L}_Z(f) = \mathcal{L}_Y(e) = a$, $\mathsf{i}(f) = m$, and $\mathsf{t}(f) = p$ if and only if $\mathsf{t}(e) = i_p$. Thus, $\mathcal{L}_Z(\varpi f) = wa$ and so $\mathcal{B}(Y) \subseteq \mathcal{B}(Z)$. Similarly, it can be shown that $\mathcal{B}(Z) \subseteq \mathcal{B}(Y)$. □

The proof of Theorem 6 makes use of the results from Propositions 4 and 5, and the proofs may be found in [14, Chapter 4].

Proposition 4. *Let $\mathcal{G} = (G, \mathcal{L})$ be a right-resolving labeled graph. Then $h(X_{\mathcal{G}}) = h(X_G)$.*

Proposition 5. *Let G be a graph having adjacency matrix A. Then $h(X_G) = \log \lambda_A$, where λ_A is the largest eigenvalue of A.*

Theorem 6. *The topological entropy of the geodesic code sofic-shift X is given as a function of the coding region Π by*

$$h(X) = \log\left[(4g-3) + \sqrt{(4g-3)^2 - 1}\right]. \tag{7}$$

Proof: Since the labeled graphs $\mathcal{G}_X$ and $\mathcal{G}_Y$ have the same underlying graph and are both right-resolving, it follows from Proposition 4 that $h(X) = h(Y)$. From Lemma 3, it follows that $h(Y) = h(Z)$. Thus, from Proposition 5 we may apply the adjacency matrix $A_{\mathcal{G}_Z}$ of $\mathcal{G}_Z$ to determine the topological entropy of X. Since $A_{\mathcal{G}_Z}$ has order two independent of the genus of the coding region Π and its entries are only dependent on the genus, we obtain an algebraic expression for $h(X)$ by determining $h(Z)$. From (5) and (6), $A_{\mathcal{G}_Z}$ is given by

$$A_{\mathcal{G}_Y} = \begin{bmatrix} 1 & 1 \\ 8g-8 & 8g-7 \end{bmatrix}. \tag{8}$$

The largest eigenvalue $\lambda_{A_{\mathcal{G}_Y}}$ of $A_{\mathcal{G}_Y}$ is the largest root of the characteristic polynomial $\lambda^2 + \lambda\,(6-8g) + 1$ of $A_{\mathcal{G}_Y}$. Thus,

$$\lambda_{A_{\mathcal{G}_Y}} = \frac{(8g-6) + \sqrt{(6-8g)^2 - 4}}{2} = (4g-3) + \sqrt{(4g-3)^2 - 1}.$$

Hence, $h(X) = h(Y) = \log \lambda_{A_{\mathcal{G}_Y}}$. □

5. Discussion

There exists a well defined geodesic coding process that share similarities with the UPO's coding by defining a Poincaré section on the edges of Π [11], and there exits an easy-coding procedure from the obtained symbol sequences and those obtained by the method described in Section 3. These geodesic codes are a two-fold analytical tool: allows an association of the UPO's sequences with the generators of the Fuchsian group of Γ_0 from the first coding method, and a hyperbolic measure from the second. These elements are considered in usual encoding and decoding procedures, and they may also be employed in chaotic communication systems.

6. Acknowledgments

This work has been supported by FAPESP under grants DRII-06/60976-8 and 2007/56052-8, and by CNPq under grant 306617/2007-2.

References

[1] Gilmore R and Lefranc M 2002 *The Topology of Chaos: Alice in Stretch and Squeezeland* (New York, United States: John Wiley & Sons)
[2] Gilmore R 1998 *Rev. Mod. Phys.* **70** 1455–529
[3] Plumecoq J and Lefranc M 2000 *Physica D* **144** 231–58
[4] Plumecoq J and Lefranc M 2000 *Physica D* **144** 259–78
[5] Lefranc M, Glorieux P, Papoff F, Molesti F and Arimondo E 1994 *Phys. Rev. Lett.* **73** 1364–67
[6] Lefranc M 2007 The topology of deterministic chaos: Stretching, squeezing and linking *Physics and Theoretical Computer Science: From Numbers and Languages to (Quantum) Cryptography* ed Gazeau J, Nesetril J and Rovan B (Amsterdam, The Netherlands: IOS Press) pp 17–29
[7] Casetti L, Pettini M and Cohen E G D 2000 *Phys. Rep.* **337** 237–341
[8] Pettini M and Valdettaro R 1995 *Amer. Inst. Phys.* **5** 646–52
[9] Stillwell J 1992 *Geometry of Surfaces* (New York, United States: Springer-Verlay)
[10] Firby P A and Gardiner C F 1992 *Surface Topology* 2nd ed (New York, United States: Ellis Horwood)
[11] Adler R L and Flatto L 1991 *Amer. Math. Soc.* **25** 229–334
[12] Series C 1981 *Acta Math.* **146** 103–28
[13] Bowen R and Series C 1979 *Inst. Hautes Études Sci. Publ. Math.* **50** 153–70
[14] Lind D and Marcus B H 1995 *An Introduction to Symbolic Dynamics and Coding* (Cambridge, United Kingdom: Cambridge University Press)

Dynamic Days South America 2010 IOP Publishing
Journal of Physics: Conference Series **285** (2011) 012036 doi:10.1088/1742-6596/285/1/012036

Neural networks for emulation variational method for data assimilation in nonlinear dynamics

Helaine Cristina Morais Furtado

Haroldo Fraga de Campos Velho

Elbert E. N. Macau

Laboratório Associado de Computação e Matemática Aplicada, São José dos Campos, Brasil

E-mail: helaine.furtado@lac.inpe.br, haroldo@lac.inpe.br, elbert@lac.inpe.br

Abstract. Description of a physical phenomenon through differential equations has errors involved, since the mathematical model is always an approximation of reality. For an operational prediction system, one strategy to improve the prediction is to add some information from the real dynamics into mathematical model. This aditional information consists of observations on the phenomenon. However, the observational data insertion should be done carefully, for avoiding a worse performance of the prediction. Technical data assimilation are tools to combine data from physical-mathematics model with observational data to obtain a better forecast. The goal of this work is to present the performance of the Neural Network Multilayer Perceptrons trained to emulate a Variational method in context of data assimilation. Techniques for data assimilation are applied for the Lorenz systems; which presents a strong nonlinearity and chaotic nature.

1. Introduction

Data assimilation is nowadays prevailing process in the Numerical Weather Prediction (NWP)[1, 2], ocean circulation forecasting [3], air monitoring [4] and space weather models [5, 6]. Technical data assimilation can be defined as a set of methods that can be used to combine data from physical-mathematics model with observational data to obtain a better forecast.

From filtering point of view, data assimilation process can be represented by:

$$w^a = w^f + Kp[y - H(w^f)] \tag{1}$$

where w^a is the value of the analysis; w^f is the forecasting (from the mathematical model, also known as background field); K is the weight matrix; y denotes the observation; H represents the observation system; the difference $\{y - H(w^f)\}$ is the innovation; and $p[.]$ is a discrepancy function. Another approach for solving the data assimilation is by computing the minimum solution for the cost function:

$$J(w) = \langle (w - w^f), C(w - w^f)\rangle + \langle [y - H(w^f)], S[y - H(w^f)]\rangle \tag{2}$$

Published under licence by IOP Publishing Ltd

where C and S denote the inverse covariance matrices of the background (in general such matrix is represented by B or Q for the data assimilation or for the control theory communities, respectively) and of the measurement errors (denoted by R), and $\langle u, v\rangle$ expresses the internal product between vectors u and v. For the Kalman filter (KF), the matrix K (Kalman gain) at equation (1) is computed from a formula involving matrices B and R. Variational scheme is linked to the optimization problem with the objective function given by expression (2).

In general, it is not easy the process of estimating the background matrix B (also called the modeling error covariance matrix). There are many techniques to estimate this matrix: using some type of parameterization [7], employing ensemble KF [8], or a Fokker-Planck equation [9, 10]. For applying the KF, in general some linearization is necessary, and the random variables are assumed to have Gaussian distribution.

The process of data assimilation can be seen as a problem of determining the initial condition (i.c.) and defined as: "*The science of having an appropriate combination of data from a mathematical model with observation data to determine the data analysis (or i.c.)*" [11]. Therefore, a better agreement with the reality for the i.c. implys in a better forecasting. Figure 1 illustrates the technique of data assimilation. Note that whilst observations are inserted into the system, the dynamics of the estimation (red curve) is near to the truth (blue curve), since the data assimilation process is interrupted, there is a decoupling of the dynamics. Therefore, observations, when combined effectively with data from mathematical model, serve to bring the estimation of the initial condition to the desired value. In this work, synthetic observations are determined by integrating the Lorenz system adding some noise.

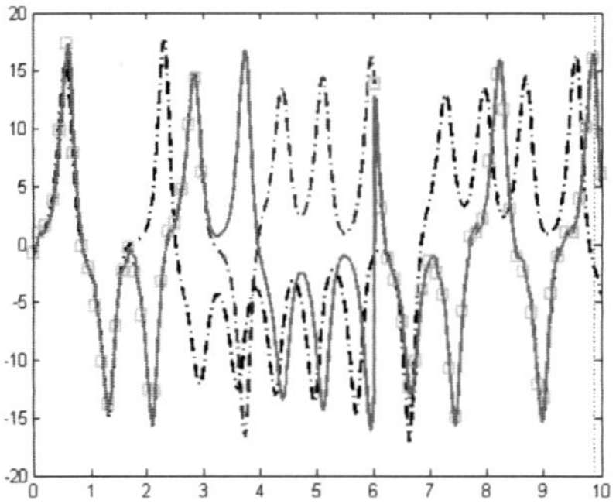

Figure 1. Time series of x component Lorenz's system. Blue curve: truth; green squares: observations; red curve: estimate and black curve: background. Adapted [12]

For considering a more general problem (nonlinear models and non-Gaussian distribution) the particle filter technique has been proposed [13]. However, this approach has a greater computational complexity than extended KF (EKF). The latter technique has a similar complexity to the 4-dimensional variational (4D-Var) scheme described in this work at Section 3.1.

Recently, we have introduced a new approach based on artificial neural network (ANN) for data assimilation [14, 15, 16, 17]. In this former articles, the ANN emulates the KF. For the present paper, the data assimilation by variational method is emulated by a supervised artificial neural networksb[12]. It is important to mention that after training ANN has a lower computational cost than extended and linear KF, variational method, and particle filter. The Lorenz system under chaotic regime will be used to illustrate the method.

The Lorenz system under chaotic regime is described in Section 2 will be employed for testing the assimilation schemes.

2. Testing model: Lorenz system

The dynamical Lorenz system is a simplified Saltzman's model [18]. The celebrated Lorenz model has been employed as a standard test for examining the performance of data assimilation methods. The Lorenz nonlinear system is expressed as:

$$\frac{dx}{d\tau} = -\sigma(x - y) \tag{3}$$

$$\frac{dy}{d\tau} = \rho x - y - xz \tag{4}$$

$$\frac{dz}{d\tau} = \rho xy - \beta z \tag{5}$$

where $\tau \equiv \pi^2 H^{-2}(1+a^2)\kappa t$ is the non-dimensional time (being H, a, κ, and t layer height, thermal conductivity, wave number (diameter of the Rayleigh-Bérnard cell), and time, respectively), $\sigma \equiv \kappa^{-1}\nu$ is the Prandtl number (ν is the kinematic vicosity); $\beta \equiv 4(1+a^2)^{-1}$. The parameter $\rho = R/R_c \propto \Delta T$ is the Rayleigh number (T is the temperature), and R_c is the critical Rayleigh number. The variables x, y and z are the coefficients of time-dependent quantities from a partial differential equation of the atmosphere under heat source (convective motion, and the temperature gradients in the horizontal and vertical directions, respectively). The numerical solution this model results on the attractor shown by Figure 2.

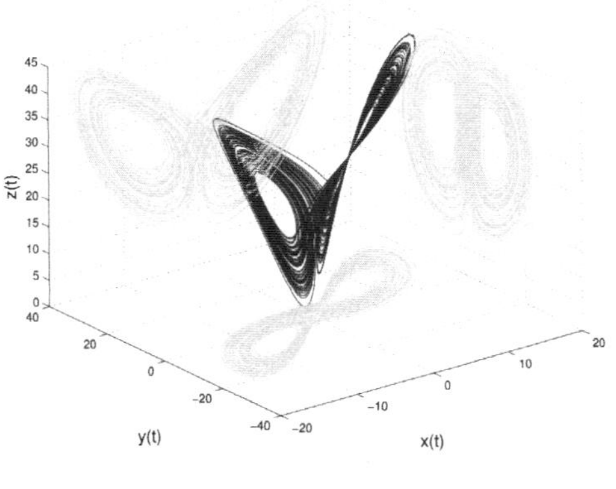

Figure 2. Lorenz attractor with initial conditions $(-1.5, 1.5, 20)$ and its projections on the axes xy, xz and yz.

3. Data assimilation schemes

3.1. Variational method

This scheme focuses to compute the solution of an optimization problem. The objective function includes a term measuring the distance to the background at the beginning of the interval, and a summation over time of the objective function for each observation increment computed with respect to the model integrated to the of the observation [2]:

$$J[\mathbf{x}(t_0)] = \underbrace{\frac{1}{2}[\mathbf{x}(t_0) - \mathbf{x}^b(t_0)]^T \mathbf{B}_0^{-1}[\mathbf{x}(t_0) - \mathbf{x}^b(t_0)]}_{J_b} + \underbrace{\frac{1}{2}\sum_{i=0}^{N}[H(\mathbf{x})_i - \mathbf{y}_i^o]^T \mathbf{R}_i^{-1}[H(\mathbf{x})_i - \mathbf{y}_i^o]}_{J_o} \tag{6}$$

The control variable $\mathbf{x}$ is the initial state of the model. The time integration of the dynamical system from the initial condition up to the time t_n could be represented by $\mathbf{x}(t_n) = M_0[\mathbf{x}(t_0)]$. In this work, the analysis solution should satisfy the model equations. The variational method seeks an initial condition such that the forecast has the best fit with the observation within the assimilation interval.

The first variation of the objective function is given by

$$\delta J = J[\mathbf{x}(t_0) + \delta\mathbf{x}(t_0)] - J[\mathbf{x}(t_0)] \approx \left[\frac{\partial J}{\partial \mathbf{x}(t_0)}\right]^T .\delta\mathbf{x}(t_0)] \tag{7}$$

where the gradient of the cost function $[\partial J/\partial \mathbf{x}(t_0]$ is a column vector. To obtain the minimum functional J by an iterative method, it is necessary to integrate the adjoint model $(M)_i^t$, which is the tangent linear model implemented ($\mathbf{M}_i$), described in section 3.2

The gradient of the first term of functional J_b of 6 with respect to $\mathbf{x}(t_0)$ is given by:

$$\frac{\partial J_b}{\partial \mathbf{x}(t_0)} = \mathbf{B}_0^{-1}[\mathbf{x}(t_0) - \mathbf{x}^b(t_0)] \; . \tag{8}$$

The gradient of the second term J_o of functional 6 is more complicated due $\mathbf{x}_i = M_i[\mathbf{x}(t_0)]$, where M represents the model prediction. If we introduce a disturbance to the initial state, then $\delta\mathbf{x}_i = \mathbf{L}(t_0, t_i)\delta\mathbf{x}_0$, so that:

$$\begin{aligned}\frac{\partial(H(\mathbf{x}_i) - \mathbf{y}_i^o)}{\partial \mathbf{x}(t_0)} &= \frac{\partial H}{\partial \mathbf{x}_i}\frac{\partial M}{\partial \mathbf{x}_0} \\ &= \mathbf{H}_i\mathbf{L}(t_0, t_i) \\ &= \mathbf{H}_i \prod_{j=i-1}^{0} \mathbf{L}(t_j, t_{j+1}) \; .\end{aligned} \tag{9}$$

The arrays $\mathbf{H}_i$ and $\mathbf{L}_i$ are linearized Jacobian $\frac{\partial H}{\partial \mathbf{x}}_i$, $\frac{\partial M}{\partial \mathbf{x}}_0$. $\mathbf{L}$ is the matrix that propagates an initial disturbance to final time of integration (tangent linear model) [19].

Therefore, the gradient of the term of observation is given by:

$$\left[\frac{\partial J_o}{\partial \mathbf{x}(t_0)}\right] = \sum_{i=0}^{N} \mathbf{L}(t_i, t_0)^T\mathbf{H}_i^T\mathbf{R}_i^{-1}[\mathbf{H}(\mathbf{x}_i) - \mathbf{y}_i^o] \; . \tag{10}$$

Equation (10) shows that each iteration of the minimization of this functional requires the calculation of the gradient, i.e., to calculate the increment $[H(\mathbf{x}_i) - \mathbf{y}_i^o]$ at the observation time t_i during the advanced integration, multiplying them by $\mathbf{H}_i^T\mathbf{R}_i^{-1}$ and integrating these weighted increments back to the initial time using the adjoint model (transposed tangent linear model). Since the adjoint integration is common in several time intervals, the sum 10 can be rearranged more conveniently (see [2]).

3.2. Tangent linear model and adjoint model

Consider a nonlinear model. Since this model must be discretized in space using finite differences, the model can be written as a set of n coupled differential equations:

$$\frac{d\mathbf{x}}{dt} = \mathbf{F}(\mathbf{x}), \quad \mathbf{x} = \begin{bmatrix} x_1 \\ \vdots \\ x_n \end{bmatrix} \quad \text{e} \quad \mathbf{F} = \begin{bmatrix} F_1 \\ \vdots \\ F_n \end{bmatrix} \tag{11}$$

This is the model in differential form. An atmospheric model consists of one system of difference equations which, for example, using the Crank-Nicholson method can be described of the form:

$$\mathbf{x}^{n+1} = \mathbf{x}^n + \triangle t\mathbf{F}\left(\frac{\mathbf{x}^n + \mathbf{x}^{n+1}}{2}\right) \tag{12}$$

A numerical solution of 11 starting from an initial time t_0 can be obtained by integrating the model numerically using 12 between initial time and a final time. This provides us a nonlinear model solution that depends only on the initial conditions:

$$\mathbf{x}(t) = M[\mathbf{x}(t_0)] \tag{13}$$

where M is the time integration of the numerical scheme from the initial condition up to time t. A small perturbation $\mathbf{y}(t)$ can be added to the basic model integration $\mathbf{x}(t)$:

$$\begin{aligned} M[\mathbf{x}(t_0) + \mathbf{y}(t_0)] &= M[\mathbf{x}(t_0)] + \frac{\partial M}{\partial \mathbf{x}}\mathbf{y}(t_0) + O[\mathbf{y}(t_0)^2] \\ &= \mathbf{x}(t) + \mathbf{y}(t) + O[\mathbf{y}(t_0)^2] \end{aligned} \tag{14}$$

At any given time, the linear perturbation of $\mathbf{y}(t)$ will be given by:

$$\frac{d\mathbf{y}}{dt} = \mathbf{J}\mathbf{y} \tag{15}$$

where $J = \partial\mathbf{F}/\partial\mathbf{x}$ is the Jacobian of $\mathbf{F}$.

This system of linear ordinary differential equations is the tangent linear model in differential form. Its solution between t_0 e t can be obtained by integration of (15) in time using the same time difference scheme used in the nonlinear model (11):

$$\mathbf{y}(t) = \mathbf{L}(t_0, t)\mathbf{y}(t_0) \tag{16}$$

Here $\mathbf{L}(t_0, t) = \partial M/\partial\mathbf{x}$ is an $m \times m$ matrix known as the matrix of the tangent linear model, it propagates an initial perturbation at time t_0 into the final perturbation at time t. Lorenz (1965) [19] introduced the concept of tangent linear model of an atmospheric model, obtained from 14, neglecting quadratic or higher order terms in the perturbation $\mathbf{y}$:

$$M[\mathbf{x}(t_0)] + \mathbf{L}(t_0, t)\mathbf{y}(t_0) = \mathbf{x}(t) + \mathbf{y}(t) \approx M[\mathbf{x}(t_0) + \mathbf{y}(t_0)]\ . \tag{17}$$

Adding a small perturbation of size ε along the vectors $\mathbf{y}_i(t_0) = \varepsilon\mathbf{e}_i$ and applying (17) to each of these perturbations and subtracting (13) obtain the matrix that defined the tangent linear model:

$$L(t_0, t)[\varepsilon e_1, \ldots, e_n] = \varepsilon\mathbf{L}(t_0, t) = [\mathbf{y}_1(t), \ldots, \mathbf{y}_n(t)]\ . \tag{18}$$

The Euclidean norm of a vector is the inner product of the vector with itself:

$$||\mathbf{y}||^2 = \mathbf{y}^T\mathbf{y} = \langle\mathbf{y}, \mathbf{y}\rangle\ . \tag{19}$$

The Euclidean norm of $\mathbf{y}(t)$ is related to the initical perturbation by:

$$||\mathbf{y}(t)||^2 = (\mathbf{L}\mathbf{y}(t_0))^T\mathbf{L}\mathbf{y}(t_0) = \langle\mathbf{L}\mathbf{y}(t_0), \mathbf{L}\mathbf{y}(t_0)\rangle = \langle\mathbf{L}^T\mathbf{L}\mathbf{y}(t_0), \mathbf{y}(t_0)\rangle \tag{20}$$

The *adjoint* of an operator $\mathbf{K}$ is defined by the property $\langle\mathbf{x}, \mathbf{K}\mathbf{y}\rangle \equiv \langle\mathbf{K}^T\mathbf{x}, \mathbf{y}\rangle$. In this case of a model with real variables, the *adjoint* tangent linear model $\mathbf{L}(t_0, t)$ is simply the *transpose* of the tangent linear model.

Now assume that we separate the interval (t_0, t) into two successive time intervals. For exemple, if $t_0 < t_1 < t$,

$$\mathbf{L}(t_0, t) = \mathbf{L}(t_1, t)\mathbf{L}(t_0, t_1)\ . \tag{21}$$

Since the adjoint of tangente linear model is the transpoe of it, the propperty of the transpose of a product is also valid:

$$\mathbf{L}^T(t_0, t) = \mathbf{L}^T(t_0, t_1)\mathbf{L}^T(t_1, t)]\ . \tag{22}$$

Equation (21) shows that the tangent linear model can be computed with a product of the tangent linear model matrices corresponding to short integrations. From equation (22), the adjoint model can be separated into single time steps, but they are executed backwards in time, starting from the last time step at t, and ending with the first time step at t_0. For more explanations (see [2]).

3.3. Artificial neural networks

Artificial Neural Networks (ANN) have became important tools for information processing [20]. Much research has been conducted in pursuing new neural network models and adapting the existing ones to solve real life problems, such as those in engineering [20]. ANN are made of arrangements of processing elements called neurons. The artificial neuron model basically consists of a linear combiner followed by an activation function, Figure 1 (left side), given by:

$$y_k = \varphi\left(\sum_{j=1}^{n} w_{kj}x_j + b_k\right) \tag{23}$$

where w_{kj} are the connection weights, b_k is a threshold parameter, x_j is the input vector and y_k is the output of the k_{th} neuron.

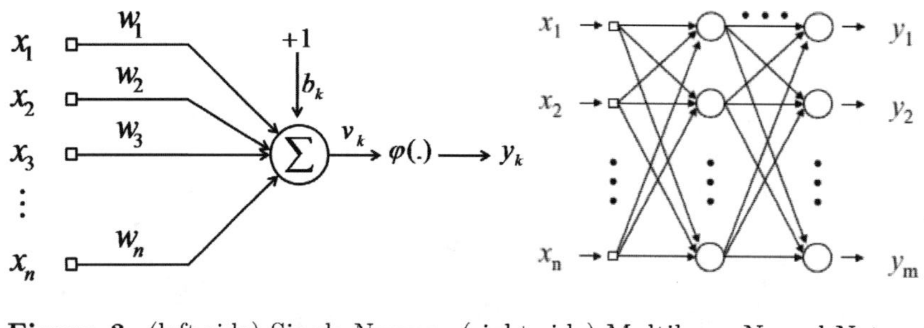

Figure 3. (left side) Single Neuron, (right side) Multilayer Neural Network

Arrangements of such units form the ANN that are characterized by:

- Very simple neuron-like processing elements;
- Weighted connections between the processing elements;
- Highly parallel processing and distributed control;
- Automatic learning of internal representations.

ANN aim to explore the massively parallel network of simple elements in order to yield a result in a very short time slice and, at the same time, with insensitivity to loss and failure of some of the elements of the network. These properties make artificial neural networks appropriate for application in pattern recognition, signal processing, image processing, financing, computer vision, engineering, etc.

There are different architectures of ANN that are dependent upon the learning strategy adopted. This paper briefly describes the Multilayer Perceptron (MLP) with error backpropagation learning. Detailed introductions on ANN can be found in [20] and [21]. MLP with backpropagation learning algorithm, are feedforward networks composed of an input layer, an output layer, and a number of hidden layers, whose aim is to extract high order statistics from the input data [20]. Figure 1(right side) depicts a multilayer neural network with a hidden layer. Functions $\varphi(.)$ provide the activation for the neuron. Neural networks will solve nonlinear problems, if nonlinear activation functions are used for the hidden and/or the output layers. From several activation functions, the sigmoid are commonly used:

$$\text{bipolar function } \varphi(v) = \frac{1 - \exp(-av)}{1 + \exp(-av)} \tag{24}$$

A feedforward network is a non-linear mapping to compute the output vector from an input vector. The connections among the several neurons (Fig. 1(b)) have associated weights that are adjusted during the learning process, thus changing the performance of the network. Two distinct phases can be devised while using ANN: the training phase (learning process) and the run phase (activation of the network). The training phase consists of adjusting the weights for the best performance of the network in establishing the mapping of many input/output vector pairs. Once trained, the weights are fixed and the network can be presented to new inputs for which it calculates the corresponding outputs, based on what it has learned.

4. Numerical results

The Lorenz systems was integrated using a second order Runge Kutta methods, with $\triangle t = 10^3$, with initial condition for the Lorenz system: $w_0 = [x_0 y_0 z_0]^T = [1 : 5088701 : 53127125 : 46091]^T$. For training data set, 2000 data are considered, and 333 data are used for cross validation. In order to test multilayer perceptron ANN architectures, for emulating a variational method in data assimilation, momentum constant we used for hidden layer α_h and $\alpha_n (n = x; y; z)$ for the output layer. A similar feature is employed for learning rates: η_h for hidden layer, and $\eta_n (n = x; y; z)$ for the output layer. The numerical values for these parameters are shown in table 1.

Table 1. Neural network parameters: one hidden layer

neurons	α_h	α_x	α_y	α_z	η_h	η_x	η_y	η_z
2	0.6	0.6	0.6	0.6	0.001	0.001	0.001	0.001

The application of this technique uses the variational functional 6, called objective function, containing the term observation J_o and the term modeling J_b. In this work we used only the term of observation J_o, ie the functional form:

$$J(\mathbf{x}(t_0)) = \underbrace{\frac{1}{2} \sum_{i=0}^{N} (\mathbf{y}_i^o - \mathbf{H}(\mathbf{x}_i))^T \mathbf{R}^{-1} (\mathbf{y}_i^o - \mathbf{H}(\mathbf{x}_i))}_{J_o} \tag{25}$$

where $\mathbf{R}$ is the covariance matrix of observation error, $\mathbf{y}$ the vector of observations, $\mathbf{H} = \mathbf{I}$ and $\mathbf{x}$ is the vector of state variable dynamical system. The method determines the best fit between the model data and observation data for the period of assimilation. Figure 4 shows the results of assimilation with this technique for the components x and z, respectively. The observations were inserted at every 12 time steps. Figures 5 and 6 show the results with the assimilation

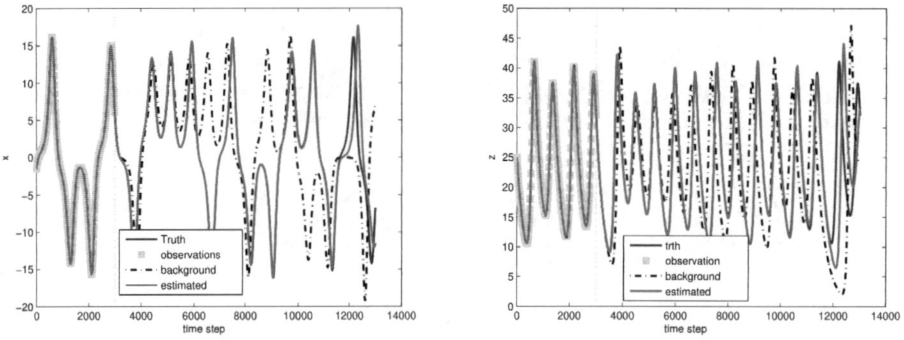

Figure 4. (left side) Time serie of x component; (right side) Time serie of z component. Data assimilation every 12 time steps. Blue curve: truth; green squares: observations; red curve: estimate and black curve: background. Estimated by variational method.

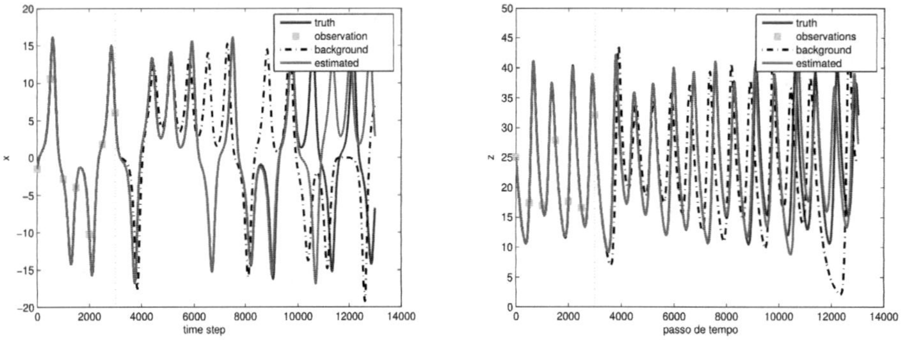

Figure 5. (left side) Time serie of x component; (right side) Time serie of z component. Data assimilation every 500 time steps. Blue curve: truth; green squares: observations; red curve: estimate and black curve: background. Estimated by variational method.

performed every 500 time steps. In Figures 4 and 5, the assimilation window corresponding to timestep 3000 thereafter we have the forecast. Note that for the assimilation window with observations inserted every 12 time steps (Figure 4), it was obtained a good prediction around 12000 time steps, while for the window of assimilation with observations every 500 time steps, we have obtained good prediction up to 10000 time steps. This indicates that the frequency of observations is relevant to the quality of the forecast. In Table 2, the erros for data assimilation process are shown for three frequencies of observations. The artificial neural network was efficient for emulate the variational method, as we can see by the graph of the error in Figure 7, which shows the absolute difference between truth and estimated, and Figure 8 show this error scaled in semi-log.

5. Conclusion

Neural network can be successfully used as a data assimilation method. The Multilayer Perceptron technique was tested in nonlinear systems described by Lorenz equations, which have chaotic dynamics. As neural network employed supervised training, where the target of

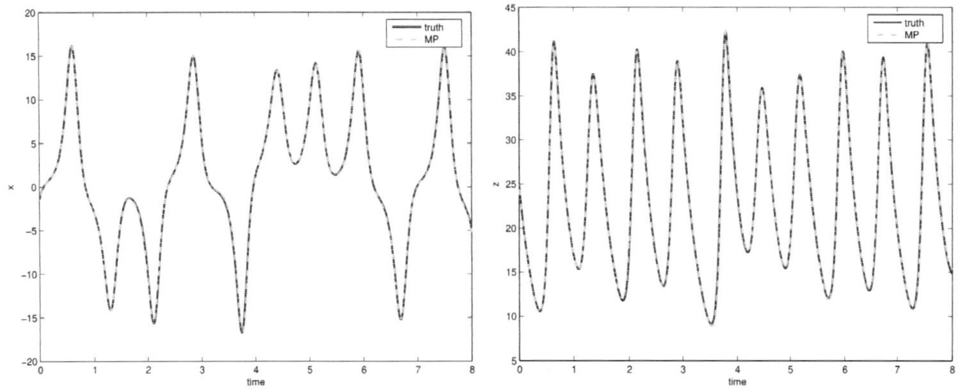

Figure 6. (left side) Time serie of x component; (right side) Time serie of z component. Data assimilation every 500 time steps. Blue curve: truth; green curve: estimated. Estimated by multilayer perceptron (MP).

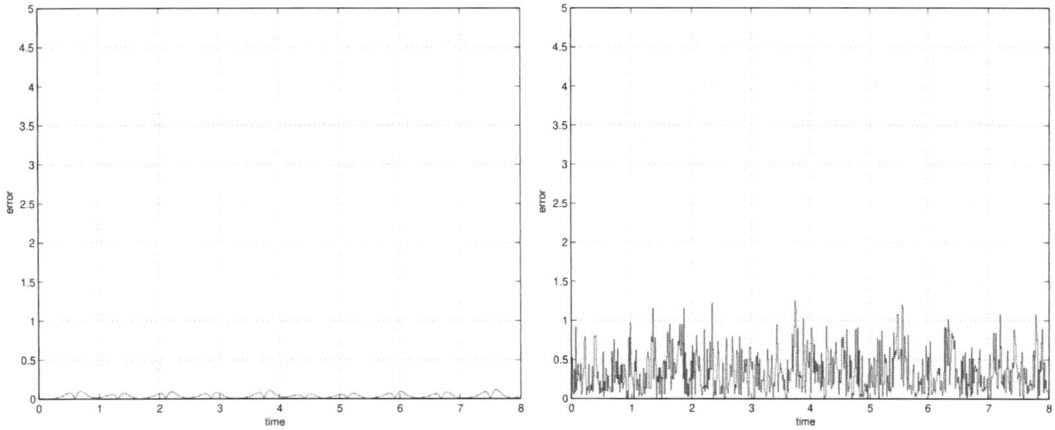

Figure 7. Graph of the error of x components Lorenz's model for assimilation performed every 12 time steps, (left side): variational method; (right side): multilayer perceptron.

Table 2. Erros: Variational Method and Multilayer Perceptron.

	Variational Method				Multilayer Perceptron			
freq	x	y	z	$(x+y+z)/3$	x	y	z	$(x+y+z)/3$
25	0.0011	0.0015	0.0018	0.0015	0.3186	0.3403	0.3251	0.3280
50	0.0017	0.0024	0.0029	0.0070	0.3087	0.3450	0.3118	0.3218
100	0.0020	0.0025	0.0030	0.0025	0.4106	0.5609	0.4937	0.4884
500	0.0061	0.0085	0.0113	0.0086	0.4341	0.6385	0.7928	0.6218

network was determined by a variational method. This ANN presents a lower computational complexity than the Particle Filter, and variational approach [22] or Kalman filter [16].

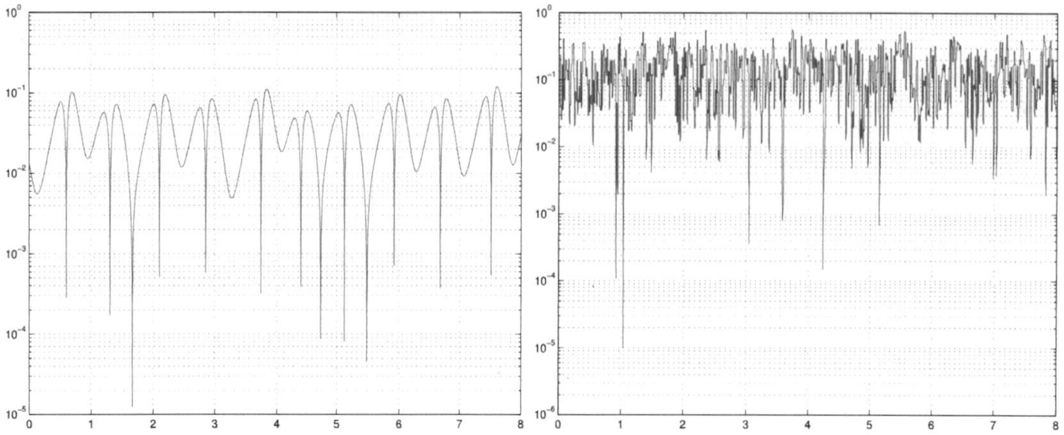

Figure 8. Erros at log scale; (left side): variational method; (right side): multilayer perceptron.

References

[1] Daley D 1993 Atmospheric Data Analysis (Cambridge University Press).

[2] Kalnay E 2003 Atmospheric modeling, data assimilation and predictability (Cambridge University Press).

[3] Bennett A F 2002 Inverse Modeling of the Ocean and Atmosphere (Cambridge University Press).

[4] Zannetti P 1990 Air pollution modeling: theories, computational methods and avaible software (Computational Mechanics Publications)

[5] Garner T W, Wolf R A, Spiro R W, Thomsen M F 1999 First attempt at assimilation data to constrain a magnetospheric model Journal of Geophysical Research 104, 25145-25152.

[6] Schunk R W, Scherliess L, Sojka J J, Thompson D C, Anderson D N, Codrescu M, Minter C, Fuller-Rowell T J, Heelis R A, Hairston M, Howe B M 2004 Global assimilation of ionospheric measurements (GAIM00). Radio Science, 1, RS1S02.

[7] Jazwinski A H 1970 Stochastic Processes and Filtering Theory (Academic Press).

[8] Evensen G 2007 The Ensemble Kalman Filter (Springer Verlag).

[9] Belyaev K, Meyers S, O'Brien J J 2000 Fokker-Planck equation application to data assimilation into hydrodynamic models J. Math. Sci., 99(4) 1393-1403.

[10] Belyaev K P, Tanajura C A S, O'Brien J J 2001 A data assimilation method used with an ocean circulation model and its application to the tropical Atlantic Appl. Math. Modelling 25 655-670.

[11] Campos Velho H F, Cintra R S, Furtado H C M. Introduo a assimilao de dados 2007. Available:¡www.lac.inpe.br/ haroldo/Curso-DataAssimilation/$Curso_Assim_Dados - INPE - 2$.pdf¿. Accessed: September 22, 2010.

[12] Furtado H C M, Campos Velho H F, Macau E E N "Redes neurais e diferentes métodos de assimilação de dados em dinâmica não linear."125 p. (INPE-15235-TDI/1322). Dissertação (Mestrado em Computação Aplicada) – Instituto Nacional de Pesquisas Espaciais, São José dos Campos. 2008. Available: `http://urlib.net/sid.inpe.br/mtc-m17@80/2008/02.07.10.49`

[13] Nakano S, Ueno G, Higuchi T 2007 Merging particle filter for sequential data assimilation Nonlinear Processes in Geophysics 14 395-408.

[14] Nowosad A G, Rios Neto A, Campos Velho H F 2000 Data assimilation in chaotic dynamics using neural networks. In: International Conference on Nonlinear Dynamics, Chaos, Control and Their Applications in Engineering Sciences 6 212-221 (Associação Brasileira de Ciências Mecânicas).

[15] Härter F P, Campos Velho H F 2005 Recurrent and feedforward neural networks trained with cross correlation applied to the data assimilation in chaotic dynamic Brazilian Journal of Meteorology 20 411-420.

[16] Härter F P, Campos Velho H F 2008 New approach to applying neural network in nonlinear dynamic model Applied Mathematical Modelling 32 2621-2633 - DOI 10.1016/j.apm.2007.09.006 - ISSN: 0307-904X.

[17] Härter F P, Rempel E L, Campos Velho H F, Chian A 2008 Application of artificial neural networks in auroral data assimilation. Journal of Atmospheric and Solar – Terrestrial Physics, 70 1243-1250, - DOI 10.1016/j.jastp.2008.03.018.

[18] Lorenz E N 1963 Deterministic Nonperiodic Flow Journal of the Atmospheric Sciences 20(2) 130-141.

[19] Lorenz E N 1965 A study oh the predictability of a 28-variable model Journal Tellus 17

[20] Haykin S 1993 Neural Networks: A Comprehensive Foundation (New York: Prentice Hall).
[21] Nadler M and Smith E P 1993 Pattern Recognition Engineering (New York: John Wiley & Sons)
[22] Furtado H C M, Campos Velho H F, Macau E E N, Härter F P 2007 Diferent data assimilation methods applied to the Lorenz dynamical system 6th Brazilian Conference on Dynamics, Control and Their Applications (DINCON-2007) May 22-25 (UNESP São José do Rio Preto (SP) Brazil) Proceedings in CD-Rom 1030-1035 - in Portuguese.

A novel cell-coupling leading to nonlocal interactions

David Ciro[†] **and Boris A. Rodríguez**[‡]
[†,‡] Instituto de Física
Universidad de Antioquia
Medellín, A.A. 1226 Colombia

E-mail: [†]davidcirotaborda@pegasus.udea.edu.co
[‡]banghelo@fisica.udea.edu.co

Abstract. It is shown that nonlocal interactions and phenomena can be achieved through local considerations, in which the departure of some scalar field from an harmonic one in a point changes as a function of the field itself in such point. After discretizing the equation of motion, it is shown that the shape of the nonlocal interaction function depends deeply on the choice of boundary conditions. As a physical implementation, the found interactions describes the evolution for the inductively coupled nonlinear networks. A qualitative analysis suggests that under such interactions the system self-organizes quite naturally, finally this is evidenced through the numerical solution of the equations of motion in the case of local cubic nonlinearities for two different boundary conditions.

1. Introduction

In the investigation of spatiotemporal behavior in arrays of coupled nonlinear oscillators, it has been very common to use local interaction functions. This leads to equations with diffusive structure that produce naturally the phenomena of propagation and the emergence of Turing patterns. The locality of the equations arises from the nearest neighbors interaction which is used for modeling coupled nonlinear networks (CNNs) [1]. This causes the dynamic of the system to be almost insensitive to the choice of boundary conditions, except at the borders; such locality can be achieved through the resistively coupling of cells [2] in electronic implementations. In this paper we show, that there is a straightforward way to induce nonlocal interactions through local considerations, but at variance with local systems, the interaction functions and the behavior of the system will be highly dependent on boundary conditions. Nonlocality can be a key idea in the electronic implementation of neural networks with no need of a large number of connections. However, a deeper knowledge of the implications on different kinds of local considerations in the resulting nonlocal system must be developed. For electronic implementations we show that inductive coupling of cells leads to such nonlocality.

The paper is organized as follows: In section 2 it is shown how to find nonlocal interactions through local considerations from a mathematical point of view; later in section 3, we generalize the interactions found in section 2, and show how those depends on boundary conditions; in section 4 we present a physical realization of such interactions through electronic devices and introduce the local dynamical systems; finally, section 5 is devoted to study self-organization under the found nonlocal interactions and numerical results are also presented.

2. A Source for Nonlocal Behavior

To begin with, we consider the scalar field $\xi(\boldsymbol{r},t)$, defined in some domain $\Omega \in M$, where M is an Euclidean Space. From this, let's define the *harmonicity* field $\Upsilon(\boldsymbol{r},t) = \nabla^2\xi(\boldsymbol{r},t)$ that says how far is ξ from being an harmonic function in the point $\boldsymbol{r}$ at the moment t, this also tell us about the "concentration" of the scalar field $\xi(\boldsymbol{r},t)$ in each point and time. Now we introduce the following rule, or dynamical system

$$\frac{\partial}{\partial t}\Upsilon(\boldsymbol{r},t) = \Im[\xi(x,t)], \tag{1}$$

where the *Harmonizing Function* $\Im(x)$ is a one-variable function that says how Υ changes in time depending on the value of ξ; i.e. in this dynamical system the "harmonicity" of some scalar field ξ changes as a function of the field itself. Note that equation (1) is local, in the sense that $\Upsilon(\boldsymbol{r},t)$ depends only on value of field in a neighborhood (see fig. 1), and this is also true for $\Im[\xi(\boldsymbol{r},t)]$.

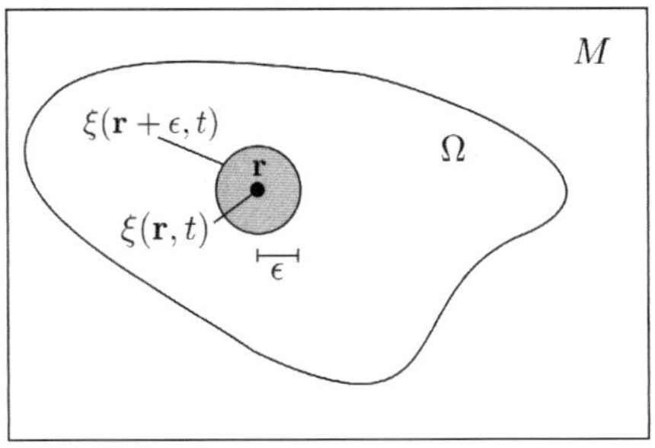

Figure 1. The spatial locality of eq. (1) comes from the fact that to define the *harmonicity* we need to know ξ in a arbitrarily small ϵ-ball around $\boldsymbol{r}$, and of course the quantity $\frac{\partial}{\partial t}\Upsilon$ is local in the time, because it does not involve future or past values of $\Upsilon(\boldsymbol{r},t)$, only depends on the actual one.

Although eq. (1) may look simple, if we substitute the *harmonicity* function we get:

$$\frac{\partial}{\partial t}\nabla^2\xi(\boldsymbol{r},t) = \Im[\xi(\boldsymbol{r},t)], \tag{2}$$

a third order partial differential equation, in which both time and space derivatives are to be applied to the same term. In the following we exhibit a procedure that renders such *local* equations as a nonlocal one.

2.1. Discretization and Standardization

We start considering the one-dimensional case, $\boldsymbol{r} = r$, where the ϵ-ball (fig. 1) becomes an interval and after discretizing the differential operators, we get:

$$\nabla^2\xi(\boldsymbol{r},t) = \frac{\partial^2}{\partial r^2}\xi(r,t) = \lim_{\epsilon\to 0}\frac{x_{n+1}(t) - 2x_n(t) + x_{n-1}(t)}{\epsilon^2}, \tag{3}$$

where the substitution $\xi(r,t) = \xi(n\epsilon,t) = x_n(t)$, was used. Taking this to eq. (2), and defining $f = \epsilon^2\Im$ we get the discrete dynamical equations associated to the continuous field equation (2):

$$\frac{d}{dt}(x_{n+1} - 2x_n + x_{n-1}) = f(x_n). \tag{4}$$

Observe that eq. (4) does not have the usual structure for a dynamical system, i.e. $\dot{x} = f(x,t)$, but we can do some algebra to give it such form. In eq. (4) there are N variables x_n, then collecting all terms in vectors and defining $\boldsymbol{x} = (x_1, x_2, ..., x_N)^T$ and $\boldsymbol{F}(\boldsymbol{x}) = -(f(x_1), f(x_2), ..., f(x_N))^T$, we can write eq. (4) in vector form like:

$$\frac{d}{dt}\mathbb{A}\boldsymbol{x} = \boldsymbol{F}(\boldsymbol{x}), \tag{5}$$

where the Coupling Matrix $\mathbb{A}$, is the matrix containing the weights of the corresponding neighbors for every independent equation. In the following sections the structure of $\mathbb{A}$ will be made precise depending on the boundary conditions. Now suppose for a moment that the coupling matrix can be inverted, and as it is time independent we have

$$\dot{\boldsymbol{x}} = \mathbb{A}^{-1}\boldsymbol{F}(\boldsymbol{x}). \tag{6}$$

This last expression has the usual form of a Dynamical System, so we will name it *The Standard Form* of eq. (5). If $\mathbb{A}$ can not be inverted, it is also possible to write the Standard Form by removing the *zero frequency* modes in the system [3].

2.2. Nonlocal Interaction

Looking the structure of eq. (4), we can see that the Coupling Matrix is a *tridiagonal Toeplitz matrix* [4] in most cases; but this does not means that $\mathbb{A}^{-1}$ is also tridiagonal. In the following sections it will be shown that the inverse of the Coupling Matrix is a dense matrix, so that the product $\mathbb{A}^{-1}\boldsymbol{F}(\boldsymbol{x})$ involves all inputs of the vector $\boldsymbol{F}(\boldsymbol{x})$, this means that the n'th element in equation (6) is:

$$\dot{x}_n = \sum_{m=1}^{N} c_{n,m} f(x_m), \tag{7}$$

where the $c_{n,m}$ are generally different from *zero* and represent the *discrete interaction function.* Therefore we have a nonlocal interaction, given that the evolution of the n'th variable involves the actual values of all the other elements in the network. This appears to be a simple consequence of the proposed coupling, but is remarkable that through *local considerations* and even *local equations* we obtained nonlocal interactions.

3. The Nonlocal Interaction Functions

In order to find the Standard Form equation, we will use the usual procedure for decomposing systems of linear second order coupled differential equations in its *normal oscillating modes* [5]. $\mathbb{A}$ is always an *hermitic matrix*, then defining

$$\mathbb{B} = (\boldsymbol{a}_1, \boldsymbol{a}_2, \cdots, \boldsymbol{a}_N), \tag{8}$$

as the matrix whose columns are the eigenvectors $\{\boldsymbol{a}_n\}$ of $\mathbb{A}$ and after premultiplying by $\mathbb{B}^T$, eq. (5) can be written as:

$$\mathbb{A}_D\mathbb{B}^T\dot{\boldsymbol{x}} = \mathbb{B}^T\boldsymbol{F}(\boldsymbol{x}), \tag{9}$$

where $\mathbb{A}_D$ is the diagonal form of $\mathbb{A}$. Then we can suppress in eq. (9) those rows corresponding to *zero* eigenvalues, and define the *non-square matrix* $\tilde{\mathbb{B}}$ as the obtained after suppressing from $\mathbb{B}$ the eigenvectors related to *null* eigenvalues. We finally get:

$$\dot{\boldsymbol{x}} = (\tilde{\mathbb{B}}\tilde{\mathbb{A}}_D^{-1}\tilde{\mathbb{B}}^T)\boldsymbol{F}(\boldsymbol{x}) = \mathbb{C}\boldsymbol{F}(\boldsymbol{x}), \tag{10}$$

in which $\tilde{\mathbb{A}}_D^{-1}$ is the diagonal matrix containing the inverse of no-null eigenvalues. This is the needed result, given that it does not carry with singularity problems produced by a *non-invertible* Coupling Matrix. Note also that eq. (10) is well defined, and is a generalization of eq. (6), where $\mathbb{C}$ is the interaction function matrix.

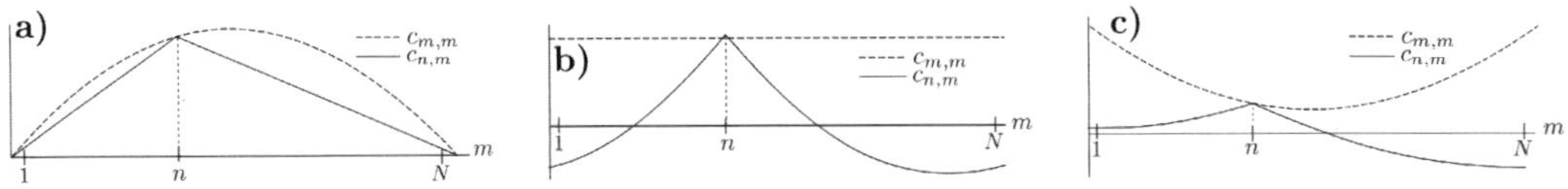

Figure 2. Interaction function matrix elements $c_{n,m}$ for different boundary conditions. a) For fixed boundaries: $c_{n,m} = \min(n,m) - \frac{mn}{N+1}$, which has a maximum in the position of the affected cell. The plot of $c_{m,m}$ shows that the strength of the interactions vanishes to the borders, so the dynamics are richest in the center. b) For periodic boundaries: $c_{n,m} = |n-m|\frac{|n-m|-N}{2N} + \frac{N^2-1}{12N}$, that again is maximum in the affected cell. Now $c_{m,m}$ shows that the strength in this case does not change, as is expected in a periodic system. c) For free borders: $c_{n,m} = \frac{n^2+m^2-n-m}{2N} - \max(n,m) + \frac{(2N+1)(N+1)}{6N}$, that depends strongly on the cell's neighbors, but now the interaction strength grows in the borders.

3.1. Dependence of Interaction Functions on Boundary Conditions

Contrary to the case of local interactions, here the choice of boundary conditions have deep consequences in the shape of the interacting functions and the resulting behavior. The boundary conditions, in some cases can be included completely on the Coupling Matrix, but in other ones auxiliary conditions on the eigenvectors are needed (see Table 1). Due to the discretization of the Laplace operator, our Coupling Matrix is:

$$\mathbb{A} = \begin{pmatrix} 2 & -1 & 0 & \cdots & -1(0) \\ -1 & 2 & -1 & \cdots & 0 \\ 0 & -1 & 2 & \cdots & 0 \\ \vdots & \vdots & \vdots & & \vdots \\ -1(0) & 0 & 0 & \cdots & 2 \end{pmatrix}, \tag{11}$$

where the top-right and down-left corner inputs "$-1(0)$" are related to boundary conditions like "periodic (fixed, free)" respectively. In this case, it is easy to show that the eigenvectors are *discrete harmonic functions* [7] for *all* boundary conditions, and that the eigenvalues have the structure:

$$a_m = 4\sin^2\left(\frac{mk}{2}\right), \; m \in \{1, 2, \cdots, N\}, \tag{12}$$

with k depending on the boundary conditions.

Table 1. Dependence of k on the boundary conditions.

	Fixed	Free	Periodic
k :	$\frac{\pi}{N+1}$	$\frac{\pi}{N}$	$\frac{2\pi}{N}$
Conditions:	$x_0 = 0$ $x_{N+1} = 0$	$x_0 = x_1$ $x_N = x_{N+1}$	$x_0 = x_N$ $x_{N+1} = x_1$

After solving the eigenvalue problem, and suppressing all trivial modes, we perform the matrix products in (10) to find the coefficients $c_{n,m}$ in eq. (7), obtaining as can be seen in figure 2 quite different interaction functions [7].

Those are long-range interactions, and their strength does not necessarily decreases for far cells, so that this kind of systems are strongly coupled and can not be reduced through local approximations to simpler systems. There is something interesting about this coupling, namely for free and periodic borders the interaction becomes negative at the far cells, likewise in the diffusive local interactions where is negative in the first neighbors and positive in the affected cells.

3.2. The Continuum Limit

Now that we have successfully inverted the equation (5) we are in position to solve the problem numerically, provided the system is already in standard form. But first, it is worth making some observations about the continuum limit of such dynamical system. Let's write eq. (7) in the special case of the periodic boundaries:

$$\dot{x}_n = \sum_{m=1}^{N} \left(|n-m| \frac{|n-m| - N}{2N} + \frac{N^2 - 1}{12N} \right) f(x_n), \tag{13}$$

using $x_n(t) = \xi(n\epsilon, t)$, $r = n\epsilon$,$r' = m\epsilon$, and $f(x_n) = \epsilon^2 \Im[\xi(r,t)]$ we have:

$$\frac{\partial}{\partial t}\xi(r,t) = \int_0^L \Im[\xi(r',t)] \left(\frac{L}{12} - |r - r'| \frac{L - |r - r'|}{2L} \right) dr', \tag{14}$$

where we made $N \to \infty$, $\epsilon N = L$, and the limit $\epsilon \to 0$ was taken. This is an integro-differential equation, that puts in evidence the non-local character of this interaction. Generally, for any boundary condition the problem may be transformed into:

$$\frac{\partial}{\partial t}\xi(r,t) = \int_0^L K(r,r') \Im[\xi(r',t)] dr', \tag{15}$$

with $K(r,r')$ a well defined kernel emerging from the found discrete non-local interactions. Note that this last equation is the formal solution of eq. (2), but now is clear that the scalar field $\xi(r,t)$ changes in relation to the current state of the field in every point of the domain Ω, in this case the interval $(0, L)$. This procedure is equivalent to applying the inverse operator of ∇^2 in eq. (2), and this will be always possible for any local consideration by the presented analysis. Something important to say is that coupling matrices similar to $\mathbb{A}$ with no vanishing values near the main diagonal will generally be dense, then local considerations similar to those used for defining eq. (1) will produce nonlocal interactions. Observe that this procedure give us consistent interaction functions depending on boundary conditions and the coupling matrix, then is not necessary to introduce by hand boundary conditions in the numerical simulations.

4. Devising nonlocality

Here we want to show that equations with the structure in (2) can be easily achieved through the *Inductive Coupling* of electronic nonlinear cells. In figure 3 nonlinear Current Controlled Voltage Sources (CCVS elements) [6], represented by rectangular shapes, are connected in a one-dimensional array containing inductors as the inter-cell coupling element.

Applying the first Kirchhoff law in the nodes A and B, and the second to the n'th loop we obtain the equation:

$$\frac{d}{dt}(i_{n+1} - 2i_n + i_{n-1}) = L^{-1} v(i_n) \tag{16}$$

where $v_n = v(i_n)$ is the n'th potential drop produced by the n'th current i_n. This equation is identical to eq. (4). Therefore for this structure we have nonlocal interactions, or in other words

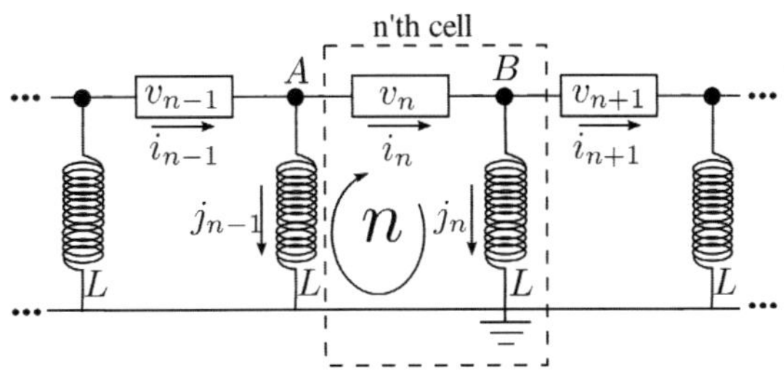

Figure 3. Inductive Coupling of Nonlinear Cells.

each cell interact with all others *directly* without the extensive connectivity of a neural network. Then, for each cell we can write:

$$\frac{d}{dt} i_n = -L^{-1} \sum_{m=0}^{N} c_{n,m} v_m \tag{17}$$

4.1. The Case of Dynamic CCVS Elements

Consider now that in the figure 3 the CCVS elements are replaced with dynamic elements, i.e. the rectangular shape has its own dynamics and the current entering to the element is just one of the dynamical variables that evolves in time. The most simple case would be

$$\dot{v}_n = F(v_n, i_n). \tag{18}$$

We can think of v_n as a *local variable*, given that its evolution only depends on the cell's state; and i_n as a global variable, because its evolution depends on the current value of *all* local variables.

In figure 4 we show such replacement, where a Chua's diode [8] has been used to introduce a "cubic" nonlinearity in the local function $F(v, i)$. Nondimensionalizing the current ($i_n \to y_n$), voltage ($v_n \to x_n$) and making $g(v_n) \to f(x_n)$

$$\dot{x}_n = f(x_n) + \epsilon y_n, \tag{19}$$

$$\dot{y}_n = -\delta \sum_{m=0}^{N} c_{n,m} x_m, \tag{20}$$

With

$$f(x_n) = \begin{cases} -mx - 1 & , \quad x < -1 \\ mx & , \quad -1 \le x \le 1 \\ -mx + 1 & , \quad x > 1 \end{cases} \tag{21}$$

This structure allows us to consider the global variable y_n as a small perturbation on the *local system* ,eq. (19), given that ϵ can be made as small as we want.

Locally we have three fixed points, two stable ones (x_-, x_+), and one unstable (x_0) disposed has in the figure 5. Observe that the effect of the global perturbation is to *move* the cubic shape up or down. This means that the basins of attraction of the stable fixed points may change under such influence, even, if y_n is big enough ($|\epsilon y_n| > h$) it may produce *local* saddle-node bifurcations, so that just one stable fixed point survives.

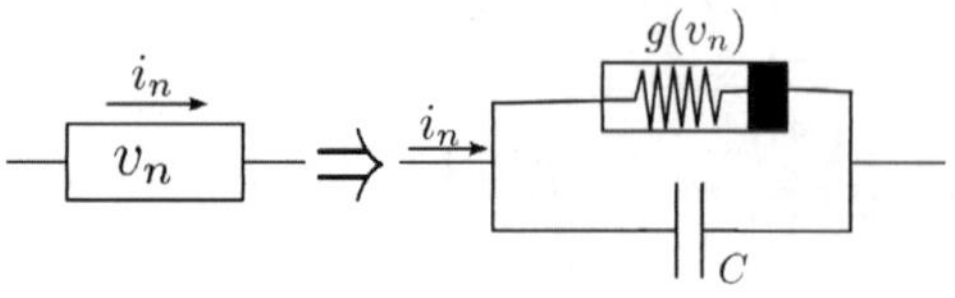

Figure 4. Replacement for the CCVS elements.

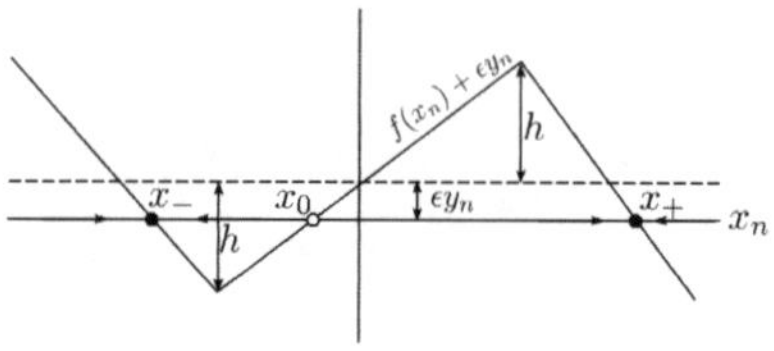

Figure 5. Nonlinear local function for the n't cell.

5. Self-organization Under Nonlocal Interactions

Now that we have introduced the basic structure of our non-local system (and its electronic realization), and convert it to the standard form, it is possible to solve numerically some problems. As we have said, different boundary conditions produces different nonlocal interactions, then we need to consider separately each one.

5.1. The evolution for a Periodic System

First of all let's consider the system with periodic boundaries. In this case, the Coupling Matrix $\mathbb{A}$ is singular because there exist a *zero* eigenvalue. Such eigenvalue is related to an eigenvector in which all entries have the same value ($N^{-\frac{1}{2}}$, after normalization), i.e. an homogeneous state. This means that in this case $\tilde{\mathbb{B}}$ is the $(N-1)\times N$ matrix $(\boldsymbol{a}_1, \boldsymbol{a}_2, ..., \boldsymbol{a}_{N-1})$. with $\boldsymbol{a}_N$ the homogeneous vector. Now, if we assume that the system is in a global equilibrium state we have:

$$\mathbb{C}\boldsymbol{x} = \boldsymbol{0} \Rightarrow \tilde{\mathbb{B}}\mathbb{A}_{\mathbb{D}}{}^{-1}\tilde{\mathbb{B}}^T\boldsymbol{x} = \boldsymbol{0}, \tag{22}$$

which can be reduced to:

$$p_1 a_1^{-1}\boldsymbol{a}_1 + p_2 a_2^{-1}\boldsymbol{a}_2 + \cdots + p_{N-1}a_{N-1}^{-1}\boldsymbol{a}_{N-1} = \boldsymbol{0}, \tag{23}$$

where a_n and $\boldsymbol{a}_n$ are the no-null eigenvalues and nonsingular eigenvectors of $\mathbb{A}$ respectively, and $p_n = \boldsymbol{x}\cdot\boldsymbol{a}_n$ is the projection of the *local* variables vector over the eigenvector $\boldsymbol{a}_n$. Due to the orthogonality of the set $\{\boldsymbol{a}_n\}$, it is clear that this equation is satisfied only if $p_n = 0, \forall n < N$, but this only happens if $\boldsymbol{x}$ is proportional to the singular eigenvector $\boldsymbol{a}_N$. The last means that for periodic boundaries only *homogeneous* states can be the equilibrium states. The stability of such states is determined by the stability of the local system, whose fixed points satisfies $\epsilon y_n + f(x_n) = 0$, but due to the homogeneity: $x_n = x^0, \forall n$, which means $y_n = -f(x_0), \forall n$. This can only be true if $f(x^0) = 0$, because we have to accomplish $\sum_{n=0}^{N} y_n = 0$ due to the application of the second Kirchhoff's law to the global loop containing all currents.

In conclusion, there are three global attractors in this system, in which all internal variables have the same value x_-, x_+ or x_0, and all currents are null; the stability for such attractors are the same as in the local system, and of course the homogeneous asymptotic state (see figures 6 and 7) is sensitive to initial conditions.

5.2. Phenomena for Fixed Boundaries

For fixed boundaries $\mathbb{A}$ is fully invertible, then there are no trivial modes or null eigenvalues, however, we can make some qualitative observations after joining equations (19) and (20).

$$\ddot{x}_n - f'(x_n)\dot{x}_n + \epsilon\delta\sum_{m=1}^{N} c_{n,m}x_m = 0, \tag{24}$$

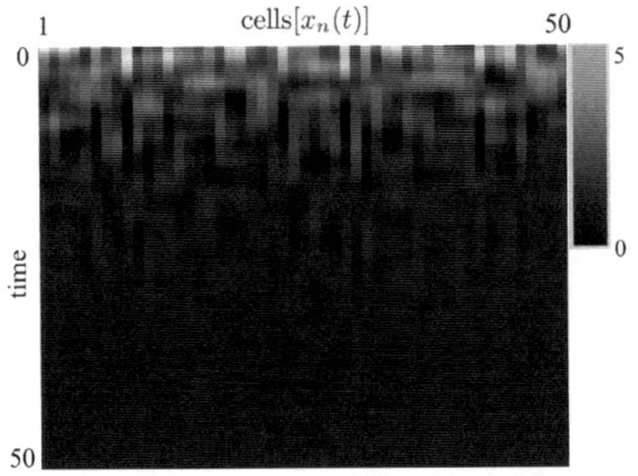

Figure 6. Evolution from random initial condition ($x_{\pm} = \pm 2, x_0 = 0$).

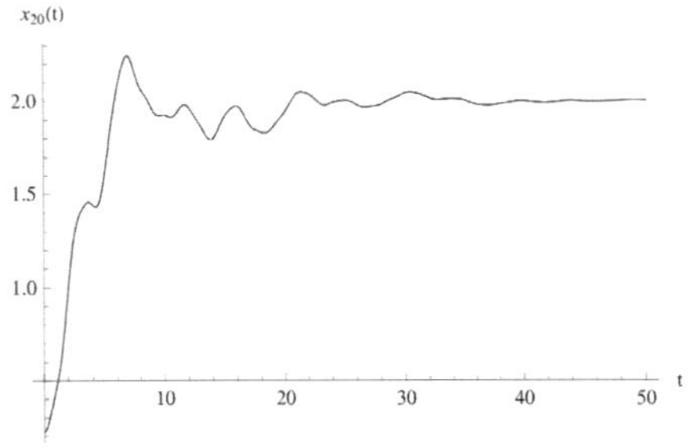

Figure 7. Evolution of an internal variable ($x_{20}(t)$).

that can be written as:

$$\ddot{x}_n - f'(x_n)\dot{x}_n + c_{n,n}\omega_0^2[x_n - x_n^0(t)] = 0, \tag{25}$$

with $\omega_0^2 = \epsilon\delta$ and $x_n^0(t) = -c_{n,n}^{-1}\sum_{m\neq n} c_{n,m}x_m$. Equation (25), says that x_n behaves like a damped oscillator whose equilibrium point moves like $x_n^0(t)$, and also whose damping ratio $f'(x_n)$ can take three different values (m_-, m_0 and m_+), depending on the x_n value. Note that the *new* equilibrium point contains the non-local effects.

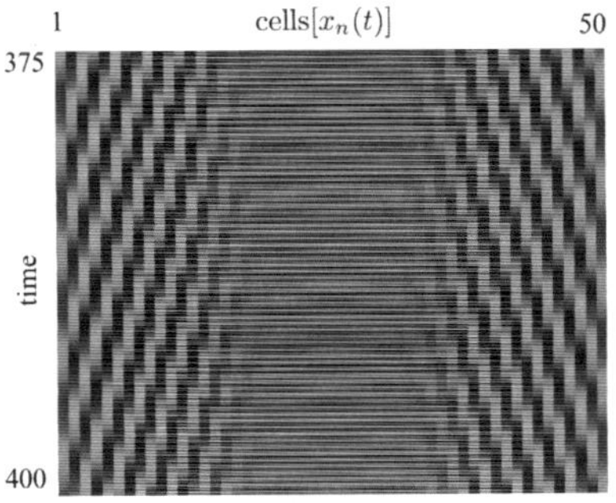

Figure 8. Evolution from random (> 0) initial condition ($x_{\pm} = \pm 2, x_0 = 0$).

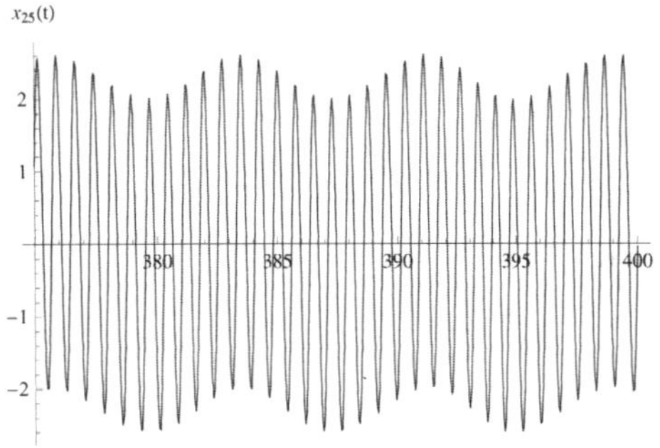

Figure 9. Evolution of the internal variable $x_{25}(t)$, for $t \in [375, 400]$.

Given the definition of $x_n^0(t)$, and due to the large number of cells N, it is possible to think that $x_n^0(t) \approx x_m^0(t), m \neq n$, so that the whole system affects each cell in a similar way. Now, as the whole network is acting as a forcing agent over the cells, this may lead to self-organization produced by the robustness of $x_n^0(t)$. In figure 8 we can see the asymptotic state for the system if it starts from random positive initial conditions. The time series for one variable ($x_{25}(t)$) shows that there are two embedded dynamics: one fast and the other slow. The fast dynamics frequency is close to the predicted natural one $\omega_n^2 = c_{25,25}\epsilon\delta$, then we can think that the slow one corresponds to the net influence of the whole system over the cell; such influence is similar for every cell, but differs in the scaling factor, that's why at the center slow oscillations happens with a smaller amplitude that in the borders (figure 10). The Fourier transform in figure 11 shows the natural larger frequency for $x_1(t)$, but also shows that the net influence of the network is not harmonic. After building a delay plot, we can see that the system may live inside a torus and is cuasiperiodic.

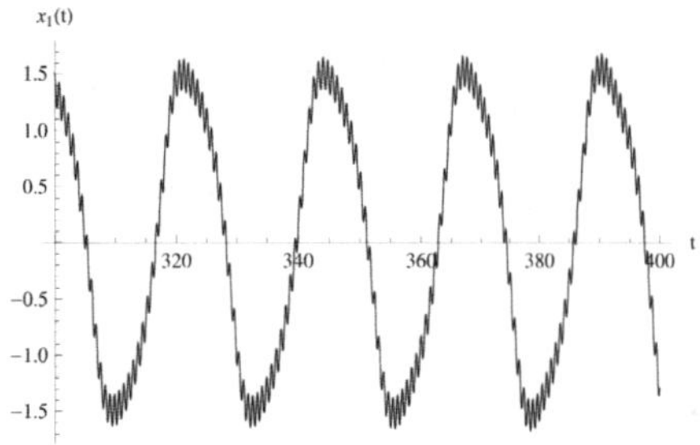

Figure 10. Evolution for the internal variable $x_1(t)$, for $t \in [300, 400]$.

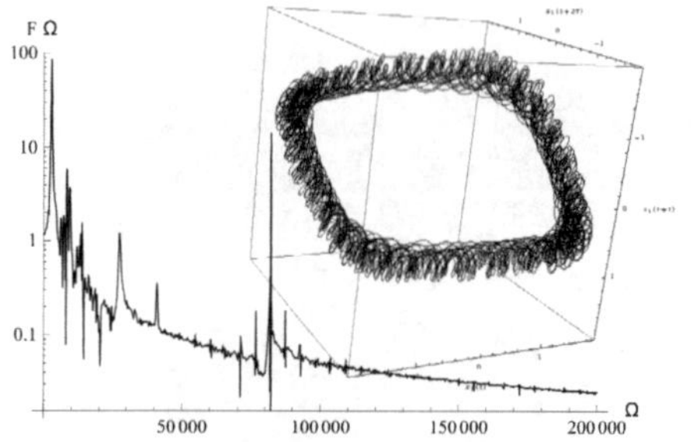

Figure 11. Fourier transform for $x_1(t)$, and delayed diagram.

In this system we also found stronger forms of self-organization, in which it evolves to a global limit cycle, as can be seen from figures 12 and 13. Such asymptotic state is reached through collisions of "structures" that propagates with constant speed until collide with another to form a new structure with a different speed. In every simulation it was observed that in the asymptotic state only one structure survives with null velocity. Obviously this behavior escapes from the scope of this paper.

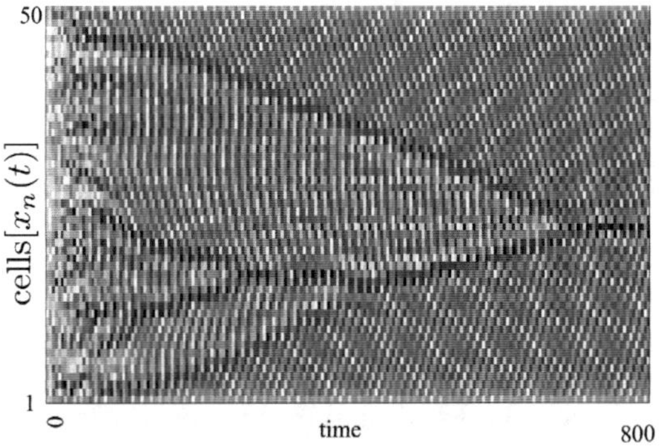

Figure 12. Self-organization for random initial conditions

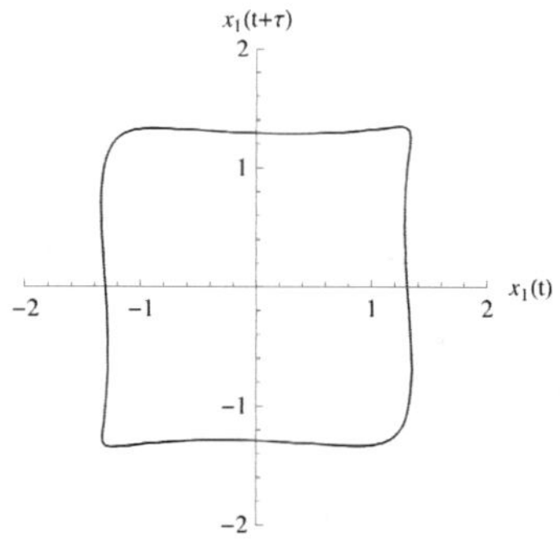

Figure 13. Delayed diagram for $x_1(t)$.

6. Conclusions

First of all, we demonstrated that through local considerations based in how close is a scalar field to be an harmonic one, it is possible to build evolution rules that explicitly involves the whole scalar field in the domain, i.e. a nonlocal interaction. By the introduction of physical implementations, we shown that it is possible, through *local connections*, to build a system endowed with nonlocal interactions. In other words, it is possible to build a highly connected system by using only local connections. When the connected cells contains dynamic elements, we shown that for periodic boundaries, for each stable fixed point in the local system exists a global attractor consisting of an homogeneous state. In the special case of cubic-like local nonlinearities, for random positive initial conditions, the system self-organizes through the robust behavior of the global system reaching an asymptotic cuasiperiodic state. But in the case of uniform random initial conditions a stronger type of self-organization by means of propagating structures leads the system to a global limit cycle.

References

[1] Chua L O and Yang L 1988 *IEEE Trans. Circuit Syst.* **35** 1257
[2] Muruganandam P, Murali K and Lakshmanan M 1999 *Int. J. Bif. Chaos* **9** 805
[3] Goldstein H 1980 *Classical Mechanics, Second Ed.* (Addison-Wesley)
[4] Bottcher A and Grudsky S 2000 *Toeplitz Matrices, Asymptotic Linear Algebra, and Functional Analysis* (Birkhauser)
[5] Guerrero A 2005 *Oscilaciones y Ondas* (Bogotá: Editorial Universidad Nacional)
[6] Chua L O, Desoer C A and Kuh E S *Linear and Nonlinear Circuits* (McGraw-Hill)
[7] Ciro D 2010 Chaos & Complexity in non-Linear Oscillator Networks: Nonlocal systems *Preprint* (Undergraduated Thesis), Instituto de Física, Universidad de Antioquia
[8] Chua LO 1993 *Proc. Int. Symp. NOLTA (Hawaii)* Part 2

Dynamic Days South America 2010 IOP Publishing
Journal of Physics: Conference Series **285** (2011) 012038 doi:10.1088/1742-6596/285/1/012038

Stochastic cellular automata model for wildland fire spread dynamics

Rodolfo Maduro Almeida

Programa de Pós-graduação em Computação Aplicada, Instituto Nacional de Pesquisas Espaciais, Avenida dos Astronautas, 1758, 12.227–010, São José dos Campos, SP, Brazil

E-mail: `rodolfo@lac.inpe.br`

Elbert E. N. Macau

Laboratório Associado de Computação e Matemática Aplicada, Instituto Nacional de Pesquisas Espaciais, Avenida dos Astronautas, 1758, 12.227-010, São José dos Campos, SP, Brazil

E-mail: `elbert@lac.inpe.br`

Abstract. A stochastic cellular automata model for wildland fire spread under flat terrain and no-wind conditions is proposed and its dynamics is characterized and analyzed. One of three possible states characterizes each cell: vegetation cell, burning cell and burnt cell. The dynamics of fire spread is modeled as a stochastic event with an effective fire spread probability S which is a function of three probabilities that characterize: the proportion of vegetation cells across the lattice, the probability of a burning cell becomes burnt, and the probability of the fire spread from a burning cell to a neighboring vegetation cell. A set of simulation experiments is performed to analyze the effects of different values of the three probabilities in the fire pattern. Monte-Carlo simulations indicate that there is a critical line in the model parameter space that separates the set of parameters which a fire can propagate from those for which it cannot propagate. Finally, the relevance of the model is discussed under the light of computational experiments that illustrate the capability of the model catches both the dynamical and static qualitative properties of fire propagation.

1. Introduction

The wildland fire spread is a combustion reaction where the necessary ingredients for its occurrence are: the vegetation, which provides the combustible source for the reaction; the oxygen in the air, which actuates as an oxidizing agent; and a heat source responsible by the initiation and the self-sustainability of the reaction [5]. The fire spreads across the landscape consuming the vegetation and this process can be decomposed into four combustion phases, the so called: pre-heating, ignition, combustion and extinction [5]. The fire front is the region of intense flaming combustion where a large quantity of heat is released. Part of this heat is transmitted to the vegetation not yet burning, and heating it up to the ignition temperature. When this happens, the flames rise and the fire front occupies a new position ahead. The flames remain as the vegetation is burnt out.

In this work a simple model for wildland fire dynamics under flat terrain and no-wind conditions is proposed and its dynamics is analyzed. The model formulation is based on stochastic cellular automata and its dynamics is analyzed qualitatively and quantitatively. Cellular automata are models that assume space, state and time discrete [1, 6]. A square lattice represents the space and each element that constitutes the lattice is called cell. Each cell has a neighborhood, set of internal states variables, and a local rule that describes the transition between the states variables and define the future state as a function of the cell present state and the neighborhood present states. In stochastic cellular automata the rules for cell states transition are performed by means of probabilities. Stochasticity is used to represent the vegetation heterogeneity and to include random component in the dynamics of the vegetation combustion and ignition process during the fire spread.

The paper is structured as follow. In the Section 2 the modeling approach and the model parameters are described. In the section 3 the model dynamics is characterized and analyzed. Finally, in the last section, the model relevance is discussed.

2. Model description

The model is based on the spatially explicit representation and the landscape is depicted as a square and two-dimensional lattice $\mathcal{L}$ of dimensions $L_x \times L_y$. Each cell is defined by:

- its discrete position $(i, j) \in \mathbb{Z}^2$ in the lattice, where $i = 1, \ldots, L_x$ is the column and $j = 1, \ldots, L_y$ is the row;
- the finite set of internal states variables that describes the possible behavior of the cells in a given time step t which are $S^t_{(i,j)} \in \{E, V, F, O\}$ where: E is an empty cell, which denotes unburnable cells or without vegetation; V is a vegetation cell, with denote cells with potential to burn; F is burning cell, which denotes a cell whose the vegetation in its inside is burning; and O is a burnt cell, which denote vegetation cell that was burned by the fire;
- the finite set of neighborhood cells $N(i, j)$, where the Moore neighborhood, as illustrated in the Figure 1(a), illustrate the neighborhood relations in the model and comprises the eight cells surrounding (i^*, j^*) of a central cell (i, j) according with the definition $N(i, j) = \{(i^*, j^*) : |i - i^*| \leq 1, |j - j^*| \leq 1\}$;
- the local rule that determines the future cell state as a function of the present cell state and present neighborhood cell states $f : S^t_{(i,j)} \times S^t_{N(i,j)} \mapsto S^{t+1}_{(i,j)}$, where the time t is also represented by discrete values or time steps. Thus, the time evolution of the model is driven by the interaction between the cell states and the cell neighborhood states. Starting from a given configuration of cells initial states, the cellular automaton self-replicates the sequent cell states. The cellular automata model is stochastic because the state transition function is performed according to probabilities values.

The heat transfer from burning regions to non-burning regions governs the fire spread. Thus, the fire spread is modeled as a set of ignitions of vegetation cells as the burning cells persist. Stochasticity is used to include the heterogeneity of spatial conditions present in real vegetation patterns and to include a random component to the combustion dynamics and ignition process [3, 4, 2]. The dynamics of fire spread is modeled as a stochastic event with an effective fire spread probability S which is as a function three probabilities, which are:

- the probability D, that represents the heterogeneity of fuel conditions along the lattice and determines the proportion of cells with vegetation across the lattice. Thus, for each cell in $t = 0$, there is a probability D for its state to be vegetation cell and the probability $1 - D$ to it is empty cell;
- the probability B, that models the combustion process, where, in each time step, a burning cell has a probability B to change its state to burnt cell;

- the probability I, that models the ignition process, where, there is a probability I for the fire spreads from a burning cell to a neighboring vegetation cell.

The local rule for perform the transition between the states are assessed according to these probabilities values. The cell state transition diagram is showed in the Figure 1(b). An empty cell is unchangeable and always remains in this state. The fire spread is considered a diffusion contagious process and the fire can spreads only from a burning cell to a neighbor vegetation cell. Thus, the transition $V \Rightarrow F$ is conditioned for a vegetation cell that has at least one burning cell in this neighborhood. Given two neighbors cells, one burning cell and the other a vegetation cell, in each time step, there is a probability I for the burning cell ignites the neighbor vegetation cell. Once ignited, in each time step, there is a probability $1 - B$ for a burning cell remain burning, otherwise its state changes to burnt cell, which is the transition $F \rightarrow O$.

The model input parameters are the probabilities D, B and I, the lattice size, and the maximum time step $NSTEPS$. A complete visit to all cells of the lattice is called a sweep. A simple simulation is performed in the two stages: initialization and fire spreading algorithm. A complete visit to all cells of the lattice is called a sweep. The initialization stage includes: (1) define the model input parameters; (2) execute a sweep and for each cell and change its state to vegetation cell with probability D or empty cell with probability $1 - D$; and (3) select one or more vegetation cells and change its state to burning cell. In the fire spreading algorithm, for each time step $t = 1, \ldots, NSTEPS$ execute a sweep and: (1) for each burning cell, evaluate the transition $B \rightarrow O$; (2) for each neighbor of a burning cell evaluate the transition $V \Rightarrow F$. Time simulation stop if $t = NSTEPS$ or if there are none burning cell in a given time step.

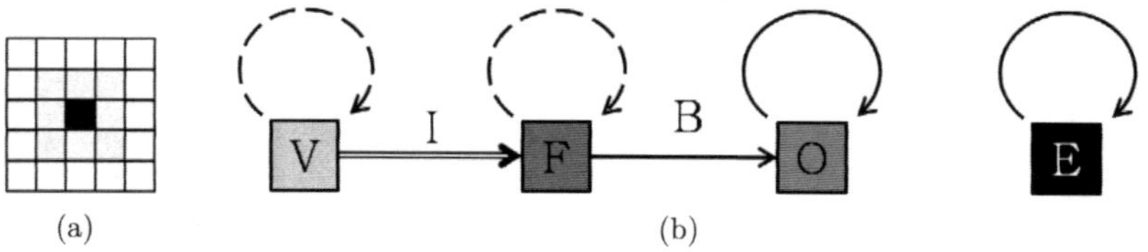

Figure 1. (a) The Moore neighborhood comprises eight cells (yellow cells) which surround the central cell (black cells). (b) In the cell state transition diagram, arrows indicate the state transitions paths. The double arrow indicates that the transition depends on the neighbor cell state. The round dashed arrows indicate that the state transitions are conditioned by the values of other probabilities.

3. Simulations and results

3.1. Qualitative analysis of the fire patterns

The effective fire spread probability S describes the fire behavior across the lattice as a function of the probabilities D, B and I. Different fire patterns, with different size and shape, can be obtained varying the values of these probabilities. The Figure 2 characterizes some fire patterns using different values of D, B and I, for a lattice of size 201×201 and the fire starting from a cell at the middle of the lattice $S^0_{(100,100)} = F$. Each cell state along the lattice is represented by colors that which are, empty cell (black), vegetation cell (green), burning cell (red) and burnt cell (gray).

The proportion of vegetation cells across the lattice determines the spatial distribution of available fuel along the lattice. Higher values of D implies in more quantity of available fuel along the landscape and therefore the fire propagates with more facility. This effect can be observed comparing the Figures 2(d) and 2(f), which can be observed that in the Figure 2(f) the burned area is larger than in the Figure 2(d).

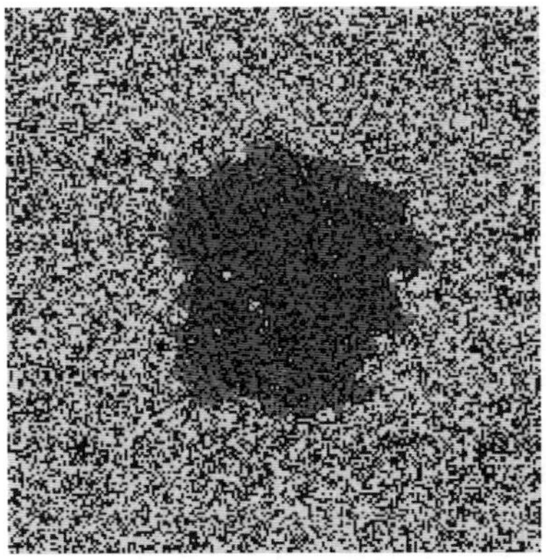

(a) $D = 0.6$, $B = 0.1$, $D = 0.3$

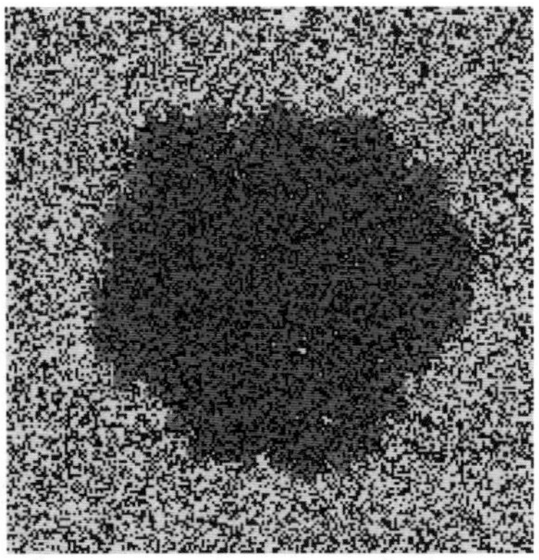

(b) $D = 0.6$, $B = 0.1$, $D = 0.4$

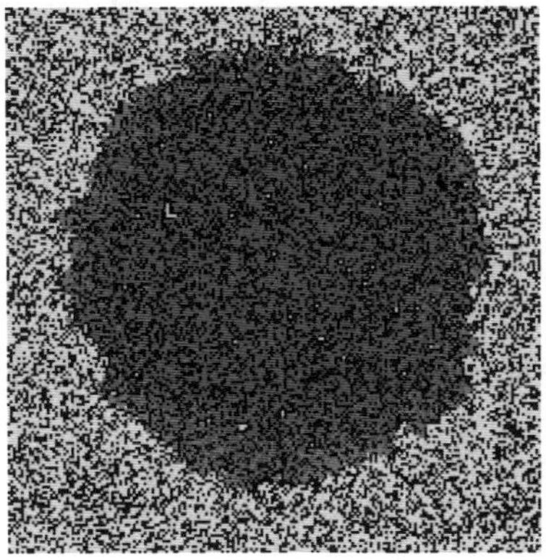

(c) $D = 0.6$, $B = 0.1$, $D = 0.5$

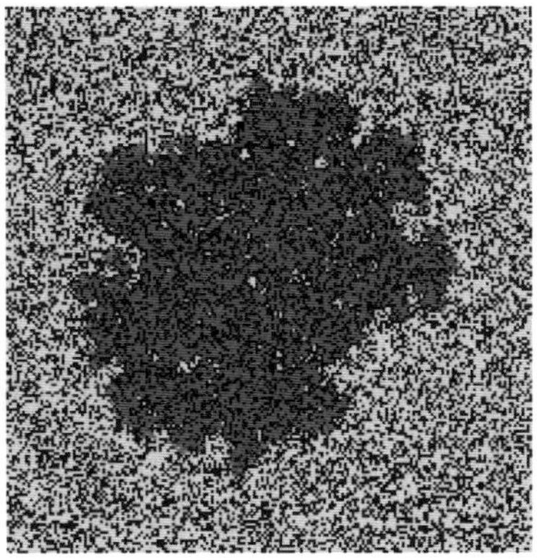

(d) $D = 0.6$, $B = 0.5$, $D = 0.5$

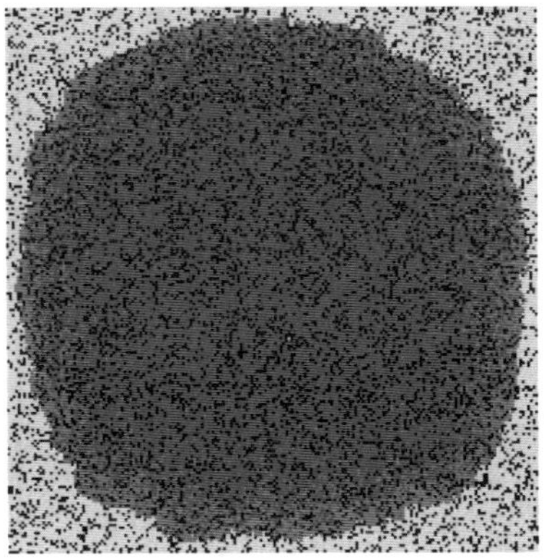

(e) $D = 0.8$, $B = 0.1$, $D = 0.5$

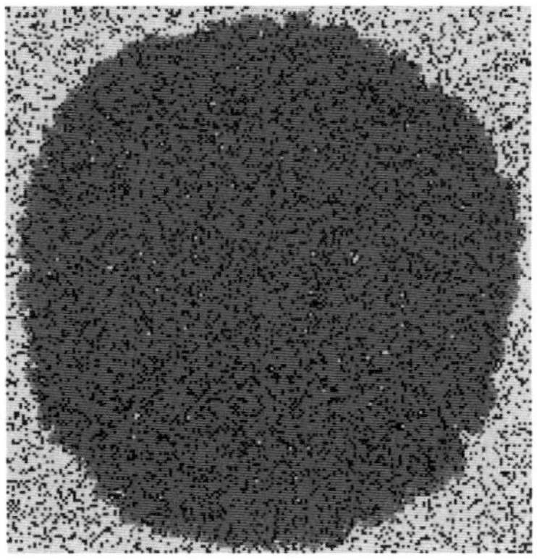

(f) $D = 0.8$, $B = 0.5$, $D = 0.5$

Figure 2. Different fire patterns for $t = 100$ using a lattice with size 201×201 and the fire starting from the middle cell positioned at $(i, j) = (100, 100)$. The parameters values are showed immediately bellow the figures.

The probability B asserts the combustion latency for a burning cell (B^{-1} is a conception for the mean reaction time). A burning cell with a high value of B burns most quickly (i.e., in less time steps) than those that have a low value. This behavior can be observed comparing the

Figures 2(e) and 2(f). Furthermore, how smaller is the values of B, more high is the possibility of the fire spread from the burning cell to neighbors vegetation cells, because the cells remain burning in more time steps. This effect can be observed comparing the Figure 2(c) and 2(d).

The ignition probability I determines how lightly the fire spreads along the lattice. The effects of different fire rates of spread can be observed by the different burned areas comparing the Figures 2(a), 2(b) and 2(c). Higher values of I are related with fire fronts which spread most quickly.

3.2. Monte-Carlo simulations

Because the model stochasticity, a same set of parameters values can generate fire patterns that are slightly different, according showed in the Figure 3(a), 3(b) and 3(c). Thus, is necessary to obtain the mean behavior computed during N simulations based on different sequence of generated random numbers. This is the objective of the Monte-Carlo simulations (MCS). For a given set of model input parameters, a large quantity of simulations are carried out and for each cell is computed the number of times that it burns. The number of time that a cell burn divided by the total number of MCS is the estimative of the cell burning risk. The Figure 3(d) shows the cell burning risk computed by a MCS for $D = 0.6$, $B = 0.5$, $I = 0.5$ and $N = 100$.

3.3. Evaluating the value of S

The fire spreads along the lattice following a pathway of interconnected cells which varies as a function of the effective probability S. Studies in percolation theory [7] corroborate that there is a critical value S^*, called percolation threshold, so that when $S > S^*$ always there is the pathway for the fire spreads from any starting cell to some other point inside the lattice. The main question here is how to characterize the probability S for the model. The existence of the percolation threshold and the consequent description of the critical line as a function of the probabilities D, B and I are investigated using MCS.

A set of N MCS are performed using identical lattices and different values of D, B and I. The fire starts at the left border of the lattices and during the N simulations is computed the number of times that the fire reaches the right border of the lattice. If the fire propagates from one side to the other then the fire percolate the lattice. Thus, the approximation of S, denoted by $\langle S \rangle$, is calculated as:

$$\langle S \rangle = \frac{1}{N} \sum_{j=1}^{N} C_j \, ,$$

where $C_j = 1$ if the fire percolate the lattice and $C_j = 0$ otherwise.

The Figure 4 characterizes the values of $\langle S \rangle$ for different values of D varying the values of B and I. For each set of parameters values a number of $N = 1000$ MCS are carried out using one lattice of size 52×52. The color map displays values varying from $\langle S \rangle = 0$ (black) to $\langle S \rangle = 1$ (white). If $\langle S \rangle = 0$ the fire not percolates the lattice, in other words, the fore extinction regime predominates. Otherwise, if $\langle S \rangle = 1$, the fire propagation regime predominates and the fire spreads incessantly across the lattice. The Figs. 4(a)–4(f) indicate the existence of a critical line defining a division on the model parameter space that separates the set of parameters for which a fire can propagate from those for which it cannot.

3.4. The fire rate of spread

The fire rate of spread R is a common measure of the fire behavior. To compute its value and its relation with the model parameters, a MCS is used to compute the mean fire front edge position, computed in the central cells in the direction of spread, for fires initiated at the left edge of the lattice. The time evolution of the fire front edge position is compatible with a linear

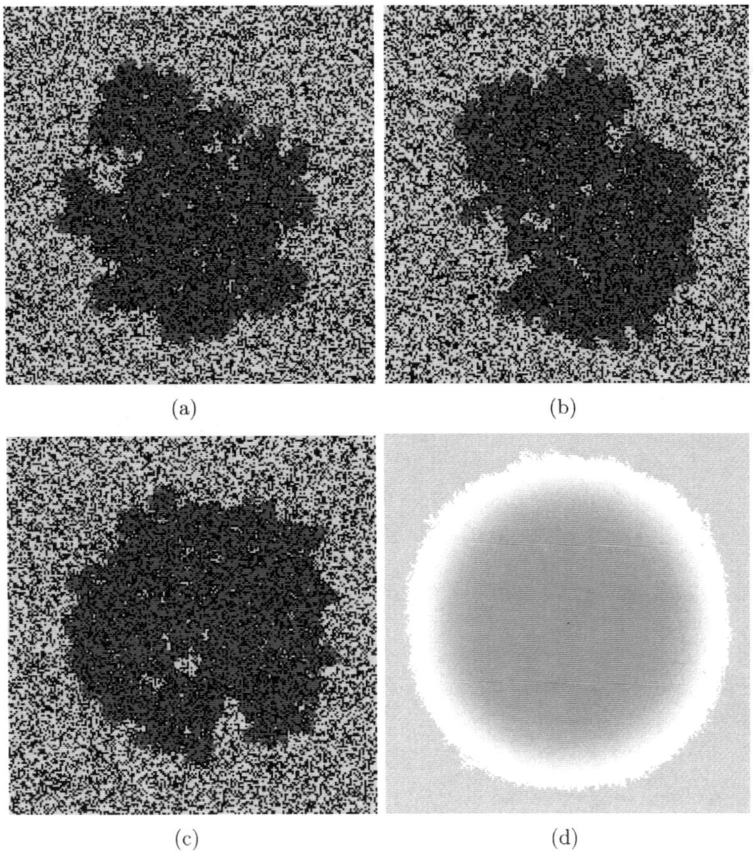

Figure 3. (a)–(c) Fire patterns slightly different for a lattice with size 201×201, $D = 0.6$, $B = 0.5$, $I = 0.5$, $t = 100$ and the fire starting from the middle cell positioned at $(i, j) = (100, 100)$. (d) Burning risk for $N = 10000$ Monte-Carlo simulation. The color map in the figure varies from 0 (black) to 1 (white), and the cells that not burn (green cells).

fit, and thus a rate of spread can be defined as the angular coefficient of the linear fit. The rate of spread unit is calculated in cell dimensions per time steps, assuming that the cells have the same dimensions.

The fire pattern generated by the model is resultant of two coupled processes: fire diffusion and fire persistence. The fire diffusion process is the dynamics of fire transmission between the cells, and it is governed by the ignition probability I. The fire persistence process supports the fire remaining in a burning cell and is responsible by the dynamics of fire pattern formation during the time and drives the fire diffusion process.

The combination of these two processes can generate a large quantity of different fire patterns and the values of I and B are the weights of each one. A high fire persistence and one low fire diffusion provides fire fronts that propagate slowly during the time. Keeping constant the persistence and increasing the diffusion, the fire rate of spread increases. This effect can be observed comparing the dependence of rate of spread on I in the Figs. 5(a) – 5(d).

The probability D is also important to the fire pattern formation. De facto, the diffusive fire spread require a continuous pathway or clusters of vegetation cells and the spatial distribution of vegetation cells across the lattice depends on D. Lower values of D generate a fuel distribution

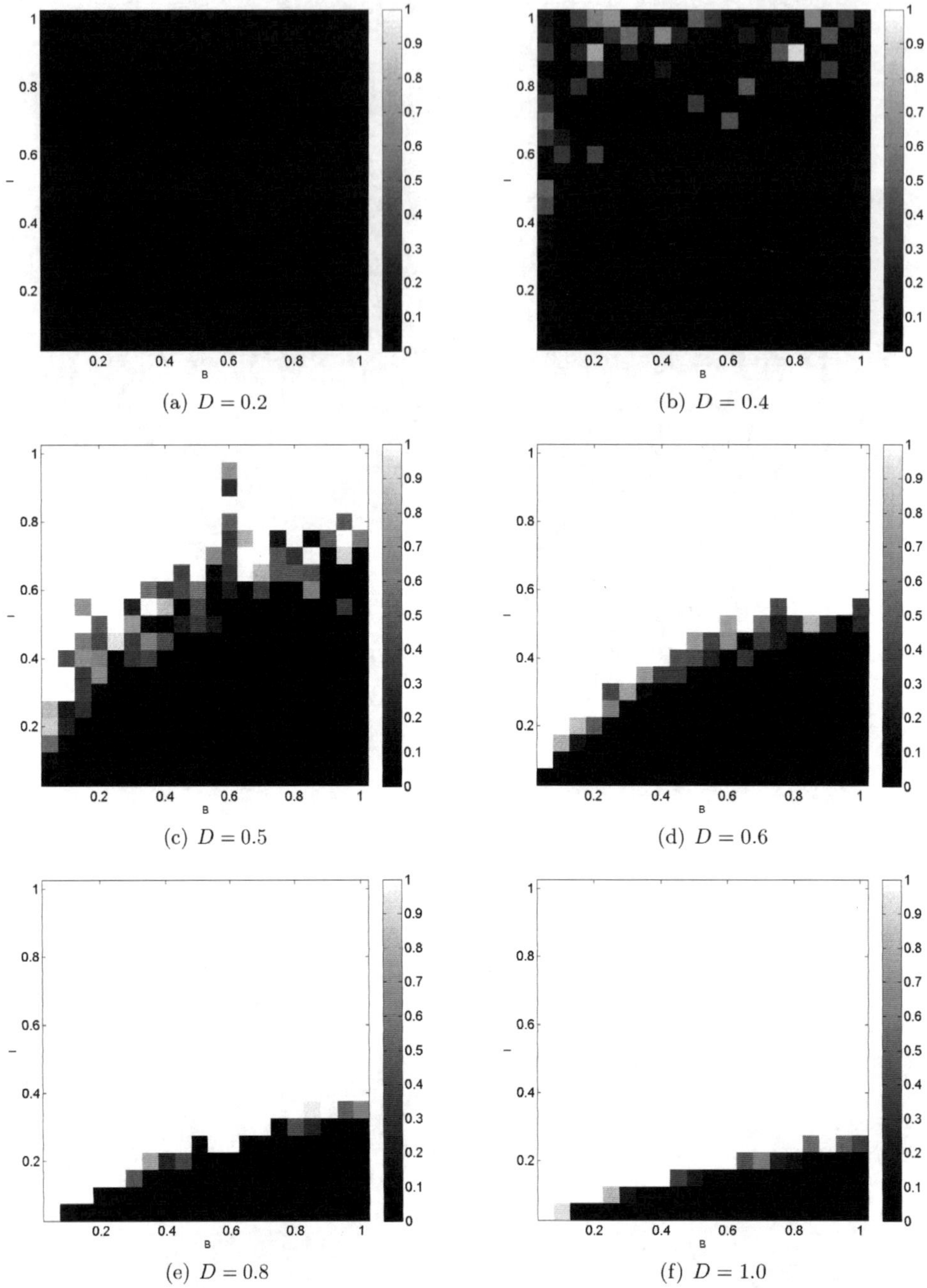

(a) $D = 0.2$ (b) $D = 0.4$

(c) $D = 0.5$ (d) $D = 0.6$

(e) $D = 0.8$ (f) $D = 1.0$

Figure 4. Values of $\langle S \rangle$ for different values of D varying the values of B (x-axis) and I (y-axis). A total of N=1000 Monte-Carlo simulations are carried out for each set of values D, B and I using a lattice of size 52×52.

that is sparse enough to the fire not propagate across it, and that can be observed in the Fig. 4(a). As D is increased, the vegetation cells cluster take place and there is fuel enough to support the fire spread. Thus, there is a critical value D^*, such that for $D >= D^*$ there is the consequent formation of a percolation cluster of neighbor vegetation cells. This percolation cluster not indicate that the fire can propagate uninterruptedly because the fire spread probability is also conditioned to the values of B and I according the critical line characterized in the Figure 4. In the Figs. 4(a)–4(f) can be observed that when D increases, also increases the proportion of the set of model parameters for which the fire can propagate.

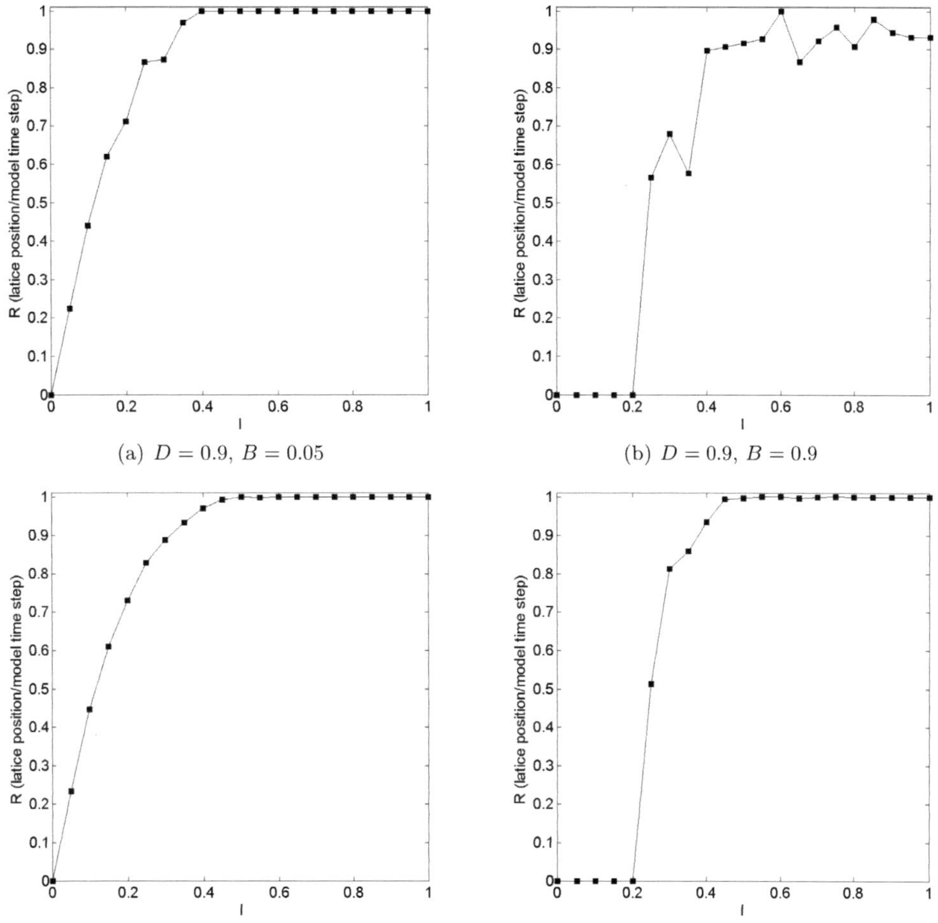

(a) $D = 0.9$, $B = 0.05$ (b) $D = 0.9$, $B = 0.9$

(c) $D = 1.0$, $B = 0.05$ (d) $D = 1.0$, $B = 0.9$

Figure 5. Dependence of the rate of spread on I for different values of D and B.

4. Final considerations

Although the model formulation include only fire spread dynamics under flat terrain and no-wind conditions, the qualitative and quantitative analysis performed in this paper indicate that this model constitutes a qualitative framework for wildland fire spread dynamics simulation. However, for further ecological applications of this model, the relation of the model parameters with meteorological, vegetation and topographical factors remain to be quantitatively established.

The effects of wind and slope on ignition process can be represented according to the incorporation of a directional bias that proportions an anisotropic diffusive process across the lattice [3]. Different values of vegetation density D can be used to represent several phyto-physiognomy vegetation clusters. The probability B can be adjusted to include different values of fuel load over the surface and different fuel moisture conditions. This model parameterization consists in finding an explicit expression between the model parameters and the environmental conditions of historical and documented forest fires.

Acknowledgments

This work was partially supported by CAPES, CNPq and FAPESP (Brazilian financial agencies).

References

[1] Chopard B and Droz M 1998 *Cellular Automata Modeling of Physical Systems* (Cambridge University Press)
[2] Favier C 2004 Percolation model of fire dynamics *Physics Letters* A **330** 396–401
[3] Hargrove W W, Gardner R H, Turner M G, Romme W H and Despain D G 2000 *Ecological Modelling* **135** 243–263
[4] Nahmias J, Téphany H, Duarte J, Letaconnoux S 2000 *Canadian Journal Forest Research* **30** 1318–1328
[5] Pyne S J, Andrews P L and Laven R D 1996 *Introduction to wildland fire* (New York, USA: John Wiley and Sons)
[6] Schiff J L 2007 *Cellular Automata: A Discrete View of the World* (Hoboken, New Jersey: Wiley Series in Discrete Mathematics & Optimization)
[7] Stauffer D and Aharony A 1992 *Introduction to Percolation Theory* 2nd ed. London: Taylor & Francis.

The initial inhomogeneity and halo formation in intense charged particle beams

L. C. Martins[1], **and R. P. Nunes**[2]

[1]Centro de Ciências Tecnológicas, Universidade Estadual de Santa Catarina, Campus Universitário Prof. A. Marcante, s/n°. CEP: 89223-100. Joinville, SC, Brasil.

[2]Instituto de Física e Matemática, Universidade Federal de Pelotas, Campus Universitário, s/n°. Caixa Postal 354. CEP: 96010-900. Pelotas, RS, Brasil.

Email: dfi2lcm@joinville.udesc.br and/or roger.pizzato@ufpel.edu.br

Abstract. Although undesired in many applications, the intrinsic and spurious spatial inhomogeneity that permeates real systems is the forerunner instability that leads high-intensity charged particle beams to its equilibrium. In general, this equilibrium is reached in a particular way, by the development of a tenuous particle population around the original beam, conventionally known as the halo. In this direction, the purpose of this work is to analyze the influence of the magnitude of the initial inhomogeneity over the dynamics of quasi-homogeneous mismatched beams. For that, all beam constituent particles, which are initially disposed in an equidistant form, suffer a progressive perturbation through a noise of a variable amplitude. Beam quantities are quantified as functions of the noise amplitude, which indirectly is assumed a consistent measure of the initial beam inhomogeneity. The results have been obtained by the means of full self-consistent N-particle beam numerical simulations and seem to be an important complement to the investigations already carried out in prior works.

1. Introduction

It is well known that perfectly round and homogeneous beams do not suffer any thermalization effect under its excursion inside the magnetic focusing system. This means that if the beam is assumed initially cold (all beam particles have initially a negligible velocity), during its linear path inside the magnetic focusing channel, the beam remains cold (particles' velocity in magnitude are still unworthy). In essence, while confined inside the accelerator structure, the particles that compose the beam individually cannot earn extra mechanical energy through the increasing of its own kinetic energy. Using the jargon of the field, the beam particles are not heated as propagate inside the channel.

All of this is readily possible to show analytically by the calculus of the beam emittance. The emittance [1] is a macroscopic beam quantity that involves the velocity of each beam constituent. Emittance in average depends on a monotonic way of particle velocities: if the particle velocities in average increase during the confinement process, beam emittance also increases; if the particle velocities in average decrease, beam emittance in the same way also decreases. Analytical calculations show that perfectly round and homogeneous beams do not experiment any heating process. If the beam is initially cold, which implies emittance to be initially zero, after the beam confinement process the emittance conserves its initial value and thus remains zero. With this in mind, since emittance is a good indicator of the macroscopic increasing of beam kinetic energy, it is possible to say that the particles that compose pretty homogeneous beams in average do not have their velocities increased.

However, one will find a much different situation if, instead of considering an initial completely homogeneous beam, one considerer a quasi-homogeneous beam. By a quasi-homogeneous beam it should be understood the situation in which the beam particles are not disposed equidistantly one from the other anymore. The beam becomes quasi-homogeneous because the equidistance property (between beam particles) is weakly broken. The direct result of this is that for a given spatial region of the beam transversal section, particles can be slightly closer when compared with other neighbor beam spatial regions, in which the particles can be slightly farther. The beam particles are not dispersed homogeneously anymore, giving rise to some kind of charge concentration and/or accumulation. This tenuous inhomogeneity introduced implies that the forces inside the beam become weakly nonlinear. Under the action of this kind of force, the beam particles pass to oscillate with a frequency that is thus dependent of its initial coordinate [2]. As a natural consequence of this new situation, synchronization problems can potentially occur inside the beam: during their dynamics, some beam particles can eventually collapse their orbits, being expelled and/or ejected from the beam core. From the beam phase space picture, this ejection looks like the breaking of density waves [3][4].

Once out of the core, the ejected particles become susceptible to the influences of the envelope mismatch. These particles are driven by the oscillation of the core and, as a consequence, progressively have their velocity improved. As the beam core one should understand the set of beam particles that have not been ejected. Since the velocity of the ejected particles increase by the excitation imposed by the oscillatory movement of the core, the beam emittance also increases. In this way, due to the initial small inhomogeneity, initially quasi-homogeneous beams suffer some heating and present an emittance growth that cannot be neglected. Note: the core movement just can excite beam particles because they are out of the beam [5]. And for the particles get out the beam, a nonlinear, even weak, force must act over them inside the beam. For this reason, one can say that while the beam envelope mismatch scales the emittance growth, the initial inhomogeneity is the forerunner mechanism that allows the emittance to increase [6]. The greater is the initial beam envelope mismatch, the greater is the emittance growth. But one must say that even in initially matched beams the inhomogeneity induces an emittance growth that should be considered.

Perfectly homogeneous beams are such an idealization pretty hard to achieve in the engineering sense. Always some inhomogeneity will be present and its influences over the beam dynamics are then something to be better investigated. Following this reasoning line, the purpose of this work is to evaluate how the time scale of the emittance growth depends of the magnitude of the initial inhomogeneity introduced in the beam density. In prior works [2][6], it has been found that the time scale of emittance growth has contributions from the initial inhomogeneity and from the beam envelope mismatch. Also, it is of interest to verify quantitatively how the beam quantities behave at equilibrium as a function of the initial inhomogeneity.

The present work is organized in the following way. In the section 2, some aspects related with the system of interest in this work are discussed. The scheme for generation of the initial beam density is shown. In the section 3, the approach developed to produce – in a controlled manner – the inhomogeneity is presented. The section 4 contains the results obtained with the analysis. Opportunely, this section also describes the method of numerical simulation employed. Finally, in the section 5, the conclusions and the perspectives of future works are addressed.

2. The system of interest

The system considered here is an initially cold and azimuthally symmetric beam of charged particles evolving in linear path inside an accelerator structure. To keep the particles confined, a constant magnetic field is applied in the direction of propagation of the beam. For mechanical and electrical reasons, a conducting pipe surrounds the linear channel. The beam particles propagate along the longitudinal direction, which is aligned with the pipe symmetry axis. The z axis of the coordinate system considered here is exactly over the pipe symmetry axis. The motion of beam particles transversal to this longitudinal direction is accounted by the radial coordinate $r = r(x, y)$. The time is referred as the quantity s hereafter.

Before starting the discussion about quasi-homogeneous beams, it is necessary to analyze the completely homogeneous case. The first step related to the problem is to generate the initial beam distribution for the numerical simulation. Considering that the beam is initially cold, the numerical generation of the initial beam distribution becomes a little bit simplified since just the initial position of each particle has to be specified. The velocity of all the beam particles is initially zero.

Suppose that the beam transversal section is composed by N particles disposed equidistantly one from the other. The beam transversal section is circular with an initial radius r_0. Using jargons of the field, the quantity r_0 is the initial beam envelope. For this case, the beam density n_b can be written in the form

$$n_\mathrm{b} = \frac{N}{\pi r_0^2}. \tag{1}$$

Note that $n_\mathrm{b} = \text{constant}$. Observe also that n_b is a superficial density. In this way, to compose n_b, roughly the particles would have to be numerically distributed in the plane. However, due to the cylindrical symmetry of the problem, it is pretty easer to generate a homogeneous linear density λ and then find a mathematical relation between n_b and λ to achieve the desired superficial beam density. This is much easier because numerically it is simpler to generate homogeneous linear densities than the superficial ones: instead of 2, it is just necessary to specify 1 spatial coordinate for each particle.

The relation between n_b and λ can be directly obtained by particle conservation. Suppose a small ring of width dr at the radial coordinate r of the beam superficial density n_b. For the superficial density to be compatible with the linear density, the number of beam particles inside this ring must be the same that the one found in an element of length dr_g of the linear density λ. In this way, one finds

$$n_\mathrm{b} \cdot (2\pi r dr) = \lambda dr_\mathrm{g} \tag{2}$$

which simplifies to the discrete version

$$r_i = r_0 \sqrt{r_{\mathrm{h}i}}, \tag{3}$$

considering that the coordinates $r_{\mathrm{g}i}$ can be expressed as $r_{\mathrm{g}i} = r_0 r_{\mathrm{h}i}$ with $0 \leq r_{\mathrm{h}i} \leq 1$. The subscript i is an integer and indexes each particle that composes the beam, $1 \leq i \leq N$. Observe that the set of $r_{\mathrm{h}i}$ coordinates are homogeneously generated in the interval $[0; 1]$ while r_i is the associated coordinate each particle i must have for its dispersion satisfies a homogeneous superficial density n_b.

3. The approach

The key point to achieve the goal is how to turn the beam density inhomogeneous in a controlled way. That is, how to progressively introduce inhomogeneity in the beam density in a way that its effects over the beam dynamics could be identified and then quantified.

One interesting approach is to suppose that initially all the beam particles are disposed in an equidistant form. However, each beam particle at radial coordinate r_i starts to be affected by a noise ξ_i. As a consequence, the beam as a whole starts to be a little bit inhomogeneous. The generated noise ξ_i, comprised in the interval $[-1; 1]$ with some probability distribution $P(\xi_i)$, has its intensity controlled by the parameter A, here defined as the noise amplitude. To assure the noise ξ_i affects each particle with the same intensity, the noise amplitude A is considered a constant for all the beam particles.

From the Equation (3), it is possible to observe that the coordinate r_i is obtained from the homogeneous generation of coordinate $r_{\mathrm{h}i}$. Thus, the noise ξ_i only equally affects all beam particles if it acts over the quantity $r_{\mathrm{h}i}$. Thence, to make the beam quasi-homogeneous, it is possible to substitute $r_{\mathrm{h}i}$ by $r_{\mathrm{nh}\,i}$, being $r_{\mathrm{nh}\,i}$ a quantity computed through the following equation

$$r_{\mathrm{nh}\,i} = (i + A\xi_i)\bar{r}\,. \tag{4}$$

The quantity $\bar{r} = 1/N$ is a characteristic spatial scale associated with the regular distance that initially each one of the beam particles is disposed from the others. For any particle, parameter A satisfies the condition $A\xi_i\bar{r} \ll 1$. If $A = 0$, in fact the noise ξ_i does not impacts over each particle coordinate r_i and they are homogeneously distributed, composing what is called, for compactness, a crystalline beam. Nevertheless, if $A \neq 0$, the initial noise starts to perturb the spatial coordinate r_i of each particle, imposing that it fluctuates around $i\bar{r}$. Once probability distribution $P(\xi_i)$ of ξ_i is not completely uniform and N — although usually large — is finite, the noise introduced inside the beam does not affect randomly each one of its particles. The generation of ξ_i is a little bit biased.

The noise breaks the equidistance characteristic between particles existent before, given rise to the concentration or accumulation of charge mentioned in the previous section. The beam loses its initial crystalline structure and starts to be inhomogeneous. Since parameter A controls the noise intensity, and the noise induces spatial fluctuations, indirectly A is a significant measure of the magnitude of inhomogeneity that exists in the beam density. As greater is A, more inhomogeneity permeates the beam. Although the beam cannot be considered homogeneous anymore, because in fact $A \neq 0$, it is possible to say that the beam is quasi-homogeneous, since the condition $A\xi_i\bar{r} \ll 1$ is yet satisfied.

4. The results

The influences of the inhomogeneity over the beam dynamics have been analyzed through full self-consistent N-particle beam numerical simulations. The method employed to simulate numerically the system is based on Gauss' Law: particles interact with each other just by the means of the generated electromagnetic fields. No collisions exist and thus just collective effects are accounted, such as desired. The angular momentum of each beam particle is conserved and thus only the radial set of ODEs must be integrated. All the results have been obtained considering a total number of $N = 10000$ beam particles, which has proportioned convergence to the results. The beam particles follow the ODE below [2]

$$\frac{d^2}{ds^2}r + \kappa r - \frac{K}{r}Q(r) = 0. \tag{5}$$

The quantity κ is the coefficient of magnetic focusing, which accounts how strength is the force originated by the applied magnetic field. The quantity K is known as the beam perveance, which measures how intense are the interactions between beam particles. $Q(r) = (1/N)\int n_{\mathrm{b}} r dr d\theta$ is the fraction of charge trapped by a Gauss surface at r. Note that in fact $Q(r)$ accounts the effects of all the other particles with coordinate smaller than r in that at r.

The initial beam envelope has been set to $r_0 = 1.5$ for all the numerical simulations carried out. This means a 50% mismatch, once the beam envelope of equilibrium has been adjusted to $r_{\mathrm{eq}} = 1$, by the scaling of the set of ODEs. The equilibrium radius r_{eq} can be readily found imposing $d^2r/ds^2 = 0$ in Equation (5). Fixing the beam envelope mismatch, it is expected that any influence of the envelope mismatch is suppressed. This allows that one can better quantify the effects of the inhomogeneity induced by the noise over the initial beam density. The only parameter that is varied between all the numerical simulations is the amplitude of noise A. As desired, with this approach, only the magnitude of the initial inhomogeneity that permeates n_{b} is changed.

The Figure 1 presents results that exemplify how the fluctuations introduced by the noise acts over the initial beam superficial density n_{b}. The Panel (a) of this figure shows the result for the completely homogeneous – the crystalline beam – case given by $A = 0$. As expected, no spatial fluctuations can be seen, and n_{b} is strictly constant. However, as noise amplitude A is increased from zero, spatial fluctuations in n_{b} arise. The beam becomes weakly inhomogeneous. The effects of the introduced noise over the superficial beam density n_{b} are shown in Figure 1b. The noise amplitude employed is $A = 100$. Through a direct inspection of Figure 1b, the fluctuations along the radial coordinate r of the superficial density n_{b} can be clearly perceived. Both panels of Figure 1 show the superficial beam density n_{b} when $s = 0$.

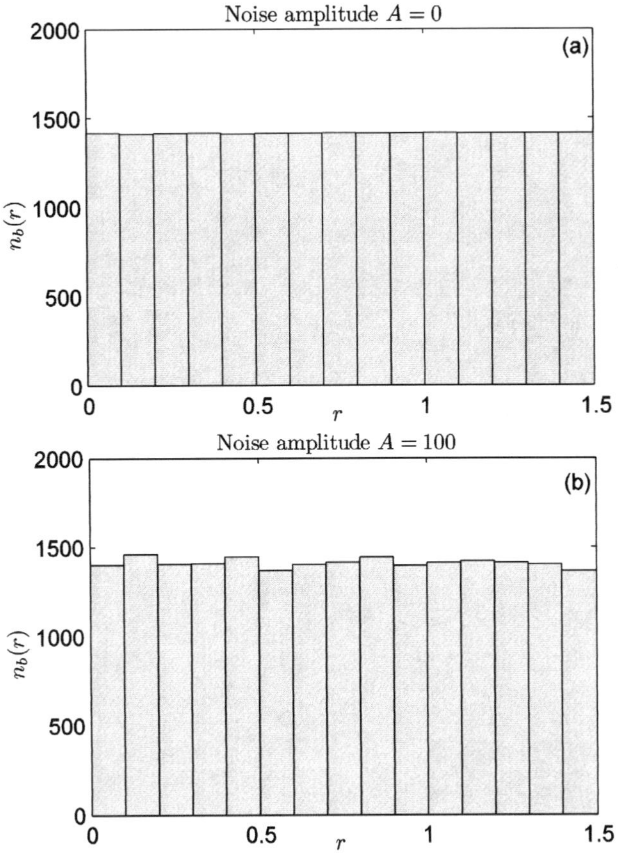

Figure 1. The initial density of beam particles $n_b(r, s = 0)$ for two distinct values of noise amplitude A. In panel (a) $A = 0$, the crystalline beam case, and in panel (b) $A = 100$, a typical quasi-homogeneous beam. Note the spatial fluctuations in n_b induced by the introduced noise.

The influences of the noise amplitude in the density n_b can be also visualized in Figure 2, which presents the initial beam transversal section for $A = 100$. The noise perturbs the beam crystalline structure so particles accumulate in some regions, living gaps in its original unperturbed positions, and consequently giving rise to the desired inhomogeneity in the superficial beam density n_b. This is exactly the phenomenon of charge concentration and/or accumulation previously commented.

The envelope r_b and emittance ϵ of the beam as it evolves inside the focusing channel is respectively presented in the panels (a) and (b) of Figure 3. While in black it is shown the results of the beam envelope and emittance for the noise amplitude $A = 0$, in blue the results of the same quantities for $A = 0$ is plotted. The values of the noise amplitude are $A = 0$, for the black curves, and $A = 100$, for the blue curves. It is important to note the pretty different behavior of both quantities when de noise amplitude is increased from $A = 0$ to $A = 100$. While for the crystalline case the beam oscillates with an invariant profile, for the quasi-homogeneous situation the beam suffers a pretty perceptible decay. From Figure 3a, for the $A = 100$ case, it is possible to observe that the beam envelope r_b starts to decrease when time $s \cong 400$. Concatenated with the envelope decay, considering also the situation for $A = 100$, approximately at this same time, from Figure 3b, the beam emittance suffers a sharp increasing. There is no envelope decay and emittance growth if one analyses the respective curves in Panels (a) and (b) of Figure 3 for the $A = 0$ situation. The numerical simulations have been carried out from $s = 0$ until $s = 800$. The initial beam envelope is $r_0 = 1.5$.

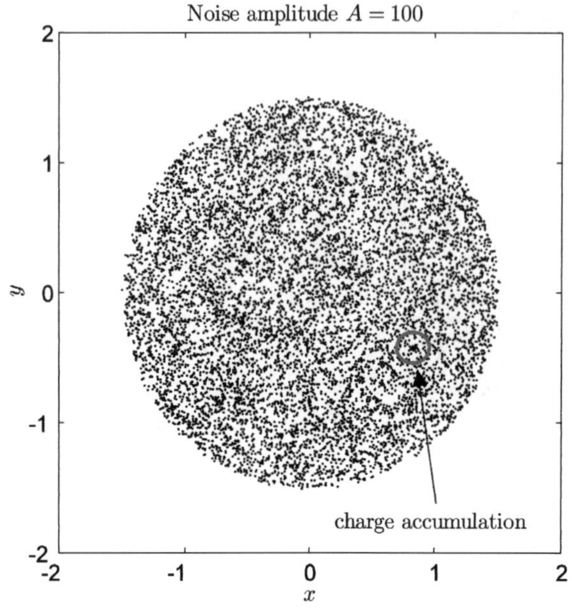

Figure 2. The spatial appearance of the initial beam density $n_b(r, s = 0)$. The amplitude of the noise that perturbs the beam crystalline structure is $A = 100$. A red circle detaches the process of charge accumulation induced by the noise.

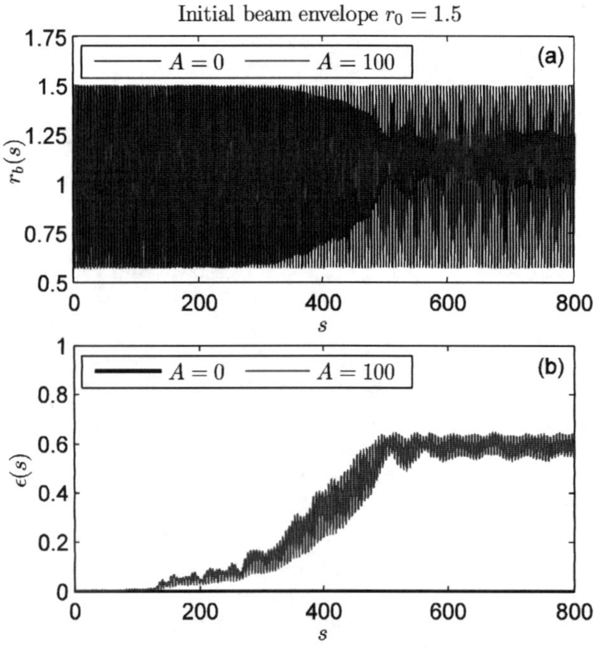

Figure 3. The beam (a) envelope r_b and the beam (b) emittance ϵ as functions of the time s for $A = 0$ and $A = 100$. The results pertain to the interval $0 \leq s \leq 800$.

As commented before, the noise introduced in the crystalline structure of the beam perturbs each one of its initial radial coordinates so that the beam density becomes a little bit inhomogeneous. Inhomogeneity implies that the forces inside the beam become nonlinear and the oscillation frequency of each particle become dependent of its initial position. In this way, in a finite time, many particles' orbits will collapse. These particles will be violently ejected from the beam core and then driven by the envelope oscillations. From the energy conservations point of view, the macroscopic potential energy associated with the envelope oscillations is transformed into kinetic energy of this small group of ejected particles, known as the beam halo. Energy flows from the beam core to this small amount of ejected particles. Although the group of ejected particles is small, since the coupling of the halo particles with beam core is resonant, a very important amount of energy is transferred from the core to the halo. This is the reason by which the beam envelope decay and consequently the emittance growth are so prominent as the Figure 3 shows.

In the Figure 3b, one can observe that the beam emittance grows from zero to some equilibrium value. It is interesting to investigate how this equilibrium value of emittance depends of the magnitude of the initial beam inhomogeneity. With this objective in mind, numerical simulations have been carried out for a set of distinct initial beam inhomogeneity magnitudes, specified by the noise amplitude $A = \{0,5,25,125,625\}$. The results obtained are compactly presented in Figure 4. It is possible to observe that as greater is A, as earlier the beam emittance increases. However, although apparently the noise amplitude A affects the time scale in which the emittance grows, the noise amplitude does not influence in its equilibrium value. In fact, the time scale of the emittance growth has contributions from the initial inhomogeneity that permeates the beam and from the initial beam mismatch [2]. When the noise amplitude A is changed, exactly the contribution from the initial inhomogeneity is changing. Even in the presence of the envelope mismatch, if the beam is pretty homogenous, no emittance growth is observed. For this reason the initial beam inhomogeneity arises as the forerunner mechanism of emittance growth in nonneutral charged particle beams.

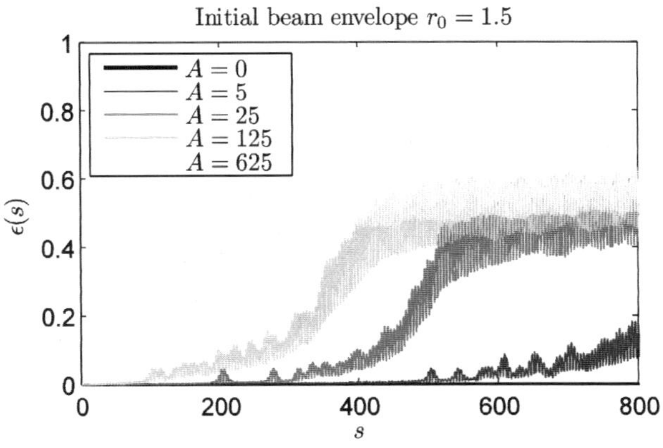

Figure 4. The emittance $\epsilon(s)$ for different values of noise amplitude A. Observe how the time scale of the emittance growth strongly depends of the noise amplitude A.

To finish the current analysis, in Figure 5 the equilibrium values of the beam envelope and emittance are shown for the same set of noise amplitude A employed before. The beam envelope and emittance are respectively plotted in red and blue lines in Figure 5. It is promptly possible to find that the noise amplitude does not impact in the equilibrium values of these beam quantities. This is what one would expect, since the equilibrium values of beam quantities are related just with topological aspects of the problem. Although the beam envelope has a coadjutant participation in the time scale of the emittance growth, it has greater importance and is fundamentally associated with the values of the beam quantities at the equilibrium. Large beam envelope mismatches does not only imply that the energy stored in the beam oscillations is large, but also that the interactions between the beam and individual particles are intense. This clearly appears if one analyses the phase space orbit of a particle excited by the beam oscillations: large envelope mismatches induces the formation of large resonant islands. When in this large resonances, some of beam particles can earn much kinetic energy from the core oscillations. Great initial envelope mismatches implies in great emittance growths and envelope decays.

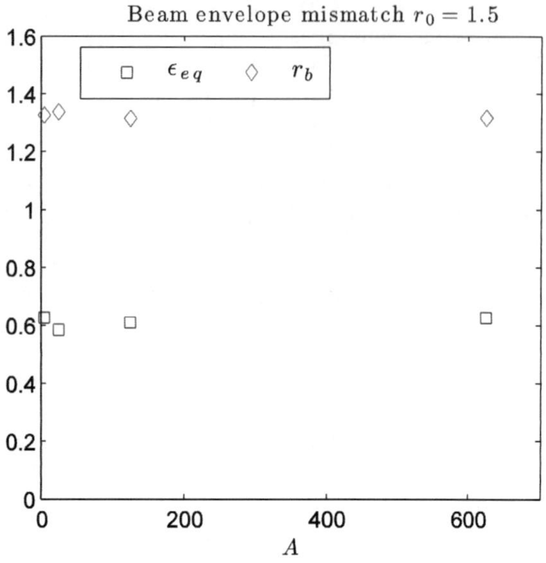

Figure 5. The equilibrium values of the emittance ϵ and beam envelope r_b for different values of noise amplitude A. Observe that the equilibrium values of both quantities do not depend of the of the noise amplitude A.

5. Conclusions and future works

The initial inhomogeneity is determinant for the emittance growth observed in charged particle beams and has impact over its time scale. However, no influences over beam macroscopic quantities such as envelope and emittance have been observed at the equilibrium. In fact, beam quantities at equilibrium strong depend of the initial beam envelope mismatch. Future works will contain more information about the issue.

Acknowledgements

This work has been done with the aid of the National Centre of Supercomputation of the Brazil's South Region (CESUP), at the Federal University of Rio Grande do Sul (UFRGS). L. C. Martins thanks the support provided by Universidade Estadual de Santa Catarina (UDESC), Brazil. The author R. P. Nunes has received financial support from the Conselho Nacional de Desenvolvimento Científico e Tecnológico (CNPq), and Fundação de Amparo à Pesquisa do Estado do Rio Grande do Sul (FAPERGS), both from Brazil.

References

[1] A. Cuchetti, M. Reiser, and T. Wangler, in Proceedings of the Invited Papers, 14th Particle Accelerator Conference, San Francisco, California (IEEE, New York, 1991), pp. 251.
[2] R. P. Nunes. PhD Thesis. UFRGS (2008).
[3] R. P. Nunes, R. Pakter, F. B. Rizzato, A. Endler, E. G. Souza, Phys. Plasmas 16, 033107 (2009).
[4] R. P. Nunes et al., in Proceedings of the 2008 European Particle Accelerator Conference, Genoa, Italy. (EPS-AG, Genoa, 2008), pp. 3218.
[5] R. P. Nunes, R. Pakter, and F. B. Rizzato, J. Appl. Phys. 104, 013302 (2008).
[6] R. P. Nunes, and F. B. Rizzato, in Proceedings of the 2009 Particle Accelerator Conference, Vancouver, Canada. (IEEE, Vancouver, 2009).

A minimalist model of calcium-voltage coupling in GnRH cells

Martin Rennie[1], **Rossanna Chan**[1], **David Schneider**[2], **Wen Duan**[1] **and James Sneyd**[1]

[1]Department of Mathematics, University of Auckland, Auckland 1142, New Zealand
[2]Departamento de Física, Comisión Nacional de Energía Atómica. Av. del Libertador 8250, 1429 Buenos Aires, Argentina

E-mail: schneide@tandar.cnea.gov.ar

Abstract. We present a minimalist model to describe the interplay between burst firing and calcium dynamics in Gonadotropin-releasing hormone (GnRH) cells. This model attempts to give a qualitative representation of Duan's model [3], and it comprises two FithzHugh-Nagumo (FHN) coupled systems describing the dynamics of the membrane potential and calcium concentration in the GnRH cells. Within the framework of our minimalist model, we find that the calcium subsystem drives burst firing by making the voltage subsystem to undergo a Hopf bifurcation. Specifically, fast relaxation oscillations occur in a specific region of the $c-z$ plane (c being the calcium concentration, and z a calcium-dependent gating variable). Slow calcium oscillations, instead, are carried by the voltage subsystem by successive shifts of the calcium steady state, and have the net effect of an external perturbation. The full comprehension of the phase-plane of the voltage subsystem and the 3-dimensional phase-space of the calcium subsystem ultimately allows us to study the behaviours of the entire model under the change of certain parameters. Those special parameters do not necessarily follow realistic assumptions, but merely intend to mimic some pharmacological tests which have been performed experimentally and also simulated by Duan's model under the corresponding physiological considerations.

1. Introduction

GnRH cells are located in the basal hypothalamus, and are responsible for the production of the GnRH decapeptide. The episodic release of this hormone into the portal system stimulates the pituitary to release gonadotropins, LH and FSH, which drive sexual developments and control the female menstrual cycles and male spermatogenesis. The mechanisms governing GnRH release dynamics are still unknown, and the investigation of those mechanisms could provide some clues to understand central aspects of life such as maturation and fertility.

Based on experimental data recorded by the group of Herbison in GnRH-Pericam transgenic mice [1, 2, 3], Duan *et al* constructed a detailed model for the behaviour of GnRH cells [3, 4]. Duan's model relies on the specific molecular components participating in the control of the membrane potential and the intracellular (and endoplasmic reticulum (ER)) calcium concentrations, and the model accurately reproduces the results obtained by Herbison *et al.* Moreover, the model was tested for different pharmacological conditions, and it was used to predict the existence of a slow activated long-duration potassium current (UCL2077 sensitive sI_{AHP}), which was later identified experimentally.

Published under licence by IOP Publishing Ltd

However, the extremely high dimensionality of the system (due to the many gating variables and variables describing the states of the different channels and pumps), makes it cumbersome to undertake a dynamical analysis of the model in a straightforward way[1].

In this work, we attempt to approach Duan's model starting from an essentially simplified framework. In order to capture only the excitable features of both the membrane potential and calcium concentration dynamics, and the fact that spiking activity in GnRH cells not only drives calcium transients (as is known to be the case) but it is itself driven by the latter, we put aside the detailed underlying biology and represent both burst firing and intracellular calcium concentration via two coupled FHN dynamical systems, and one (calcium-dependent) gating variable controlling voltage.

For such a simplified model, we can set the time scales according to the experimental findings (a fast evolution of the voltage, superimposed on a slow evolution of the calcium concentration). The remaining parameters are set so as to mimic the qualitative features of both the calcium concentration and the membrane potential, as obtained in the experiments or under Duan's model simulations. Nevertheless, the main goal of this paper is to achieve a full comprehension of the dynamics of the two subsystems, and the interplay between these two dynamics. Such a characterization of the voltage subsystem phase-plane, and the influence of that on the 3-dimensional phase-space associated to the calcium concentration, will eventually allow us to explain the behaviour of the complete model under the change of certain parameters. Those parameter modifications do not necessarily follow realistic assumptions, but are merely an attempt to mimic some pharmacological tests which were performed experimentally to recognize the key molecular components controlling the voltage and the intracellular calcium concentrations in the GnRH cells[2].

2. The minimalist representation of Duan's model

The full minimalist model for the activity of the GnRH cells is given by two FHN subsystems. The first subsystem concerns the fast voltage dynamics, and it reads,

$$\begin{aligned} \epsilon_1 \frac{\mathrm{d}v}{\mathrm{d}t} &= v(\theta_1 - v)(v - \alpha_1) - w_1 + I(c, z) \\ \frac{\mathrm{d}w_1}{\mathrm{d}t} &= \beta_1(v - \gamma_1 w_1). \end{aligned} \tag{1}$$

The second subsystem represents the slow dynamics of the calcium concentration, and it involves an extra variable (denoted by 'z') which plays the role of a delay variable,

$$\begin{aligned} \epsilon_2 \frac{\mathrm{d}c}{\mathrm{d}t} &= (c - 0.1)(\theta_2 - c)(c - \alpha_2) - w_2 + q(v) \\ \frac{\mathrm{d}w_2}{\mathrm{d}t} &= \beta_2(c - 0.1 - \gamma_2 w_2) \\ \tau_z(c)\frac{\mathrm{d}z}{\mathrm{d}t} &= z_\infty(c) - z. \end{aligned} \tag{2}$$

The fast voltage-subsystem and the slow calcium-subsystem are coupled by means of "current-like" terms, which are as follows,

[1] For a discussion of the bifurcation analysis of Duan's model, see [4].

[2] The simulations by means of Duan's model, on the contrary, obey more physiological considerations.

$$
\begin{aligned}
I(c, z) &= z\left(1 - \frac{c^{n_1}}{Kd_1{}^{n_1} + c^{n_1}}\right) \\
\text{and} & \\
q(v) &= h_2\left(\frac{v^{n_2}}{Kd_2{}^{n_2} + v^{n_2}}\right).
\end{aligned}
$$

The description is complete with the addition of the calcium-dependent functions governing the delay variable evolution,

$$
z_\infty(c) = \exp\left(-\frac{(c-p_2)^2}{b_2}\right) \quad \text{and} \quad \tau_z(c) = f_2 \exp(-d_2\, c).
$$

The time scales of both subsystems are set by choosing suitable values for the parameters ϵ_1, ϵ_2, β_1, and β_2. The full list of the parameters of the model is given in the Table 1.

Table 1. List of the model parameters.

index	α	β	γ	ϵ	θ	n	Kd	b	d	f	h	p
1	0.1	1.0	0.5	0.01	1.0	4	0.5	-	-	-	-	-
2	0.18	5.95×10^{-4}	1.0	0.2	0.52	4	0.5	5×10^{-4}	20.0	1.4×10^{3}	0.025	0.132

Figure 1 displays the simulations (for the standard experimental setup) of Duan's model and those of the minimalist model.

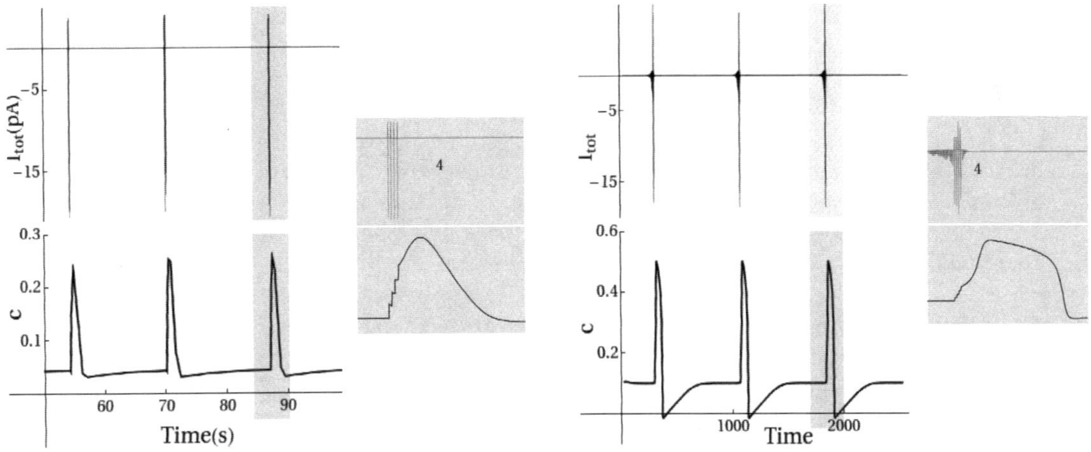

Figure 1. For the regular conditions of Herbison's experimental setup (left) Duan's model simulations and (right) the double FHN model simulations.

Even though the shapes of the calcium transients do not match perfectly well, the qualitative features are properly accomplished[3]. Namely, intracellular calcium displays slow-scale transients,

[3] In the simulations corresponding to the minimalist model the total current $I_{\rm tot}$ is computed as the time derivative of the voltage.

whereas the voltage displays fast-scale burst firings composed of 3 or 4 spikes per burst perfectly synchronized with the beginning of the calcium transients. There is only one feature which our model does not account, which is the randomness in the interval between two burst firings. This interburst interval setting is an important issue not only because it is correlated with the amplitude of the next calcium transient (see figure 1), but because is one of the main subjects under the control of the two afterhyperpolarization potassium currents studied in [3]. That failure of our model is a consequence of its fully deterministic character.

However in the double FHN model the relative time scales are controlled so as to reproduce the time scales of the experiments, the absolute value of the time remains measured in arbitrary units. The units of the calcium concentration and the current are arbitrary as well.

3. Dynamical analysis of the double FHN model[4]

The slow subsystem controls the fast subsystem by means of the term $I(c,z)$ (equation (1)). As depicted in figure 2, the effect of this "current" is a shift up and down of the cubic shaped nullcline of the fast subsystem (the v-nullcline). Now, for a specific value $I = I_{HB}$ the intersection of the v and w_1-nullclines happens exactly at the minimum of the former. Lower values of I mean an intersection to the left of that minimum, whereas higher values mean an intersection to the right. As an intersection to the left corresponds to a stable fixed point and an intersection to the right to an unstable one surrounded by a limit cycle, this special pair of "parameters" c and z (which make the w_1-nullcline to intersect the v-nullcline at its minimum) represents a Hopf bifurcation of the fast subsystem (see reference [5] for a further explanation). In fact, there is not one pair, but a continuous curve (in the $c-z$ plane) for which $I(c,z) = I_{HB}$, so we have actually a two parameter-dependent Hopf bifurcation separating two specific regions in the parameter-plane (figure 2, right). Region I corresponds to a steady state-regime, and region II to a relaxation oscillations-regime. Most of the time the fast system stays in the region I (close to the steady state), it eventually undergoes the Hopf bifurcation and enters region II (turning around the limit cycle four times), and finally it enters again region I (and approaches asymptotically to the steady state). The bottom panel of figure 2 displays the time series of the voltage during the incursions of the fast subsystem in the region II of the parameter-plane. These short term incursions driven by the slow subsystem are actually the mathematical substrate of the burst firing.

When viewed on the scenario of the slow subsystem, the two parameters (c and z) defining the Hopf bifurcation of the fast subsystem have to be regarded as dynamical variables. However, the calcium submodel involves three differential equations (see equation (2)), so now we have to visualize the $c-z$ parameter space of the voltage submodel (figure 2, top-right) as a single piece of a 3-dimensional phase-space, namely, the section $w_2 =$ constant. Figure 3 displays the trajectory of the slow subsystem in the full $c-w_2-z$ phase-space. 2-dimensional graphs depict the projections of the trajectory on the planes $c-w_2$ (left) and $c-z$ (right).

Even though the effect of the fast submodel on the slow one can be regarded also as a shift of the cubic-shaped nullcline (represented now by the c-nullcline of the calcium 3-dimensional submodel), the way as slow oscillations are generated is different. As depicted in figure 3, the nullclines governing the dynamics in the directions c and w_2 keep intersecting to the left of the minimum of the former even when the "current" q saturates near its maximum value h_2. As ϵ_2 is small enough for being under the FHN's model conditions (which means a fast evolution in the c direction and a slow evolution in the w_2 direction), the previous remark implies that the fixed point doesn't change its stability. Yet, when the slow subsystem is perturbed by means of a strong current q (near the top of the voltage spikes), the c-nullcline is momentarily

[4] The analysis performed in this section is based on the well known time scale separation techniques and bifurcation analysis. For a detailed overview of the subject, see references [5, 6].

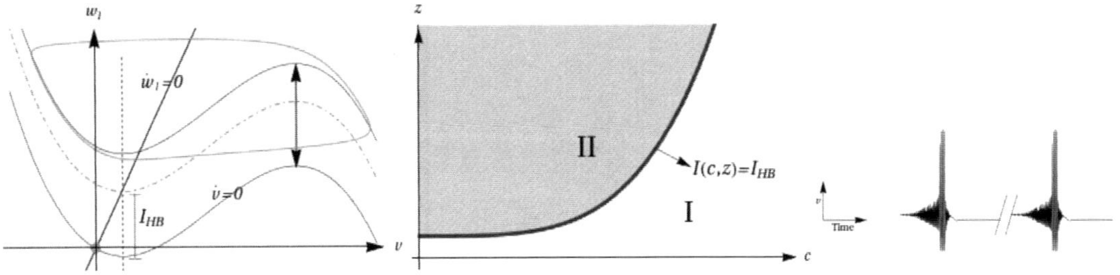

Figure 2. Effect of calcium concentration on the voltage. (Left) For low values of $I(c, z)$ the fast subsystem stays close to a stable fixed point. Beyond the threshold given by $I(c, z) = I_{HB}$, it starts turning around a limit cycle. The corresponding attractors are displayed in orange. (Middle) The Hopf bifurcation curve in the parameter-plane separating two different behaviour regions. Region I corresponds to a steady state regime where the membrane potential is at rest. Region II corresponds to the relaxation oscillations-regime. Short term incursions of the fast subsystem into the region II are the responsible for the spiking activity of the membrane potential (right).

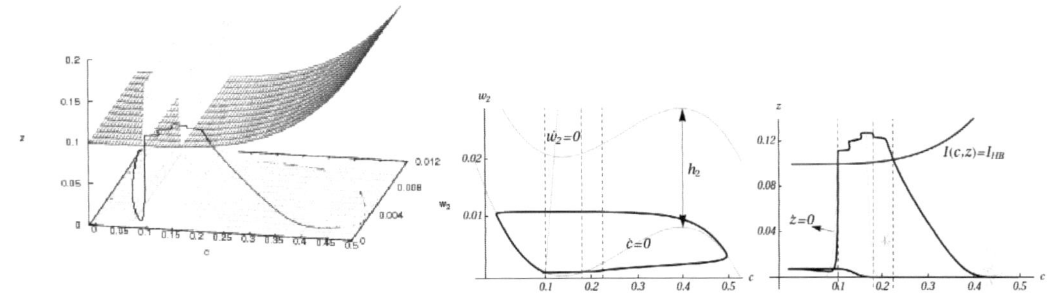

Figure 3. (Left) 3-dimensional phase-space of the slow-subsystem displaying the trajectory of a regular calcium transient (black), the Hopf bifurcation surface (blue) and the z-nullcline (green). (Middle and right) Projections of the trajectory on the $c{-}w_2$ and the $c{-}z$ planes. The most left and the most right vertical dashed lines enclose the bursting region. In between, a third vertical dashed line denotes the point where the trajectory intersects the resting c-nullcline.

'replaced' by a positive cubic-shaped $\dot{c}$ curve, which gives rise to an hypothetical trajectory with the typical shape of that of a perturbed excitable system (namely, an excitable system which has been perturbed enough and consequently performs a large excursion in the phase-space). This mechanism is illustrated in figure 4, where we superimpose to the phase-space (the $c{-}w_2$ plane only) the contours of the c component of the velocity field. We stress that this trajectory is hypothetical in the sense that the slow system doesn't complete it as in the case of an usual excitable system, but it follows it only for the very short time for which the current q is large enough, namely, the duration of a single voltage spike. Between spikes, the current q vanishes, the c-nullcline returns to its original position, and the $c{-}w_2$ phase-plane looks again as at the beginning of the burst firing (figure 4, left). Moreover, in these interspike intervals $\dot{c}$ becomes negligible and z starts to be the only relevant direction of movement (figure 3, bottom-right), until a new spike takes place and the process repeats again[5].

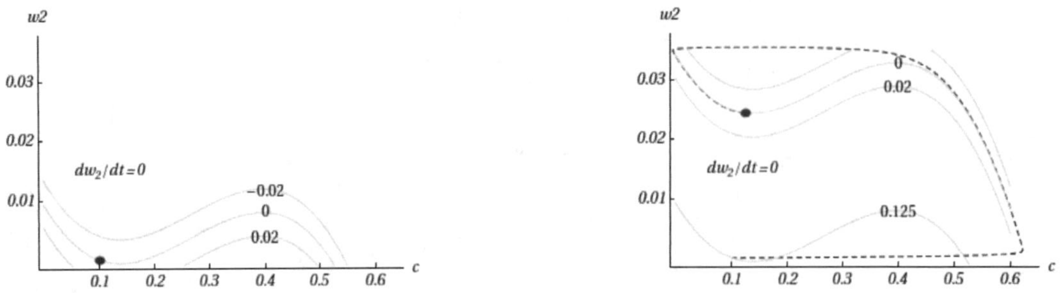

Figure 4. Superimposed to the $c-w_2$ plane-space of the slow subsystem, the contours of the c component of the velocity field (light blue curves). (Left) Interburst intervals and interspike intervals within a burst firing: $v \simeq 0$, so that $q \simeq 0$ and therefore the c-nullcline is at rest. (Right) Near the top of the first spike: $v \gg Kd_2$, so that $q \simeq h_2$ and therefore the c-nullcline is replaced by a positive $\dot{c}$ cubic-shaped curve. The dashed curve represents the trajectory followed by the slow subsystem when the voltage is at the top of the spike. In both panels, the purple disks represent the $c-w_2$ projection of the corresponding steady state.

The mechanism described in the previous paragraph repeats several times (as many times as spikes are in one burst), and it is the responsible for the 'jumps' of the trajectory in the c-direction during the burst firing (see figure 3, bottom-right. In this picture the most left and the most right vertical dashed lines depict the values of c which enclose the burst region, which corresponds to the portion of the trajectory above the Hopf bifurcation surface of the fast system). However, we stress that this mechanism, being necessary, is not enough by itself to generate the experimental large-amplitude calcium transients that the model is capable to simulate. A second mechanism, which has again the feature of the perturbation of an excitable system (but that works in a different manner) is required. In both the left and right panels of figure 3 we plot three vertical dashed lines, the most left and the most right being the specific values of c for which the trajectory intersects the Hopf bifurcation of the fast system. The middle vertical dashed line, on the other hand, depicts the value of c for which the trajectory intersects the c-nullcline. This specific value of c, which not necessarily coincides with the end of the burst firing region, corresponds to the real slow system's stability threshold. Namely, once the c variable goes beyond this value, $\dot{c}$ becomes positive regardless the value of the current q. Later on, a short term increase of the velocity has a negligible effect on the future trajectory, and the large excursion in the slow phase-space happens regardless what the fast system is doing. The fast system might actually keep spiking, but eventually it will cross again the Hopf bifurcation surface and leave the burst firing region (region II in the central panel of figure 2).

To sum up, fast oscillations are driven by the slow subsystem by means of a calcium-dependent current which set the fast system into the relaxation oscillations-regime. Slow oscillations of large amplitude, instead, are a consequence of the interplay of two similar, but qualitatively different

[5] As a matter of fact, between spikes new hypothetical trajectories emerge with almost negligible (although negative) initial values of $\dot{c}$. This effect is more evident during the intermediate interspike intervals, and it is a consequence of the fact that, during these intervals the trajectory is slightly above the middle branch of the c-nullcline (figure 3, bottom-left panel). Coming again to the comparison with an usual excitable system, this situation would be equivalent to a perturbation below the stability threshold (which in this case is in addition of a short term), and is therefore not relevant for the outcome of the present analysis. Yet, the negative component of the velocity in the c direction is still noticeable in the $c-z$ projection of the trajectory (figure 3, bottom-right panel.)

mechanisms. Both mechanisms are comparable to that of an usual FHN system which, being out of the relaxation oscillations-regime, is perturbed in a certain way. One of these mechanisms is voltage-dependent and it involves a short term increase of the c component of the velocity. The second mechanism is voltage-independent and means a simple increase of the variable c (not its time derivative) beyond the middle branch of the FHN's cubic shaped nullcline, which is actually the result of several realizations of the first mechanism.

4. The behaviour of the minimalist model beyond the regular conditions

In this section we will give three examples which illustrate how the mechanisms controlling the burst firing and the calcium oscillations might fail, and what happens to both subsystems in each situation. Those examples, being the outcome of the modification of some specific parameters, are connected with different pharmacological perturbations of the standard experimental setup. For a detailed discussion of those pharmacological perturbations and the corresponding Duan's model realizations see [3, 4].

4.1. The Tetrodotoxin test

Tetrodoxin (TTX) is a well known sodium current blocker, and it has therefore the effect of abolishing action potentials in excitable cells. In [3] TTX has been supplied to test whether burst activity initiated the high amplitude calcium transients. As the outcome of the experiment was a total suppression of both burst firing and calcium oscillations, the conclusion was that it actually did.

To prevent the fast subsystem of the minimalist model to oscillate, we only need to avoid the incursions of the fast subsystem trajectory into the relaxation oscillations region of the $c-z$ parameter-plane (region II in figure 2, bottom right panel). An easy way to accomplish this task, is by increasing the numerical value of I_{HB}, which has the effect of displacing the Hopf bifurcation upwards in the parameter-plane. A direct examination of the left panel of figure 2 immediately suggests that this can be done by simple making the parameter γ_1 equal to zero, as this parameter inversely controls the slope of the w_1-nullcline. As a result, the w_1-nullcline acquires a vertical slope, and the minimum of the c-nullcline can not intersect it regardless the value of the actual current $I(c, z)$. Hence, the region of relaxation oscillations disappears from the parameter-plane. Figure 5 displays the behaviour of the model for such conditions. The limitation of the fast subsystem to reach the relaxation oscillations-regime prevents the slow subsystem to be perturbed by means of the mechanisms described in the previous section, and hence it moves to its steady state causing calcium oscillations to be completely suppressed.

4.2. The Cyclopiazonic Acid test

The roles of calcium internal stores were evaluated by Herbison *et al* by treating GnRH neurons with 2-aminoethoxydiphenylborate (2-APB) and cyclopiazonic acid (CPA); antagonists at IP_3R and the sarcoplasmic reticulum calcium-ATPase pump, respectively. They found that 2-APB blocked both calcium transients and burst firings, whereas CPA had the effect of suppressing calcium transient amplitude and making burst firing disorganized[3].

To suppress calcium transients in our minimalist model (allowing voltage activity), we have to focus on the first mechanism controlling the slow subsystem dynamics. By eliminating that mechanism (which in turn causes the suppression of the second mechanism, however this is an indirect effect -we will see shortly the outcome of a direct suppression of the second mechanism), we prevent at all calcium oscillations keeping unaffected the chance of the fast subsystem to develop action potentials. A possible parameter to be perturbed to accomplish such a behaviuor is the parameter which controls the amplitude of the current $q(v)$, namely, h_2. Figure 6 displays the model simulations for the normal conditions ($h_2 = 0.025$), followed by an interval of 'low extracellular calcium' ($h_2 = 0.005$). During the top of the spike, the current $q(v)$ makes calcium

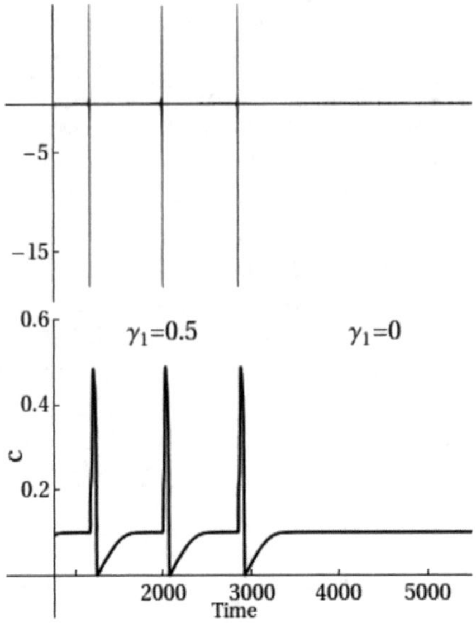

Figure 5. Simulations of the TTX test.

to increase even for small value of $q(v)$ (see figure 6, right panel). However, as the current is very week, the positive velocity which calcium acquires during the spikes is comparable to the negative velocity it acquires during the interspike intervals. As a result, calcium performs very tiny, constant average, tent-like oscillations around the baseline, and the trajectory looks as it reaches a steady state in the 3-dimensional phase-space (figure 6). Moreover, as in our simulations the parameter h_2 changes abruptly, during a short time-window just after the change z turns out to be the main direction of motion (this is because $\dot{c}$ is small both when it is negative and when it is positive), and therefore the system oscillates around the z-nullcline until the actual intersection of the three nullclines is enough approached.

4.3. The Free Calcium test

To analyse the role of external calcium in burst firing generation, Herbison *et al* replaced calcium in the normal artificial cerebral spinal fluid with equimolar magnesium. As a result they observed a gradual suppression of calcium transients correlated with a failure of the normal burst firing patterns. The reduction of calcium transients correlated with a decrease of interburst intervals, an increase in burst duration and an increase in spikes per burst[3].

Most of free calcium experiment findings are accomplished by the minimalist model when the second mechanism controlling the slow subsystem dynamics is directly suppressed. We want now the first mechanism to work properly, so we will keep the parameter h_2 somehow close to the standard conditions (however we will decrease it slightly to enlarge the spikes per burst). As discussed in section 3, the second mechanism means a perturbation of the slow subsystem beyond the threshold given by the middle branch of the c-nullcline; it is a voltage independent mechanism (namely, it makes the c component of the velocity positive even if the burst firing have finished), and is the responsible for the high amplitude of the calcium transient. A direct way for suppressing this mechanism is by displacing the middle branch of the c-nullcline to the right. This can be done easily by increasing the parameter α_2, which defines the position of the middle root of the c-nullcline. The results of the simulations are depicted in figure 6.

We stress that in this scenario calcium oscillations actually occur, however the slow system

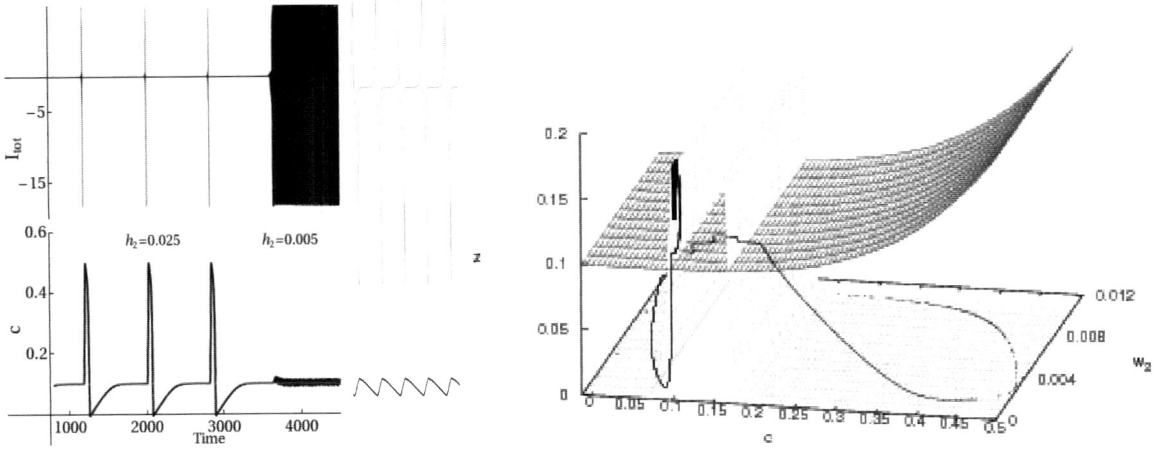

Figure 6. (Left) Simulations of the CPA test (in the region $h_2 = 0.005$, a small time-window is squeezed to emphasize the behaviour of the model). (Right) Phase-space of the slow subsystem (one regular calcium transient is included before the CPA simulation).

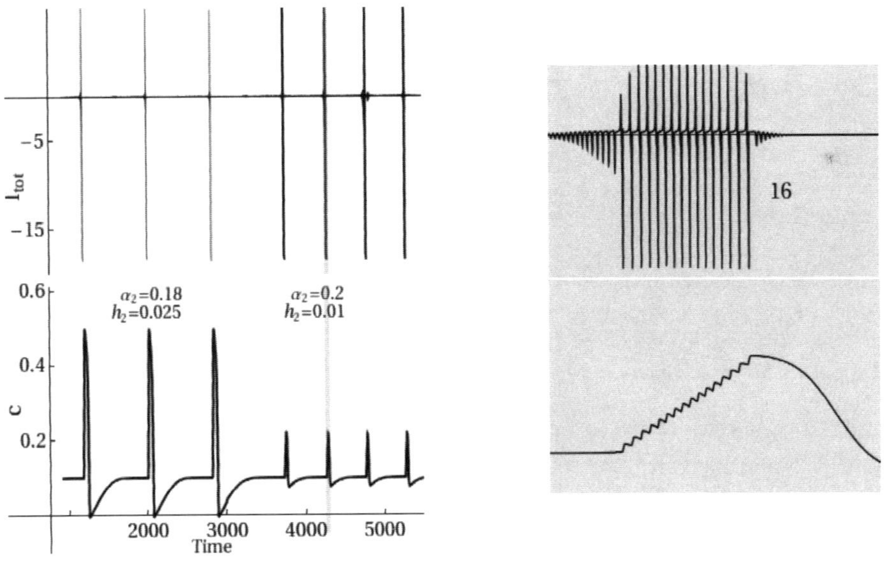

doesn't reach the threshold represented by the middle branch of the c-nullcline neither. This fact is depicted in 7, where we display the trajectory of the slow subsystem in the 3-dimensional phase-space and the corresponding projections in the $c-w_2$ and the $c-z$ planes. Unlike the case of the regular bursting, the trajectory does not intersect now the resting cubic shaped nullcline, and therefore the oscillation of the slow subsystem does not mean a big excursion in the phase-space (compare with figure 3).

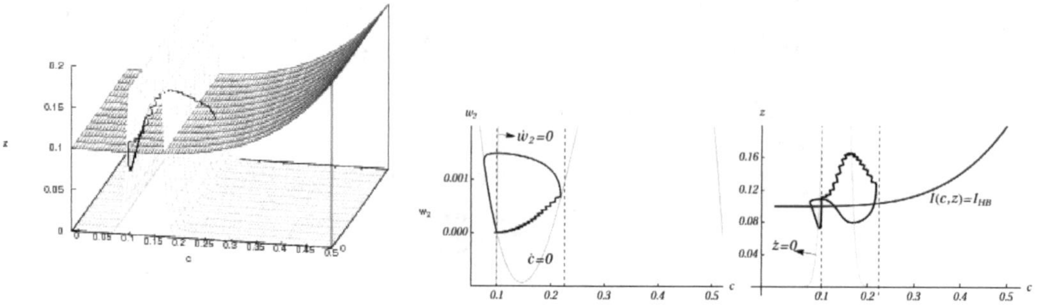

Figure 7. (Left) Phase-space of the slow subsystem for the free Calcium test (compare with figure 3). (Middle and right) Projections of the trajectory on the c–w_2 and the c–z planes. As in figure 3, the most left and the most right vertical dashed lines enclose the bursting region. The trajectory does not intersect here the resting c-nullcline, and thus the middle vertical dashed line is not present.

5. Discussion

We have constructed a minimalist model to reproduce the qualitative features of voltage-calcium coupling in GnRH cells, and have analysed its dynamical behaviour. As a result, we observed that calcium drives burst firing by setting the voltage subsystem to a relaxation oscillation-regime. In turn, the fast voltage oscillations are responsible for the initiation of the calcium transients, however the high amplitude of the transients is a consequence of the capability of the slow subsystem to cross a threshold which is not voltage dependent.

As we stated previously, the parameters of our model are not related to the physiological properties of the GnRH cell. This fact makes it hard to take advantage of the results to extract realistic conclusions. However, as a next step we are focused in linking both Duan's and the double FHN model further. This will be accomplished by comparing the dynamical properties of the two models. For instance, we are concerned to establish the normal form of the bifurcation of Duan's fast submodel when burst firing is developed. In our model, this process is associated to a Hopf bifurcation, so we expect to observe this type of bifurcation for the fast piece of Duan's model. If this assertion turns out to be true, a suitable fitting between both models would revel the physiological value of the parameters . An approach to the bifurcation structure of Duan's model is accomplished in [4].

Ideally, this procedure would provide a better understanding of the interplay between the calcium and the membrane voltage activity at a molecular level, in the sense that it would tell us which machineries of the cell are actually crucial to develop the observed behaviours.

Acknowledgments

This work is supported by the Marsden Fund of the Royal Society and the Conicet.

References

[1] Jasoni C L, Todman M G, Strumia M M and Herbison A E 2007 *The Journal of Neuroscience* **27** 860-7
[2] Xinhuai L, Lee K and Herbison A E 2008 *Endocrinology* **149** 4605-14
[3] Lee K, Duan W, Sneyd J and Herbison A E 2010 *The Journal of Neuroscience* **30** 6214-24
[4] Duan W, Lee K, Herbison A E and Sneyd J. To appear in the *Journal of Theoretical Biology*
[5] Keener J and Sneyd J 2009 *Mathematical Physiology* Second ed (New York: Springer)
[6] Izhikevich E M 2007 *Dynamical Systems in Neuroscience: The geometry of excitability and bursting*

Squeezing of thermal noise in a parametrically-driven oscillator

Adriano A. Batista

Departamento de Física, Universidade Federal de Campina Grande, Campina Grande-PB, Brazil

E-mail: adriano@df.ufcg.edu.br

Abstract. In this article we report a theoretical model based on Green's functions and averaging techniques that describes the dynamics of parametrically-driven mechanical resonators under the action of thermal noise. Quantitative estimates for quadrature thermal noise squeezing near the first parametric instability zone of the oscillator are given. Furthermore, the parameter space where these phenomena occur is presented. Very good agreement between analytical estimates and numerical results is achieved.

1. Introduction

The field of microelectromechanical devices (MEMs) has been developing rapidly since the advent of the atomic force microscope in mid 80's [1]. MEMs have been used primarily for measuring small forces and as ultrasensitive mass detectors. Further detection techniques have been developed in the early 90's that use mechanical parametric amplification (before transduction) to improve the sensitivity of measurements. This amplification method works by driving the parametrically-driven oscillator on the verge of parametric unstable zones. Rugar and Grütter [2] have shown ways, using this method, to obtain linear parametric gain. They were also looking for ways of reducing noise and increasing precision in a detector for gravitational waves, when they experimentally found classical thermomechanical squeezing. DiFilippo et al. [3, 4] have proposed explanations for the noise squeezing phenomenon, but their models do not treat noise directly in the equations of motion. They use a Gaussian (thermal) distribution of initial states and evolve it deterministically. In this evolution amplitude squeezing of the signal was observed, at the same time higher dephasing was obtained. In Ref. [4], the authors proposed a theoretical model for explaining the cause of thermomechanical squeezing in which the system parameters are chosen to be inside the first zone of parametric instability. Such approach is not correct, since, with this choice, the solutions diverge exponentially. One would have to use a nonlinear model, based on the Duffing oscillator [5, 6], instead of the linear model used to prevent the divergence. Furthermore, detuning, dissipation, and stochastic dynamics are all neglected. Basicaly, there is no treatment of the dynamics due to noise, apart from the probabilistic initial state distribution. In Ref. [7], a nonlinear noise squeezing procedure is experimentally observed in a driven Duffing oscillator, again the theoretical model used does not treat noise dynamically.

Here, we investigate the effect of adding thermal noise to the fundamental mode of a doubly-clamped beam resonator that is axially loaded (i.e., its first normal mode is equivalent to a parametrically-driven oscillator) with the objective of explaining thermomechanical noise squeezing. This model can also be applied to the linear response of non-linear driven oscillators, in which, due to the linearization, the response is given by parametrically-driven oscillations. With our model we obtain analytical quantitative estimates of the amount of quadrature thermal-noise squeezing. In the following section we use a Green's function approach to solve the Langevin equation and, aligned with averaging techniques, we obtain analytical estimates of the oscillator fluctuations showing how thermal-noise squeezing occurs.

2. Theoretical model

The equation for the parametrically-driven oscillator (in dimensionless format) is given by

$$\ddot{x} + \omega_0^2 x = -\gamma \dot{x} + F_p \cos(2\omega t)\; x, \tag{1}$$

in which γ and $F_p \sim O(\varepsilon)$, where $\varepsilon << 1$. Since we want to apply the averaging method (AM) [8, 9] to situations in which we have detuning, it is convenient to rewrite Eq. (1) in a more appropriate form with the notation $\Omega = \omega_0^2 - \omega^2$, where we also have $\Omega \sim O(\varepsilon)$. With this substitution we obtain $\ddot{x} + \omega^2 x = -\Omega x - \gamma \dot{x} + F_p \cos(2\omega t)\; x$. We then rewrite this equation in the form $\dot{x} = y$, $\dot{y} = -\omega^2 x + f(x, y, t)$, where $f(x, y, t) = -\Omega x + F_p \cos(2\omega t)\; x - \gamma y$. We now set the above equation in slowly-varying form with the Van der Pol transformation

$$\begin{pmatrix} x \\ y \end{pmatrix} = \begin{pmatrix} \cos\omega t & -\sin\omega t \\ -\omega \sin\omega t & -\omega\cos\omega t \end{pmatrix} \begin{pmatrix} \mathcal{U} \\ \mathcal{V} \end{pmatrix}$$

and obtain

$$\begin{aligned} \begin{pmatrix} \dot{\mathcal{U}} \\ \dot{\mathcal{V}} \end{pmatrix} &= \begin{pmatrix} \cos\omega t & -\frac{1}{\omega}\sin\omega t \\ -\sin\omega t & -\frac{1}{\omega}\cos\omega t \end{pmatrix} \begin{pmatrix} 0 \\ f(x,y,t) \end{pmatrix} \\ &= -\frac{1}{\omega} \begin{pmatrix} \sin\omega t f(x,y,t) \\ \cos\omega t f(x,y,t) \end{pmatrix}. \end{aligned}$$

After an application of the averaging method (in which, in our case, basically, we filter out oscillating terms at and near 2ω), we obtain

$$\begin{aligned} \dot{u} &= \frac{-1}{2\omega}\left[\gamma\omega u + \left(\Omega + \frac{F_p}{2}\right) v\right], \\ \dot{v} &= \frac{-1}{2\omega}\left[\left(-\Omega + \frac{F_p}{2}\right) u + \gamma\omega v\right], \end{aligned} \tag{2}$$

where the functions $\mathcal{U}(t)$ and $\mathcal{V}(t)$ were replaced by their slowly-varying averages $u(t)$ and $v(t)$, respectively. It is known that, by integration of Eq. (2), one finds the first parametric resonance, i.e. the boundary between the stable and unstable regions. It is given by

$$(\gamma\omega)^2 = (F_p/2)^2 - \Omega^2. \tag{3}$$

This result is valid for $\omega \approx \omega_0$ even in the presence of noise. In Fig. (1) we find very good agreement between results obtained from numerical integration of Eq. (1) and the boundary given by the averaging technique.

We now will investigate the effect of noise on the parametric amplification mechanism [2, 3, 4, 7]. We start by adding noise to Eq.(1) and obtain

$$\ddot{x} = -\omega_0^2 x - \gamma\dot{x} + F_p \cos(2\omega t)\; x + R(t), \tag{4}$$

where $R(t)$ is a random function that satisfies the statistical averages $\langle R(t)\rangle = 0$ and $\langle R(t)R(t')\rangle = 2T\gamma\delta(t - t')$, according to the fluctuation-dissipation theorem [10]. T is the heat-bath temperature of the surrounding medium in which the oscillator is embedded. Once we integrate these equations of motion we can show how classical mechanical noise squeezing occurs.

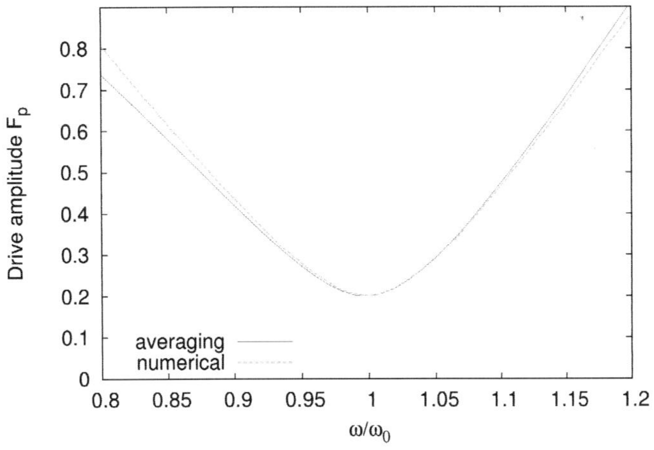

Figure 1. (color online) Comparison between numerical and averaging method predictions for the boundary of the first instability zone of the damped parametrically-driven oscillator of Eq. (1). In the region above the green line lies the unstable zone obtained by numerical computation, while the region above the red line is the analytical prediction for the unstable zone. The numerical results are obtained by numerically calculating the corresponding Floquet multipliers, when at least one of them has modulus equal to 1. The averaging predictions are given by Eq. (3). The fixed parameters of the equations of motion are $\gamma = 0.1, \omega_0 = 1.0$. These parameters were also used to obtain the results portrayed in the following figures.

2.1. Green's function method

Equation (4) can be solved exactly by using Floquet theory and Green's functions methods [11]. The equation for the corresponding Green's function is given by

$$\left[\frac{\partial^2}{\partial t^2} + \omega_0^2 + \gamma\frac{\partial}{\partial t} - F_p\cos(2\omega t)\right] G(t,t') = \delta(t-t') \tag{5}$$

Using the Green's function we obtain the solution $x(t)$ of Eq. (4) in the presence of noise $R(t)$

$$x(t) = x_h(t) + \int_{-\infty}^{\infty} dt'\; G(t,t')R(t'),$$

where $x_h(t)$ is the homogeneous solution, which in the stable zone decays exponentially with time; since we assume the pump has been turned on for a long time, $x_h(t) = 0$. By statistically averaging the fluctuations as a function of time we obtain

$$\begin{aligned}\langle x^2(t)\rangle &= \textstyle\iint_{-\infty}^{\infty} dt'\; dt'' G(t,t')G(t,t'')\langle R(t')R(t'')\rangle \\ &= 2T\gamma \textstyle\int_{-\infty}^{t} dt'\; G(t,t')^2 = 2T\gamma \int_0^{\infty} d\tau\; G(t,t-\tau)^2,\end{aligned} \tag{6}$$

where $\tau = t - t'$.

Instead of using Floquet theory, which, though exact, yields no simple analytical solutions, we obtain fairly simple analytical approximations to the Green's functions using the averaging method, in which we solve Eq. (2), where the initial conditions at $t = t'$ are given by $u(t') = -\sin(\omega t')/\omega$ and

$v(t') = -\cos(\omega t')/\omega$. The fundamental matrix of this system is given by

$$e^{A(t-t')} = e^{-\gamma(t-t')/2} \times \begin{pmatrix} \cosh(\kappa\,(t-t')) & \frac{\beta-\delta}{\kappa}\sinh(\kappa\,(t-t')) \\ \frac{\beta+\delta}{\kappa}\sinh(\kappa\,(t-t')) & \cosh(\kappa\,(t-t')) \end{pmatrix},$$

where $\kappa = \sqrt{\beta^2-\delta^2}$, $\beta = -F_p/4\omega$ and $\delta = \Omega/2\omega$. The approximate Green's function is given by

$$G(t,t') \approx e^{-\gamma(t-t')/2}\left[\cos(\omega t)\left(\cosh(\kappa\,(t-t'))u(t') + \frac{\beta-\delta}{\kappa}\sinh(\kappa\,(t-t'))v(t')\right)\right.$$
$$\left. -\sin(\omega t)\left(\frac{\beta+\delta}{\kappa}\sinh(\kappa\,(t-t'))u(t') + \cosh(\kappa\,(t-t'))v(t')\right)\right] \quad (7)$$

for $t > t'$ and $G(t,t') = 0$ for $t < t'$. In the stable zone of the parametrically-driven oscillator, when $|\beta| < |\delta|$, we can rewrite the Green's function, replacing explicitly the initial conditions and using simplifying trigonometrical identities. We obtain

$$G(t,t') \approx -\frac{e^{-\gamma(t-t')/2}}{\omega}\left\{\cos(\omega t)\left[\cos(|\kappa|\,(t-t'))\sin(\omega t') + \frac{\beta-\delta}{|\kappa|}\sin(|\kappa|\,(t-t'))\cos(\omega t')\right]\right.$$
$$\left. -\sin(\omega t)\left[\frac{\beta+\delta}{|\kappa|}\sin(|\kappa|\,(t-t'))\sin(\omega t') + \cos(|\kappa|\,(t-t'))\cos(\omega t')\right]\right\}. \quad (8)$$

The change of variables $t' = t-\tau$ results in

$$G(t,t-\tau) \approx \frac{e^{-\gamma\tau/2}}{\omega}\left\{\cos(|\kappa|\,\tau)\sin(\omega\tau) + \frac{\delta}{|\kappa|}\sin(|\kappa|\,\tau)\cos(\omega\tau) - \frac{\beta}{|\kappa|}\sin(|\kappa|\,\tau)\left[\cos(\omega\tau)\cos(2\omega t)\right.\right.$$
$$\left.\left. +\sin(\omega\tau)\sin(2\omega t)\right]\right\}. \quad (9)$$

The squared Green's function is given by

$$G(t,t-\tau)^2 \approx \frac{e^{-\gamma\tau}}{\omega^2}\left\{\left[\cos(|\kappa|\tau)\sin(\omega\tau) + \frac{\delta}{|\kappa|}\sin(|\kappa|\tau)\cos(\omega\tau)\right]^2 + \frac{\beta^2}{2|\kappa|^2}\sin^2(|\kappa|\tau)\right.$$
$$-\frac{2\beta\sin(|\kappa|\,\tau)}{|\kappa|}\left[\cos(|\kappa|\,\tau)\sin(\omega\tau) + \frac{\delta}{|\kappa|}\sin(|\kappa|\,\tau)\cos(\omega\tau)\right]\left[\cos(2\omega t)\cos(\omega\tau) + \sin(2\omega t)\sin(\omega\tau)\right]$$
$$\left. +\frac{\beta}{2|\kappa|^2}\sin^2(|\kappa|\tau)\left[\cos(2\omega\tau)\cos(4\omega t) + \sin(2\omega\tau)\sin(4\omega t)\right]\right\}.$$

To obtain the squared Green's function when $|\beta| > |\delta|$ one should replace $|\kappa|$ by $i\kappa$ in the above expression. When there is no external parametric driving, $F_p = 0$, and $\Omega = 0$, we obtain

$$G(t,t-\tau) \approx \frac{e^{-\gamma\tau/2}}{\omega_0}\sin(\omega_0\tau),$$

which is within an error of order $O(\epsilon^2)$ of the simple harmonic oscillator Green's function. Also, the average fluctuations at $F_p = \Omega = 0$ become

$$\langle x^2\rangle = T/\omega_0^2\left[1 - \frac{\gamma^2}{\gamma^2+4\omega_0^2}\right],$$

reproducing the equipartition within an error of $O(\epsilon^2)$. As we are using an approximation of the Green's function with an error of $O(\epsilon)$, as stated below Eq. (1), this result is within the expected error bounds. An estimate of the time average of the thermal fluctuations, when $|\beta| < |\delta|$, is given by

$$\begin{aligned}
\overline{\langle x^2(t)\rangle} &= \frac{2T\gamma}{\omega^2}\int_0^\infty e^{-\gamma\tau}\left\{\left[\cos(|\kappa|\tau)\sin(\omega\tau)+\frac{\delta}{|\kappa|}\sin(|\kappa|\tau)\cos(\omega\tau)\right]^2+\frac{\beta^2}{2|\kappa|^2}\sin^2(|\kappa|\tau)\right\}d\tau \\
&= \frac{2T\gamma}{\omega^2}\left\{\frac{\beta^2}{\gamma(\gamma^2+4|\kappa|^2)}+\frac{\delta\gamma}{2|\kappa|}\left[\frac{1}{\gamma^2+4(|\kappa|-\omega)^2}-\frac{1}{\gamma^2+4(|\kappa|+\omega)^2}\right]\right. \\
&\quad +\frac{1}{4}\left[\frac{1}{\gamma}+\frac{\gamma}{\gamma^2+4|\kappa|^2}-\frac{\gamma}{\gamma^2+4\omega^2}-\frac{\gamma}{2}\left(\frac{1}{\gamma^2+4(|\kappa|-\omega)^2}+\frac{1}{\gamma^2+4(|\kappa|+\omega)^2}\right)\right] \\
&\quad \left.+\frac{\delta^2}{4|\kappa|^2}\left[\frac{1}{\gamma}+\frac{\gamma}{\gamma^2+4\omega^2}-\frac{\gamma}{\gamma^2+4|\kappa|^2}-\frac{\gamma}{2}\left(\frac{1}{\gamma^2+4(|\kappa|-\omega)^2}+\frac{1}{\gamma^2+4(|\kappa|+\omega)^2}\right)\right]\right\}
\end{aligned} \tag{10}$$

An estimate of the time average of the thermal fluctuations, when $|\beta| > |\delta|$, is given by

$$\begin{aligned}
\overline{\langle x^2(t)\rangle} &= \frac{2T\gamma}{\omega^2}\int_0^\infty e^{-\gamma\tau}\left\{\left[\cosh(\kappa\tau)\sin(\omega\tau)+\frac{\delta}{\kappa}\sinh(\kappa\tau)\cos(\omega\tau)\right]^2+\frac{\beta^2}{2\kappa^2}\sinh^2(\kappa\tau)\right\}d\tau \\
&= \frac{2T\gamma}{\omega^2}\left[I_1+I_2+\delta^2 I_3+\delta I_4\right],
\end{aligned} \tag{11}$$

where the integrals are given by

$$\begin{aligned}
I_1 &= \frac{\beta^2}{2\kappa^2}\int_0^\infty e^{-\gamma\tau}\sinh^2(\kappa\tau)d\tau = \frac{\beta^2}{\gamma(\gamma^2-4\kappa^2)}, \\
I_2 &= \int_0^\infty e^{-\gamma\tau}\cosh^2(\kappa\tau)\sin^2(\omega\tau)d\tau \\
&= \frac{1}{2}\left\{\frac{1}{2\gamma}-\frac{\gamma}{2(\gamma^2+4\omega^2)}+\frac{\gamma}{2(\gamma^2-4\kappa^2)}-\frac{1}{4}\mathrm{Re}\left[\frac{1}{\gamma-2\kappa-2i\omega}+\frac{1}{\gamma+2\kappa-2i\omega}\right]\right\}, \\
I_3 &= \frac{1}{\kappa^2}\int_0^\infty e^{-\gamma\tau}\sinh^2(\kappa\tau)\cos^2(\omega\tau)d\tau \\
&= \frac{1}{\kappa^2}\left\{\frac{\kappa^2}{\gamma(\gamma^2-4\kappa^2)}+\frac{1}{8}\mathrm{Re}\left[\frac{1}{\gamma-2\kappa-2i\omega}+\frac{1}{\gamma+2\kappa-2i\omega}\right]-\frac{\gamma}{4(\gamma^2+4\omega^2)}\right\}, \\
I_4 &= \frac{1}{2\kappa}\int_0^\infty e^{-\gamma\tau}\sinh(2\kappa\tau)\sin(2\omega\tau)d\tau = \frac{\gamma}{4i\kappa}\left[\frac{1}{(\gamma^2-4(\kappa-i\omega)^2}-\frac{1}{\gamma^2+4(\kappa+i\omega)^2}\right].
\end{aligned}$$

An estimate of the thermal fluctuations is given by

$$\langle x^2(t)\rangle \approx \overline{\langle x^2(t)\rangle}+A_{2\omega}\cos(2\omega t)+B_{2\omega}\sin(2\omega t)+A_{4\omega}\cos(4\omega t)+B_{4\omega}\sin(4\omega t). \tag{12}$$

From Eqs. (6) and (7), we find that the coefficients necessary for the calculation of the fluctuations are given by

$$\begin{aligned}
A_{2\omega} &= -\frac{4\beta T\gamma}{\omega^2}\int_0^\infty d\tau\, e^{-\gamma\tau}\frac{\sinh(\kappa\,\tau)}{\kappa}\left[\cosh(\kappa\,\tau)\sin(\omega\tau)+\frac{\delta}{\kappa}\sinh(\kappa\,\tau)\cos(\omega\tau)\right]\cos(\omega\tau), \\
B_{2\omega} &= -\frac{4\beta T\gamma}{\omega^2}\int_0^\infty d\tau\, e^{-\gamma\tau}\frac{\sinh(\kappa\,\tau)}{\kappa}\left[\cosh(\kappa\,\tau)\sin(\omega\tau)+\frac{\delta}{\kappa}\sinh(\kappa\,\tau)\cos(\omega\tau)\right]\sin(\omega\tau), \\
A_{4\omega} &= \frac{\beta^2 T\gamma}{\omega^2\kappa^2}\int_0^\infty d\tau\, e^{-\gamma\tau}\sinh^2(\kappa\tau)\cos(2\omega\tau), \\
B_{4\omega} &= \frac{\beta^2 T\gamma}{\omega^2\kappa^2}\int_0^\infty d\tau\, e^{-\gamma\tau}\sinh^2(\kappa\tau)\sin(2\omega\tau).
\end{aligned}$$

The following results are found when $|\beta| > |\delta|$,

$$\begin{aligned}
A_{2\omega} &= -\frac{\beta T\gamma}{2\omega^2}\left\{\text{Im}\left[\frac{1}{\kappa}\left(\frac{1}{(\gamma-2\kappa)+2i\omega}-\frac{1}{(\gamma+2\kappa)+2i\omega}\right)\right]\right.\\
&+ \left.\frac{\delta}{\kappa^2}\text{Re}\left[\frac{1}{(\gamma-2\kappa)+2i\omega}+\frac{1}{(\gamma+2\kappa)+2i\omega}\right]+\frac{2\delta}{\kappa^2}\left[-\frac{1}{\gamma}+\frac{\gamma(\kappa^2+\omega^2)}{(\gamma^2+4\omega^2)(\gamma^2-4\kappa^2)}\right]\right\},\\
B_{2\omega} &= -\frac{4\beta T\gamma}{\omega^2}\left\{\frac{1}{2(\gamma^2-4\kappa^2)}+\frac{1}{8}\text{Re}\left[\frac{1}{\kappa}\left(\frac{1}{(\gamma+2\kappa)+2i\omega}-\frac{1}{(\gamma-2\kappa)+2i\omega}\right)\right]\right.\\
&\left.+\frac{\delta}{8\kappa^2}\text{Im}\left[\frac{1}{(\gamma+2\kappa)-2i\omega}+\frac{1}{(\gamma-2\kappa)-2i\omega}+\frac{2}{\gamma+2i\omega}\right]\right\},\\
A_{4\omega} &= \frac{\beta T\gamma}{4\omega^2\kappa^2}\text{Re}\left[\frac{1}{(\gamma+2\kappa)-2i\omega},+\frac{1}{(\gamma-2\kappa)-2i\omega}-\frac{2}{\gamma+2i\omega}\right]\\
B_{4\omega} &= \frac{\beta T\gamma}{4\omega^2\kappa^2}\text{Im}\left[\frac{1}{(\gamma+2\kappa)-2i\omega}+\frac{1}{(\gamma-2\kappa)-2i\omega}+\frac{2}{\gamma+2i\omega}\right].
\end{aligned}$$

When $|\beta| < |\delta|$, replace κ by $i|\kappa|$ in the expressions above for coefficients of the thermal fluctuations.

3. Results and discussion

In Fig. (2) we plot the Green's functions obtained directly from the numerical integration of Eq. (5) alongside analytical approximation results given by Eq. (7), if $|\beta| > |\delta|$, or by Eq. (9) if $|\beta| < |\delta|$. We obtain excellent agreement between the two methods, what implies that our analytical estimates of $\langle x^2(t)\rangle$ are accurate. The numerical integration was performed using a RK4 algorithm with a time step given by $\pi/(512\omega)$.

In Fig. (3) we show a time series of $\langle x^2(t)\rangle$ given by Eqs. (12). By looking at the time average values (dynamical temperatures) we see that cooling occurs at positive detuning (here at $\omega/\omega_0 = 1.1$), heating (at $\omega/\omega_0 = 0.9$ and 1.0), and thermal noise quadrature squeezing (at all parameters). It is noteworthy to mention that heating or cooling are related to the decay rate of the oscillations of the Green's function, as can be seen in Eq. (6). Lower decay rates imply more heating, while higher decay rates imply less heating or even cooling.

In Fig. (4), a 2D color intensity plot displays the amplitude of the thermal noise squeezing as a function of half the pump frequency, namely ω, and the pump amplitude. We notice that the closer one gets to the first instability zone the higher the amplitude of the squeezing, i.e. the amplitude of the oscillations of the thermal fluctuations $\langle x^2(t)\rangle$. The oscillations occur basically at 2ω, the superharmonics at 4ω are too small and are thus neglected in this plot. For low dynamical temperatures, the squeezing amplitude has to be small too. For higher dynamical temperatures, the squeezing amplitude tends to increase, although there is also dependence on detuning. From this figure one sees that the larger the detuning, the smaller the squeezing effect, although the effect is not symmetric as one can clearly see. This is related to terms that depend linearly on detuning in the Green's functions.

The results given by Eq. (12) imply that by varying the pump amplitude F_p and the detuning Ω, we can create a continuous family of classical thermo-mechanical squeezed states, generalizing the experimental results of Rugar and Grütter [2].

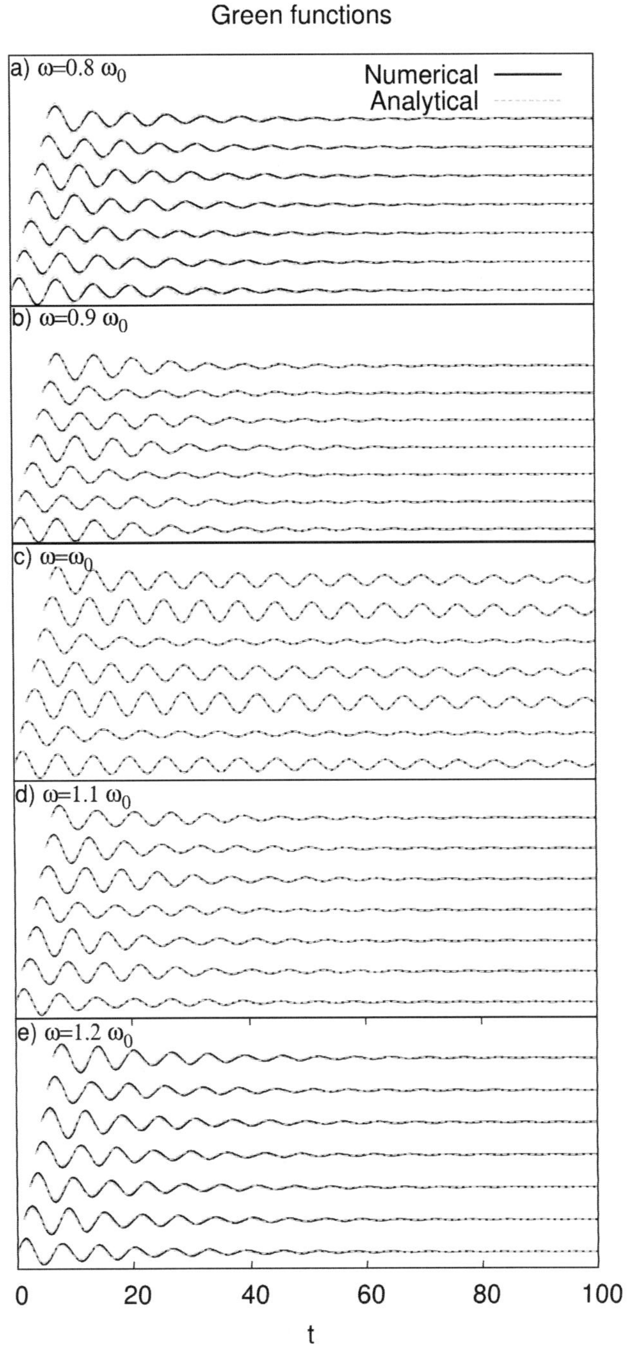

Figure 2. (color online) In the frames above we show several Green's functions with equally-spaced in time initial conditions in one given period of the parametric driving. They are vertically spaced for clarity, since all asymptotes are zero. In each frame we have a comparison between numerical results given by the numerical integration of Eq. (5) and the analytical approximate results given by Eqs. (7) or (9). We have a) $\omega = 0.9\,\omega_0$, b) $\omega = 1.0\,\omega_0$, and c) $\omega = 1.1\,\omega_0$. The initial values of the Green's functions are $G(t, t') = 0$ and $\frac{\partial}{\partial t}G(t, t') = 1.0$ when $t = t' + 0^+$. The pump amplitude used was $F_p = 0.15$.

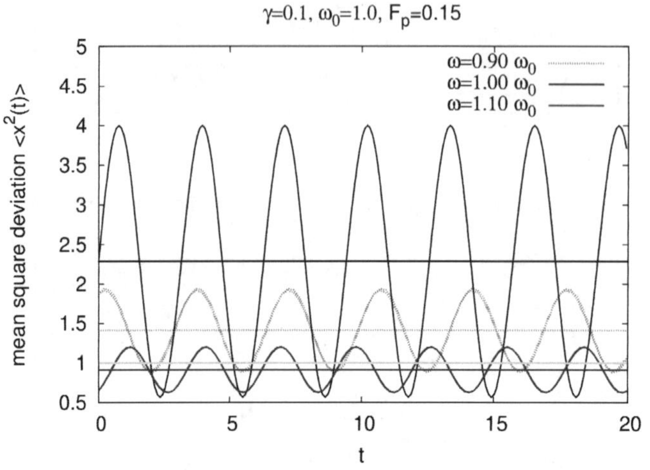

Figure 3. (color online) Steady-state time evolution of the mean square displacement $\langle x^2(t)\rangle$ versus time as defined in Eq. (6), with the Green function given approximately by Eqs. (7) or (9). The same results are obtained by using Eqs. (12), indicative that the expressions for the analytical estimates of $\langle x^2(t)\rangle$ are correct. We obtain cooling when $\omega = 1.1\omega_0$ and heating at the other frequencies. The horizontal solid green line is obtained from the equipartition theorem, it indicates the expected equilibrium value of the fluctuations (for $F_p = 0$) when the heat-bath temperature is at $T = 1.0$, while the other straight lines represent the corresponding dynamical temperatures, i.e. time-averaged values $\overline{\langle x^2(t)\rangle}$ as given by Eq. (10) or by Eq. (11) for $F_p = 0.15$.

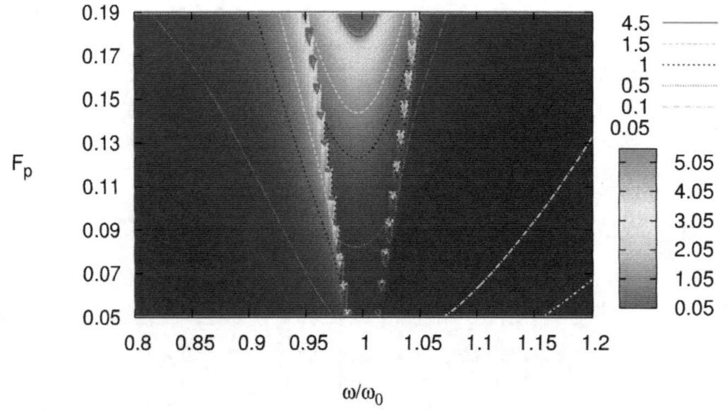

Figure 4. (color online) Contour plot of the amplitude of squeezing at 2ω of the mean square displacement $\langle x^2(t)\rangle$ as given by $\sqrt{|A_{2\omega}|^2 + |B_{2\omega}|^2}$, in which $A_{2\omega}$ and $B_{2\omega}$ are defined in Eqs. (12). The amplitude at 4ω is very small for the pump amplitudes F_p used here.

4. Conclusions

The ability to reduce thermal noise in MEMS devices, in addition to parametric amplification mechanisms, can greatly improve the accuracy and precision of measuring small masses, weak forces, and, furthermore, may contribute to the effort of measuring gravitational waves. Our main contribution resides in developing a relatively simple and general theoretical framework that provides us with quantitative estimates of cooling, heating, and thermal noise squeezing in mechanical resonators. We believe these estimates can well serve, with the appropriate modifications to adjust to particular systems, as a guide to the experimentalist in search for such phenomena. Further development to our model may be achieved by including more details about the noise model, such as memory effects [12, 13], by taking into account the coupling to a heat bath that could be made out of photons, as in radiation-pressure cooling, or via coupling to phonons. Finally, we note that here we provided a general framework for studying classical thermomechanical squeezing that could be applied to several different systems in the presence of noise, such as ac-driven doubly-clamped beams [14], ions in ion traps [15], opto-mechanical cavities [16], and to magnetic nanoparticles driven by ac magnetic fields.

Acknowledgements

The author acknowledges financial support from CNPq (Brazil) through grant numbers 481035/2009-5 and 557542/2009-0.

References

[1] Binnig G, Quate C and Gerber C 1986 *Phys. Rev. Lett* **56** 930–933
[2] Rugar D and Grutter P 1991 *Phys. Rev. Lett.* **67** 699
[3] DiFilippo F, Natarajan V, Boyce K and Pritchard D 1992 *Phys. Rev. Lett.* **68** 2859–2862
[4] Natarajan V, DiFilippo F and Pritchard D 1995 *Phys. Rev. Lett.* **74** 2855–2858
[5] B Yurke, D S Greywall, A N Pargellis, and Busch P A 1995 *Phys. Rev. A* **51** 4211
[6] Nayfeh A H and Mook D T 1979 *Nonlinear Oscillations* (New York: Wiley)
[7] Almog R, Zaitsev S, Shtempluck O and Buks E 2007 *Phys. Rev. Lett.* **98** 78103
[8] Guckenheimer J and Holmes P 1983 *Nonlinear Oscillations, Dynamical Systems, and Bifurcations of Vector Fields* (New York: Springer-Verlag)
[9] Verhulst F 1996 *Nonlinear differential equations and dynamical systems* (New York: Springer-Verlag)
[10] Kubo R 1966 *Reports on Progress in Physics* **29** 255–284
[11] Wiesenfeld K and McNamara B 1985 *Phys. Rev. Lett.* **55** 13
[12] Grifoni M and Hänggi P 1998 *Phys. Rep.* **304** 229–354
[13] Batista A A, Oliveira F A and Nazareno H N 2008 *Phys. Rev. E* **77** 066216
[14] Aldridge J S and Cleland A N 2005 *Phys. Rev. Lett.* **94** 156403
[15] Paul W 1990 *Rev. of Mod. Phys.* **62** 531–540
[16] Dorsel A, McCullen J, Meystre P, Vignes E and Walther H 1983 *Phys. Rev. Lett.* **51** 1550

Dynamic Days South America 2010 IOP Publishing
Journal of Physics: Conference Series **285** (2011) 012042 doi:10.1088/1742-6596/285/1/012042

Generalising the logistic map through the q-product

R W S Pessoa, E P Borges

Escola Politécnica, Universidade Federal da Bahia, Rua Aristides Novis 2, Salvador, Bahia, Brazil

E-mail: robsonpessoa2007@gmail.com,ernesto@ufba.br

Abstract. We investigate a generalisation of the logistic map as $x_{n+1} = 1 - ax_n \otimes_{q_{map}} x_n$ $(-1 \leq x_n \leq 1,\ 0 < a \leq 2)$ where $\otimes_q$ stands for a generalisation of the ordinary product, known as q-product [Borges, E.P. Physica A **340**, 95 (2004)]. The usual product, and consequently the usual logistic map, is recovered in the limit $q \to 1$, The tent map is also a particular case for $q_{map} \to \infty$. The generalisation of this (and others) algebraic operator has been widely used within nonextensive statistical mechanics context (see C. Tsallis, *Introduction to Nonextensive Statistical Mechanics*, Springer, NY, 2009). We focus the analysis for $q_{map} > 1$ at the edge of chaos, particularly at the first critical point a_c, that depends on the value of q_{map}. Bifurcation diagrams, sensitivity to initial conditions, fractal dimension and rate of entropy growth are evaluated at $a_c(q_{map})$, and connections with nonextensive statistical mechanics are explored.

1. Introduction

Low-dimensional non-linear maps represent paradigmatic models in the analysis of dynamic systems. The discrete time evolution and the small number of relatively simple equations make their treatment easy, without losing the richness of the behaviour, exhibiting order, chaos and a well defined transition between them (see, for example, [1, 2]).

Strongly chaotic systems are of special interest for statistical mechanics, once they feature well known characteristics: exponential sensitivity to the initial conditions, ergodicity, exponential relaxation to the equilibrium state, gaussian distributions [3].

In-between ordered systems (with negative Lyapunov exponent) and (strongly) chaotic systems (with positive Lyapunov exponent) there are those with zero maximal Lyapunov exponent. These systems are characterised by power-law sensitivity to initial conditions, instead of the exponential sensitivity, and thus are considered as weak chaotic systems. This change in the dynamics may lead to break of ergodicity, non-exponential relaxation to equilibrium and/or non-gaussian distributions. These behaviours are usually expected to be found in systems that are described by nonextensive statistical mechanics [4, 5]. Some low dimensional maps, *e.g.* the logistic map, also exhibit weak chaoticity at the edge of chaos, and hence the interest in studying them to better understand nonextensivity.

Power-law like sensitivity to initial conditions and power-law like relaxation to the attractor (more precisely a q-exponential law) have already been found in logistic-like maps [2, 6]. q-exponential function ($e_q^x \equiv [1 + (1-q)x]_+^{1/(1-q)}$, the subscript $_+$ is explained in the following) appear within nonextensive statistical mechanics and it generalises the usual exponential function (recovered as $q \to 1$). It is asymptotically a power-law (for $q > 1$ and $x < 0$ or $q < 1$ and $x > 0$). Sensitivity to initial conditions of the logistic map at the edge of chaos is identified to a q-exponential, with a specific value of the parameter q, denoted as q_{sen}. The rate of entropy growth S_q/t (S_q is the nonextensive entropy, defined later by Eq. (9), and t is time) is another parameter usually evaluated in maps. It must be finite at the macroscopic limit, and there is one special value of q denoted q_{ent} (from *entropy*) that makes $S_{q_{ent}}/t$ finite. At the edge of chaos, $q_{ent} \neq 1$. It is numerically verified that $q_{sen} = q_{ent}$ (see [5] and references therein). Relaxation of the logistic map to the attractor at the edge of chaos also follows a q-exponential behaviour, with a different and specific value of the parameter q, denoted q_{rel}. The relation between $q_{ent} \leq 1$ and $q_{rel} \geq 1$ plays a central role in the foundations of nonextensive statistical mechanics. For completely chaotic systems, these values collapse to $q_{ent} = q_{rel} = 1$ (See [5] for details).

Nonextensive statistical mechanics has lead to developments in many related areas, including generalised algebras [7, 8]. These works have introduced generalised algebraic operators, and here we are particularly interested in the q-product[1], defined as

$$x \otimes_q y \equiv \text{sign}(x)\text{sign}(y) \left[|x|^{1-q} + |y|^{1-q} - 1\right]_+^{\frac{1}{1-q}} \tag{1}$$

where the symbol $[A]_+$ means that $[A]_+ = A$ if $A > 0$ and $[A]_+ = 0$ if $A \leq 0$ (known as cut-off condition, $[A]_+ \equiv \max\{0, A\}$, for short). The limit $q \to 1$ recovers the usual product ($x \otimes_1 y = xy$). Our work consists in generalising the logistic map as

$$\begin{aligned} x_{n+1} &= 1 - a\left(x_n \otimes_{q_{map}} x_n\right) \\ &= 1 - a[2|x_n|^{1-q_{map}} - 1]_+^{\frac{1}{1-q_{map}}} \end{aligned} \tag{2}$$

($-1 \leq x_n \leq 1$, $0 < a \leq 2$). The cut-off condition implies that $[2|x_n|^{1-q_{map}} - 1]_+ = 0$, and thus $x_{n+1} = 1$, if $|x_n| \leq 1/2^{1/(1-q_{map})}$. This q-logistic map recovers the usual logistic map for $q_{map} = 1$, and also the tent map for $q_{map} \to +\infty$ (the tent map properly shifted as $x_{n+1} = 1 - a|x_n|$). At the limit $q_{map} \to -\infty$, it becomes $x_{n+1} = 1$ for $-1 < x_n < 1$ and $x_{n+1} = 1 - a$ for $x_n = \pm 1$. Some details regarding these limits are sketched in Appendix A. Figure 1 shows one iteration of the map.

The z-logistic map ($x_{n+1} = 1 - a|x_n|^z$, $z > 1$, $0 < a \leq 2$) [11, 12, 13, 14], is another generalisation of the logistic map for a general power $z > 1$ and holds some similarity to the present q-logistic map.

Bifurcation diagrams for different values of q_{map} are shown in Fig. 2. As q_{map} goes from 1 (the usual logistic map) to 2, the value of the parameter a for the first bifurcation goes from $a = 0.75$ to $a = 1$ (it approaches 1 from the left). Also the value of the parameter for the accumulation of bifurcations goes from $a_c = 1.401155189092\ldots$ to $a_c = 1$ (it approaches 1 from the right). It means that the period doubling cascade becomes narrower until it eventually disappears at $q_{map} = 2$. Similar behaviour also happens with the other windows of order inside chaos (where there is tangent bifurcation): they get narrower as q_{map} increases, and eventually disappear for $q_{map} = 2$. Complete chaos is preserved at $a = 2$, $\forall q_{map}$. The Schwarzian derivative for the q-logistic map is given by (see Appendix B)

$$(Sf)(x) = \left(q_{map} - \frac{1}{2}q_{map}^2\right)\frac{1}{|x|^2}\left(1 - \frac{4}{4 - 4|x|^{q_{map}-1} + |x|^{2(q_{map}-1)}}\right). \tag{3}$$

This expression is negative in the interval $0 < q_{map} < 2$, and positive for $q_{map} > 2$ and for $q_{map} < 0$ ($SD(f(x)) = 0$ for $q_{map} = 0$ and for $q_{map} = 2$). This means that the route to chaos for $0 < q_{map} < 2$ is by period doubling bifurcation.

[1] The q-product has been also used in the generalisation of Gauss's law of errors [9], in the formulation of the q-Fourier transform, and in the generalisation of the central limit theorem [10].

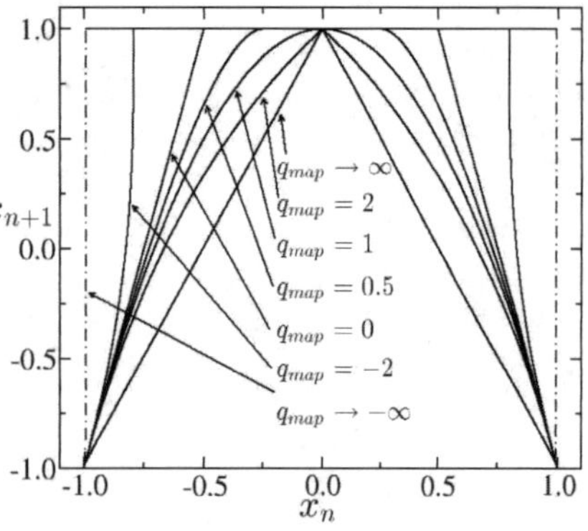

Figure 1. x_{n+1} as a function of x_n for the q-logistic map (Eq. (2) with $a = 2$). The usual parabolic behaviour is recovered at $q_{map} = 1$. Tent map is found at $q \to \infty$ (with the peak parameter set to unit). As q_{map} departs from 1 towards $-\infty$ ($q_{map} < 1$) the cut-off condition in Eq. (1) yields an increasing region in which $x_{n+1} = 1$. For $q_{map} \to -\infty$, $x_{n>0}$ alternates between 1 and $(1-a)$ $\forall x_0 \in [-1, 1]$. For $q_{map} > 1$ the map is discontinuous at $x = 0$ and for $q_{map} < 1$ the discontinuity is at $|x| = 1/2^{1/(1-q)}$.

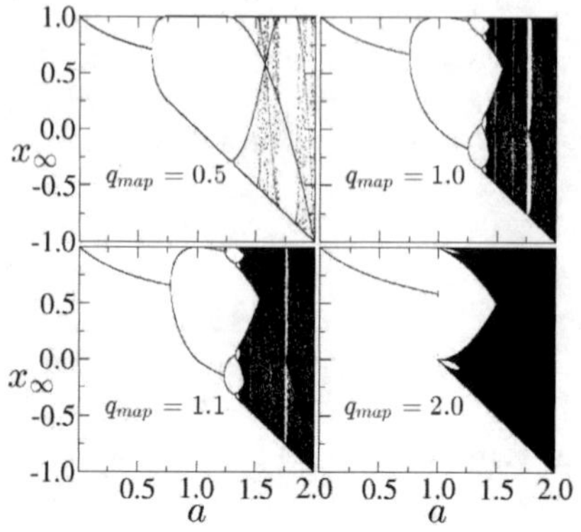

Figure 2. Bifurcation diagrams for different values of q_{map} (indicated). Windows of order inside chaos vanishes as $q_{map} \to 2$. In this paper we explore $q_{map} > 1$ but $q_{map} = 0.5$ is shown as an instance just to give an idea of the scenario for $q_{map} < 1$: the regions of chaos become narrower and the regions of order become dominant.

2. Sensitivity to initial conditions

The maximal Lyapunov exponent may be evaluated at the edge of chaos according to (see, for instance, [3])

$$\lambda_{max} = \lim_{N\to\infty} \frac{1}{N} \sum_{i=0}^{N-1} \ln |f'(x_i)| \tag{4}$$

where $f'(x)$ is the derivative of the q-logistic map,

$$f'(x) = -|x|' 2a|x|^{-q_{map}} \left[2|x|^{1-q_{map}} - 1\right]_+^{\frac{q_{map}}{1-q_{map}}}. \tag{5}$$

$|x|'$ is the derivative of the absolute value function: $|x|' = 1$ for $x > 0$ and $|x|' = -1$ for $x < 0$. For values of the control parameter a different from critical ones, the sensitivity to initial conditions are characterised by exponential divergence at regions of chaos and exponential decay at regions of order, *i.e.*, positive or negative Lyapunov exponent λ_1 (subscript 1 will be clear in the following) in

$$\xi(t) = \lim_{\Delta x(0)\to 0} \frac{\Delta x(t)}{\Delta x(0)} = e^{\lambda_1 t}. \tag{6}$$

At the edge of chaos it was proposed that the divergence follows an asymptotic power-law (in fact a q-exponential law) [2] characterising a slow dynamics,

$$\xi(t) = e_{q_{sen}}^{\lambda_{q_{sen}} t} = (1 + (1 - q_{sen})\lambda_{q_{sen}} t)^{\frac{1}{1-q_{sen}}}, \tag{7}$$

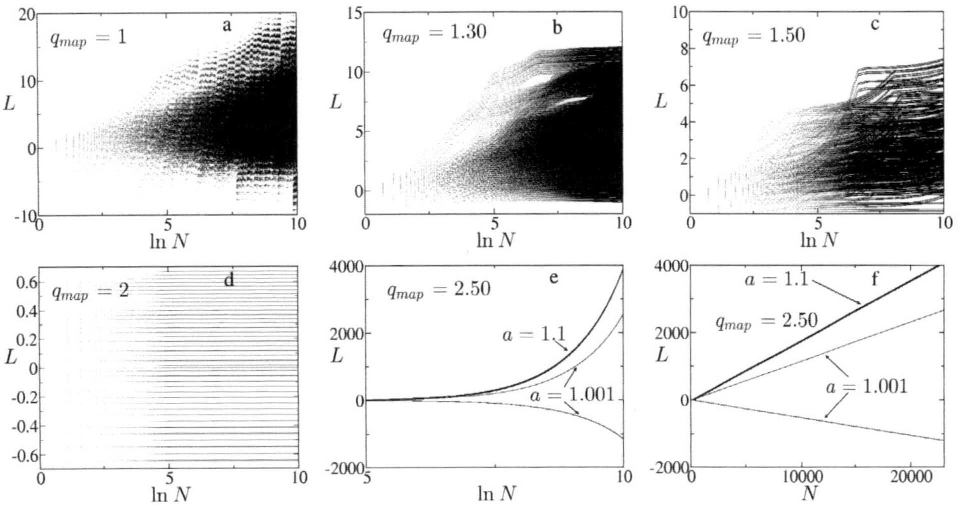

Figure 3. Sensitivity to initial conditions as defined by variable L (Eq. (8)) for different values of q_{map} (indicated). 40 initial conditions are used. For $q_{map} = 2.5$ (figures e and f) two values of the control parameter a are presented. It can be seen that it is possible to have coexistence of attractors for a certain ensemble of initial conditions (for $a = 1.001$), one with $L > 0$ and the other with $L < 0$. The same value of $q_{map} = 2.5$ and with $a = 1.1$, the initial conditions always lead to $L > 0$. Fig. f shows linear dependence of L with N. Slopes (in absolute values) for $a = 1.001$ for $L > 0$ and $L < 0$ differ.

where $\Delta x(0)$ represents the distance between two neighbouring initial conditions and *sen* stands for sensitivity to initial conditions ($q_{sen} \leq 1$). Eq. (6) is recovered at $q_{sen} = 1$ (see Appendix A) and this is the reason for the subscript 1 in λ_1, Eq. (6). As a graphical representation of the sensitivity, it can be defined the variable L as [2]:

$$L = \sum_{i=0}^{N-1} \ln |f'(x_i)|. \tag{8}$$

Figure 3 shows five instances of L vs. $\ln N$. For $q_{map} = 2$ (Fig. 3d) the dependence of L on N is very slow and cannot be seen up to $N = 2^{15}$ (the upper limit of the figures). For $q_{map} > 2$ it is possible to have coexistence of attractors according to the control parameter a: depending on the initial conditions, L may be increasingly positive or decreasingly negative. Fig 3f shows that there is a linear dependence of L on N but with different slopes for the cases $L > 0$ and $L < 0$. For values of $a > 1.07499\ldots$, $L < 0$ is never exhibited. Coexistence of attractors was also found in [15] for a different deformation of the logistic map (the authors also call their deformation as q-logistic map).

Lyapunov exponents are displayed in Fig. 4 showing transitions from order to chaos. The figures show that these transitions become sparse as q_{map} departures from unit, and for $q_{map} = 2$ there is only one transition (robust chaos)[16]. Fig. 4f shows coexistence of attractors for $q_{map} = 2.5$.

Figure 5 illustrates the first point of accumulation of period doubling bifurcation a_c as a function of q_{map}. For $1 < q_{map} < 2$, the behaviour is ordinary in the sense that for $a < a_c$, $L < 0$, and for a slightly greater than a_c, $L > 0$. For $q_{map} > 2$ a different behaviour appears: order is found for $a < 1$ (region below the solid line), while chaos is found above the dashed

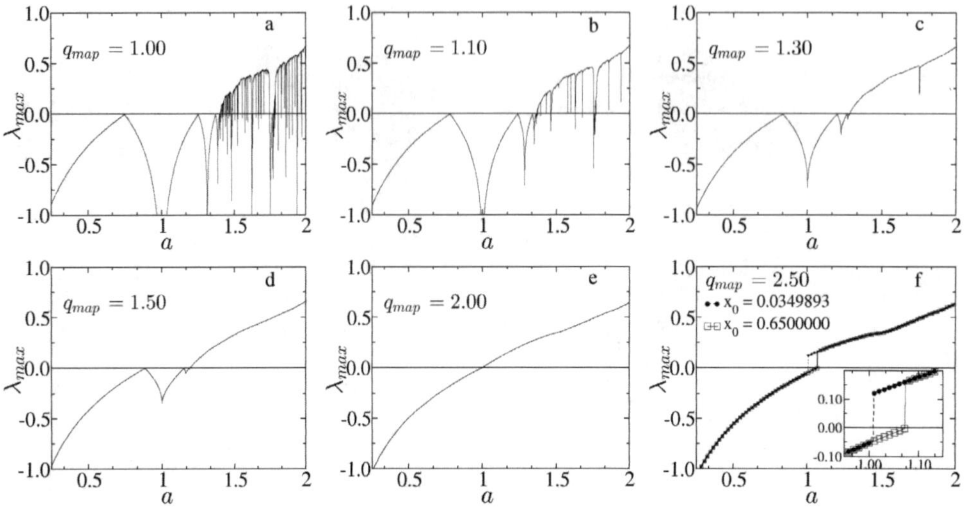

Figure 4. Lyapunov exponent as a function of the control parameter a for different values of q_{map} (indicated). Up to the time limit used in Eq. (4), the almost vertical line in Fig. c at $a = 1.7477$ does not crosses zero, but it does present negative values if the calculation is done with a greater precision (it was used $t = 2^{18}$ and $\Delta a = 2 \times 10^{-4}$). Fig. f exhibits coexistence of attractors for two different initial conditions. It's inset is an amplification of the region of coexistence of attractors for two initial conditions (indicated as full circles and open squares).

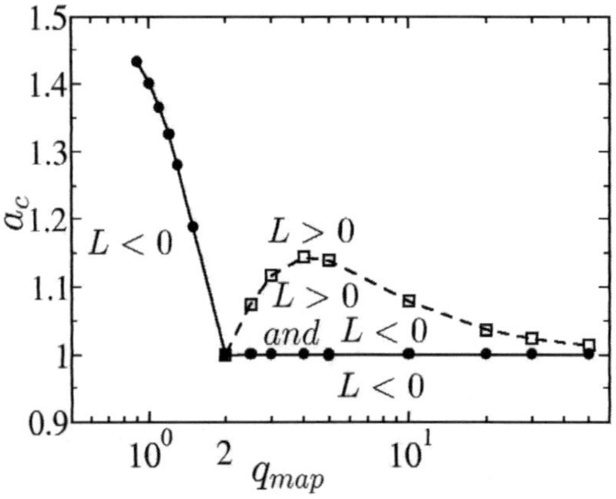

Figure 5. Dependence of the first point of accumulation of period doubling bifurcation a_c on q_{map}. For $q_{map} > 2$ there is a region that exhibits coexistence of attractors depending on the initial conditions. Values of L for this figure were calculated without transient time and final time $t_{end} = 2^{18}$, and increments for the control parameter $\Delta a = 10^{-7}$. Critical point is taken as that for which $0 < \lambda_{max} < 5 \times 10^{-5}$.

line. The region in-between presents coexistence of ordered and chaotic behaviours, depending on the initial conditions (see figures 3e, 3f and 4f).

Table 1 shows the range of the critical points a_c (first point of accumulation of bifurcations)

Table 1. Critical points

q_{map}	a_-	a_+	a_c
1.00	1.40115518	1.40115520	1.401155189092
1.01	1.3977569	1.3977571	1.397757026
1.02	1.3943177	1.3943179	1.394317802
1.03	1.3908370	1.3908372	1.390837098
1.04	1.3873144	1.3873146	1.387314512
1.05	1.3837496	1.3837498	1.383749669
1.10	1.3652805	1.3652807	1.365280586
1.20	1.3250906	1.3250908	1.325090670
1.30	1.2811360	1.2811362	1.281136143
1.40	1.2353387	1.2353389	1.235338767
1.45	1.2125150	1.2125152	1.21251512
1.50	1.1900820	1.1900822	1.1900822

for different values of q_{map}. The value of a_c is between a_- and a_+. The fourth column shows adopted value for a_c. For the evaluation of a_c listed in Table 1, the Lyapunov exponents were calculated with a transient time of $t_{trans} = 2^{23}$ and a final time of $t_{end} = 2^{23}$ (the values of the Table are more accurate than those of Fig. 5). Initial condition was fixed in $x_0 = 0.65$ for all cases.

In Fig. 6 we show the cycle 3 window for $q_{map} = 1.25$. This window of order inside chaos (as well as all the others) becomes narrow: in order to identify the Lyapunov exponent in this instance it was necessary to give increments of $\Delta a = 1.0 \times 10^{-8}$ and $t_{end} = 2^{23}$, with a transient $t_{trans} = 2^{23}$. Identification of windows of order in this q-logistic map is computationally time consuming as q_{map} departs from unit.

3. Entropy production

Parameter q_{ent} (from *entropy*) in Tsallis entropy [4]

$$S_q = k\frac{1 - \sum_{i=1}^{W} p_i^q}{q-1} \tag{9}$$

(W is the number of microstates and p_i is the probability of microstate[2] i; we use $k = 1$ without loss of generality) is estimated according to the method developed in [17]. It consists in calculating the rate of increase of entropy, which must be *finite* for a large (virtually infinite) system[3]. The phase space ($x_n \in [-1, 1]$) is divided into W cells, with N points (initial conditions) inside one of them. As time evolves, Tsallis entropy S_q is calculated for many values of q. Calculation is done again with N points inside another initial cell, and this process is repeated for each cell of a certain ensemble called "best initial condition cells" (the definition of "best initial condition cells" is: the integrated number of occupied cells must be greater than a certain (arbitrary) threshold, the most visited cells). Then it is taken the average of entropy for each time step, and this average is finally plotted against time, for various values of q (see Fig. 7

[2] In statistical mechanics, for a given macrostate (that is, macroscopically measured state) there are a number W of compatible configurations of the constituent elements, or microscopic states (microstates, for short). A macroscopic variable is, in fact, a measure of the average of all compatible microscopic states.

[3] By large system we mean a macroscopic system. The density variable of a quantity F, F/N, must remain finite as the number of microscopic constituents N reaches the thermodynamical limit $\lim_{N\to\infty}$.

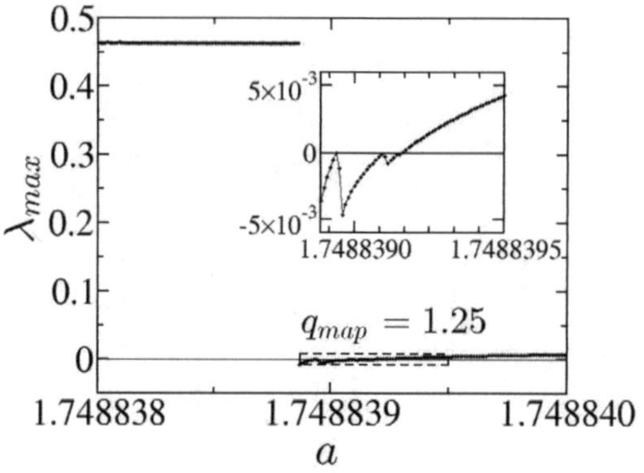

Figure 6. Maximal Lyapunov exponent for $q_{map} = 1.25$ at the vicinity of the largest window, of cycle 3, with increments of $\Delta a = 1.0 \times 10^{-8}$. It was used a transient time of $t_{trans} = 2^{23}$ and the time used to estimate the Lyapunov exponent $t_{end} = 2^{23}$ (with lower time intervals the transition from negative to positive Lyapunov exponent cannot be seen). Inset corresponds to the dashed rectangle in the main panel and shows points of zero Lyapunov exponents that corresponds to period doubling bifurcation.

for an instance). There is only one special value of q for which the increase of S_q is linear in the macroscopic limit, that is, the production of entropy $\kappa_q = \lim_{t\to\infty} \lim_{W\to\infty} \lim_{N\to\infty} S_q/t$ remains finite. This special value is identified with q_{ent} (see [17] for details; in that paper, q_{ent} is called q^*). For values of the control parameter a that corresponds to chaotic behaviour $q_{ent} = 1$, so ordinary Boltzmann-Gibbs-Shannon entropy is the proper one to be used. This is not the case at the edge of chaos ($a = a_c$), which we are interested in this work, and the value of q_{ent} which yields linear growth of entropy (and consequently finite rate at the macroscopic limit) is smaller than one (for the ordinary logistic map, $q_{map} = 1$, at the edge of chaos, $q_{ent} = 0.2445...$). We applied this procedure to the q-logistic map. The result of this procedure for various values of q_{map} is presented in Fig. 8.

The integrated number of occupied cells, used to identify the best initial condition cells, is shown in Fig. 9. We see a reduction of the number of visited cells as q_{map} goes from one to two and a change in the displayed pattern. Fig. 10 shows number of cells with visits greater than 5000 as a function of q_{map}.

4. Relaxation to the critical attractor

Relaxation to the critical attractor was presented at [6] and basically consists in the division of the phase space [-1,1] into W cells and take an ensemble of N initial conditions uniformly distributed in the entire phase space (thus maximal entropy). Time evolution leads to the decreasing of the number of occupied cells $W_{occ}(t)$ according to a power-law with log-periodic oscillations ($W_{occ}(0) = W$). It is supposed [6] that the power-law is the asymptotic limit of a q-exponential, and the parameter is denoted as q_{rel}, for *relaxation* ($q_{rel} > 1$):

$$W_{occ}(t) = (1 + (1 - q_{rel})K_{q_{rel}}t)^{\frac{1}{1-q_{rel}}}, \tag{10}$$

$K_{q_{rel}}$ is the inverse of a characteristic time. Figure 11 presents the results for the fraction of occupied cells $W_{occ}(t)/W_{occ}(0)$ for different values of q_{map} at their corresponding critical

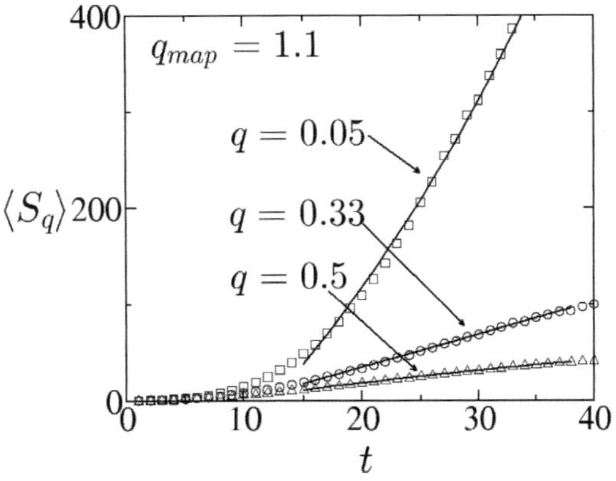

Figure 7. Rate of increase of entropy for $q_{map} = 1.1$. We found $q_{ent} = 0.33$. This value is found by fitting a parabola $\langle S_q(t)\rangle = a + bt + ct^2$ for the time interval $15 \leq t \leq 38$ (the same used in [17]). Initial times are always left out of calculation because the concavity is always positive at that region. For $q < q_{ent}$ the concavity of the curve is positive (squares, $c > 0$) and thus the rate of growth of entropy diverges at the macroscopic limit, which is unphysical. For $q > q_{ent}$, the concavity is negative (triangles, $c < 0$), thus the rate of growth of entropy is zero at the macroscopic limit, which is also unphysical. $c = 0$ is shown as circles.

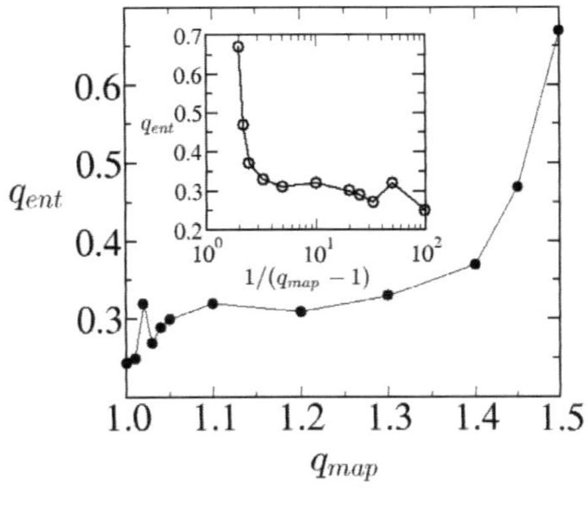

Figure 8. q_{ent} as a function of q_{map}. Inset shows abscissa in a different scale (note that abscissa is in log scale). The value at $q_{map} = 1.02$ may have numerical imprecision. Lines are only guide to the eyes.

points. Log-periodic oscillations present increasing periods for $q_{map} \rightarrow 2$. The slope in the log-log plot (Fig. 11a) at the region of the log-periodic oscillations is used to estimate q_{rel} as slope $= 1/(1 - q_{rel})$. Fig. 11b presents q_{rel} as a function of q_{map}.

The procedure for estimating q_{rel}, that is, the shrinking of the number of occupied cells, is also used to estimate the fractal dimension d_f at the edge of chaos (it is a kind of box counting method). $W_{occ}(\infty)$ increases with $W_{occ}(0)$ according to $W_{occ}(\infty) \propto [W_{occ}(0)]^{d_f}$. Fractal dimension decreases with q_{map} as shown in Fig. 12.

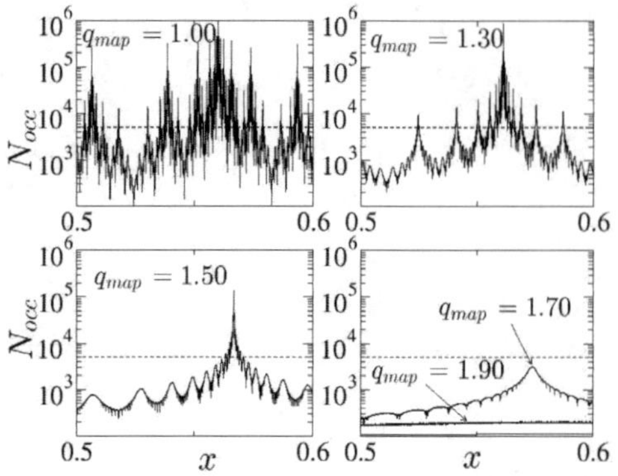

Figure 9. Integrated number of visits per cell for five cases ($q_{map} = 1, 1.30, 1.50, 1.70, 1.90$). Phase space is divided into $W = 10^5$ cells, each one contains 10^6 points uniformly distributed. Time evolves up to 50 iterations. Figures show $x \in [0.5, 0.6]$ just for better visualisation.

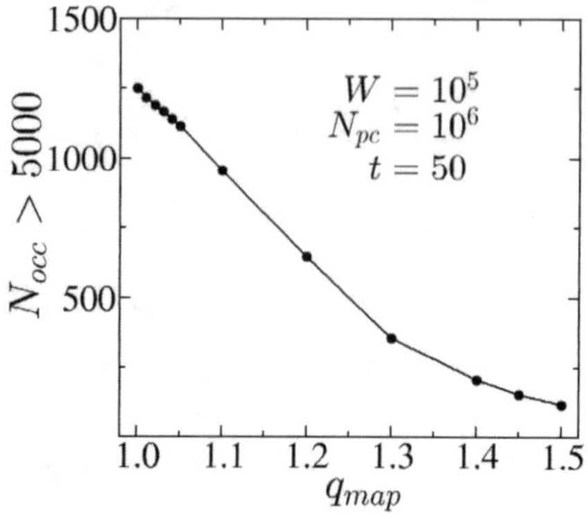

Figure 10. Number of cells with more than 5000 visits in a run of 50 iterations as a function of q_{map}. $N_{pc} = 10W$ is the number of initial conditions per cell. Dashed horizontal line in Fig. 9 indicates the threshold of 5000 visits per cell.

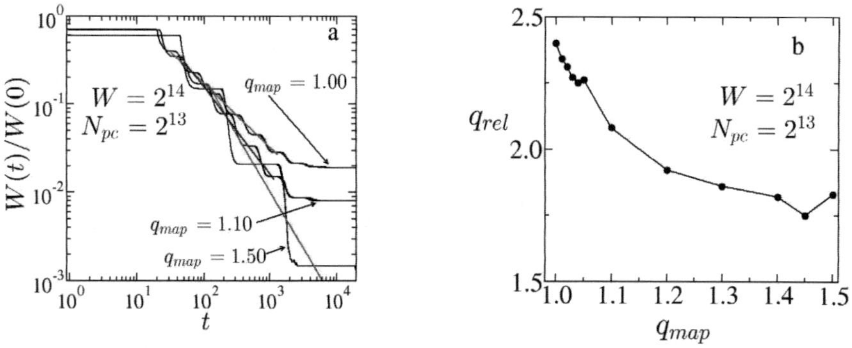

Figure 11. Relaxation to the critical attractor at the edge of chaos. Fig. a is a log-log plot of $W_{occ}(t)/W_{occ}(0) \times t$ for three different values of q_{map}. It can be seen the results of regression of power-laws, whose slopes are identified as $1/(1 - q_{rel})$. Fig. b shows q_{rel} as a function of q_{map}. It was used $W = 2^{14}$ and number of initial conditions per cell $N_{pc} = 2^{13}$. Lines are only guide to the eyes. Fluctuations in the curve of Fig. b are due to inaccuracy — the results are very sensitive to small changes in the characteristic parameters.

5. Final remarks

The generalisation of the logistic map by means of the q-product introduces some interesting features in its dynamical behaviour. Our analysis is focused on $q_{map} > 1$. In this region, the windows of order inside chaos become narrower as q_{map} increases until all of them disappear and the map becomes the tent map. For $q_{map} < 1$ (not analysed in this paper) the opposite behaviour occurs: the regions of chaos become narrower and order dominates the scenario. A remarkable feature of the q-logistic map is to continuously pass from a map with a variety of

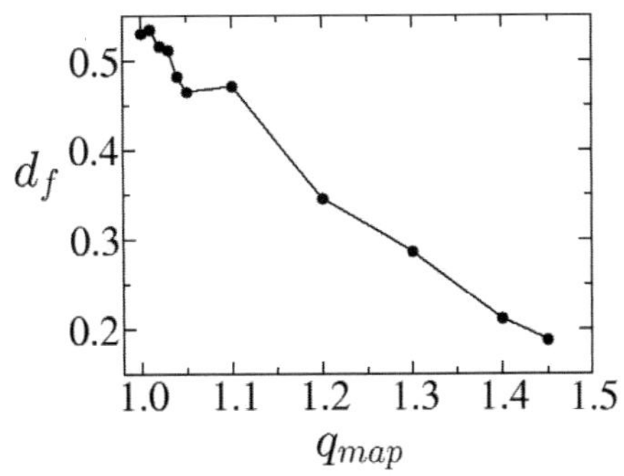

Figure 12. Fractal dimension at the edge of chaos as a function of q_{map}. For the usual logistic map ($q_{map} = 1$), $d_f(a_c) = 0.53665$. Lines are only guide to the eyes. Fluctuations in the curve are due to inaccuracy — the method is very sensitive to small changes in the characteristic parameters.

behaviours, such as period doubling bifurcation, multifractality and power-law like sensibility to initial conditions at the edge of chaos (the logistic map), to a robust map (the tent map). We have calculated the sensitivity to initial conditions, the rate of entropy growth, the relaxation to the critical attractor and the fractal dimension at the edge of chaos. These methods permit to estimate the parameters q_{ent} and q_{rel} for different values of q_{map}. The entropy parameter and the relaxation parameter are two indices that appear within the context of nonextensive statistical mechanics and the understanding of their dependence on the control parameters of the system may lead to their *a priori* determination. Some other evaluations for this q-logistic map remains to be done, *e.g.* the dependence of q_{rel} on the coarse graining W and its relation to q_{sen} (as in [12] for the z-logistic map, and as in [18, 19] for the Hénon map), the probability distributions of sums of iterates as in [20, 21], multifractality as it was done in [22], and tangent bifurcations.

Acknowledgements
This work was partially supported by FAPESB (Fundação de Amparo à Pesquisa do Estado da Bahia). We thank FESC and GSUMA (research groups of the Institute of Physics of UFBA) for using their computational resources.

Appendix A. Special limits for the q-product

The limit $q \to v$ where v is one of the following $\{-\infty, 1, \infty\}$ for the q-product $x \otimes_q x$ as it appears in the q-logistic map, Eq. (2) and also in Eq. (1) with $y = x$, leads to $\{0^0, 1^\infty, \infty^0\}$ respectively. The indeterminates can easily be solved by means of a simple trick that is to rewrite the q-product as

$$x \otimes_q x = [2|x|^{1-q} - 1]_+^{\frac{1}{1-q}} = h(q)^{g(q)} \tag{A.1}$$

with $h(q) = [2|x|^{1-q} - 1]_+$ (note that $h(q) \geq 0$ due to the cut-off condition in Eq. (1)) and $g(q) = \frac{1}{1-q}$. Of course h is a function of q and x but for now we are not interested in the dependency on x, so we omit it. We also omit the subscript of q for the sake of brevity. Then

$$\lim_{q\to v}[2|x|^{1-q}-1]_+^{\frac{1}{1-q}} = e^{\lim_{q\to v} g(q)\ln h(q)}. \tag{A.2}$$

Straightforward application of L'Hospital rule leads to (we remind the reader that $|x| \leq 1$)

$$\lim_{q\to v} x \otimes_q x = \begin{cases} \left.\begin{array}{ll} 1, & \text{if } |x| = 1 \\ 0, & \text{if } |x| < 1 \end{array}\right\} & \text{for } v = -\infty, \\ x^2, & \text{for } v = 1, \\ |x|, & \text{for } v = \infty. \end{cases} \tag{A.3}$$

L'Hospital rule must be applied twice in the case $v = \infty$. A similar procedure applied to Eq. (7), now with $h(q) = 1 + (1 - q_{sen})\lambda_{q_{sen}} t$ and $g(q) = \frac{1}{1-q_{sen}}$, leads to Eq. (6).

Appendix B. Schwarzian derivative for the q-logistic map

The Schwarzian derivative is defined by

$$(Sf)(x) = \frac{f'''}{f'} - \frac{3}{2}\left(\frac{f''}{f'}\right)^2. \tag{B.1}$$

The function $f(x) = 1 - a\,(x \otimes_{q_{map}} x)$ represents the q-logistic map and it may be written as $f(x) = 1 - ah(x)^{g(q)}$ with $h(x)$ and $g(q)$ given by Eq. (A.1) — now we are interested in the dependency of h on x. The three first derivatives of $f(x)$ are given by

$$\begin{aligned} f'(x) &= -a\tfrac{1}{1-q} h^{\frac{q}{1-q}} h' \\ f''(x) &= -a\tfrac{1}{1-q}\left(\tfrac{q}{1-q} h^{\frac{2q-1}{1-q}}\left(h'\right)^2 + h^{\frac{q}{1-q}} h''\right) \\ f'''(x) &= -a\tfrac{1}{1-q}\tfrac{q(2q-1)}{(1-q)^2} h^{\frac{3q-2}{1-q}}\left(h'\right)^3 \\ &\quad -a\tfrac{1}{1-q} 3\tfrac{q}{1-q} h^{\frac{2q-1}{1-q}} h' h'' \\ &\quad -a\tfrac{1}{1-q} h^{\frac{q}{1-q}} h''' \end{aligned}$$

with $h' = |x|'2(1-q)|x|^{-q}$, $h'' = -2q(1-q)|x|^{-1-q}$ and $h''' = |x|'2q(1-q)(1+q)|x|^{-2-q}$. Substitution in Eq. (B.1) leads to

$$\begin{aligned} (Sf)(x) &= \frac{4q(2q-1)}{|x|^{2q}[2|x|^{1-q}-1]_+^2} - \frac{6q^2}{|x|^{1+q}[2|x|^{1-q}-1]_+} + \frac{q(1+q)}{|x|^2} \\ &\quad -\frac{3}{2}\left(\frac{2q}{|x|^q[2|x|^{1-q}-1]_+} - \frac{q}{|x|}\right)^2 \end{aligned} \tag{B.2}$$

This expression may be rearranged as in Eq. (3) that is more convenient to analyse its signal. The Schwarzian derivative for the q-logistic map presents a divergence at $x = 0$ for $1 \leq q < 2$ and a divergence at $x = 0$ and at $x = \pm 1/2^{1/(1-q)}$ for $q < 1$. The case $q_{map} = 1$ corresponds to the Schwarzian derivative for the usual logistic map, $(Sf)(x) = -3/(2x^2)$, as can be easily verified.

References

[1] Hilborn R 2000 *Chaos and Nonlinear Dynamics: An Introduction for Scientists and Engineers* (Oxford University Press, USA)
[2] Tsallis C, Plastino A R and Zheng W M 1997 *Chaos Sol. Fract.* **8** 885–891
[3] Beck C and Schlögl F 1995 *Thermodynamics of Chaotic Systems: An Introduction (Cambridge Nonlinear Science Series)* (Cambridge University Press)
[4] Tsallis C 1988 *J. Stat. Phys.* **52** 479–487
[5] Tsallis C 2009 *Introduction to Nonextensive Statistical Mechanics: Approaching a Complex World* 1st ed (Springer)
[6] de Moura F, Tirnakli U and Lyra M L 2000 *Phys. Rev. E* **62** 6361–6365
[7] Nivanen L, Le Méhauté A and Wang Q A 2003 *Rep. Math. Phys.* **52** 437–444
[8] Borges E P 2004 *Physica A* **340** 95–101
[9] Suyari H and Tsukada M 2005 *Information Theory, IEEE Transactions on* **51** 753–757
[10] Umarov S, Tsallis C and Steinberg S 2008 *Milan Journal of Mathematics* **76** 307–328
[11] Costa U M S, Lyra M L, Plastino A R and Tsallis C 1997 *Phys. Rev. E* **56** 245–250
[12] Borges E P, Tsallis C, Añaños G F J and de Oliveira P M C 2002 *Phys. Rev. Lett.* **89** 254103
[13] da Silva C R, da Cruz H R and Lyra M L 1999 *Braz. J. Phys.* **29** 144–152
[14] Tirnakli U and Tsallis C 2006 *Phys. Rev. E* **73** 037201
[15] Jaganathan R and Sinha S 2005 *Phys. Lett. A* **338** 277–287
[16] Banerjee S, Yorke J and Grebogi C 1998 *Phys. Rev. Lett.* **80** 3049–3052
[17] Latora V, Baranger M, Rapisarda A and Tsallis C 2000 *Phys. Lett. A* **273** 97–103
[18] Borges E P and Tirnakli U 2004 *Physica D* **193** 148–152
[19] Borges E P and Tirnakli U 2004 *Physica A* **340** 227–233
[20] Tirnakli U, Beck C and Tsallis C 2007 *Phys. Rev. E* **75** 040106
[21] Tirnakli U, Tsallis C and Beck C 2009 *Phys. Rev. E* **79** 056209
[22] Lyra M and Tsallis C 1998 *Phys. Rev. Lett.* **80** 53–56

Dynamic Days South America 2010 IOP Publishing
Journal of Physics: Conference Series **285** (2011) 012043 doi:10.1088/1742-6596/285/1/012043

Locking-time and Information Capacity in CML with Statistical Periodicity

R M Szmoski, R F Pereira, F A S Ferrari, S E de S Pinto

Departamento de Física, Universidade Estadual de Ponta Grossa, 84030-900, Ponta Grossa, PR, Brazil.

E-mail: desouzapinto@pq.cnpq.br

Abstract. In this work we address the statistical periodicity phenomenon on a coupled map lattice. The study was done based on the asymptotic binary patterns. The pattern multiplicity gives us the lattice information capacity, while the entropy rate allows us to calculate the locking-time. Our results suggest that the lattice has low locking-time and high capacity information when the coupling is weak. This is the condition for the system to reproduce a kind of behavior observed in neural networks.

1. Introduction

There are low-dimensional dynamical systems whose density evolution does not converge to an invariant state. Instead, it becomes periodic in time with well defined oscillation period. Such behavior is called statistical periodicity [1]. Dynamical systems with statistical periodicity, although possess an invariant density, do not evolve to it [2, 3]. Once in physical world the asymptotic equilibrium is rather an exception than the rule, systems with statistical periodicity are essentials for understanding nonequilibrium phenomena.

One-dimensional maps, such as the logistic and tent ones, exhibit statistical periodicity for certain values of the control parameter. For tent map it was demonstrated [2] there is statistical periodicity for $(\sqrt{2})^{\frac{1}{T}} < a \leq (\sqrt{2})^{\frac{1}{T-1}}$, where $T = 1, 2, \cdots$ is the oscillation period and a the map control parameter. In general, both logistic map and tent one exhibit statistical periodicity just at the parameter values where they generate banded chaos or quasiperiodicity.

In this work, we study a coupled map lattice (CML) composed by N maps with statistical periodicity. CMLs are suitable mathematical models for spatially extended systems and spatiotemporal chaos [4]. Furthermore, they are considered useful to investigate neural networks [5, 6], because they can reproduce some experimentally observed features such as: a rapid response to stimuli, an irregular or chaotic behavior of individual elements and a great variability of patterns.

The dynamic of a CML is, in general, governed by two competing mechanisms: a local reaction, which is given by map, and interaction due to coupling. Here, using the local-type coupling we shall try to answer two questions: a) How does the coupling influence the asymptotic behavior? b) Is there any value interval of the coupling parameter from which the statistical periodicity is no longer observed?

CMLs with statistical periodicity were studied in [6, 7] driven by experimentally observed features in neural network. Hauptmann *et al* [7] show that the CML can reproduce such features when the coupling is weak. However, they determined the time for the lattice to achieve the permanent periodic state or locking-time using a somewhat primitive method. Here, we propose an accurate method to determine such quantity, which is based on the system entropy rate. Conversely, we show that there exist a dependence between locking-time and accessibility of effective binary states. Also we calculate the conditional entropy as information capacity measure of the CML. This paper is organized as follows: section 2 describes the coupling-type and map used; section 3 presents the periodicity detection method proposed; section 4 the findings and discussions and, finally, the conclusions.

2. The lattice

We consider that system state at each instant of time n is completely defined by the vector $\mathbf{x}_n = (x_n^{(1)}, x_n^{(2)}, \cdots, x_n^{(N)})^T$, in which the components evolve according to the equation

$$x_{n+1}^{(i)} = (1-\varepsilon)f(x_n^{(i)}) + \frac{\varepsilon}{2}\left[f(x_n^{(i-1)}) + f(x_n^{(i+1)})\right], \tag{1}$$

where $\varepsilon \in [0,1]$ is the coupling strength between the sites of the lattice and $f(x)$ is local dynamics.

Following [7] we consider

$$f(x) = \begin{cases} ax & , \quad 0 \le x < \frac{1}{2} \\ a(1-x) & , \quad \frac{1}{2} \le x \le 1 \end{cases}, \tag{2}$$

with $a = \sqrt{2}$. For this parameter value the density flow exhibits a period-2 oscillation in time. Figure 1(a) shows the evolution of a ensemble with 100,000 maps of the form (2), in which $a = \sqrt{2}$ and $T = 2$. This periodicity can be viewed as the alternation between light and dark colors. We can see a frontier line on distribution separating it into two regions. This line represents the unstable fixed point $x^* = a/(a+1)$ of the map.

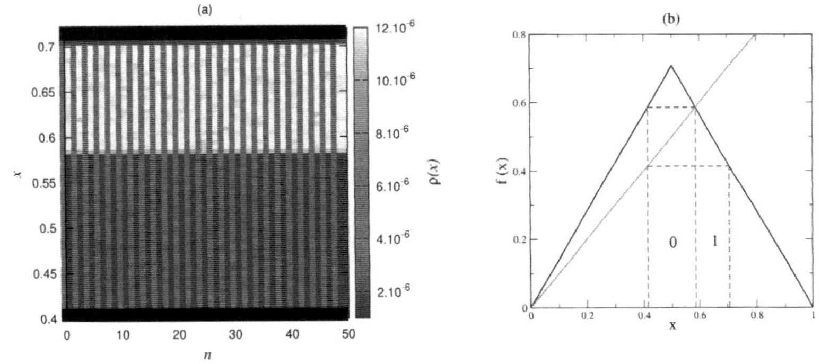

Figure 1. (a) Density evolution for an ensemble of 10^5 uncoupled tend maps with $a = \sqrt{2}$. The states (time) correspond to vertical (horizontal) axis while the colors code the state density. Color intense (yellow online) corresponds to higher density, while dark color (red online) lower density. (b) $F_a(x)$ on intervals $[0,1]$ for $a = \sqrt{2}$ and partition generated by mapping the density boundaries.

The periodic behavior of the density becomes clearer if we realize a proper partitioning of phase space. Using the fixed point of the map as a boundary we split the phase space into two regions. We call these regions 0 and 1, as shown in Fig. 1(b). We observe that all points in 0 (1) are mapped to 1 (0). It is easy to verify that any point outside that region is eventually mapped to it.

Now we present an alternative procedure in order to identify the statistical periodicity for the binary pattern of the lattice. We define the binary pattern regarding the fixed point of the map: we assign the value 1 if $x_n^{(i)} > x^*$ or 0 if $x_n^{(i)} \leq x^*$ [7] . To avoid finite size problems, we assume periodic boundary conditions. The initial conditions are uniformly distributed on the interval $[a(1-a/2), a/2]$, in which the invariant density has non-null supports.

3. The entropy rate

The entropy rate is defined [8] as the average number of bits to encode one additional state when we know all previous states of the system. Denoting for $x'^{(i)}$ the binary state of the $i-$th site of the lattice, the entropy rate of the system is given by

$$h = -\sum_{i=1}^{N} p(x'^{(i)}_{n+1}, x'^{(i)}_{k}) \ln p(x'^{(i)}_{n+1} | x'^{(i)}_{k}) \tag{3}$$

where $p(x'^{(i)}_{n+1}, x'^{(i)}_{k})$ is the joint probability, $p(x'^{(i)}_{n+1}|x'^{(i)}_{k})$ the conditional probability and $x'^{(i)}_{k}$ a shorthand notation for $(x_n^{(i)}, x_{n-1}^{(i)}, \cdots, x_{n-k+1}^{(i)})$. Thus, the entropy rate measures the uncertainty degree of the state at instant $n+1$ with respect to k previous states.

Note that, for systems whose the asymptotic pattern presents statistical periodicity with a arbitrary period T, we have $p(x_{n+1}^{(i)}|x_{T-1}^{(i)}) = 1 \ \forall \ i$, which implies $h = 0$. Therefore, since the entropy is null only if the pattern has periodicity, this quantity can be used to calculate the locking-time. Figure 2 presents the entropy rate $h(k=2)$ and the binary pattern as a function of the time for the lattice (1), for two different initial distributions. We can see that the entropy rate abruptly decays to zero if period-2 oscillations occurs. We also observe that there are transient states which oscillate, before to reach the asymptotic behavior. In Fig. 2(a), for instance, we note a oscillation state, with period-2, between $n = 6$ and $n = 10$, which coincides with $h = 0$. After this interval we note a small fluctuation in the pattern, and only at $n = 13$ the asymptotic state is reached. The similar behavior is observed for the case in which the initial conditions are distributed on the interval $[0, 1]$ (Fig. 2 (b)).

To ensure that the system reached the steady state we perform a temporal average for h on the time interval τ. In the inset of Fig. 2 we plot $h(\tau = 20)$ for two different sets of initial conditions. We see that, in this case, $h(\tau = 20)$ goes to zero only if the system reaches the asymptotic periodic state. We also note that the convergence of the pattern depends on the initial conditions. So, for a reasonable estimate of the locking-time is appropriate to consider the means on the ensemble of initial conditions. Thus, we rewrite the equation (3) as follows

$$\langle h \rangle = -\sum^{\#}\sum^{\tau}\sum^{N} p(x'^{(i)}_{n+1}, x'^{(i)}_{T-1}) \ln p(x'^{(i)}_{n+1}|x'^{(i)}_{T-1}), \tag{4}$$

where the first summation refers to initial conditions, the second one to time interval, and the last one takes into account the elements of the lattice. The dotted curve in the inset of the Fig. 2 indicates $\langle h \rangle$, which is calculated with 1,000 random initial distributions.

4. Results

4.1. The locking-time

In this work, our first analysis about the influence of coupling on the periodic behavior was related with the locking-time, which is the needed time for the systems to present a permanent

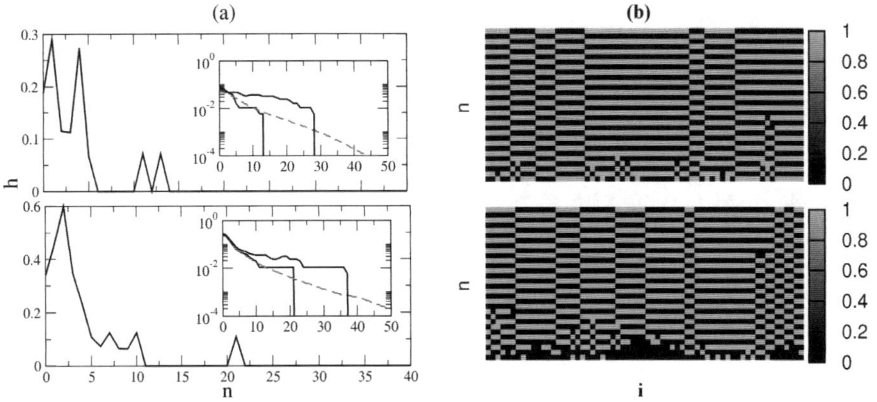

Figure 2. Entropy rate (a) and binary activity pattern (b) as a function time for $N = 64$ and $\varepsilon = 0.20$. The inset in (a) shows the mean entropy rate on interval $\tau = 20$ for two different sets of initial conditions. The dashed curve (red online) corresponds to mean entropy rate on 1,000 sets of initial conditions. In (b) the space (time) correspond to the horizontal (vertical) axis while the colors code the binary states. Color intense (yellow online) correspond to 1 and dark black 0. Figures appear at the top correspond to random initial conditions distributed on $[a(1-a/2), a/2]$ and those below on intervals $[0, 1]$.

behavior [7] . As the periodicity of the binary pattern implies $\langle h \rangle = 0$, we initially investigated the behavior of $\langle h \rangle$ as a function of time for different values of ε and initial distributions, in order to identify the values for the locking-time. Figure 3(a) exhibits $\langle h \rangle$ calculated on a ensemble of 20,000 random initial distribution, with $\tau = 100$, $\varepsilon = 0.26$, $\varepsilon = 0.27$ and $\varepsilon = 0.28$. We observe that at the interval of time in the figure $\langle h \rangle$ follows an exponential law whose slope depends on ε, *i.e.*, $\langle h \rangle \propto n^{\beta(\varepsilon)}$. For $\varepsilon = 0.26$ and $\varepsilon = 0.27$ the coefficients are negative, $\beta(0.26) \approx -10^{-2}$ and $\beta(0.27) \approx -10^{-3}$, while for $\varepsilon = 0.28$ we obtained $\beta(0.28) \approx 10^{-4}$.

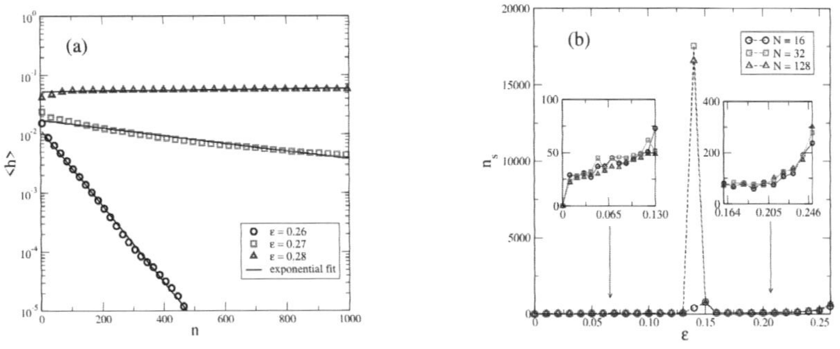

Figure 3. (a) Mean entropy rate as a function of the time for $N = 128$ and 20,000 sets of initial conditions. (b) Locking-time as a function of the coupling strength for $N = 16, 32, 128$, $\tau = 100$ and $n_{max} = 10^7$ iterates. The initial conditions were randomly distributed on the interval $[a(1-a/2), a/2]$.

The qualitative change in the slope for the $\langle h \rangle$ plot in Fig 3(a) suggests the existence of a critical value of ε, above that periodicity will no longer be observed. To go into the details in this question, we calculate the mean locking-time n_s for different values of N. We consider a maximum time observation of 10^7 iterations, and 10^{-6} as a threshold values for $\langle h \rangle$, below that we consider the system is in the periodic permanent state. Figure 3(b) shows the dependence of n_s on ε for $N = 16$, 32 and 128. Since the distribution of initial conditions respected the density interval for the map, we obtained a locking-time equals 1 with $\varepsilon = 0$, for $\varepsilon > 0$ the locking-time is directly proportional to ε. Besides, the curve presents a peak divergence in $\varepsilon \approx 0.15$ for all values of N, and for $\varepsilon > 0.27$ we did not obtain any value for n_s until 10^7 iterations.

Although we have not found any values of n_s for $\varepsilon > 0.27$, we cannot assert that they do not exist. We have three possibilities: a) $n_s > 10^7$ iterations; b) the period of oscillation is greater than 2; c) the binary pattern ceases to exist for $\varepsilon > 0.27$. Once we determined the locking-time from an ensemble of initial distributions, some of them can take more than 10^7 iterations to reach the stationary state, and contributed significantly for the calculation of $\langle h \rangle$. To investigate this hypothesis we verify the portion of initial distribution that satisfy $\langle h \rangle < 10^{-6}$ for $n \leq 10^7$. Figures 4(a) and 4(b) present the graphical verification of the hypothesis for a ensemble of 10,000 initial conditions and $N = 16$, the observation time being equal to 10^7. The mean locking-time is small – $\mathcal{O}(10^2)$ iterations – for $\varepsilon < 0.27$, exhibits a maximum value in the order of 10^5 iterations for $\varepsilon \in (0.27, 0.34)$, and a irregular oscillation after that interval. The number of initial distribution that has not reached the periodic state until 10^7 iterations is equals to zero when $\varepsilon < 0.27$, however, for values greater than the former, nearly 70% of the distributions have not converged yet. Therefore, our results suggest that the oscillation of period two ceases to be the only typical behavior if $\varepsilon > 0.27$. We should pay attention to the fact that this critical value do not depends on the size of the lattice.

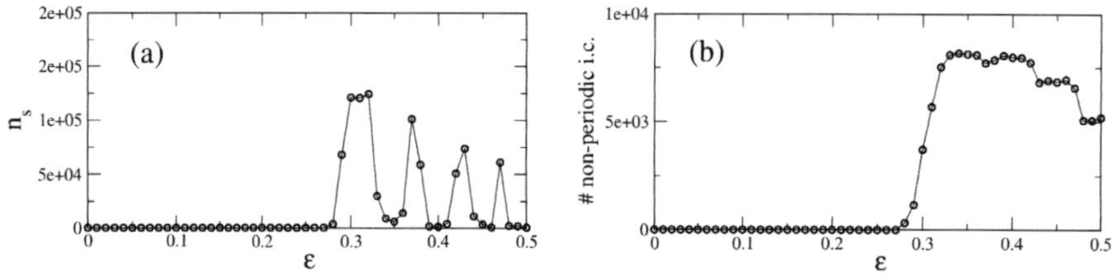

Figure 4. Mean locking-time (a) and number of non-periodic initial condition sets (b) as a function of the coupling strength. The mean locking-time was weighed by distributions for which $\langle h \rangle_\tau < 10^{-6}$ in 10^7. We consider $N = 32$, $\tau = 100$ and 10.000 random initial distributions on intervals $[a(1 - a/2), a/2]$.

Aiming to examine the peak and the divergence of the n_s in Fig. 3(b), we studied the density of the effective binary states as the parameter ε suffers variation. We define and label each effective binary state s in relation to the number of 0's and 1's that appears in the binary pattern after a transient time, *i.e.*

$$s \equiv \begin{cases} \frac{1}{2}|N_0 - N_1| & \text{if } N \text{ is even} \\ \frac{1}{2}(|N_0 - N_1| - 1) & \text{if } N \text{ is odd} \end{cases} \tag{5}$$

in which, $N_0(N_1)$ is the number of 0's(1's) in the asymptotic binary pattern. From this definition follows that a lattice with N elements have $\frac{N}{2} + 1$ effective binary states if N is even and $\frac{N+1}{2}$ states if N is odd. Figure 5 shows s for $N = 16$ and $N = 32$, where we used 20,000 initial distributions e considered 100 iterations after a transient of 10^6. In Fig. 5(a) we see that when $\varepsilon \approx 0.15$ the density of some states s abruptly varies, which occurs for the states 6, 7 and 8. More precisely, we see that while the occupation probability of the states 6, and 8 increases, for the state 7 it goes to zero. Because of this the locking-time diverges in that point. We note another abrupt change in the probability at $\varepsilon \approx 0.27$, for which three new states becomes unlikely. Therefore, we attribute the divergence of time-locking to the fact that some states have almost null probability of being occupied.

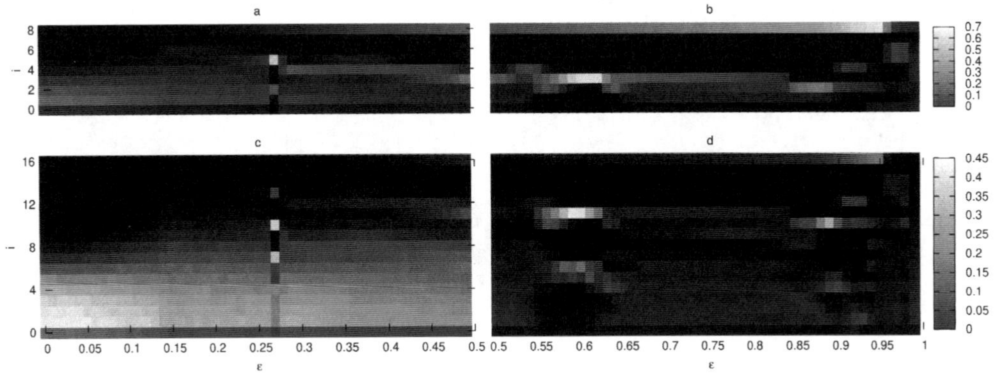

Figure 5. Density of effective binary states as a function the coupling strength. (a) and (b) refers to $N = 16$, (b) and (c) $N = 32$. The states (coupling strength) correspond to vertical (horizontal) axis while color code the state density. Color intense (yellow online) correspond to higher density and dark color lower density. The density were weighted on $20,000$ initial distribution and 100 iterates, considering 10^6 of transient.

4.2. The information capacity

In the previous section we see that the intensification of the coupling implies destruction of binary patterns. Since the multiplicity of the patterns is related to information capacity [7], we have that the stronger the coupling, the lower the information capacity in the lattice. This means that the intensification of the interaction between sites decreases the uncertainty about the asymptotic states. A way to quantify such uncertainty is through the conditional entropy given by

$$H(x^{(i)}|y^{(i)}) = -\sum_{x^{(i)}}\sum_{y^{(i)}} p(x^{(i)}, y^{(i)}) \ln p(x^{(i)}|y^{(i)}), \tag{6}$$

where $x^{(i)}$ is the binary state of the $i-$th element, $y^{(i)}$ are the binary states of the next-neighbors of i, $p(x^{(i)}, y^{(i)})$ is the joint probability and $p(x^{(i)}|y^{(i)})$ is the conditional probability.

The conditional probability is one that concerns the element be "1" or "0" if its neighbors, to which it is connected , are in a certain configuration. Such probability is given by

$$p(x^{(i)}|y^{(i)}) = \frac{p(x^{(i)}, y^{(i)})}{\sum_x p(x^{(i)}, y^{(i)})} \tag{7}$$

in which the summation for x is over the two possible states of $i-$th element.

Using Eqs. (7) and (6) we evaluate the conditional entropy of the binary pattern which corresponds to the lattice (1) as a function of ε on the interval $[0.00, 0.27]$. Figure 6(a) shows the dependence of $H(x^{(i)}|y^{(i)})$ on ε for different values of N. When $\varepsilon = 0.0$ the entropy is equal to $\ln 2$. Until $\varepsilon \approx 0.15$, $H(x^{(i)}|y^{(i)})$ presents a more pronounced decay than those for $\varepsilon > 0.15$.

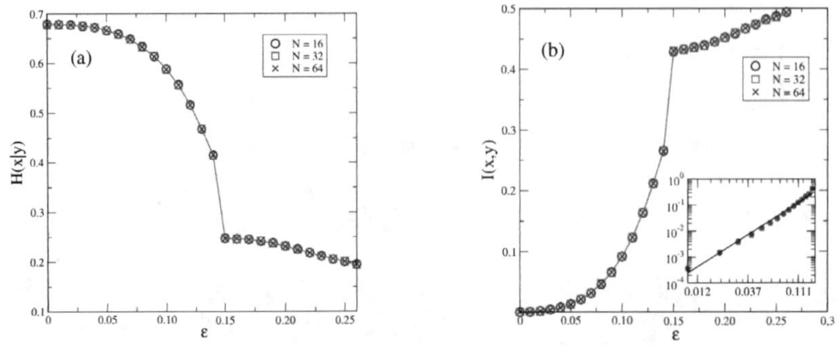

Figure 6. Conditional entropy or entropy per element (a) and mutual information (b) as a function of coupling strength. These quantities were evaluated on the permanent binary pattern of 20,000 initial condition sets. We consider the CML on permanent binary pattern when $h < 10^{-12}$.

The curves For the three different values of N the curves coincide. This indicates that the conditional entropy do not depends on the size of the lattice.

Conditional entropy, Eq. (6), also can be used to obtain the mutual information. This quantity corresponds to information obtained about the state of one element by knowing the states of its neighbors. Thus, we have

$$I(x, y_i) = H(x) - H(x|y_i) \tag{8}$$

where $H(x) \approx \ln(2)$ is the entropy of x.

In Fig. 6(b) we show the behavior of $I(x, y_i)$ as a function of ε. We note that the mutual information increases with the coupling parameter. The fit curve for $\varepsilon \in (0, 0.15)$ (inset) shows that it follow a power-law,*i.e.* $I \propto \varepsilon^{\alpha}$, being $\alpha \approx 2.6$. As for the entropy, an abrupt changes in behavior of $I(x, y_i)$ occurs at $\varepsilon \approx 0.15$, which the first maximum of the locking-time and where one effective binary state becomes unlikely. With this variation the mutual information (conditional entropy) assumes a larger (smaller) value from its mean value and therefore the state of an arbitrarily chosen element is known for his neighborhood more than 50% the uncoupled case one.

On the multiplicity of binary patterns in response to the variability of initial distributions, our results show that it becomes smaller increasing the interaction between network sites. For $\varepsilon = 0$, for example, about $2^{NH/\ln 2} \approx 2^N$ while for $\varepsilon = 0.25$ it is reduced to approximately $2^{N/3}$ possible patterns. Therefore, both characteristics of neural networks, optimal capacity information and smaller locking-time, are verified when the network elements interact only weakly.

5. Conclusion

We studied the information capacity of a coupled map lattice with statistical periodicity considering the multiplicity of the patterns into periodic state. We propose a method to determine the locking-time of systems with periodic states and we apply it to CML under study. We found that periodic behavior and oscillation period of the uncoupled map are preserved by CML for weak coupling strength. However, we observed that the average locking-time and the number of initial patterns that not achieve the periodic state is large for strong coupling. Our results, supported by the entropy rate and conditional entropy, suggest that the information capacity decreases and the locking-time increase as the interaction between the sites of the CML becomes more intense. Therefore, if the interest is to study neural networks using coupled map lattices, weak coupling are required. On the other hand, if the objective is more efficient communication, it should be used strong coupling strength.

Acknowledgments
This work was made possible through partial financial support from the following Brazilian research agencies: Fundação Araucária, CAPES and CNPQ.

References
[1] Lasota A and Mackey M 1985 *Probabilistic Properties of Deterministic Systems* (Cambridge University Press: Cambridge)
[2] Provatas N and Mackey M C *Physica D* **53** 295
[3] Provatas N and Mackey M C *Journal of Statistical Physics* **63** 585
[4] Kaneko K 1993 *Theory and Applications of Coupled Map Lattices* (Chichester: Wiley)
[5] Dhamala M, Jirsa V K and Ding M 2004 *Physical Review Letters* **92**
[6] Milton J G and Mackey M C *Journal of Physiology* **94** 489
[7] Hauptmann C, Touchette H and Mackey M C 2003 *Physical Review* E **67** 026217
[8] Schreiber T *Physical Review Letters* **85** 461

Dynamic Days South America 2010 IOP Publishing
Journal of Physics: Conference Series **285** (2011) 012044 doi:10.1088/1742-6596/285/1/012044

Model for neural signaling leap statistics

Martine Chevrollier and Marcos Oriá

Laboratório de Física Atômica e Lasers
Departamento de Física, Universidade Federal da Paraíba
Caixa Postal 5086
58051-900 João Pessoa, Paraíba, Brazil

E-mail: oria@otica.ufpb.br

Abstract. We present a simple model for neural signaling leaps in the brain considering only the thermodynamic (Nernst) potential in neuron cells and brain temperature. We numerically simulated connections between arbitrarily localized neurons and analyzed the frequency distribution of the distances reached. We observed qualitative change between Normal statistics (with T = 37.5°C, awaken regime) and Lévy statistics (T = 35.5°C, sleeping period), characterized by rare events of long range connections.

1. BASIC INFORMATION

Everyone knows about the "weirdness" of our dreams taking place in REM period, the last stage of each sleeping cycle. These sleep cycles have different stages, well characterized by physiologic signatures, as brain waves, eye movements and brain temperature variations [1, 2]. It is also known that the associative capacity of our brain increases during its sleep period [3], resulting in better learning and memory fixation [4], as well as possibly increasing our capacity for solving problems. As the energy used by the brain is a big fraction (one fifth) of the whole energy used by the body [5, 6] if compared with their mass ratio (the body mass is about fifty times the brain mass), it is necessary that each small thermodynamic mechanism be a result of evolutionary optimization leading to better information processing. Therefore we propose that these behaviors and improvements are facilitated by the increase in the likeliness of long-distance neuron connections, as a result of different thermodynamic regimes between the two phases of the circadian cycle, sleep and wakefulness.

The very simple mathematical model we propose here is grounded on the expected change in the likeliness of long-distance transmissions between neurons due to the drop in brain temperature and slight changes in the neural electrochemical potential while sleeping [7]. We calculated the distance traveled by "spikes" and obtained that in certain conditions the statistical properties of the leap length distribution changes from Normal (Gaussian) to Superdiffusive (Lévy) statistics [8]. In other words, the connection leaps may evolve from a well determined averaged value toward a more unpredictable one, characterized by very long individual steps, as illustrated in Figure 1. We would like to emphasize the distinction between the problem of neuronal transmission discussed here and the known property of networks whose connections follow power laws as an intrinsic consequence of net topologies [9].

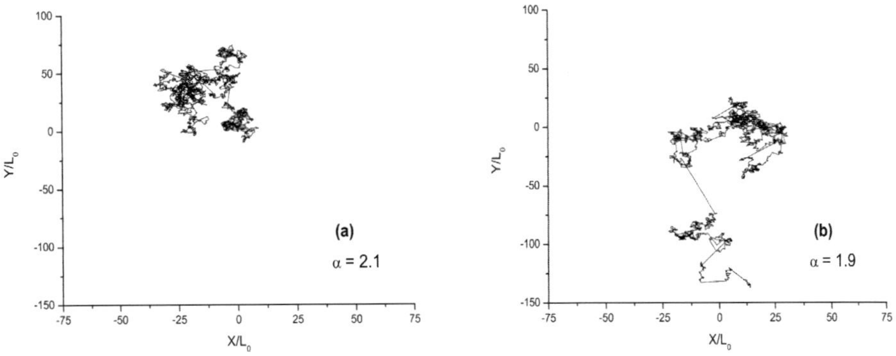

Figure 1. 2D propagation of a single neural spike whose single-leap length is determined by equations (1) and (2) (see text). The propagation regime changes at a critical value $\alpha_c = 2$, where $\alpha = \frac{\beta kT}{V_0}$ (a) Normal (diffusive) regime(T = 37.5 °C, $\alpha = 2.1$) and (b) Levy (superdiffusive) regime (T = 35.5 °C, $\alpha = 1.9$). In both cases, the number of leaps is 5000.

To explain the difference between Normal and Superdiffusive regimes, we investigated the idea that the neural signal leap length may exhibit an Arrhenius-cascade-like behavior as a function of temperature. Let us then remember the Arrhenius cascade model [10], used to analyse the total time it takes for a particle to go down an inclined, randomly corrugated surface (washboard), jumping from valley to valley. The well depths are distributed according to $p(U) = U_0^{-1} exp(-U/U_0)$ and the time spent in each well depends on U according to the Arrhenius law: $\tau = \tau_0 \; exp(U/kT)$, with k the Boltzmann constant, T the temperature and τ_0 the high temperature limit of τ. For a large number of explored wells, the distribution of individual dwell time $p(\tau)$ follows a power law $p(\tau) = 1/\tau^{(1+\alpha)}$, where $\alpha = kT/U_0$. The behavior of the total time qualitatively changes as α goes through a critical value $\alpha_c = 2$.

2. THE MODEL

We started from the Nernst thermodynamic potential and considered a variation of 2°C in brain temperature between sleep and wakefulness or, more precisely, between REM and NREM sleep stages [11]. We adjusted the parameters in order to get correspondingly normal ($\alpha \geq 2.0$) and superdiffusive ($\alpha < 2$) behaviors for the neuron connection leaps. Likewise we could have taken measured temperatures for sleep and awake regimes and estimated an effective potential between connected points. In this stage of model building, we did not consider the neural cell structure (cell body, dendrites, axons, synapses,...) but took interconnected points uniformly distributed in the considered volume (brain volume).

We therefore began with a very simplified model [12] for the membrane potential distribution $f(V)$, as well as for the leap length dependence at this potential, $l(V)$. We write

$$f(V) = \frac{1}{V_0} e^{-(V+V_m)/V_0} \tag{1}$$

and

$$l = l_0 e^{\frac{V}{\beta kT}}, \tag{2}$$

with values of V_m, V_0, l_0 and β to be adjusted, β being a temperature-dependent coefficient. Eqs. (1) and (2) lead to a leap length distribution:

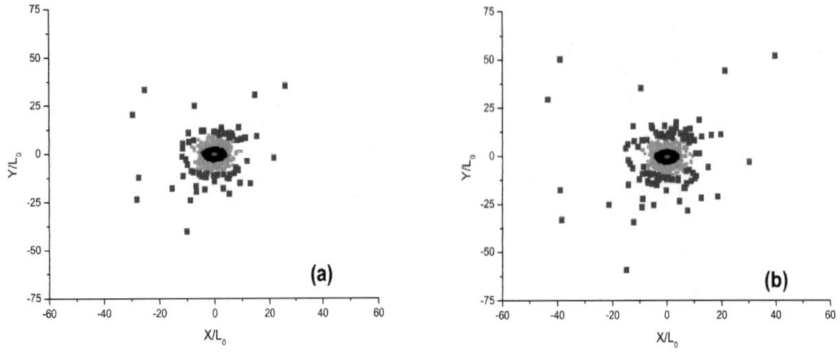

Figure 2. Spatial distribution of final position of 10^4 spikes, all beginning at the same (0,0) position and propagating according to Eqs. (1) and (2) (see text and Figure 1). The stop condition is a voltage threshold (V_c). (a) $\alpha = 2.1$; (b) $\alpha = 1.9$.

$$f(l) = \alpha l_0^{\alpha} e^{-V_m/V_0} \frac{1}{l^{1+\alpha}} \tag{3}$$

with $\alpha = \frac{\beta k T}{V_0}$.

For $\alpha \geq 2$, the diffusion of leaps is normal and the total displacement distribution, $f(r)$, converges to a Gaussian. However, if $\alpha < 2$ the r follow a stable Lévy distribution, as predicted by the Generalized Central Limit Theorem [13]. This superdiffusive behavior is characterized by occasional long leaps that dominate the system's dynamics. We thus claim that the leap length statistics would be Gaussian in the high temperature regime ($\beta T \geq 2V_0/k$) and Lévy in the low temperature regime ($\beta T < 2V_0/k$).

We assume that a transition between the normal and the Lévy regimes occurs between the temperatures $T_1 = 35.5°C$ and $T_2 = 37.5°C$, with corresponding α parameters $\alpha_1 = 1.9$ and $\alpha_2 = 2.1$ [14]. This allows us to determine the β coefficient at these temperatures. We simulate for these temperatures the spatial propagation of a spike, based on a randomly chosen value for the potential at each neuron position reached and the value of the next leap length determined from Eq. (2), with a cut at $V = V_c$ (i.e. if $V < V_c$, the spike is stopped). The final spatial distribution of 10^4 spikes starting at the same position is shown in Figure 2 for the two regimes.

3. ANALYSIS

Notice that in this oversimplified model we neglected the difference between electrical and chemical signal transport, the intermediary synapses between cells, and details of propagation through axons, dendrites, etc. Even the notion of potential is not well defined here, which is not necessarily a drawback because a threshold potential is not needed to get train spikes from neurons [15]. Moreover, a higher sensitivity to noise was observed for neurons firing at low temperatures [16], which seems to be valid even when the temperature drops only a fraction of degree. Let us remember that noise has been shown to stabilize spike timing [17], as observed from neocortical neurons [18].

In fact this model allows one to speculate about behaviors, like yawning [19] or taking a deep breath before falling asleep, being physiological procedures for basal brain cooling. Also, one can ask why Gaussian statistics would be more adapted to our vigilant stage, and what the advantage of long connections during sleeping is. Notice that *Gaussian noise* is known to be an important ingredient in models of nonlinear systems described as limit-cycle oscillators [20] and that, recently, Ermentrout and collaborators [21] generalized the validity of limit-cycle oscillators to model sustained oscillations submitted to colored and non-Gaussian noise. Considering this general behavior of nonlinear systems, we point out that the potential distribution $f(V)$ we worked with, which may be generated by thermal noise, may result in oscillations of neurons firing.

As a conclusion, we have presented a speculative model in which the thermodynamical properties of the action potential in neurons may determine a qualitative changing in the statistical properties of the spikes propagation between neurons. Therefore we hope this model can serve as a starting point for new investigations looking for signatures of short and localized communication range in the awake regime, and superdiffused images of the sleeping time (during certain periods). We expect that an experimental confirmation of this connectivity between neurons will help to build a more elaborate model for connections and networking in the brain [22].

References

[1] Parmeggiani P L 2007 *Archives Italiennes de Biologie* **145** 13
[2] Imeri L, Bianchi S and Mancia M 1997 *Am. J. Physiol. Regulatory Integrative Comp. Physiol.* **273** 1663
[3] Stickgold R 2005 *Nature* **437** 1272
[4] Talamini L M, Nieuwenhuis I L C, Takashima A and Jensen O 2008 *Learn. Mem.* **15** 233
[5] Magistretti P J 2009 *Science* **325** 1349
[6] Alle H, Roth A and Geiger J R P 2009 *Science* **325** 1405
[7] H. Kawamura H and Sawyer C H 1965 *Science* **150** 912
[8] Tsallis C, Levy S V F, Souza A M C and Maynard R 1995 *Phys. Rev. Lett.* **75** 3589
[9] Barabási A L and Albert R 1999 *Science* **286** 509
[10] Bardou F 2000 Cooling gases with Lévy flights: using the generalized central limit theorem in physics (*Preprint* arXiv:physics/0012049)
[11] Moser E, Mathiesen I and Andersen P 1993 *Science* **259** 1324
[12] Gluss B 1967 *Bull. of Math. Biophys.* **29** 233
[13] Lévy P 1937 *Théorie de l'Addition des Variables Aléatoires* (Paris: Gauthier-Villiers)
[14] The 2 degrees of temperature difference are not essential in this model because we made the transition between the two thermodynamic regimes to occur at the chosen brain temperatures. Experimental observation of such transition will allow to adjust the model constants. See, for instance, Boudreau P, Shechter A, Dittmar A, Gehin C, Delhomme G, Nocua R, Dumont G and Boivin D B 2008 *Proc. 30th Annual International IEEE EMBS Conference (Vancouver)* p.4856
[15] Izhikevich E 2007 *Dynamical systems in neuroscience: the geometry of excitability and bursting* (Cambridge: MIT Press)
[16] Du Y, Lu Q and Wang R 2010 *Cogn. Neurodyn.* **4** 199
[17] Pei X, Wilkens L and Moss F 1996 *Phys. Rev. Lett.* **77** 4679
[18] Mainen Z F and Sejnowski T J 1995 *Science* **268** 1503
[19] Gallup A C and Gallup Jr G G 2008 *Physiology & Behavior* **95** 10
[20] Pikovsky A S 1984 *Radiophys. Quantum Electron.* **27** 390
[21] Goldobin D S, Teramae J, Nakao H and Ermentrout G B 2010 *Phys. Rev. Lett.* **105** 154101
[22] Jia H, Rochefort N L, Chen X and Konnerth A 2010 *Nature* **464** 1307

Dynamics of the Brazil-Malvinas Confluence: Energy Conversions

C P F Francisco[1], I C A da Silveira[2] and E J D Campos[2]

[1]Universidade Federal do ABC, Avenida dos Estados 5001, Santo André, São Paulo, Brazil.
[2]Universidade de São Paulo, Praça do Oceanográfico 191, São Paulo, São Paulo, Brazil.

E-mail: cayo.francisco@ufabc.edu.br

Abstract. In this work, we investigated the mesoscale dynamics of the Brazil-Malvinas Confluence (BMC) region. Particularly, we were interested in the role of geophysical instability in the formation and development of the mesoscale features commonly observed in this region.

We dynamically analyzed the results of numerical simulations of the BMC region conducted with 'Hybrid Coordinate Ocean Model" (HYCOM). We quantified the effect of barotropic and baroclinic energy conversions in the modeled flow and showed the dominance of the latter in the region.

1. Introduction

The confluence of the warm waters of the Brazil Current (BC) flowing S-SW with the cold waters of the Malvinas Current (MC) flowing N-NE generates a region of strong potential vorticity gradient around 36°S, in the Atlantic Basin, as can be seen from AVHRR images obtained by [1], containing meanders, eddies and filaments, known as the Brazil-Malvinas Confluence (BMC).

Figure 1 shows a series of sea surface temperature images from the BMC region, obtained by [3]. We can clearly observe the signature of the confluence region in the thermal front, as also the formation of coherent structures in the flow.

The processes that lead to the formation of the retroflection and eddy-shedding patterns are akin to occur via geophysical instability processes. Moreover, analytical contour dynamics models constructed by [2], allowing only the baroclinic instability mechanism, exhibited the development of both retroflection pattern and baroclinic vortical dipole formation.

These authors verified that dipoles were pinched off from either the retroflection lobe (i.e., the primary crest of the wave train) or the primary trough when the baroclinically unstable current system was perturbed. Particularly, their results showed that dipole formation occurs at the vicinities of the continental boundary when unstable waves propagate phase westward.

The processes responsible for the formation and development of these instabilities, although, are not yet fully understood.

2. Model formulation.

As the theoretical results obtained by [2] pointed that the baroclinic instability effect could be the main source of meander growing in the BMC region, we performed, in this study, an energy conversion analysis of the flow using the numerical simulation outputs of HYCOM. Our aim was

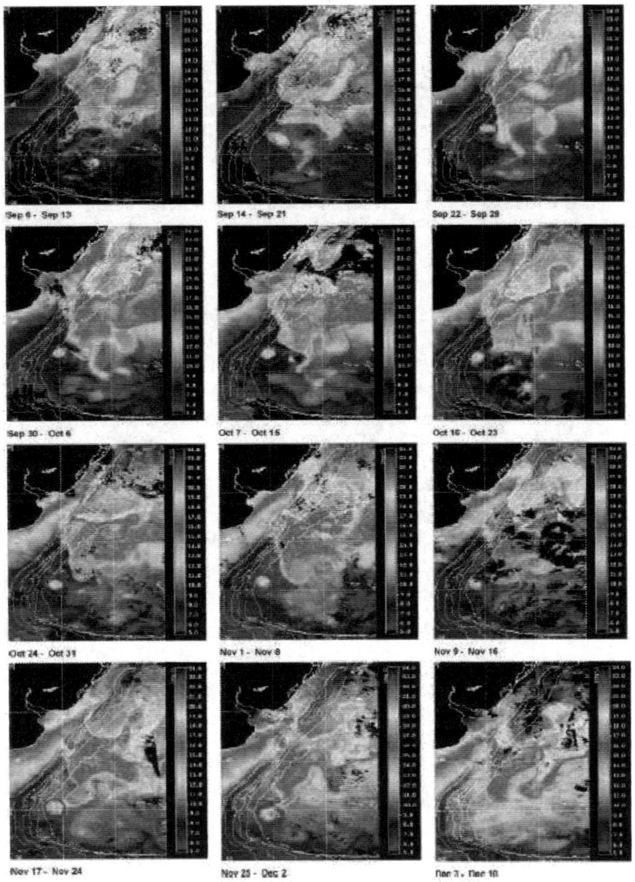

Figure 1. Images obtained by [3], showing weekly sea surface temperature averages, between 09-06-2002 and 12-10-2002. The red circle identifies a warm ring that was generated and absorbed by the current.

to account for the effects of both baroclinic and barotropic instability, not possible in [2] more idealized study.

The implementation of HYCOM, for the BMC region, followed [4]. First, a 21-layers coarser resolution (1/3-degree) version for the entire Atlantic Basin, from 65°S to 60°N, was forced with monthly mean winds, obtained from COADS base, and initialized with climatological hydrography from [5]. This initial configuration was run for 25 years. After such period a one-way nesting with 1/12-degree resolution and 21 layers was implemented for the western South Atlantic region delimited by the latitudes 30°S and 45°S and the longitudes 60°W and 45°W, as showed in figure 2.

The nested model was run for one full year, forced with the same COADS climatological data and lateral boundary conditions provided by the coarse resolution model, using a 20 grid-points relaxation.

The BMC flow was obtained using the nested modeled year average and the corresponding 26th year data of simulation of the whole domain generated by HYCOM, for the fields of velocity, temperature and salinity. Those fields compared well with observed data.

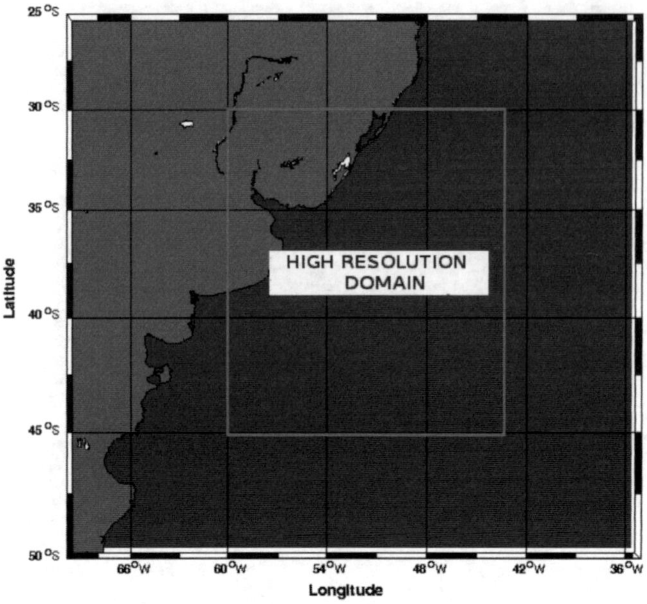

Figure 2. High resolution domain, delimited by the red square.

The surface long-term velocity field is showed in figure 3, in IS units.

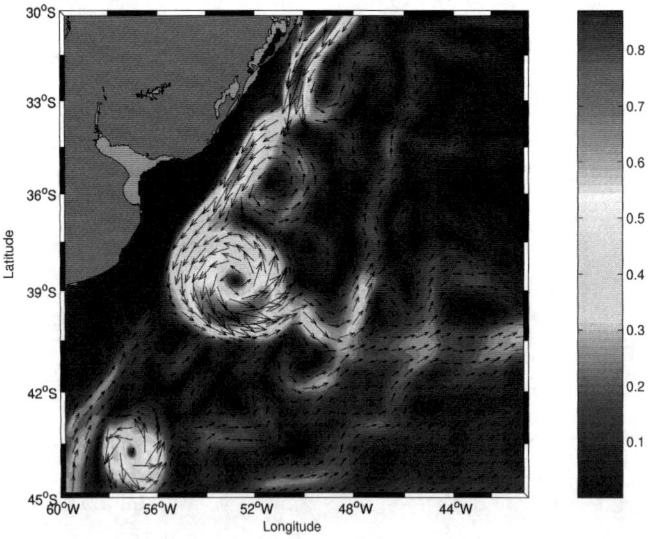

Figure 3. Mean surface velocity field, in IS units. The color scale represents the absolute values of velocities.

We clearly observe the BMC front located around 41°S, in accordance with observations performed by [6], as a result of the confluence between the BC flowing in the S-SW direction and the MC flowing N-NE. We also identify the zonal current extension, corresponding to the

South Atlantic Current (SAC).

The maximum absolute value of the velocities in the BC flow occur around 0,55 ms^{-1}, closely to the ones observed by [7] at 28°S. Maximum velocities associated with the MC flow are approximately 0,3 ms^{-1}, with mean velocity values around 0,17 ms^{-1}, higher than the geostrophic values measured by [8] at 36,5°S.

The flow associated with the BC weakens with increasing depth due to its baroclinicity, while the MC flow, dominated by the barotropic dynamics, presents small variability. These characteristics can be observed in figure 4, where we present the results for the velocity field around 650m.

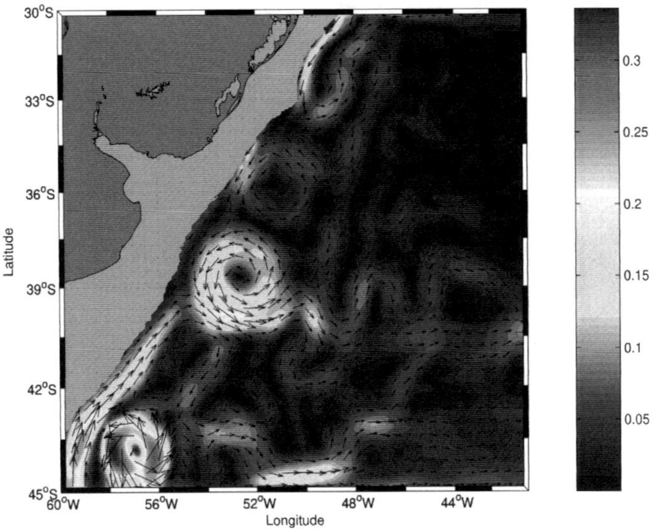

Figure 4. Velocity field, in IS units, for the 650m depth. The color scale represents the absolute values of velocities.

Particularly, the velocity field calculations clearly show the intense vortical activity associated with the BMC front, showing the presence of two large anticyclonic structures in both sides of the confluence.

With the dynamical fields of the modeled BMC we performed the energy conversion calculations utilizing the method developed by [9], obtaining both baroclinic and barotropic energy conversion maps for all depths of the flow.

3. Energy conversions

To perform the energy conversion calculations we used the quasigeostrophic framework, following [9] and decomposed the three-dimensional flow into a leading order geostrophic flow component $\mathbf{u} = (\mathbf{u}, \mathbf{v}, \mathbf{0})$ and a small ageostrophic component $\mathbf{u_a} = (\mathbf{u_a}, \mathbf{v_a}, \mathbf{w})$.

In order to calculate the eddy energetics we need to decompose the instantaneous fields into mean and eddy fields, which is done by defining the mean field as the long term average and the eddy field as the deviations from it.

According to [9] we can define the barotropic energy conversion, representing momentum advection associated with the flow, as

$$BT = -\overline{u'u'}\frac{\partial \bar{u}}{\partial x} - \overline{u'v'}\left(\frac{\partial \bar{u}}{\partial y} + \frac{\partial \bar{v}}{\partial x}\right) - \overline{v'v'}\frac{\partial \bar{v}}{\partial y}, \tag{1}$$

where the mean terms are denoted by an over bar and the eddy terms are denoted by primes.

The baroclinic energy conversion, which is relevant to baroclinic instability, representing a mechanism that lowers the center of mass of the fluid is defined as

$$BC = \frac{g\alpha}{\theta_z}\left(\overline{u'T'}\frac{\partial \bar{T}}{\partial x} + \overline{v'T'}\frac{\partial \bar{T}}{\partial y}\right) , \tag{2}$$

with α being the specific volume and θ a z coordinate function representing a purely vertical temperature field given by $T = \theta(z)+\delta T(x,y,z,t)$, with $\delta T(x,y,z,t) = \Delta T(x,y,z)+T'(x,y,z,t)$ and $\bar{T} = \theta(z) + \Delta T(x,y,z)$.

The energy conversion rates were calculated for all vertical levels of the simulation, showing the same dynamical characteristics, being less effective with increasing depths. Furthermore, it is important to note that only positive values of the conversion rates represent truly energy conversions.

The surface barotropic and baroclinic energy conversion rates are showed in figure 5 and figure 6, respectively.

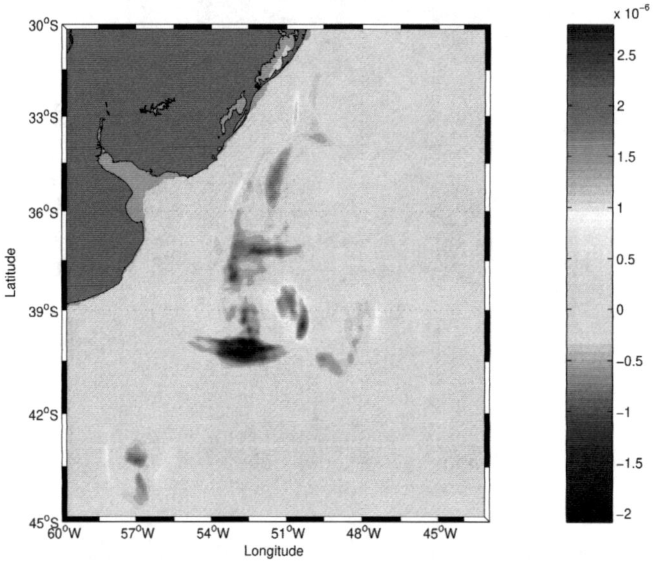

Figure 5. Surface barotropic energy conversion rate, in m^2 s^{-3}.

We can observe in figure 5 that regions where barotropic conversions occur are associated with the retroflection region within the boundaries of the flow and the boundaries of the anticyclones. The maximum values of the barotropic conversions reach 2×10^{-6} m^2 s^{-3}, located around 39°S and 51°W.

In figure 6 we observe that the baroclinic energy conversions are also more effective in the retroflection region. Maximum values of baroclinic conversion are 14×10^{-6} m^2 s^{-3}. The zonal flow extension, corresponding to the SAC, also presents positive baroclinic energy conversions corresponding due to the meandering formation in this current.

The energy conversion rates associate with the BMC are of the same order of those calculated for the Gulf Stream by [9], although in the BMC the conversion process decays quickly with increasing depth.

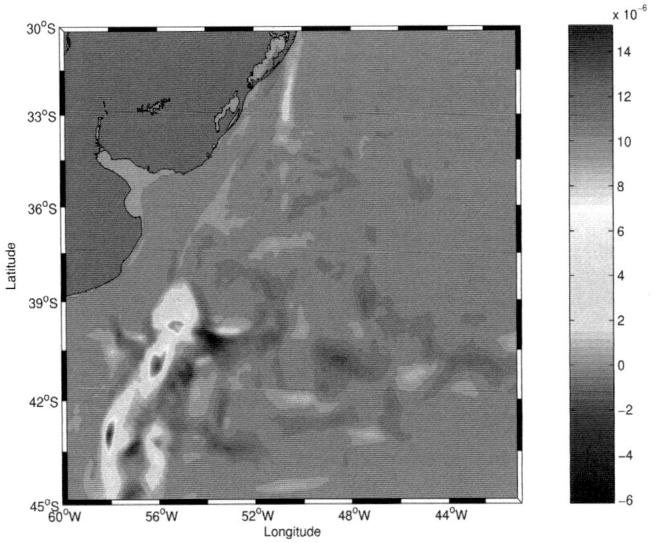

Figure 6. Surface baroclinic energy conversion rate, in $m^2\ s^{-3}$.

4. Conclusions

We reproduced, in this work, the dynamic flow patterns observed in the BMC region, using an hybrid coordinate ocean model (HYCOM), to study the origin and development of the intense eddy activity commonly observed in this region. We were particularly interested in quantifying the importance of geophysical instability in the dynamics of the flow.

This was achieved through the calculation of the energy conversion rates, following [9] quasigeostrophic eddy energetic formulation, using HYCOM output fields for the region delimited by the latitudes between 30°S and 45°S and the longitudes between 60°W and 45°W.

The results of the energy conversion calculations showed the dominance of the baroclinic conversions over the barotropic ones, although these process do not always occur at the same locations.

The dominance of baroclinic energy conversion mechanism suggest that heat fluxes are more efficient to destabilize the flow than are the momentum fluxes. The conversion of the available potential energy to kinetic energy in the BMC seems to be the main cause for the growing of instabilities, while the barotropic conversions seems to be important in regions of intense shear flow, particularly in the boundaries of the anticyclones.

References

[1] Olson D, Podestá G P, Evans R H and Brown O B 1988 *Deep Sea Res.*, **35** 12
[2] Francisco C P F and Silveira I C A 2004 *Braz. J. Gephys.* **22** 2
[3] Buss R S., Mata M M, Garcia C A E, Kampel M, Oliveira E N and Lorenzetti J A 2006 *Rem. Sens. Envir.*, **100** 52
[4] Campos E J D 2006 *Geophys. Res. Let.*, **33**
[5] Levitus S and Boyer T P 1998 *World Ocean Atlas 1994, Technical Report, National Oceanographic Data Center* **4** 117.
[6] Boebel O, Schmid C, Podestá G and Zenk W 1999 *J. Geophys. Res.*, **104** (C9) 21
[7] Muller T J, Ikeda Y, Zangenberg N and Nonato L V 1998 *J. Geophys. Res.* **103** 5429
[8] Garzoli S L 1993 *Deep Sea Res.* **40** 1379
[9] Cronin M and Watts D R 1996 *J. Phys. Oceanogr.* **26**, 2107

Dynamic Days South America 2010 IOP Publishing
Journal of Physics: Conference Series **285** (2011) 012046 doi:10.1088/1742-6596/285/1/012046

Detection and Alert of muscle fatigue considering a Surface Electromyography Chaotic Model

V Herrera[1], J F Romero[1] and M Amestegui[2]

[1] Engineering, Modeling and Applied Social Sciences Center, ABC Federal University, Santo Andr - SP - Brazil
[2] Engineering Faculty, Electronics Engineering, Universidad Mayor de San Andrés, La Paz, Bolivia

E-mail: victoria.herrera@ufabc.edu.br

Abstract. This work propose a detection and alert algorithm for muscle fatigue in paraplegic patients undergoing electro-therapy sessions. The procedure is based on a mathematical chaotic model emulating physiological signals and Continuous Wavelet Transform (CWT). The chaotic model developed is based on a logistic map that provides suitable data accomplishing some physiological signal class patterns. The CWT was applied to signals generated by the model and the resulting vector was obtained through Total Wavelet Entropy (TWE). In this sense, the presented work propose a viable and practical alert and detection algorithm for muscle fatigue.

1. Introduction

Paraplegia deprives individuals of movement and sometimes sensitivity. Paraplegic patient should attend sessions of electro-therapy, not to regain mobility but to avoid muscles to suffer other damage due to lack of movement. However, if fatigue is sustained more than necessary during therapy, the muscle can also be damaged. The therapy main drawback is lack of equipment to identify accurately when to stop the stimulation as each patient reacts differently to electrical stimulation therapy. Most of researches, focused in this kind of signal, work with Fourier Transform FT, as we can see in [1], [2], [3] and [4] although recently literature shows that SEMG signal has been characterized as chaotic [5], [6]. The present work proposes a detection algorithm for muscle fatigue in paraplegic patients undergoing electro-therapy sessions by means of surface electromyographic (SEMG) signal processing considering a chaotic model approach of SEMG. In section 2, the main definitions related to Wavelet Transform and chaos theory, which are used throughout the work, are presented. In section 3 is presented a chaotic model that emulates the electro-therapy process. Section 4 describes the pattern recognition technique, based on a continuous Wavelet Transform, proposed in this work. Aiming to exhibit the main functionalities of the proposed methodology, several numerical results are presented in section 5. Finally, in section 6, the main conclusions of the work are presented.

2. Nonlinear Chaos theory and Wavelet Transform

Nonlinear chaos theory and wavelet transform are applied in modeling of an electro-therapy process. In the following sections the main tools and methods used in the proposed recognition algorithm are briefly described.

2.1. Logistic Map

In this section is briefly described the logistic map, which is the basis of the proposed chaotic generator. The logistic map equation is represented by (1) [7]:

$$x_n = rx_n(1 - x_n) \tag{1}$$

Where parameter r defines the complexity of the system by means of the following relations [7]:
When $r < 1$ the equilibrium point is zero.
For $1 < r < 3$ there are one equilibrium point.
For $3 < r < 4$ there are more than one equilibrium points.
When $r > 4$ the system is not observable.

2.2. Chaos Theory

Chaotic systems are used to modeling sequences of disorderly or random events by means of inherent rules that determine some characteristics of their behavior [7]. The sensitivity to initial conditions suggests that, the evolution of two states, in the same neighborhood, subject to the same law of evolution, can be completely different. The existence of deterministic laws, but unpredictability at the same time, is perhaps the most striking feature of chaos. Chaotic systems are characterized by the following topics: Sensitivity to initial conditions, irregular behavior (not predictable for long periods of time), not correlated signs and wide bandwidth [8].

2.3. Wavelet Transform

A wavelet is a time-limited signal whose average value is zero. Comparing Wavelets with sinusoidal functions (basis of Fourier analysis), it is possible to note that the main difference between them is that sinusoidal signals are no time-limited, since they extend from $-\infty$ to ∞ [1]. Wavelet analysis represents a logical next step to Short-time Fourier Transform STFT, more precisely, it considers windows with regions of variable size [8]. Wavelet analysis allows the use of large intervals of time in those segments that require more precision in low frequency, and smaller intervals of time in regions where information is required in high frequency. Similarly to FT, signal analysis using WT decomposes the signal, displaced (in time) and scaled (in frequency), into versions of the original wavelet, known as Mother Wavelet. Since the range of frequencies that corresponds to each signal is known, it is possible to group them to make a graph in three dimensions, being the axis: time, frequency and wavelet coefficients. Thus, it is possible to observe what frequencies occur at what time. An important advantage provided by WT is its ability to analyze localized areas of large signals [9], [8].

2.4. Continuous Wavelet Transform.

There are two types of WT, Continuous Wavelet Transform (CWT) and the Discrete Wavelet Transform (DWT). CWT is optimal for purposes of extracting intuitive characteristics [9], therefore, in this work, CWT is applied for SEMG signal analysis in coordinate time-scale. Equation (2) describes WT, as equation (3) represents the Mother Wavelet, which displacement and scaling depends on τ and s, respectively.

$$C_{(\tau,s)} = \int_{-\infty}^{+\infty} f(t)\psi^*{}_{\tau,s}(t)\, dt \tag{2}$$

$$\psi^*{}_{\tau,s}(t) = \frac{1}{\sqrt{|s|}}\psi\left(\frac{t-\tau}{s}\right) \tag{3}$$

To analyze a signal by means of WT the next steps must be followed [8]:

1º The wavelet function, which will be the Mother Wavelet, should be chosen. This function will serve as a prototype for all windows that are used in the process.

2º After selecting the mother wavelet, equation (2), is applied along the signal with a scale factor determined (constant).

3º Then, the scaling factor should be varied in (3) in order to reduce or enlarge the size of windows

4º Equation (2) should be applied again along the signal.

5º Repeat steps 3 and 4 until attain the amount of information required for the analysis.

2.5. *Total Wavelet Entropy.*

Total Wavelet Entropy (TWE) is based on the principle of Shannon Entropy. TWE is defined in (4), [10]:

$$S_{wt} = -\sum_{n=0}^{n} p_j \ln p_j \tag{4}$$

Where, p_j is the probability distribution or Relative Wavelet Energy, at a given scale. Then, the total wavelet entropy is defined as:

$$TWE = \frac{S_{wt}}{\ln N} \tag{5}$$

Where N is the maximum scale used in the analysis. This measure indicates the level of energy contained in the signal [10].

3. Chaotic Signal Generator.

This section presents the Chaotic Signal Generator proposed for Surface Electromyographic signals and its validation. It must be noticed, that this stage has the main goal of simulate data acquisition stage.

3.1. *Chaotic Signal Generator.*

The chaotic signal generator is based on the chaotic model of the logistic map, which has been modified according to the characteristics of the Surface Electromyogram (SEMG) signal of a paraplegic patient.

To perform the signal generator, the time segment of the logistic map was modified in order to obtain characteristics of SEMG signal showed in [5]. The most important considerations, in order to develop the Chaotic Generator are listed following.

- The main variables considered are r and x_0, system complexity and initial condition, respectively.
- A constant value of r is used in the suggested model, aiming minimal variations of SEMG signals complexity, [5] [6].
- r must be greater than 3.89 once there is a non chaotic region near from this value [7], and simultaneously it must be less than 4 because from this value the system is not observable. The value r = 3.99 is selected for all numerical simulations.
- Initial condition value varies from 0 to 1.
- Changing initial condition value, SEMG signals of different paraplegic patients, or the same patient at different sessions, are simulated .

- The time segment of the logistic map was divided in three parts to perform the signal generator.

Summarizing, the time segment of the logistic map was divided in three parts where f_{si} is the scale factor that modifies each interval of logistic mapn ($f_{s1} \neq f_{s2} \neq f_{s3}$). Aiming to guarantee the process convergence in the greatest chaotic region, and considering several initial conditions, a constant value of parameter r was used in this work. Equations (6), (7) and (8) presents the proposed chaotic model of the SEMG signal generator based on the modified logistic map:

$$\text{for } 1 < i < \frac{n}{6} \qquad x_{i+1} = rx_i(1 - x_i) + f_{s1} \tag{6}$$

$$\text{for } \frac{n}{6} < i < \frac{n}{4} \qquad x_{i+1} = rx_i(1 - x_i) + f_{s2} \tag{7}$$

$$\text{for } \frac{n}{6} < i < n \qquad x_{i+1} = rx_i(1 - x_i) + f_{s3} \tag{8}$$

The main characteristics of the signal generated by the proposed model, are similar to real physiological signals as highlighted bellow:

a *Signal duration.* The signal generated, during 350 sec, simulates the signal of a patient which has reached maximum fatigue at time of 350(s). It is a plausible assumption considering that each patient could became fatigued at different times due to external and internal factors to the muscle [11].

b *Time behavior.* As can be observed in Fig.1, SEMG signal amplitude tends to increase over time. This feature can also be noticed in reconstruction of face portraits, as presented in [5] and [6], since the attractor increases its range at different time intervals. This amplitude increases due the increment in number of motor units activated by the fatigue and due other physical phenomenas. According to several studies based in real physiological signals, [12],[11] and [5], it has been defined that amplitude increases faster when the greater the effort.

Assuming an initial condition x_0=0.71, a generated temporal data example is presented in Fig.1.

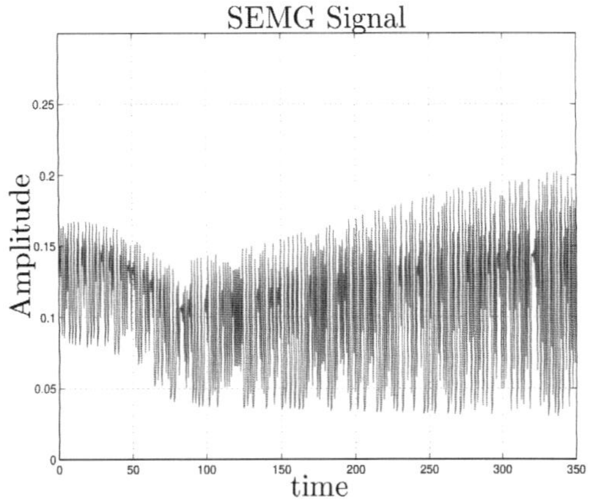

Figure 1. Chaotic Generator of a emulated SEMG

3.2. Validation of the proposed Chaotic Signal Generator.
The model based data shows that signals obtained from chaotic generator could be validated as chaotic through the following criteria [7]:

- *Sensitivity to little variations in initial conditions:* varying initial condition by 0.01, the generated signal is totally different, but maintain similar trajectories.
- *Strange attractor:* reconstruction of the phase space results in a strange attractor, which tends to grow at passage of time, as can be seen in Fig.2.
- *Correlation dimension:* the correlation dimension presents a non integer value (≈ 0.51), this indicates that the signal has a strange attractor and therefore is a chaotic signal, as was previously observed.

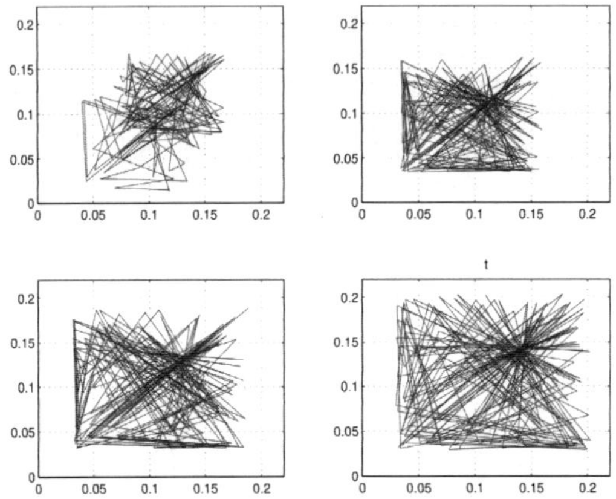

Figure 2. SEMG signal Phase Portraits, generated at four intervals of time

4. Wavelet Transform based analysis.
Once emulated a statistically uncorrelated group of signals, CWT should be applied in order to extract the main characteristics. Then, features observed through CWT portrait, are quantified applying TWE.

4.1. Continuous Wavelet Transform.
Initially, the Daubechies4 (db4) function is selected as a mother wavelet. Applying the CWT transform is verified that all signals presents the same characteristics. As an example, results obtained for a signal with an initial condition $x_0 = 0.71$ are shown in Fig.3. Values of the coefficients are represented in color scale, from dark brown (lowest coefficient values), to white (highest coefficient values).

For low scales (high frequencies) it can be observed that the passage of time do not undergo in significant changes in coefficient values, as highlighted with a blue line in Fig.3. This analysis also shows that wavelet coefficients values grow over time on high scales (low frequencies), as highlighted with a red circle. Accordingly, as the fatigue increases over time, the concentration of coefficients at low-frequencies increases proportionally. This feature supports previous analysis obtained through the Fourier Transform, which indicates that because of increased fatigue over time, there is a decrease in the Mean Frequency and Median [12].

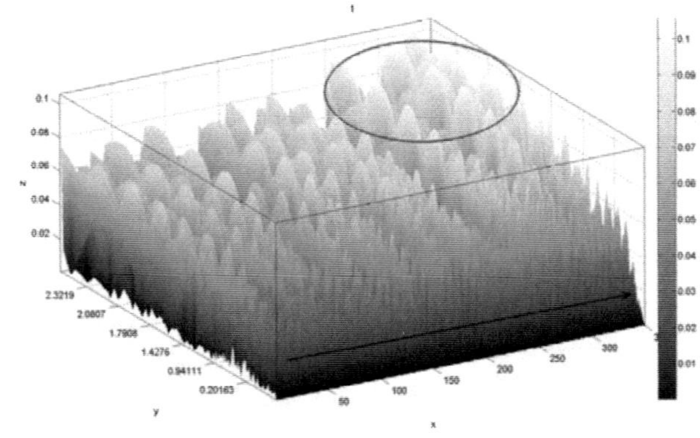

Figure 3. Continuous Wavelet Transform

As a direct proportional increase in signal power and muscle fatigue are observed, by means of CWT, a pattern recognition method could be proposed to determine thresholds values for fatigue phenomena. In order to realize this procedure, in this work a Total Wavelet Entropy is applied. This function enables summarizes the pattern characteristics in the Wavelet frequency analysis [13]

4.2. Total Wavelet Entropy.

Continuous Wavelet Transform and Total Wavelet Entropy (TWE) were calculated for several signals. Both of them present the same feature: an increment over time, in low frequencies. Total Wavelet Entropy values obtained, from five different signals are presented in Fig.4 and these values were calculated at eight different intervals of the SEMG signal sequence. It could be observed that TWE values increase over time for those five signals, tests realized in several signals presents the same pattern.

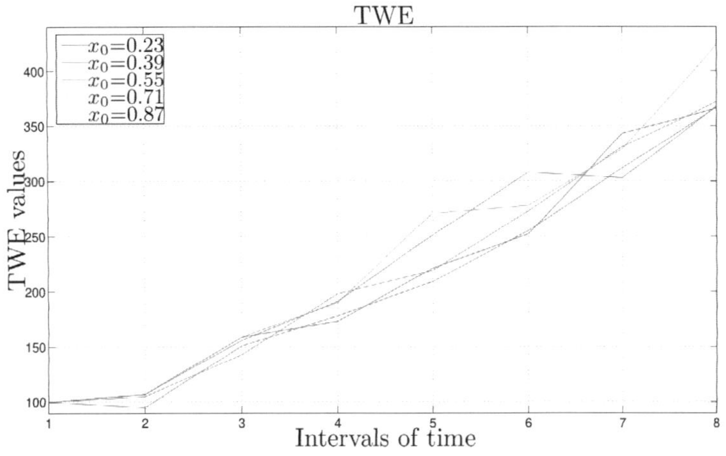

Figure 4. Total Wavelet Entropy

5. Resume of the Proposed Algorithm.

The algorithm detection and alert of fatigue proposed in this work, can be summarized in the following steps:

A. Data acquisition.
B. Calculation of the CWT (each 25 sec).
C. Calculation of the TWE (of wavelet coefficients each interval).
D. Calculation of growth percentage of TWE in 25 sec intervals (in relation to the first 25 sec).

Where:
A: A chaotic signal generator of non-variable complexity, from a scaled logistic map, that simulates data acquisition step. This generator provides several signals that emulate the behavior of electromyographic signals obtained from paraplegic patients undergoing electro-therapy sessions. These signals vary according to the initial condition, which permits simulate signals from different patients.
B: CWT shows the signal behavior in frequency domain. CWT should be calculated every 25 sec.
C: To quantify increment of energy the TWE function has been chosen, as it provides better results than other two quantifiers of wavelet energy [14].
D: Alert levels are defined based on the percentage growth of TWE, as shown in Table 1. Electrical stimulation should be stopped at Alert 2.

Table 1. Thresholds of Fatigue

Alerts	% Growth	Meaning
OK	1% -136%	Without Fatigue
Alert 1	137%-199%	Initial Fatigue
Alert 2	200%-280%	Transition
Alert 3	$\geq$ 281%	Maximum Fatigue

The proposed algorithm is validated in Matlab®7.1 platform, considering several tests, working at the most chaotic region of logistic map. As Table. 1 presents, all results obtained returns four main thresholds of fatigue from reported growth rates.

6. Conclusions

The purpose of this work consists in improve one of the most important drawbacks in physiotherapy which is referred to electro-therapy dosage. The Chaotic Signal Generator provides the enough quantity of data to realize this work. The main contribution of the work consists in the pattern recognition procedure proposed, by means of the CWT. This tool has permitted to identify a direct relationship between fatigue increment and wavelet energy increment at low frequencies. This characteristic was quantified by means of TWE. Numerical results have permitted to obtain thresholds of fatigue as shown in Table 1 and enable stopping stopping stimulation before patient allows maximum fatigue. Implementation of the algorithm will be important to both the patient, since it prevents the muscle achieve maximum fatigue during electro-therapy, and the therapist, giving adequate information about when electrical stimulation should be stopped. As a future work, the algorithm would be implemented on a Digital Signal Processor using a development kit and data acquisition devices. .

Acknowledgements
The first and second authors thanks the Universidade Federal do ABC (UFABC) for the financial support. First author thanks to Conselho Nacional de Desenvolvimento Científico e Tecnológico (CNPq) for the scholarship finantial through process. The third author thanks the Universidad Mayor de San Andrés (UMSA) for the support.

References

[1] Betancourt G, Giraldo E and Franco J 2004 *Scientia et Technica X* **Vol. 26** pp. 53–58
[2] Cai L Y, Wang Z Z and Zhang H H 2000 A surface emg signal identification method based on short-time fourier transform Tech. rep. Shanghai Jiao Tong University.
[3] Luca G D 2003 Fundamental concepts in emg signal acquisition Tech. rep. Delsys Inc
[4] Sparto P J, Parnianpour M, Barria E A and Jagadeesh J M 2000 *IEEE Transactions on Rehabilitation Engineering No. 3* **Vol. 8** pp. 433–436
[5] Erfabian A, Chizek H J and Hashemi R M 1996 *18th Annual International Conference of the IEEE Engineering in Medicine and Biology Society.*
[6] Padmanabhan P and Puthusserypady S 2004 *26th Annual International Conference of the IEEE EMBS. San Francisco, CA, USA.*
[7] Clayton K 1997 *Basic Concepts in Nonlinear Dynamics and Chaos* Society for Chaos Theory in Psychology and the Life Sciences Marquette University, Miwaukee, Wisconsin
[8] Martinez J and de Castro R M 2002 *Análisis de la teoría de ondículas orientada a las aplicaciones en Ingeniería, Eléctrica: Fundamentos* ETSII - Universidad Politcnica de Madrid
[9] Kilby J 2005 *Wavelet Analysis and Classification of surface Electromyography Signals* Master's thesis Engineering University of Technology Auckland New Zealand
[10] Zunino L, Pérez D G and Garavaglia M 2006 *Physica A: Statistical Mechanics and its Applications* **Vol. 364** pp. 79–86
[11] Martin J M R 2004 *Electroterapia en Fisioterapia* (Mdica Panamericana)
[12] Nanthavanija S and Deivanayagamb S 1989 *International Journal of Industrial Ergonomics* **Vol. 4 Issue 3** pp. 213–224
[13] Andrade A, Amestegui M and Romero J 2008 *Congresso Brasileiro de Automatica, 2008.*
[14] Herrera V and Amestegui M 2008 Diseño del algoritmo de detección y alerta de fatiga muscular en pacientes parapléjicos sometidos a sesiones de electroterapia. Tech. rep. Proyecto de Grado: Universidad Mayor de San Andres. Carrera de Ingeniera Electr nica. La Paz, Bolivia.

AUTHOR INDEX

AUTHOR INDEX